THE
INTERNATIONAL SERIES
OF
MONOGRAPHS ON PHYSICS

GENERAL EDITORS
W. MARSHALL   D. H. WILKINSON

H. S. W. MASSEY, E. H. S. BURHOP
AND H. B. GILBODY

ELECTRONIC AND IONIC IMPACT PHENOMENA

SECOND EDITION
IN FIVE VOLUMES

VOLUME V

Slow positron and muon collisions—
and notes on recent advances

OXFORD
AT THE CLARENDON PRESS
1974

Oxford University Press, Ely House, London W. 1

GLASGOW NEW YORK TORONTO MELBOURNE WELLINGTON
CAPE TOWN IBADAN NAIROBI DAR ES SALAAM LUSAKA ADDIS ABABA
DELHI BOMBAY CALCUTTA MADRAS KARACHI LAHORE DACCA
KUALA LUMPUR SINGAPORE HONG KONG TOKYO

ISBN 0 19 851283 X

© *Oxford University Press 1974*

All rights reserved. No part of this publication may be reproduced, stored in a retrieval system, or transmitted, in any form or by any means, electronic, mechanical, photocopying, recording, or otherwise, without the prior permission of Oxford University Press

*Printed in Great Britain
at the University Press, Oxford
by Vivian Ridler
Printer to the University*

PREFACE TO THE SECOND EDITION

THE immense growth of the subject since the first edition was produced has raised many problems in connection with a new edition. Apart from the sheer bulk of new material, the interconnections between different parts of the subject have become very complex, while the sophistication of both theoretical and experimental techniques has greatly increased.

It became clear at the outset that it was no longer possible to attempt a nearly comprehensive treatment. As sources of data on cross-sections, reaction rates, etc., for use in various applications are now available and are becoming more comprehensive, it seemed that in the new edition the emphasis should be on describing and discussing experimental and theoretical techniques, and the interpretation of the results obtained by their use, rather than on compilation of data. Even so, a great selectivity among the wide range of available material has been essential. Within these limitations the level of the treatment has been maintained roughly as in the first edition, although some allowance has been made for the general increase in the level of sophistication.

When all these considerations were taken into account it became clear that the new edition would be between four and five times larger than the first. To make practicable the completion of the task of writing so much against the rate of production of new results, it was decided to omit any discussion of phenomena occurring at surfaces (Chapters V and IX of the first edition) and to present the new edition in four volumes, which has now grown to five, the correspondence with the first edition being as follows·

Second edition	*First edition*
Volume I	Chapters I, II, III
Volume II	Chapters IV, VI
Volume III	Chapter VII
Volume IV	Chapters VIII, X
Volume V	New subjects and notes on recent advances

In this way Volumes I and II deal with electron impact phenomena: Volume I with electron–atom collisions and Volume II with electron–molecule collisions. Volume II also includes a detailed discussion of photo-ionization and photodetachment which did not appear in the first edition.

Volumes III and IV deal, in general, with collisions involving heavy particles. Thus, Volume III is concerned with thermal collisions involving neutral and ionized atoms and molecules, Volume IV with higher energy collisions of this kind. In addition, recombination is included in this volume. Collision processes involving slow positrons and muons, which did not appear in the first edition, are treated in Volume V, together with notes on recent advances over the whole field covered by the book.

Because of the complicated mesh of cross-connections many difficult decisions had to be made as to the place at which a new technique should be described in detail. Usually it was decided to do this in relation to one of the major applications of the technique rather than attempting, in a wholly artificial way, to avoid forward references at all stages.

In covering such a wide field in physics, and indeed also in chemistry, acute difficulties of notation are bound to arise. The symbol k for wave number is now so universally used in collision theory that we have been so impious as to use κ instead of k for Boltzmann's constant. Unfortunately, k is also used very widely by physical chemists to denote a rate constant. In some places we have adhered to this but elsewhere, to avoid confusion with wave numbers, we have substituted a less familiar symbol. Another unfamiliar usage we have employed is that of f for oscillator strength to distinguish it from f for scattered amplitude. Again we have been unfashionable in using F instead of E for electric field strength because of risk of confusion with E for energy.

No attempt has been made to adopt a set of symbols of universal application throughout the book, although we have stuck grimly to k for wave number and Q for cross-section as well as to e, h, κ and c.

We have tried not to be too pedantic in choice of units though admitting to a predilection for eV as against kcal/mole. When dealing with phenomena of strongly chemical interest we have at times used kcal/mole, but always with the value in eV in brackets.

The penetration of the work into chemistry, though perhaps occurring on a wider front, is no deeper than before. The deciding factor has always been the complexity of the molecules involved in the reactions under consideration.

In order to complete a volume it was necessary, at a certain stage, to close the books, as it were, and turn a blind eye to new results coming in after a certain date—unless, of course, they rendered incorrect anything already written. The closing date for Volumes I and II was roughly early 1967, for Volume III mid-1968, and for Volume IV about one year later. Notes on later advances over the whole field are included in Volume V.

London
February 1969
Revised May 1972

H. S. W. M.
E. H. S. B.
H. B. G.

PREFACE TO THE FIRST EDITION

THERE are very many directions in which research in physics and related subjects depends on a knowledge of the rates of collision processes which occur between electrons, ions, and neutral atoms and molecules. This has become increasingly apparent in recent times in connection with developments involving electric discharges in gases, atmospheric physics, and astrophysics. Apart from this the subject is of great intrinsic interest, playing a leading part in the establishment of quantum theory and including many aspects of fundamental importance in the theory of atomic structure. It therefore seems appropriate to describe the present state of knowledge of the subject and this we have attempted to do in the present work.

We have set ourselves the task of describing the experimental techniques employed and the results obtained for the different kinds of collision phenomena which we have considered within the scope of the book. While no attempt has been made to provide at all times the detailed mathematical theory which may be appropriate for the interpretation of the phenomena, wherever possible the observations have been considered against the available theoretical background, results obtained by theory have been included, and a physical account of the different theories has been given. In some cases, not covered in *The theory of atomic collisions*, a more detailed description for a particular theory has been provided. At all times the aim has been to give a balanced view of the subject, from both the theoretical and experimental standpoints, bringing out as clearly as possible the well-established principles which emerge and the obscurities and uncertainties, many as they are, which still remain.

It was inevitable that some rigid principles of exclusion had to be practised in selecting from the great wealth of available material. It was first decided that phenomena involving the collisions of particles with high energies would not be considered, and that other phenomena associated with the properties of atomic nuclei such as the behaviour of slow neutrons would also be excluded. It was also natural to regard work on chemical kinetics as such, although clearly involving atomic collision phenomena, as outside the scope of the book, but certain of the more fundamental aspects are included. Phenomena involving neutral atoms or molecules only have otherwise been included on an equal footing with those involving ions or electrons. A further extensive class of phenomena has been excluded by avoiding any discussion of collision processes occurring within solids or liquids, confining the work to processes occurring in the gas phase or at a gas–solid interface. Among the latter phenomena electron diffraction at a solid surface has been rather arbitrarily excluded as it is a subject already adequately dealt with in other texts. Secondary electron emission and related effects are, however, included.

By limiting the scope of the book in this way it has just been possible to provide a fairly comprehensive account of the subjects involved. It is perhaps too much to hope that even within these limitations nothing of importance has been missed, but it is believed that the account given is fairly complete. Extensive tables of observed and theoretical data have been given throughout for reference purposes and the extent to which the data given are likely to be reliable has been indicated. Every effort has been made to provide a connected and systematic account but it is inevitable that there will be differences of opinion as to the relative weight given to the various parts of the subject and to the different contributions which have been made to it.

We are particularly indebted to Professor D. R. Bates for reading and criticizing much of the manuscript and for many valuable suggestions. Dr. R. A. Buckingham has also assisted us very much in this direction while Dr. Abdelnabi has checked some of the proofs. We also wish to express our appreciation of the remarkable way in which the Oxford University Press maintained the high standard of their work under the present difficult circumstances.

H. S. W. M.
E. H. S. B.

London
August 1951

ACKNOWLEDGEMENTS

Mr. and Mrs. J. Lawson, until their retirement, continued to provide the invaluable assistance they had afforded us in the preparation of previous volumes. Mrs. Harding has not only continued to deal most effectively and with great good humour with the preparation of typescript from the almost indecipherable manuscript but has also undertaken the tedious task of preparing the author index as well as assisting in checking references etc.

We are indebted to Drs. K. Canter, C. Griffiths, and G. Heyland for many useful discussions on positrons and related subjects and to Professor L. Castillejo for clarification of certain matters concerning multiphoton processes.

The Clarendon Press have maintained the high standard which was set in earlier volumes and again have been of much assistance in checking the large bulk of material.

CONTENTS

26. COLLISIONS OF SLOW POSITRONS AND OF SLOW POSITIVE MUONS IN GASES

1. Introduction	3123
2. The annihilation of free positrons	3124
3. Positronium	3125
4. The life-history of a positron in a gas	3127
4.1. Positronium formation in the Öre gap	3127
4.2. Processes leading to quenching of ortho-positronium	3129
5. Experimental methods for investigating slow collisions of positrons and of positronium in gases	3130
5.1. The measurement of delayed coincidence annihilation spectra	3133
5.1.1. Analysis of annihilation–time spectra	3134
5.2. Measurement of orthoPs concentrations by gamma ray methods	3137
5.2.1. From the energy spectrum of single annihilation quanta	3137
5.2.2. From observation of triple coincidences	3142
5.3. Angular correlation measurements	3145
6. Effect of magnetic fields on positronium formation	3148
7. The effect of electric fields on positron annihilation rates and on positronium formation	3155
7.1. The effect on the positron annihilation rate	3155
7.2. The effect on positronium formation	3156
7.3. Experimental observation of the effects on positronium formation	3161
7.4. Summary	3165
8. Summary of experimental possibilities for studying reactions of positrons and of positronium in gases	3166
9. Some historical notes	3168
10. The theory of the collisions of slow positrons with atoms	3169
10.1. Introduction	3169
10.2. Collisions with hydrogen atoms	3171
10.2.1. The zero-order phase shift	3171
10.2.2. The higher order phase shifts	3178
10.2.3. Summarizing remarks	3180

10.3.	Collisions with helium atoms	3181
	10.3.1. The zero-order phase shift	3182
	10.3.2. The higher order phase shifts	3183
	10.3.3. The momentum-transfer cross-section	3183
	10.3.4. The effective number Z_a of annihilation electrons	3186
	10.3.5. The angular correlation between two-photon annihilation quanta	3186
10.4.	Collisions with neon and argon atoms	3187
10.5.	Theoretical evaluation of cross-sections above the Ps formation threshold	3190
11. Analysis and discussion of experimental results on delayed annihilation of free positrons in gases		3191
11.1.	Helium	3191
11.2.	Argon and neon	3202
11.3.	Other gases	3209
12. The theory of the collisions of slow Ps atoms with other atoms		3213
12.1.	Introduction	3213
12.2.	Collisions with hydrogen atoms—exchange quenching	3213
12.3.	Collisions with helium atoms—pick-off quenching	3215
13. Discussion of experimental results on the quenching of orthoPs atoms in gases		3218
13.1.	Results at room temperature	3218
13.2.	Results at low temperatures—bubble formation	3220
14. Positronium chemistry		3227
15. The passage of positive muons through matter and muonium formation		3230
15.1.	Introductory	3230
15.2.	Muonium	3231
	15.2.1. Energy levels and fine structure splitting of muonium	3231
	15.2.2. Hyperfine splitting of muonium	3231
15.3.	Magnetic moment of muonium	3233
15.4.	Depolarization of muons by muonium formation	3233
15.5.	Muonium formation in presence of longitudinal magnetic field	3235
16. Direct experimental identification of muonium formation in gases		3236
16.1.	Direct observation of muonium precession frequency	3236
16.2.	Observation of micro-wave resonance spectrum of muonium	3238
	16.2.1. Observation at high magnetic fields	3240
	16.2.2. Observation at low magnetic fields	3242
	16.2.3. Pressure shift in the micro-wave resonance spectrum	3244

17.	Muonium formation in condensed materials	3248
	17.1. Direct evidence for muonium formation in condensed materials	3248
	17.2. Indirect evidence of muonium production from the study of the spin depolarization of μ^+ stopped in various materials	3253
	17.3. Measurement of μ^+ polarization in a longitudinal magnetic field	3253
	17.4. Measurement of μ^+ polarization in a transverse magnetic field	3255
	17.5. Results of muon depolarization measurements	3255
	17.6. Measurement of spin relaxation time for μ^+ stopping in a condensed medium	3255
	17.7. Interpretation of the observations on muon spin depolarization	3259
	17.8. Physical processes involved in μ^+ depolarization	3262
18.	Measurement of electron spin-exchange cross-sections for muonium	3264
	18.1. Effect of impurities on the micro-wave resonance spectrum of muonium	3264
	18.2. Effect of impurities on μ^+ spin relaxation time in argon	3266
19.	Muonium chemistry	3269

27. COLLISION PROCESSES INVOLVING MESIC ATOMS

1.	Introduction	3276
2.	Processes leading to mesic atom formation	3276
	2.1. Energy loss of fast negatively-charged particles	3277
	2.2. Capture of negative mesons	3278
	2.2.1. Capture of negative mesons by atomic hydrogen	3279
	2.2.2. Capture of negative mesons in molecular hydrogen	3284
	2.2.3. Capture of negative mesons by atoms other than hydrogen	3286
	2.2.4. Depolarization of negative muons in the capture process	3289
3.	De-excitation processes of mesic atoms	3290
	3.1. De-excitation of mesic hydrogen (Me$^-$p)	3290
	3.2. Measurement of the de-excitation time of mesic hydrogen atoms	3294
	3.3. The de-excitation time of mesic atoms in liquid helium	3298
	3.4. De-excitation of heavier mesic atoms	3299
	3.5. The polarization of muons decaying at rest	3300
	3.6. Exchange collisions of (μ^-p) in hydrogen	3304
4.	Cross-sections and collision rates of (μ^-p) and (μ^-d) atoms obtained from the muon-catalysed fusion process, and related phenomena	3306
	4.1. The observation of muon-catalysed fusion	3306

4.2. Experimental study of the process	3308
4.2.1. The bubble chamber method	3308
4.2.2. The diffusion chamber method	3310
4.2.3. Study of γ-ray emission in muon-catalysed fusion	3312
4.3. Derivation of cross-sections for basic processes important in muon-catalysed fusion	3312
4.3.1. Description of the basic processes	3312
4.3.2. The yield of γ-rays and re-emitted muons	3314
4.3.3. Time distribution of γ-ray emission	3315
4.3.4. Influence of hyperfine structure effects on fusion rates	3318
4.3.5. Influence of muon recycling and trapping on fusion rate	3321
4.4. Impurity method of studying rates for collision processes of (μ^-p) and (μ^-d) atoms	3323
4.4.1. Rates for muon transfer and muonic molecule formation	3323
4.4.2. Transfer rate of negative pions from gaseous hydrogen to argon	3328
4.4.3. Scattering cross-sections of (μ^-p) and (μ^-d) atoms	3330
5. Theory of muonic atom collision processes	3336
5.1. The general theory	3336
5.2. States of muonic molecules	3339
5.3. Collisions of (μ^-p) and (μ^-d) in hydrogen and deuterium	3339
5.3.1. Scattering phases and cross-sections	3343
5.3.2. Hyperfine effects in (μ^-p) and (μ^-d) scattering	3345
5.3.3. Transfer of μ^- from hydrogen to heavier elements	3347
5.4. Formation of muonic molecules	3349
5.5. Nuclear fusion rate in muonic molecules	3351
NOTES ON RECENT ADVANCES	3353
ERRATA	3687
AUTHOR INDEX	*1*
SUBJECT INDEX	*5*

of methods for obtaining information about the rates of collision processes involving slow positrons and their theoretical interpretation.

The phenomena observed in the passage of positively charged μ mesons (μ^+) in matter are closely analogous to those observed for positrons. Methods are available for observing the capture of electrons by μ^+ mesons to form muonium (Mu) and its subsequent re-ionization. These methods depend on the existence of the magnetic moment of both the μ^+ and the electron, the fact that the μ^+ is an unstable particle, which decays into an e^+ and neutrinos, with a mean life of the order of about two microseconds, and the fact that the probability of electron emission is not symmetrical about a plane perpendicular to the spin axis of the muon. Mu formation can therefore be monitored by studying changes in μ^+ spin orientation inferred from observation of the asymmetry in the distribution of decay positrons relative to the direction of an applied magnetic field.

The study of the cross-section for capture of electrons by μ^+ to form muonium, and its subsequent re-ionization, as well as the cross-sections for different types of chemical reactions involving slowly moving Mu, will be discussed in the second part (§§ 15–19) of this chapter.

The following chapter will deal with slow collisions of negative particles, particularly negative muons, although some reference will be made, also, to antinucleons and negatively charged pions and strange particles.

2. The annihilation of free positrons

According to relativistic quantum theory, a positron and electron of opposite spin can mutually annihilate each other on collision to produce two gamma quanta. On the other hand, if their spins are parallel, annihilation can only take place through the much less probable process of emission of three gamma quanta. If the mean concentration of electrons at the position of the positron, with opposing spin, is n_s/cm^3 the chance of annihilation per second with emission of two γ-quanta is given by[†]

$$\lambda_s = 4\pi r_0^2 c n_s, \qquad (1)$$

where r_0 ($= e^2/mc^2$) is the classical radius of the electron and c is the velocity of light. Since the chance of annihilation per second with emission of three quanta is so small (of order 1/370 of that for the two quantum process) it follows that, if n is the total electron concentration

[†] DIRAC, P. A. M., Proc. Camb. phil. Soc. math. phys. Sci. **26** (1930) 361.

26

COLLISIONS OF SLOW POSITRONS AND OF SLOW POSITIVE MUONS IN GASES

1. Introduction

CONSIDERABLE interest is attached to the study of the rates of slow collision process in gases which involve rare particles such as positrons, antinucleons, and the strange particles. These particles, either because of the sign of their charge, or their mass, can be expected to behave, in gaseous collisions, in rather different ways from the electrons and neutral and ionized atoms and molecules which we have been discussing in earlier chapters.

It is a very severe test of an approximate theory of scattering of slow electrons by atoms that it should immediately become applicable to positrons by merely changing the sign of the charge. If experimental information about positron collisions can be obtained then it should be well worth while to test the available theory of electron scattering and, if necessary, extend it so that it is capable of reproducing and interpreting the observed behaviour of positrons.

At first sight it would seem unlikely that much could be done experimentally to determine cross-sections for slow positron collisions with atoms and molecules, partly because available positron fluxes are so small, and partly because, in matter, at pressures of an atmosphere or so, the lifetime of a positron against annihilation in impact with an electron is very small, of the order of nanoseconds. However, because of the existence of the annihilation process which produces gamma rays of sufficient energy to be detected in coincidence counters, the lifetimes of individual positrons may be determined. Many collisions with gas molecules can take place before annihilation so that observation of the lifetime as a function of age for positrons in a gas under different conditions can yield a remarkable amount of useful information about collision rates. Information may also be obtained from the details of the annihilation events. There is even the possibility, by use of time-of-flight techniques, of obtaining direct information about total cross-sections for positrons, in different gases, as a function of positron velocity.

The first part of this chapter (§§ 1–14) will be devoted to a description

so that $n_s = \tfrac{1}{4}n$, the annihilation rate is given by

$$\lambda_s = \pi r_0^2 \, cn. \tag{2}$$

We may also define an annihilation cross-section Q_a, where

$$Q_a = \pi r_0^2 \, c Z_a / v, \tag{3}$$

v being the positron velocity and Z_a the effective number of annihilation electrons per atom or molecule in the gas. Z_a will be of the order of the total number Z of atomic electrons but may be greater or less than this under different circumstances (see § 10). If the positrons and electrons are essentially at rest the two annihilation quanta, which are of equal energy 511 keV, will be emitted in opposite sense. A distribution of relative velocity, about a most probable value v, will produce an angular spread of order v/c about the 180° angle between the directions of emission of the quanta.

3. Positronium

An electron and positron can form a bound pair with allowed energies close to one-half of the corresponding values for atomic hydrogen. This is because the reduced mass of the relative motion is $\tfrac{1}{2}m$, where m is the electron mass. Thus the ground state has a binding energy of 6·8 eV, the 2-quantum states 1·7 eV, and so on. A bound electron–positron pair is referred to as positronium and denoted by the symbol Ps. Although there is only a small energy difference between the singlet and corresponding triplet states, they differ greatly in mean lifetime before mutual annihilation of electron and positron occurs.

The singlet S states may decay through emission of two γ-quanta. For the ground singlet state, referred to as paraPs, the lifetime is given by[†]

$$\tau_p = 1\cdot 25 \times 10^{-10} \text{ s},$$

while for higher states $(1s\,ns)\,^1S$ the lifetime is n^3 times greater. On the other hand for the triplet S states 2-quantum annihilation is completely forbidden and decay occurs with emission of three quanta. This increases the lifetime of the ground triplet state, referred to as orthoPs, to[‡]

$$\tau_0 = 1\cdot 41 \times 10^{-7} \text{ s}.$$

Again this is n^3 times longer for a $(1s\,ns)\,^3S$ state.

As far as decay of other excited states of Ps is concerned, the 2-quantum process is forbidden for all states in which the total angular momentum quantum number (J) is 1, and also for states with odd J and odd parity.

[†] DE BENEDETTI, S. and CORBEN, H. C., *A. Rev. nucl. Sci.* **4** (1954) 191.
[‡] ÖRE, A. and POWELL, J. L., *Phys. Rev.* **75** (1949) 1696.

For other states with orbital angular momentum number L the decay rate is of order $(e^2/\hbar c)^{2L}$ times that for the S state with the same value of n. The three quanta are emitted in directions which are coplanar in the centre of mass system for the positronium and, while their energies add to 1·022 MeV, the energy of any one has the continuous statistical distribution shown in Fig. 26.1.

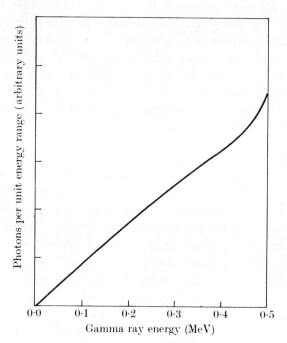

Fig. 26.1. Energy distribution of gamma rays produced in three-quantum annihilation of positrons.

The probability of an azimuthal distribution of the direction of emission of the three quanta in which these directions make angles α, β, γ with each other (see Fig. 26.2) is proportional to[†]

$$\{(1-\cos\alpha)^2+(1-\cos\beta)^2+(1-\cos\gamma)^2\}\sin\alpha\sin\beta\sin\gamma\times$$
$$\times\{\sin\alpha+\sin\beta+\sin\gamma\}^{-2}. \quad (4)$$

The relatively long life of triplet Ps, as well as the distinctive nature of the annihilation process, makes it possible to study many of the features of Ps formation and reactions in gases, as we shall see below.

[†] Siegel, R. T., Thesis, Carnegie Inst. Tech. (1952); Siegel, R. T. and de Benedetti, S., *Phys. Rev.* **87** (1952) 235.

4. The life history of a positron in a gas

A newly emitted positron with an energy of order of a few keV will slow down rapidly in a gas due to inelastic collisions which lead to ionization and excitation. The ionization cross-section for such positrons will nearly be the same as for electrons. Thus, for 100 eV positrons, the ionization cross-section will be around 10^{-16} cm² (see Chap. 3, Table 3.2). On the other hand the annihilation cross-section given by eqn (3) is only $1\cdot2Z_a \times 10^{-23}$ cm². It follows that the positrons will be reduced to an energy equal to the ionization energy E_i of the atoms or molecules in the gas before they will have any appreciable chance of annihilation. The same considerations apply as the positron energy is further reduced until it is equal to the minimum electronic excitation energy E_{ex} of the gas molecules. If the gas is monatomic, further moderation of the positron energy can then take place only through elastic collisions. The relevant comparison is now between $2mQ_d/M$ and Q_a, where m, M are the respective masses of a positron and of a gas molecule, and Q_d is the momentum-transfer cross-section for positron–atom collisions. In argon, for 1 eV positrons, taking Q_d as 10^{-15} cm²,

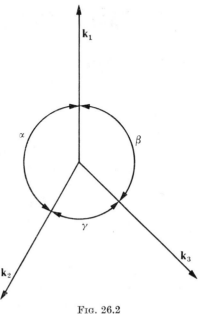

Fig. 26.2

$$2mQ_d/M = 2\cdot5 \times 10^{-20} \text{ cm}^2, \qquad Q_a = 1\cdot2Z_a \times 10^{-22} \text{ cm}^2,$$

so that the two cross-sections are becoming comparable. Even so, it appears that a considerable fraction of positrons will be brought to thermal equilibrium before annihilation.

In polyatomic gases moderation to thermal energies will proceed faster due to additional energy loss by excitation of molecular vibration and rotation.

4.1. *Positronium formation in the Öre gap*

So far we have ignored the possibility of positronium formation. In order that a positron can capture an electron from a gas molecule, it must possess initially a kinetic energy at least as great as $E_i - E_{ps}$, where E_{ps} is the dissociation energy of Ps, 6·8 eV. Also, if positrons possessing an energy $> E_i$ capture electrons, they will produce ground-state Ps possessing a kinetic energy $> E_{ps}$. Such Ps atoms will, on the average, be broken up again in collisions with gas molecules so that permanent positronium formation will not occur unless the positron energy $E < E_i$. However, most positrons with energies such that $E_i > E > E_{ex}$ will suffer inelastic collisions producing electronic excitation of the gas molecule, so degrading their energies below the positronium formation threshold before they can capture an electron.

According to these considerations one would expect that 'permanent' positronium will only be produced by positrons with energies in the range $E_{ex} > E > E_i - E_{ps}$, the so-called 'Öre gap', as its existence was first pointed out by Öre† in 1949. Assuming that, following the last ionizing collision, there is a uniform distribution of positron kinetic energies between zero and E_i, rough limits may be placed on the fraction (ϕ) of positrons which form Ps. Thus

$$\frac{E_{ex} - (E_i - E_{ps})}{E_{ex}} < \phi < \frac{E_i - (E_i - E_{ps})}{E_i} = \frac{E_{ps}}{E_i}. \tag{5}$$

Whilst an Öre gap exists for all the rare gases, for most molecules the lowest electronic excitation energy is often so low that $E_i - E_{ps} > E_{ex}$, and no Öre gap exists, in the strict sense. However, in many such cases the chance of excitation, by positron impact, of the low-lying molecular electronic levels, which are normally metastable, may be very small, so that an effective Öre gap does exist, and an appreciable fraction of positrons form positronium in the gas concerned.

The ultimate fate of the free positrons, which have passed through the Öre gap, and of the orthoPs atoms, which have been formed in the gap, depends very much on the nature of the gas. Eventually, the free positrons will be annihilated by a molecular electron, with a probability which will be much enhanced if the positron can be attached to a molecular fragment, or form a complex with a whole molecule having a lifetime greater than the mean annihilation time. A striking example is provided by annihilation of thermalized positrons in freon. Even in monatomic gases there is the possibility of complex formation in narrow energy ranges due to resonance capture, which may appreciably enhance the annihilation rate.

† ÖRE, A., *Naturvidenskap Rikke No. 9*, University of Bergen, Årbok (1949).

4.2. *Processes leading to quenching of ortho-positronium*

In addition to the relatively slow rate of three-quantum annihilation of orthoPs atoms there are a number of other ways in which the bound positron may be annihilated. These orthoPs quenching processes may be classified as follows.

(*a*) *Electron exchange in a molecular collision.* If the ground state of the molecule possesses a non-zero total electron spin, electron exchange between the molecule and an orthoPs atom, on collision, may convert the ortho- to paraPs, which will then annihilate promptly. The simplest theoretical case is that of exchange of electrons of opposite spin in a collision with an H atom. In practice, the most important example is that of quenching in nitric oxide. The ground state of the NO molecule is a doublet ($^2\Pi$), as with atomic hydrogen, and it converts orthoPs to paraPs very efficiently, on impact. This strong quenching property was put to good use in the pioneer experiments of Deutsch[†] in order to establish the existence of Ps (see § 9), and it remains a valuable tool in the study of positron and positronium behaviour (see p. 3137).

The O_2 molecule, with a triplet ($^3\Sigma$) ground state, can also convert ortho- to paraPs through electron exchange on collision, but it is much less effective than NO.

(*b*) *Pick-off annihilation.* Just as a free positron, the positron in orthoPs may annihilate on collision with one of the molecular electrons. The chance of this occurring, per impact with a particular molecule, will not be the same as for a free positron, because the presence of the Ps electron will modify the motion of the positron relative to the molecule. The simplest theoretical example in which this occurs is in helium, in which exchange quenching is not possible.

(*c*) *Quenching due to chemical reactions.* In a collision of a Ps atom with a molecule XY a reaction of the form

$$\text{Ps} + \text{XY} \to \text{XPs} + \text{Y}, \tag{6a}$$

analogous to
$$\text{H} + \text{XY} \to \text{XH} + \text{Y}, \tag{6b}$$

may occur. Because a reaction such as (6a) will maintain the positron in the close proximity of the electrons of the atom or radical X it will increase the probability of annihilation, an effect which will appear as orthoPs quenching.

A further possibility is a reaction in which the positron is freed, analogous to the reaction

$$\text{H} + \text{oxidant} \to \text{H}^+ + \text{reduced product},$$

[†] DEUTSCH, M., *Phys. Rev.* **82** (1951) 455; **83** (1951) 866.

i.e. $\quad$ Ps+oxidant $\to$ e$^+$+reduced product.

These reactions are relatively rare in the gas phase, but are important in liquid, solution, and solid phases.

(*d*) *Quenching due to radiative or three-body capture.* If the energy of the complex XPs is less than that of the ground states of X+Ps, a Ps atom may be captured by a molecule X with emission of radiation. The radiative capture cross-section will at least be comparable with cross-sections for pick-off annihilation. As for free positrons, capture of orthoPs in three-body collisions will be even more important at pressures much greater than atmospheric. In either case once the complex is formed annihilation becomes much more probable.

It is important to distinguish between *quenching*, in which orthoPs once formed is caused to annihilate rapidly through one of the processes listed above, and *inhibition* of orthoPs formation, in which the effect of the presence of a particular gaseous species is to reduce the chance that Ps formation occurs at all. Thus the addition of a polyatomic gas to a rare gas will tend to reduce the probability of Ps formation, by reducing the time spent by positrons in the Öre gap, the rate of energy loss being enhanced by inelastic collisions leading to vibrational and rotational excitation. The fractional yield of Ps will also be reduced if positrons with energies within the Öre gap are captured with appreciable probability either to form long-lived complexes, or stable compounds, with molecular radicals. The captured positrons will be annihilated promptly, and so be lost as producers of Ps. In some experiments it is difficult to distinguish between quenching and inhibition, but with the range of experimental possibilities available, it is usually possible to identify the general nature of the processes involved.

The scope of experimental study of the different processes concerned in the life history of positrons and positronium can be extended very substantially by changing certain external variables such as the gas temperature and applied electric, and magnetic, fields. We next outline the principles involved in the different types of experiments which are practicable, and discuss the nature of the information which can be derived from them about the nature and rates of the basic reactions involved.

5. Experimental methods for investigating slow collisions of positrons and of positronium in gases

There are three chief methods for obtaining information relating to the collision interactions of slow positrons and Ps atoms in gases. They may be summarized as follows:

(a) The measurement of the delayed annihilation–time spectrum of positrons. In this type of experiment the rate of annihilation is measured, with high time resolution, as a function of the time since emission from the source. These spectra may be taken at different pressures of the main or admixed gases, at different temperatures, and in the presence of electric fields.

(b) The observation of γ-rays which are produced exclusively by annihilation of orthoPs. Identification of the nature of the source can be made either by observation of triple coincidences in counters arranged with their centres coplanar with the centre of the gas chamber or by taking advantage of the energy spectrum of the annihilation γ-rays (see Fig. 26.1). A further possibility is to monitor any change in orthoPs population from observation of the coincident signals in two counters, viewing the chamber in opposite directions. These arise from singlet annihilation but, as increase in orthoPs will be accompanied by a decrease in such annihilations, and vice versa, they serve to indicate the change in orthoPs population.

(c) Measurement of the angular correlation between the two γ-rays resulting from singlet annihilation. From these observations the mean value of a single component of the momentum of the annihilating pair may be derived.

These measurements all employ scintillation counters for detecting the γ-rays arising from positron annihilation, or from the positron source. When a beam of monochromatic γ-rays of quantum energy less than 1·024 MeV ($2mc^2$) falls on a scintillating crystal such as sodium iodide, the electrical pulses recorded in a photomultiplier coupled to the crystal have a height distribution of the form shown in Fig. 26.3. The sharp peak at the high-energy end is due to γ-rays absorbed photoelectrically, the photoelectrons, as well as any Compton electrons, being reduced to rest in the crystal so that all the energy of the incident photon is released in the scintillator. The plateau results from γ-rays, part of whose energy is carried off by particles—Compton electrons or Compton degraded photons or photoelectrons—that leave the crystal. The spectrum is more complicated for γ-rays of energy above 1·024 MeV when positron–electron pair production is possible. Either one, or both, photons of the positron annihilation radiation may leave the crystal, giving two additional peaks corresponding to one-photon escape or two-photon escape. The height of the peak corresponding to the total absorption of the incident γ-ray is measured by a pulse-height analyser and provides a measure of the γ-ray energy.

The monochromatic γ-rays, which result from singlet annihilations, will give rise to a pulse-height distribution as in Fig. 26.3 (i), but those from the triplet annihilation having wave-lengths which vary continuously (see Fig. 26.1) up to a maximum of 0·511 MeV give relatively greater

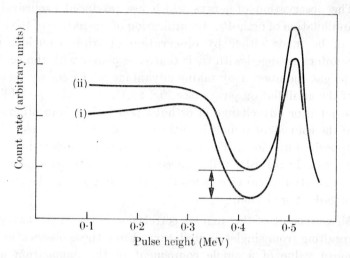

Fig. 26.3. Pulse-height spectra from annihilation radiation as recorded by a scintillation counter (i) when the radiation is monochromatic as in singlet annihilation of Ps, (ii) when the radiation includes a contribution from triplet annihilation of Ps with energy distribution as in Fig. 26.1.

contributions at lower pulse-heights, particularly in the valley region in Fig. 26.3. By selecting the pulse-height for recording, more, or less, weight may be given to singlet as compared with triplet annihilation.

A pulse-height analyser in association with scintillation counters can also be used to measure the time interval between two pulses using a time to pulse-height converter circuit. A typical circuit of this kind would consist of a condenser charged up to a certain potential. The first counter pulse triggers the discharge of the condenser through a resistance. The second counter pulse triggers the discharge of the remaining charge through the pulse-height analyser which records a height determined by the time interval between the two counter pulses.

We now consider each of these methods in turn, including an account of the technique employed with reference to particular experiments, and of the analysis, in terms of collision rates or cross-sections, of the measurements made.

5.1. *The measurement of delayed coincidence annihilation spectra*

In carrying out measurements of this kind, using positrons emitted from a ^{22}Na radioactive source, advantage is taken of the fortunate fact that a γ-ray of 1·28 MeV energy is emitted effectively simultaneously with a positron. The experiment therefore consists of observing delayed coincidences between signals recorded on scintillation counters by the 1·28 MeV γ-ray and an annihilation γ-ray.

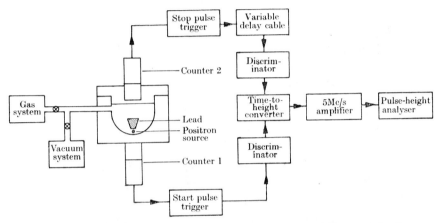

FIG. 26.4. Block diagram of the apparatus used by Duff and Heymann for the observation of delayed-coincidence positron annihilation spectra in gases.

We illustrate the techniques which have been employed by reference to the experiments of Jones and his collaborators,[†] and of Duff and Heymann.[‡] Fig. 26.4 shows diagrammatically the apparatus used by the latter. Pressures up to about 60 atm could be used in an aluminium alloy gas chamber. The positron source consisted of ^{22}Na, of strength 1 mCi, mounted on a small aluminium disc. Two plastic scintillator counters in the form of two cylinders 2 in long and 2 in in diameter were used. They were coupled to EMI 6097 photomultipliers. A lead cone screened one of the counters (2) from the source so that it picked up a signal only from the positron annihilation γ-rays while counter 1 picked up a signal from the nuclear γ-rays as well. A pulse from counter 1 signalling the emission of a nuclear γ-ray was used to start the operation of the time to pulse-height converter. A pulse from counter 2 signalling the emission of an annihilation γ-ray photon was used to stop the converter and a pulse of amplitude proportional to the time interval between

---

† ORTH, P. H. R. and JONES, G., *Phys. Rev.* **483** (1969) 7.
‡ DUFF, B. G. and HEYMANN, F. F., *Proc. R. Soc.* **A270** (1962) 517.

the arrival of the two pulses at the converter was fed into the pulse-height analyser. The converter enabled time intervals over a range of 6×10^{-7} s to be measured with a resolution of 2×10^{-9} s. The discriminators were used to avoid triggering on dark current pulses from the photomultiplier. The variable delay cable enabled a known delay time to be introduced into the stop pulse, and was used to calibrate the time scale of the

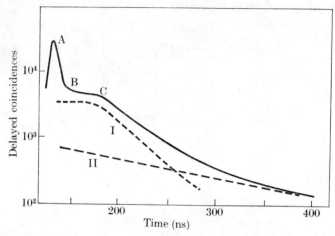

Fig. 26.5. Typical form of delayed coincidence annihilation spectrum for positrons in a rare gas. I, component arising from free positron annihilation. II, component arising from orthoPs annihilation.

converter, by noting the shift in separation between the two peaks. The linearity of the time to pulse-height conversion was checked by using two independent γ-ray sources, and placing one near each counter. This would be expected to give rise to a nearly flat spectrum, since the time separation between counts in 1 and 2 should then be random. Departures from flatness enabled linearity correction factors to be derived.

5.1.1. *Analysis of annihilation–time spectra.* A typical form of annihilation spectrum for a rare gas is illustrated in Fig. 26.5. It consists of a prompt peak A, occurring within some nanoseconds of the positron emission, followed by a shoulder (BC) and then by a smoothly decreasing curve which gradually merges with the background. The prompt peak is due to annihilation of positrons which have formed singlet positronium or have suffered annihilation in the source or walls of the experimental chamber. The existence of the shoulder indicates an annihilation rate depending on the time. If we suppose that, during the period spanned by the shoulder, the positrons are slowing down but have not yet reached thermal equilibrium with the gas atoms, then the presence of the shoulder implies that, over the velocity range it covers, the number Z_a of effective annihilation electrons is a function of positron velocity. We would expect in terms of this explanation that the width Δt of the shoulder should be inversely proportional to the gas pressure as it is found to be in practice. Beyond the shoulder the annihilation spectrum with background subtracted can be analysed in the form

$$ae^{-\lambda_1 t} + be^{-\lambda_2 t}, \qquad (7)$$

where $\lambda_2 < \lambda_1$, arising from a combination of two decay processes. The slower process is due to annihilation of orthoPs and the faster to a constant annihilation rate of free positrons. This constancy would be expected if the positrons are thermalized, but it could arise before this stage is reached if Z_a tends to a constant for positron velocities less than some epithermal value v_e, say.

Analysis of a spectrum to yield the two exponential decay rates presents difficult problems, and requires very reliable basic data. Accounts of procedures which have been adopted are given by several authors.† It is possible to assist in the isolation of the two decay rates by noting that λ_1 should be proportional to pressure, while

$$\lambda_2 = \lambda_0(1+p\mu_q/\lambda_0),$$

where λ_0 is the natural annihilation rate of orthoPs, p the pressure, and μ_q the quenching rate of ortho- to paraPs, per unit pressure. If no quenching process is operative λ_2 is constant. The analysis is simplified by adding a suitable gas which effectively eliminates one or other of the decay processes. Thus a powerful quencher of orthoPs, such as NO, can be used to remove the slow decay thus enabling an estimate of the annihilation cross-section for slow free positrons. On the other hand in a substance such as freon 12 (CCl_2F_2) there are effectively no free positron annihilations as the positrons become attached to the freon molecules and annihilate in this bound state, with a lifetime short compared with that for free positrons. Examples of the application of these procedures are given in §§ 11 and 13. λ_1 is given in terms of the effective number of annihilation electrons $Z_a(v)$ by

$$\lambda_1 = n\pi r_0^2 c \int_0^\infty Z_a(v) f(v) \, dv, \qquad (8)$$

where $f(v)$ is the Maxwellian distribution function at the gas temperature T, namely,
$$f(v) = (2m^3/\pi\kappa^3 T^3)^{\frac{1}{2}} v^2 \exp(-mv^2/2\kappa T). \qquad (9)$$

† COLEMAN, P. G., GRIFFITH, T. C., and HEYLAND, G. R., *J. Phys. E: Sci. Instrum.* 5 (1972) 376.

If λ_1 can be measured as a function of T over a wide enough range ΔT of T, it is possible to invert the relation (8) to obtain $Z_a(v)$ as a function of v over a range of order $(\kappa \Delta T/m)^{\frac{1}{2}}$.

The analysis of the shoulder region of the annihilation spectrum is less definite. In this region the rate of annihilation is given by eqn (8), but with the distribution function $f(v,t)$ depending on the time t. It will satisfy the Boltzmann equation (see Chap. 2, § 1.2)

$$\frac{\partial f}{\partial t} = \frac{n}{Mv^2} \frac{\partial}{\partial v}\left\{v^3 Q_d\left(\kappa T \frac{\partial f}{\partial v} + mvf\right)\right\} - \lambda_a f, \qquad (10)$$

where M is the mass of a gas atom and Q_d is the momentum-transfer cross-section for collisions between positrons of velocity v and the gas atoms.

λ_a is the annihilation decay constant $n\pi r_0^2 c Z_a(v)$ for positrons of velocity v. Since $\lambda_a \ll nvQ_d$, for almost all cases of importance, $\lambda_a f$ can usually be neglected in comparison with the other terms. In principle, it would seem possible to obtain Q_d by trial and error methods, determining $f(v,t)$ for any chosen form of Q_d by solving eqn (10). However, it is necessary to know, also, the initial velocity distribution $f(v,0)$ of the positrons; this is not known from experiment, and cannot be calculated. It is usual to make a similar assumption to that used in deriving eqn (10), namely, that there is initially a uniform distribution of positron kinetic energies between zero and the threshold for Ps formation.

The presence of impurities in quite small concentration can be very serious in modifying annihilation spectra in rare gases. We have already pointed out how certain gases can be used as additives to eliminate either one of the two decay rates. At working pressures an orthoPs atom will make around 10^{11} collisions per s, and thus as many as 10^4 in its natural lifetime. Since the probability of quenching per collision with a strongly quenching molecule can be of order unity, it is necessary to reduce the fractional impurity concentration of such molecules to $< 10^{-4}$ if the measured value of λ_2 is to be characteristic of the main gas. Similar requirements apply to the measurement of λ_1, but the effective molecules, which must capture positrons easily, are relatively unusual as impurities.

The shoulder width is very sensitive to polyatomic impurities of all kinds, because these can accelerate the slowing down of positrons to thermal equilibrium to a relatively great extent, through inelastic collisions which involve vibrational and rotational excitation. Even diatomic molecules such as N_2 and O_2 can be important in this regard.

A positron loses a fraction $2m/M$ of its energy on the average in elastic collisions with atoms of mass M. If the fractional energy loss in a collision with a molecule is μ, it follows that impurity concentrations of order $2m/M\mu$ will be important. In N_2 for electrons of characteristic energy 4 eV, $\mu \simeq 5 \times 10^{-2}$ (see p. 727), while for argon $2m/M \simeq 3 \times 10^{-5}$, so that fractional concentrations as low as 10^{-4} could be important.

We shall discuss the effect of a uniform electric field on the annihilation spectrum in § 7.1.

5.2. Measurement of orthoPs concentrations by gamma ray methods

5.2.1. *From the energy spectrum of single annihilation quanta.* We have already pointed out on p. 3132 how selection of the pulse-height from a scintillation counter can be used to give more, or less, weight to singlet as compared with triplet annihilation. In particular, if the pulse-height is set in the valley region shown in Fig. 26.3 the relative contribution from triplet decay events will be a maximum. As an illustration of the use of this result for quantitative observation of quenching of orthoPs by various gases, we describe the experiments of Heymann, Osmon, Veit, and Williams.† The positron source consisted of about $\frac{1}{4}$ mCi of ^{22}Na sandwiched between two mica sheets 6 mg cm^{-2} thick, and 2 cm in diameter, and held in the centre of the aluminium alloy gas chamber in the form of a cylinder 6 in long by 4 in diameter, and with a wall thickness of $\frac{1}{8}$ in. A sodium iodide (Tl) crystal 1 in thick by $1\frac{3}{4}$ in diameter, coupled optically to an EMI 6260 photomultiplier and constituting the scintillation counter, was placed 60 cm from the gas chamber, far enough for its detection efficiency (η) for an annihilation event to be effectively independent of the location of the event in the chamber. This was important to ensure that η should not depend on the gas pressure. The background pressure in the chamber, before introduction of the working gas, was less than 5×10^{-6} torr.

Fig. 26.6 shows as one curve an observed gamma ray spectrum, with pure argon in the chamber. The effect of adding 3 per cent NO to the argon, sufficient to quench the orthoPs, is shown by a second curve. It is clear from the reduced signals in the valley region, and the increased signal near the peak, that most of the gamma quanta from orthoPs decay are recorded in the former region. Addition of the NO converts all annihilation events to two-quantum processes, enhancing the counting rate near the peak at 511 MeV. Accordingly, Heymann and his

† HEYMANN, F. F., OSMON, P. E., VEIT, J. J., and WILLIAMS, W. F., *Proc. phys. Soc.* **78** (1961) 1038.

collaborators adjusted their apparatus to record only pulses in the range corresponding to 300–430 keV.

Let R_p be the counting rate in the gas at pressure p, and R_0 that in the same gas containing 3 per cent NO in addition. If λ_0 is the annihilation rate for pure orthoPs, and λ_q the quenching rate, the fraction of

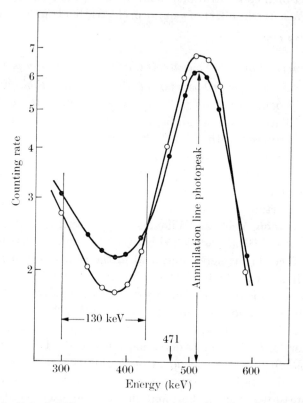

Fig. 26.6. Pulse-height spectrum of annihilation radiation from positrons in argon observed by Heymann, Osmon, Veit, and Williams, -●-●- in pure argon -○-○- in argon containing 3 per cent NO.

orthoPs atoms which will suffer three-quantum annihilation will be $\lambda_0/(\lambda_0+\lambda_q)$. Hence, if A is the number of orthoPs atoms which emerge from the Öre gap per second, the contribution to the counting rate from three-quantum events will be $\eta_3 A\lambda_0/(\lambda_0+\lambda_q)$, where η_3 is the detecting efficiency of the counter for these events. In addition, there will be a contribution from two-quantum events, for which the detecting efficiency is η_2. This will consist of an amount $\frac{1}{3}A\eta_2$ from paraPs annihilation, and $\eta_2 A\lambda_q/(\lambda_0+\lambda_q)$ from the quenching of orthoPs. We

thus have
$$R_p = \eta_3 A \frac{\lambda_0}{\lambda_0+\lambda_q} + \tfrac{1}{3}\eta_2 A\left(\frac{3\lambda_q}{\lambda_0+\lambda_q}+1\right)+C, \qquad (11)$$
where C is a background rate due to free positron annihilation in the gas and in the walls, source, and source-holder, and to the 1·28 MeV γ-rays emitted in near coincidence with each positron.

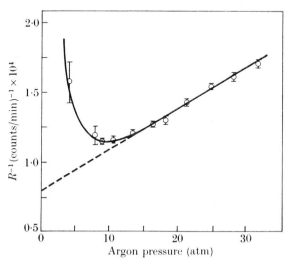

FIG. 26.7. Plot of reciprocal counting rate (R) (see eqn (14)) against argon pressure for quenching of orthoPs in argon as observed by Heymann et al.

Again, with 3 per cent NO present all annihilations proceed through the two-quantum process, so
$$R_0 = \eta_2 \frac{4A}{3}+C. \qquad (12)$$
Thus
$$R = R_p - R_0 = A(\eta_3-\eta_2)\lambda_0/(\lambda_0+\lambda_q), \qquad (13)$$
so
$$\frac{1}{R} = \frac{1}{(\eta_3-\eta_2)A}\frac{\mu_q p}{\lambda_0} + \frac{1}{(\eta_3-\eta_2)A}, \qquad (14)$$
where we have written λ_q, which is proportional to the pressure p, as $\mu_q p$. It follows from eqn (14) that a plot of $1/R$ against p will be a straight line with intercept $-\lambda_0/\mu_q$ on the pressure axis. In this way μ_q may be obtained. Fig. 26.7 shows a typical plot of this kind for argon. The departure from linearity at pressures below about 15 atm arises because, at lower pressures, not all of the positrons are stopped within the chamber. Above 15 atm the relation is accurately linear and gives
$$\mu_q = 0\cdot0352\lambda_0 \text{ per atm} \pm 2 \text{ per cent}.$$

For strongly quenching gases, such as NO and O_2, R_p referred to argon, to which a fixed percentage of the strong quencher was added sufficient to reduce the three-quantum rate to about one half (0·03 per cent NO, 1·6 per cent O_2). Figs. 26.8 and 26.9 show (R^{-1}, p) plots, for

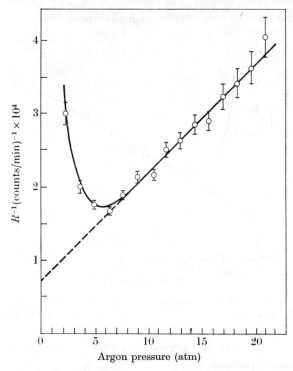

Fig. 26.8. Plot of reciprocal counting rate (R) (see eqn (14)) against argon pressure for quenching of orthoPs in a mixture of argon with 0·03 per cent of NO as observed by Heymann *et al.*

NO and O_2, from which it was found that $\mu_q = 400\lambda_0$ atm^{-1}, and $3\lambda_0$ atm^{-1} for NO and O_2, respectively. Other results obtained by this method are given in Table 26.5.

The first use of this technique for determining the fraction of positrons which form positronium was made by Pond.† Let R_m be the peak counting rate in the pure gas at a pressure in which quenching is negligible. Then if A and B are the rates at which orthoPs and free positrons pass through the lower limit of the Öre gap per second,

$$R_m = \eta_5 A + \eta_4(\tfrac{1}{3}A+B)+C, \qquad (15)$$

† Pond, T. A., *Phys. Rev.* **85** (1952) 489.

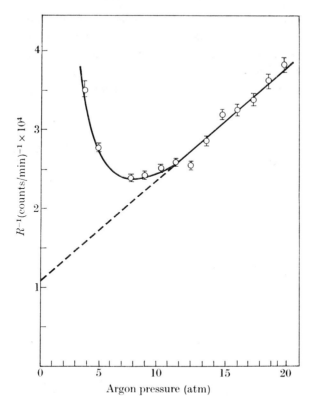

Fig. 26.9. Plot of reciprocal counting rate (R) (see eqn (14)) against argon pressure for quenching of orthoPs in a mixture of argon with 1·6 per cent of O_2 as observed by Heymann et al.

where η_4, η_5 are the detecting efficiencies, for quanta from 2- and 3-quantum annihilation processes, respectively, and C is the background count. With a sufficient pressure of NO added to quench the orthoPs, the counting rate will become $R_m(0)$, where

$$R_m(0) = \eta_4(\tfrac{4}{3}A+B)+C. \tag{16}$$

Hence
$$\frac{R_m(0)-C}{R_m-C} = \frac{\tfrac{4}{3}A+B}{\tfrac{1}{3}A+B+\eta_5 A/\eta_4}. \tag{17}$$

$R_m(0)$ and R_m may be directly measured. The estimation of the background rate C is more difficult. Hughes, Marder, and Wu† considered that an upper limit to C was given by measuring the counting rate when no gas is present. This is mainly due to positron annihilation in the walls of the chamber, the rate of which will certainly be higher than

† HUGHES, V. W., MARDER, S., and WU, C. S., Phys. Rev. **98** (1955) 1840.

when no gas is present because of the positrons which annihilate in the gas. Hughes *et al.* then took as an estimate of C, one-half of the counting rate when no gas is present. This was a correction of only 5 per cent, which itself is probably correct to about 10 per cent. A further correction is necessary because η_5/η_4 on the right-hand side, though small, is finite.

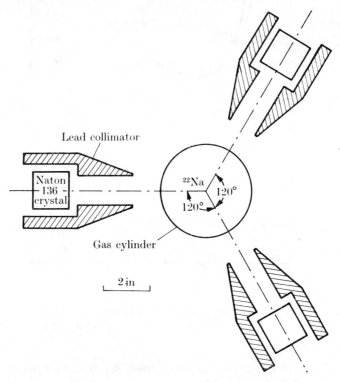

Fig. 26.10. Counter arrangement used by Celitans and Green for the detection of three-photon annihilation.

It may be estimated from the calculated energy distribution of the 3-quantum annihilation gamma rays and the geometry of the system.

5.2.2. *From observation of triple coincidences.* A further method for monitoring the three-quantum decay rate is by observation of triple coincidences. As an illustration of the application of this technique to the measurements of quenching rates for orthoPs, we describe the experiments of Celitans and Green.†

The positron source was about 0·2 mCi of ^{22}Na sandwiched between two 1 cm square mica sheets 2 mg cm^{-2} thick, mounted at the centre

† CELITANS, G. J. and GREEN, J. H., *Proc. phys. Soc.* **83** (1964) 823.

of a stainless steel gas cylinder of 5 in diameter and 8 in length. Fig. 26.10 illustrates the arrangement of the three counters placed symmetrically in a plane containing the source. The detectors were Naton 136 plastic phosphors (*p*-terphenyl in polyvinyl toluene) of 2 in diameter and 2 in

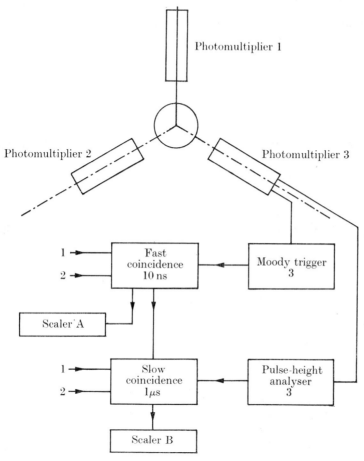

FIG. 26.11. Block diagram of the fast–slow triple coincidence system used by Celitans and Green.

length coupled to E.M.I. 6097B photomultipliers with paraffin oil. To reduce spurious coincidences, due to scattering from counter to counter, each was enclosed in a lead collimator.

Fig. 26.11 shows a block diagram of the electronics for the fast–slow coincidence method employed.† Pulses from each of the photomultipliers were fed into a fast triple coincidence circuit with resolving time 10 ns.

† BELL, R. E., GRAHAM, R. L., and PETCH, H. E., *Can. J. Phys.* **30** (1952) 35.

In addition, each pulse was passed through a channel analyser. Quadruple coincidences between the outputs from these pulse analysers, and the overall output from the fast triple-coincidence circuit, provided the signal which was taken to indicate three-photon annihilation. This quadruple-coincidence circuit had a resolving time of 1 μs, related to the slower time constants of the analyser circuits. By this means, the three-photon annihilation processes were selected on the basis of γ-ray energy in addition to the observation of the emission of three photons.

For an experiment on a particular gas, or gas mixture, the coincidence rate was determined for pressures up to 23 atm, with 2 atm intervals. To obtain a particular gas mixture the cylinder was first filled with the impurity gas and then diluted in successive stages with the background gas until the desired concentration was reached, the composition being checked by a Metropolitan-Vickers MS2 mass spectrometer. Reproducibility was found to be better than 5 per cent. The random coincidence rate was determined at each pressure by tilting one of the detectors out of the common plane by about 45°, the distance from source to crystal being adjusted to maintain the same count rate. In most experiments the random coincidence rate was about one-eighth of the true rate.

The experimental three-photon coincidence rate, after subtracting background, and correcting for the geometry of the detection system and the positron range, may be written

$$R = kf\lambda_0/(\lambda_0+p\mu_q),$$

where f is the fraction of positrons that form positronium, and k a constant depending on the positron source strength, counter efficiency and solid angle, positron absorption in the source holder, and gamma ray absorption in the gas cylinder. λ_0 is the natural annihilation rate of orthoPs and μ_q is the quenching rate per unit pressure, the assumption being made that the quenching process is proportional to the pressure as would be the case for the 'pick-off' quenching process discussed above, so that

$$R^{-1} = (kf)^{-1}(1+\mu_q p/\lambda_0),$$

i.e. R^{-1} is linear with pressure.

Using this method Celitans and Green† observed quenching rates for argon containing small admixtures of oxygen and of nitrogen. Fig. 26.12 shows plots of R^{-1} against pressure, observed for Ar containing fractional concentrations of 5×10^{-6} and 5×10^{-4} of O_2. From these and other plots, a quenching rate λ_q per atm of argon was determined as $3\cdot 5\pm 0\cdot 4\times 10^{-2}$ s^{-1}, independent of the fractional concentration in the range 5×10^{-6} to 5×10^{-4}. At higher concentrations, quenching by O_2 became important, while at concentrations of 2×10^{-6} and below, a

† loc. cit., p. 3142.

peculiar effect was found, in that the observed value of R^{-1} was found to vary quadratically with the pressure. The reason for this behaviour, which was also found in experiments by Celitans, Tao, and Green† using the lifetime method, remains obscure.

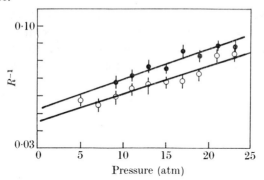

Fig. 26.12. Plots of inverse triple coincidence counting rate against pressure for orthoPs in argon containing a small admixture of O_2 as observed by Celitans and Green. ⌀: 5×10^{-6} O_2, ●: 5×10^{-4} O_2.

5.3. Angular correlation measurements

It has already been pointed out on p. 3125 that, if the centre of mass of an annihilating pair is not at rest in the laboratory system, the gamma rays from a two-quantum annihilation event will not be emitted in exactly opposite directions. If the probability distribution of the angle between the two directions can be measured information can be obtained about the velocity distribution of the centres of mass of annihilating pairs or, more correctly, about that of the component of velocity normal to the direction of observation. Thus, let **p** be the momentum of the centre of mass of an annihilating pair. If $\mathbf{P}_1$, $\mathbf{P}_2$ are the momenta of the emitted quanta,
$$\mathbf{P}_1 + \mathbf{P}_2 = \mathbf{p},$$
so that
$$\mathbf{P}_1 \times \mathbf{P}_2 = \mathbf{P}_1 \times \mathbf{p}.$$
Hence if θ is the small angle between the directions of $\mathbf{P}_1$ and $\mathbf{P}_2$ we have, since $P_1 \simeq P_2 \simeq mc$,
$$\theta \simeq p_z/mc,$$
where p_z is a component of **p** perpendicular to $\mathbf{P}_1$. Since p_z is of order $(Em)^{\frac{1}{2}}$, where E is the kinetic energy of the centre of mass of the pair
$$\theta \simeq (E/mc^2)^{\frac{1}{2}}.$$
Thus, for thermalized positronium,
$$\theta \simeq (0 \cdot 03/5 \times 10^5)^{\frac{1}{2}},$$
$$= 0 \cdot 25 \text{ mrad},$$

† Celitans, G. J., Tao, S. J., and Green, J. H., *Proc. phys. Soc.* **83** (1964) 833.

whereas for pick-off annihilation (see p. 3129), in which the positron annihilates with an outer atomic electron, $E \simeq 5$ eV and

$$\theta \simeq 3 \text{ mrad.}$$

It follows that, if measurements of angular correlation can be made, with a resolution of a milliradian, or better, very useful information can be obtained. Thus we would expect that the observations of the angular distribution in a gas should be of the form shown in Fig. 26.13 with a

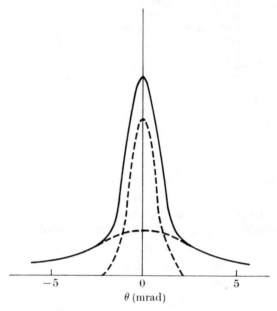

Fig. 26.13. Typical shape of an angular correlation function showing analysis into two components.

sharp peak at $\theta = 0$, with width so small as to be largely determined by the resolving power of the apparatus, superimposed on a much broader distribution, with width of order of a few milliradians, due to pick-off annihilation of free positrons and of orthopositronium, and annihilation of non-thermalized paraPs. The possibility of analysis into these two separate components can be made use of in a number of ways (see p. 3152).

Some of the earliest experiments on the annihilation of positrons in different materials were carried out by angular correlation measurements, and the technique has proved to be a very useful complement to the others we have described. As an illustration of an experiment of this kind, carried out with positrons annihilating in gases, we describe the equipment used by Obenshain and Page,† which was applied particularly to the study of the effects of electric and magnetic fields on Ps formation.

The general arrangement is shown in Fig. 26.14. A and B are the two scintillation counters each 120 cm from the centre (O) of the gas chamber. Adjacent to each counter is a pair of lead bricks forming detector slits

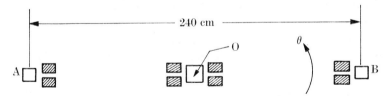

FIG. 26.14. General arrangement of the apparatus in the angular correlation experiments of Page and his collaborators.

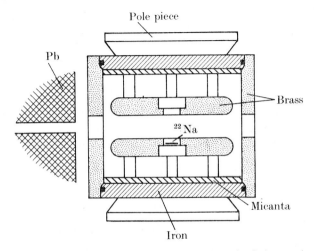

FIG. 26.15. Gas chamber used by Obenshain and Page in their experiments on the effect of strong electric and magnetic fields on Ps formation in gases.

of adjustable width. A similar pair of bricks forms a defining slit on each side of the source assembly. The defining and detector slits, as well as the counter B, are carried on a rigid arm which can be rotated about an axis through O, perpendicular to the plane of the paper.

The scintillator crystals were of dimensions $1\frac{1}{4} \times 1\frac{1}{4} \times \frac{1}{4}$ in coupled to Du Mond 6292 photomultipliers. Fig. 26.15 illustrates one of the gas chambers used, the electric field being applied between two brass electrodes, whose separations were variable up to 1 cm. One of the electrodes contained a 15 mCi ^{22}Na source deposited on a quartz backing.

† OBENSHAIN, F. E. and PAGE, L. A., *Phys. Rev.* **125** (1962) 573.

The gas pressures were such that the positrons were stopped in the gas between the electrodes. The angular resolution, determined by the slit width, corresponded to a pair of kinetic energy of about 0·5 eV. Results obtained using this equipment are described and discussed on p. 3164.

An example of high resolution angular correlation equipment is that used by Stewart and his collaborators,† who were concerned particularly with annihilation of positrons in liquid helium. The detectors were located about 20 feet away on each side of the source and subtended an angle at the liquid helium specimen of about 0·20 milliradians. This specimen was defined by a fixed lead slit and a slightly larger moving slit close to the cryostat, so that the volume of the specimen did not change with angle. With this apparatus the resolution measured by the full width, at half maximum, was 0·4 milliradians. For some experiments an even better resolution of 0·25 milliradians was achieved. Results obtained with this equipment are described and discussed on p. 3200.

6. Effect of magnetic fields on positronium formation

Application of a magnetic field to a Ps atom partially raises the degeneracy between the triplet states with $m = 0$ and those with $m = \pm 1$, m being the magnetic quantum number. In addition it mixes the triplet and singlet states with $m = 0$ so that, in a magnetic field of strength H, we have for the respective energies of these states‡

$$^3E(m = 0) = E_\mathrm{m} + \tfrac{1}{2}\Delta E(1+x^2)^{\frac{1}{2}},$$
$$^1E(m = 0) = E_\mathrm{m} - \tfrac{1}{2}\Delta E(1+x^2)^{\frac{1}{2}},$$

where E_m is the mean energy of the singlet and triplet states, and ΔE their separation in zero magnetic field. x is given by

$$x = 4\mu H/\Delta E,$$

μ being the Bohr magneton.

The wave-functions of the two states, with $m = 0$, take the form

$$\psi = a_\mathrm{s}\psi_{00} + a_\mathrm{t}\psi_{10}, \quad (18\,\mathrm{a})$$

where $\qquad a_\mathrm{s}/a_\mathrm{t} = [(1+x^2)^{\frac{1}{2}} - 1]/x.$ (18 b)

ψ_{00} and ψ_{10} are the respective zero-field wave-functions for the singlet and triplet states with $m = 0$. Because a_s is finite for both states, both will tend to annihilate through emission of two quanta. Detailed analysis shows that, for free non-interacting Ps, the fraction of annihilation events

---

† STEWART, A. T. and BRISCOE, C. V., *Positron annihilation*, edited by Stewart and Roellig, Academic Press, New York (1967), p. 383.

‡ BERESTETZKI, V. B., and LANDAU, L. D., *Zh. eksp. teor. Fiz.* **19** (1949) 673; BERESTETZKI, V. B., ibid. 1130.

which, in the zero-field limit, take place through three-quantum emission becomes, in the presence of a magnetic field (H),

$$\phi = \frac{2}{3} + \frac{1}{3}\left[1 + \left(\frac{2\mu H}{\Delta E}\right)^2 \frac{\lambda_s}{\lambda_t}\right], \tag{19}$$

where λ_s, λ_t are the singlet and triplet Ps annihilation rates. This formula includes the fact that two-thirds of the orthoPs will be in states with $m = \pm 1$, unaffected by the field.

If the Ps atoms are present in a medium in which the orthoPs quenching rate is λ_q, eqn (19) becomes

$$\phi = \frac{\lambda_t}{\lambda_q}\left[\frac{2}{3} + \frac{1}{3}\left\{1 + \left(\frac{2\mu H}{\Delta E}\right)^2 \frac{\lambda_s}{\lambda_q}\right\}\right]. \tag{20}$$

Experiments on magnetic quenching in gases were first carried out with the aim of confirming theoretical prediction of ΔE and the ratio λ_s/λ_t. With this aim it is necessary to work in a gas which has a minimal quenching effect on orthoPs, and this naturally suggests freon. The first experiments in which the effect was observed and quantitative measurements made were those of Deutsch and Dulit,† using the single-quantum energy-spectrum method to monitor the triple annihilation rate. They assumed λ_s/λ_t to have the calculated value of 1114, and derived ΔE within 10 per cent of the theoretical value. A little later Deutsch and Brown‡ measured ΔE with precision by a radio-frequency resonance technique, so later experiments in magnetic quenching were concerned with the determination of λ_s/λ_t given ΔE. Wheatley and Halliday,§ using the triple-coincidence method for monitoring the orthoPs annihilation rate, obtained a value for λ_s/λ_t of 1050 ± 140. About the same time, Pond and Dieke‖ observed magnetic quenching using the increase in two-quantum coincidences to monitor the decrease in three-quantum annihilation events.

It was pointed out a little later by Hughes, Marder, and Wu†† that, in experiments in which the single-quantum energy-spectrum method is used, allowance must be made for the angular distribution of the quanta arising from annihilation in the $m = 0$ and $m = \pm 1$ states. Thus Fig. 26.16 shows the calculated angular distributions of the relative probability of emission of annihilation radiation with energy between

---

† DEUTSCH, M. and DULIT, E., *Phys. Rev.* **84** (1951) 601.
‡ DEUTSCH, M. and BROWN, S. C., ibid. **85** (1952) 1047. See also WEINSTEIN, R., DEUTSCH, M., and BROWN, S. C., ibid. **94** (1954) 758 and **98** (1955) 223.
§ WHEATLEY, J. and HALLIDAY, D., ibid. **88** (1952) 424.
‖ POND, T. A. and DIEKE, R. H., ibid. **85** (1952) 489.
†† HUGHES, V. W., MARDER, S., and WU, C. S., *Phys. Rev.* **98** (1955) 1840.

360 and 400 keV by the different magnetic substates of orthoPs, relative to the direction of the magnetic field.† It will be seen that the distributions are quite different so that, if the quanta are observed in a particular direction, one or other of the two sets of substates will receive greater weight than the other. Thus, observations at 90° could lead to an apparent reduction by a large quenching field of more than one-third in the fraction, ϕ.

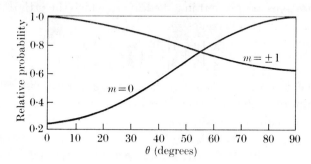

Fig. 26.16. Calculated angular distributions relative to the direction of the magnetic field of emission of annihilation radiation with energy between 360 and 400 keV by the different substates of orthoPs.

To study this effect, Hughes, Marder, and Wu carried out experiments using a 50 mCi ^{64}Cu source in the form of a disc, of diameter $\frac{1}{2}$ in and 0·001 in thickness, glued to the centre of one of the parallel ends of a cylindrical brass chamber (6 in diameter and $2\frac{1}{2}$ in long) containing freon at a pressure between 1 and 2 atm. The scintillation spectrometer consisted of a Na I (Tl) crystal coupled to a 5819 photomultiplier placed inside a lead and iron shield. It viewed the chamber through a $\frac{1}{2}$ cm hole in a block of lead shielding. A magnetic field up to 7000 G was provided by an electromagnet, of 9 in pole face and gap width about 3 in. The spectrometer detected quanta emitted at 90° to the direction of the magnetic field.

Fig. 26.17 shows typical annihilation gamma ray spectra which they obtained (a) in pure freon and zero field (b) in freon containing 4 per cent of NO and zero field and (c) in pure freon in a magnetic field of 4101 G, in each case at 2 atm pressure. Let R_a, R_b, R_c be the counting rates at the peak (485–530 keV) in each of the respective cases, and V_a, V_b, V_c the corresponding rates in the valley region (345–390 keV). Then if η_1^P, η_3^P are the detecting efficiencies at the peak region, and η_1^V, η_3^V are the

† DRISKO, R. M., Phys. Rev. A**95** (1954) 611.

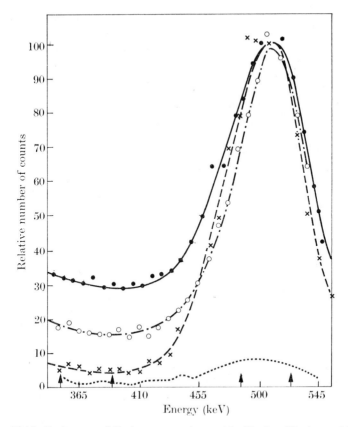

FIG. 26.17. Positron annihilation spectra observed by Hughes, Marder, and Wu in their experiments on magnetic quenching of orthoPs in freon at 2 atm. Count rates are normalized to 100 at the peak. Without magnetic field —●—, pure freon —×—, freon +4 per cent NO. With magnetic field of 4101 G –·o·– pure freon. Background counts · · · · ·. Peak region from 485 to 530 keV and valley region from 345 to 390 keV are indicated.

corresponding quantities at the valley region, for gamma rays from para- and orthoPs decay, respectively, we have

$$R_a = N_p\,\eta_1^P + N_a\,\eta_1^P + N_o\,\eta_3^P,$$
$$V_a = N_p\,\eta_1^V + N_a\,\eta_1^V + N_o\,\eta_3^V,$$
$$R_b = N_p\,\eta_1^P + N_a\,\eta_1^P + N_o\,\eta_1^P,$$
$$V_b = N_p\,\eta_1^V + N_a\,\eta_1^V + N_o\,\eta_1^V,$$
$$R_c = N_p^H\,\eta_1^P + N_a^H\,\eta_1^P + N_o^H\,\phi\eta_3^P + N_o^H(1-\phi)\eta_1^P,$$
$$V_c = N_p^H\,\eta_1^V + N_a^H\,\eta_1^V + N_o^H\,\phi\eta_3^V + N_o^H(1-\phi)\eta_1^V.$$

Here N_p, N_a, and N_o are the number of annihilations per second of paraPs, free positrons, and orthoPs, respectively, in the absence of a magnetic field, N_p^H, N_a^H, N_o^H are the corresponding quantities when the field is present. ϕ is the required fractional yield of three-quantum annihilations from orthoPs in the presence of the field. Assuming $N_o = 3N_p$ and $N_a = KN_p$, where K is a constant, and that the same relations hold when the field is present, we have

$$\phi = 1/(1+y), \qquad (21)$$

where $y = (R_a/R_b)\{(V_a/R_a)-(V_c/R_c)\}/\{(V_c/R_c)-V_b/R_b)\}$.

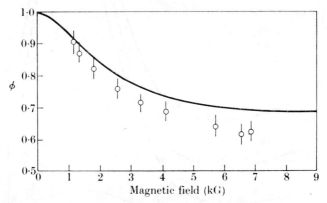

FIG. 26.18. Comparison of observed and calculated values of the function ϕ (see eqn (21)). ○ observed, —— calculated.

Fig. 26.18 shows the observed values of ϕ, with statistical accuracy indicated, as a function of the magnetic field, H. It will be seen that these values fall below the theoretical curve. However, when correction is made for the angular distribution of the annihilation radiations, there is much closer agreement. Assuming the observed accurate value for ΔE, the corrected observations give for the ratio λ_s/λ_t, 1302 ± 200, as compared with the theoretical value 1114. If allowance is made for the small quenching rate in freon, the observed value would be slightly increased.

Heinberg and Page[†] applied the angular correlation method to investigate magnetic quenching in a number of gases, using the equipment described on p. 3145. Fig. 26.19 shows the effect of a magnetic field on the angular correlation of the two-quantum annihilation gamma rays in argon at a pressure of 27 atm. In the absence of a field the angular distribution has no sharp peak at $\theta = 0$ but is relatively broad. This is,

† HEINBERG, M. and PAGE, L. A., *Phys. Rev.* **107** (1957) 1589.

presumably, because the paraPs atoms annihilate before being thermalized. In the presence of a magnetic field of 10 kG a sharp peak is produced due to two-quantum annihilation of orthoPs atoms with $m = 0$. Fig. 26.20 shows the dependence of the counting rate at $\theta = 0$ on the magnetic field. This establishes that saturation is obtained with a

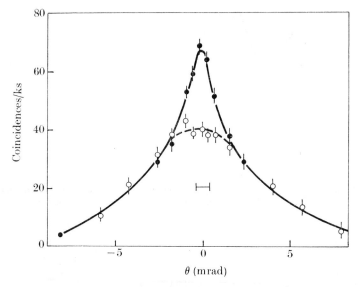

Fig. 26.19. Angular correlation of annihilation quanta for two-quantum annihilation of positrons in argon at 27 atm observed by Heinberg and Page, the slit width being 0·83 mrad. ○ observed in zero magnetic field, ● observed in a magnetic field of 10 kG, ⊢⊣ slit width.

10 kG field. The area under the sharp peak in Fig. 26.19 is about 10 per cent of that under the broad distribution obtained in zero field. According to the theory of magnetic quenching, this corresponds to one-third of the orthoPs actually present. Since the broad distribution at zero field represents only two-quantum annihilation processes, we must add a further 30 per cent for the orthoPs present if we wish to obtain the total of free positrons+paraPs+orthoPs annihilation. Of these the fraction of para-+orthoPs must be 40 per cent, since the orthoPs alone comprises 30 per cent. Hence the fraction of positrons forming Ps is $0·4/1·3 \simeq 30$ per cent, which is in reasonable agreement with that obtained by Pond, and by Hughes and Marder using other methods.

In pure oxygen an applied magnetic field has no appreciable effect on the counting rate at $\theta = 0$, as may be seen by reference to Fig. 26.21. This is to be expected because of the high quenching rate of O_2 for

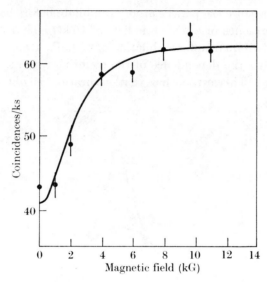

Fig. 26.20. Observed dependence of the counting rate, at $\theta = 0$, on the magnetic field, under the same experimental conditions as in Fig. 26.19.

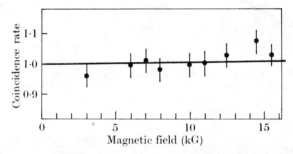

Fig. 26.21. Observed dependence of the counting rate at $\theta = 0$, on the magnetic field for oxygen. The rate normalized to 1·0 for zero magnetic field.

orthoPs, little of which should be present in the pure gas. Gittleman, Dulit, and Deutsch† have obtained some evidence which indicates that, in the presence of small concentrations of O_2, or of NO in argon, more than one-third of the orthoPs atoms may be quenched. This is consistent with the assumption that quenching by O_2 and NO involves spin re-orientation.

The fact that the effect of a magnetic field may be used to monitor orthoPs annihilation rate under different circumstances is often useful (see p. 3167). Thus, suppose that, in an experiment in a gas in which

† GITTLEMAN, B., DULIT, E. P., and DEUTSCH, M., Bull. Am. Phys. Soc. 1 (1956) 69.

the rate at which two-quantum coincidences occur is measured, a magnetic field of 5 kG is applied, so as to quench all orthoPs with $m = 0$. The coincidence rate R_H under these circumstances will be given by

$$R_H = \eta(\tfrac{1}{2}A + B),$$

where A and B are the rates at which Ps and free positrons pass out of the lower side of the Öre gap, and η is the detection efficiency. If the same measurements are made but with sufficient NO present to quench all orthoPs, the rate becomes

$$R_H(0) = \eta(A+B),$$

so that the fractional formation of positronium is given by

$$\{R_H(0) - R_H\}/R_H(0).$$

Correction must of course be made for background events.

7. The effect of electric fields on positron annihilation rates and on positronium formation

7.1. *The effect on the positron annihilation rate*

In the presence of an electric field (F) the annihilation spectrum is modified, because the field changes both the root mean square velocity, and the velocity distribution, of the positrons. Furthermore, the fraction of positrons which form positronium is changed, though this change is very small until the field exceeds a certain value depending on the momentum-transfer and energy-loss cross-sections near the positronium threshold. The effect of the field on the shorter lifetime component of the annihilation spectrum, and on the shoulder, is apparent at much smaller fields and we shall discuss it first.

Referring to eqn (26) of Chapter 2 we now have, for the velocity distribution†

$$f(v) = A \exp\left\{-(6m/M) \int_0^\epsilon \epsilon \, d\epsilon/(\epsilon_l^2 + 6m\kappa T\epsilon/M)\right\}v^2, \qquad (22)$$

where m, M are the respective masses of a positron and a gas atom, $\epsilon = \tfrac{1}{2}mv^2$, and ϵ_l is given by

$$\epsilon_l = eF/nQ_\mathrm{d}(v), \qquad (23)$$

where $Q_\mathrm{d}(v)$ is the momentum-transfer cross-section for elastic collisions

---

† In (26) of Chapter 2 $f_0(v)$ is normalized so that

$$\int f_0(v) v^2 \, \mathrm{d}v = 1,$$

whereas $f(v)$ in (22) is normalized so that

$$\int f(v) \, \mathrm{d}v = 1;$$

hence the extra factor v^2.

of the positrons with the gas molecules of concentration n. If observations of annihilation-time spectra of the type described in § 5.1 are carried out in the presence of a uniform electric field the data may be analysed as before, but now the equilibrium free-positron annihilation rate λ_1, will be given by eqn (8) but with $f(v)$ as in eqn (22) instead of eqn (9). It follows that the interpretation of results obtained for the constant free-positron decay rate in the presence of an electric field involves not only the effective number of annihilation electrons $Z_a(v)$ but also the momentum-transfer cross-section $Q_d(v)$. Unfortunately, no direct procedure can be derived to determine both these functions of positron velocity. It is necessary to adopt a trial and error procedure based, at least, on semi-empirical theory in order to obtain plausible sets of values.

A rough estimate of Q_d at the most probable velocity v_m may be made by a method suggested by Orth and Jones.† Referring to eqn (22) we see that

$$\frac{\partial f(v)}{\partial T} \bigg/ \frac{\partial f(v)}{\partial (F^2)} \simeq 3m^2\kappa v^2 n^2 Q_d^2(v)/Me^2, \tag{24}$$

so that

$$\frac{\partial \lambda_1}{\partial T} \bigg/ \frac{\partial \lambda_1}{\partial (F^2)} \simeq 3m^2\kappa v_m^2 n^2 Q_d^2(v_m)/Me^2. \tag{25}$$

The application of this relation to positrons in argon is discussed on p. 3205.

Analysis of the shoulder region may be carried out in terms of the modified Boltzmann equation

$$\frac{\partial f}{\partial t} = \frac{n}{Mv^2}\frac{\partial}{\partial v}\left[v^3 Q_d\left\{\left(\kappa T + \frac{\epsilon_l^2 M}{6m\epsilon}\right)\frac{\partial f}{\partial v} + mvf\right\}\right], \tag{26}$$

but, as in the absence of a field, uncertainty is introduced by lack of adequate knowledge of the initial velocity distribution of the positrons. An account of experimental results, and their analysis in terms of the above theory, is given for helium in § 11.1, and for argon in § 11.2.

7.2. The effect on positronium formation

The effect of an electric field on positronium formation was first noted by Deutsch and Brown,‡ and first studied in detail experimentally by Marder, Hughes, Wu, and Bennett,§ and theoretically by Teutsch and Hughes.‖ They were concerned with the increase of positronium formation due to the effect of the electric field in extending the high energy

---

† ORTH, P. H. R. and JONES, G., *Phys. Rev.* **183** (1969) 16.
‡ DEUTSCH, M. and BROWN, S. C., ibid. **85** (1952) 1047.
§ MARDER, S., HUGHES, V. W., WU, C. S., and BENNETT, W., ibid. **103** (1956) 1258.
‖ TEUTSCH, W. B. and HUGHES, V. W., ibid. 1266.

tail of the positron velocity distribution to penetrate beyond the positronium formation threshold.

Referring to the Boltzmann equation (26) we may regard the effective positron temperature T^+ in a rare gas at energies near the positronium formation threshold E_t, as given by

$$\kappa T^+ = \epsilon_l^2 M/6mE_t. \tag{27}$$

The electric field will have a significant effect only if κT^+ is comparable with or greater than E_t, i.e.

$$\frac{\epsilon_l^2 M}{6mE_t^2} > 1, \tag{28}$$

or, writing
$$\rho = \frac{\epsilon_l}{E_t}\left(\frac{M}{6m}\right)^{\frac{1}{2}} = \frac{eF}{nQ_d E_t}\left(\frac{M}{6m}\right)^{\frac{1}{2}}, \tag{29}$$

where Q_d is the momentum-transfer cross-section for collisions with a relative velocity $v_t = (2E_t/m)^{\frac{1}{2}}$, the condition is that $\rho > f$, where f is a quantity of order unity.

In a polyatomic gas an estimate of the appropriate value for ρ may be made by replacing $2m/M$ by μ, where μ is the mean fractional energy loss per collision (see Chap. 11, § 1) at a relative velocity v_t. Effects of this kind were observed with the rare gases and with N_2, H_2, and D_2, but in SF_6 only a slow decrease of the positronium yield was found as the field increased.

Later experiments showed that, in most gases, as the field increases from zero, the first effect is a gradual decrease of the yield to a minimum after which, with further increase of the field, the yield rapidly rises to a saturation value comparable with unity. The decrease is due to the electric field reducing the time a positron spends in the Öre gap.

To a close approximation we may separate the two effects by writing for the fractional yield of positronium in the presence of the electric field

$$\Phi = \phi + (1-\phi)W, \tag{30}$$

where ϕ is the contribution from formation in the Öre gap by positrons, in the course of slowing down. Because of the field, some of the fraction $(1-\phi)$ of positrons which pass through the Öre gap are subsequently accelerated by the field, so that their energies fall again within the gap. This effect is represented by the factor W. As remarked earlier, ϕ itself depends on the field. For a given gas, Φ is a function of ρ, which is proportional to the ratio of electric field F to gas pressure p, at a fixed temperature, as may be seen from eqn (29). It follows that Φ is a function of F/p.

Considering first the calculation of W, we can proceed from the Boltzmann equation (26) with the addition on the right-hand side of a term $-\lambda_f f$ in which λ_f is the rate at which positrons are lost through positronium formation. Thus, $\lambda_f = nQ_f v$, where Q_f is the corresponding formation cross-section. If

$$\lambda_f < 2mQ_d nv/M, \tag{31}$$

the added term can be treated as a perturbation and we have

$$W = \lambda_0/(\lambda_0+\lambda_a), \tag{32}$$

where λ_a is the annihilation rate and

$$\lambda_0 = \int_0^\infty \lambda_f f_0^{(0)} \, dv, \tag{33}$$

$f_0^{(0)}$ being the equilibrium distribution function in the presence of an electric field, the term $\lambda_f f$ being ignored.

Carrying out the detailed evaluation of W in this way, Brandt and Feibus[†] obtain
$$W = \Gamma(\rho)/(1+\Gamma(\rho)), \tag{34}$$
where ρ is given by eqn (29) and

$$\Gamma(\rho) = \langle Q_f/Q_a\rangle\{1+(2/\pi^{\frac{1}{2}}\rho)\exp(-1/\rho^2)-\operatorname{erf}(1/\rho)\}. \tag{35}$$

$\langle Q_f/Q_a\rangle$ is a mean value over the Öre gap energies of the ratio of the cross-section Q_f for positronium formation to the annihilation cross-section.

Teutsch and Hughes[‡] in their earlier calculation, making the approximation (31), obtained essentially the same result, except that they wrote for the formation cross-section

$$Q_f(E) = \{(E-E_t)/E\}^{\frac{1}{2}}Q'_f, \tag{36}$$

and then took Q'_f as a constant. They found

$$\Gamma(\rho) = 2^{\frac{1}{4}}\langle Q'_f/Q_a\rangle\rho^{\frac{3}{2}}\exp(-1/2\rho^2). \tag{37}$$

Fig. 26.22 (a) shows $W(\rho)$ as a function of ρ for different values of $\langle Q_f/Q_a\rangle$ according to eqn (35), and Fig. 26.22 (b) gives corresponding plots for different values of $\langle Q'_f/Q_a\rangle$ according to eqn (37). It will be seen that the two sets of curves are very similar, as they should be.

Teutsch and Hughes[‡] considered in detail the opposite approximation, that of assuming λ_f to be so large that any positron whose energy, in the electric field, crosses into the Öre gap forms positronium. This means that a total absorption barrier exists at $v = v_t$ and the problem is to solve the Boltzmann equation (26), with $\lambda_f = 0$, but with the boundary

[†] BRANDT, W. and FEIBUS, H., *Phys. Rev.* **174** (1968) 455.   [‡] loc. cit., p. 3156.

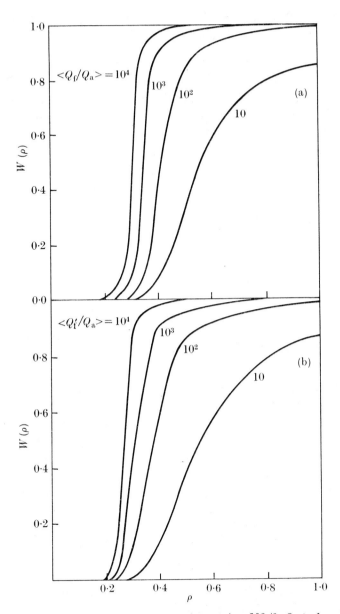

FIG. 26.22. Values of $W(\rho)$ (see eqn (34)) assuming $MQ_f/2mQ_d$ to be small. (a) calculated by Brandt and Feibus for different values of $\langle Q_f/Q_a \rangle$ as indicated. (b) calculated by Teutsch and Hughes for different values of $\langle Q_f'/Q_a \rangle$ (see eqn (36)) as indicated.

condition $f(v_t, t) = 0$. In this way $W(\rho)$ was obtained as a function of ρ for different values of a dimensionless parameter Λ_a, where

$$\Lambda_a = Q_a M / 4 m Q_d. \tag{38}$$

Typical plots of this kind are given in Fig. 26.23. On the whole, it is probable that, at least for monatomic gases, the approximation of very large Q_f is likely to be nearer the correct one than the assumption that

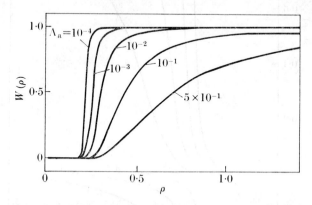

FIG. 26.23. Values of $W(\rho)$ (see eqn (34)) calculated by Teutsch and Hughes, assuming $MQ_f/2mQ_d$ to be large, for different values of Λ_a as indicated.

Q_f is small. The experimental results obtained by Marder, Hughes, and Bennett† which we discuss below are certainly interpretable in a consistent way in terms of plots such as in Fig. 26.23. In other words, the results are determined by the two cross-sections Q_d and Q_a. The field at which W first becomes appreciable is determined by Q_d, and the shape of the sigmoid curve representing the increase thereafter, is determined by the ratio Q_d/Q_a. Detailed results and their analysis in this way are discussed in § 7.3.

Brandt and Feibus‡ have evaluated the term ϕ in eqn (30) as a function of the ratio of electric field strength F to gas pressure p. They found

$$\phi(F) = \phi(0) - \{3/2(\mu-1)\}^{\frac{1}{2}} (1 - \phi(0))\rho + \text{terms in } F^2, \tag{39}$$

where $\mu = (M/2m) Q_f / Q_d$ and will usually be $\gg 1$, in which case the initial slope of $\phi(F)$ is given by

$$\frac{d\phi(F)}{dF} = -\{1 - \phi(0)\} / n E_t (Q_f Q_d)^{\frac{1}{2}}, \tag{40}$$

and so is determined by the geometric mean cross-section $(Q_f Q_d)^{\frac{1}{2}}$.

† loc. cit., p. 3156.   ‡ loc. cit., p. 3158.

Accurate measurement of $d\phi/dF$ is difficult as the slope is small. However, Marder *et al.* obtained accurate data for SF_6 at 2 atm pressure. Their results are shown in Fig. 26.24. Brandt and Feibus showed that a very good fit is obtained with eqn (40) taking

$$Q_d = 1{\cdot}1\times 10^{-14}\text{ cm}^2, \qquad Q_f = 1{\cdot}1\times 10^{-18}\text{ cm}^2.$$

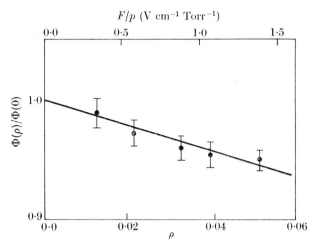

Fig. 26.24. Comparison of observed and calculated values of the fraction of positrons forming Ps in SF_6 at 2 atm pressure, as a function of the ratio of electric field F to gas pressure. ρ is the parameter defined in eqn (29). ● observed by Marder, Hughes, Wu, and Bennett; —— calculated by Brandt and Feibus.

Q_d was chosen to be the same as for electrons. The combination of the two terms in eqn (30) leads to a variation of the yield fraction Φ of the form shown in Fig. 26.25.

7.3. *Experimental observation of the effects on positronium production*

The first systematic study of the effect of electric fields on the probability of positronium formation were those of Marder, Hughes, Wu, and Bennett.† They used the single-quantum annihilation-spectrum technique described on p. 3137. We have already described how this method may be applied to determine the fraction of positrons which form positronium in zero field and there is no difficulty, in principle, in extending it to the case when a field is present.

The apparatus was much the same as in the experiments of Hughes, Marder, and Wu‡ described on p. 3141, with appropriate modification to make possible the application of a uniform electric field. This was

† loc. cit., p. 3156. ‡ loc. cit., p. 3141.

achieved by applying a static voltage between two parallel copper plates 10 cm in diameter separated by 4 cm, insulated from the gas cavity by Teflon sheets. No internal lead shielding could then be introduced because this would distort the electric field. The ^{64}Cu source was set into a depression $\frac{1}{2}$ in diameter and $\frac{1}{32}$ in deep at the centre of one of the field plates, and held in place by a copper spring. A magnetic

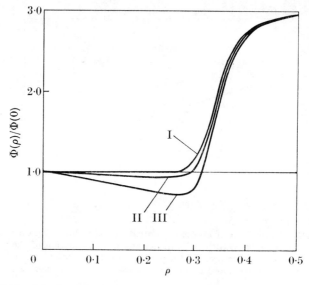

FIG. 26.25. Calculated variation of positronium yield with the parameter ρ (see eqn (29)) for $Q_f/Q_a = 10^3$, $\phi(0) = \frac{1}{3}$. Curves I, II, III refer to values of Q_f/Q_d of 6×10^{-3}, 6×10^{-4} and 6×10^{-5}, respectively.

field of between 5 and 7·5 kG in a direction parallel to the electric field was employed in most measurements because of its focusing effect which had the additional advantage of maintaining the positrons in the region of uniform electric field.

Precautions were taken to insure gas purity though in most cases commercial grade gases were used. Reproducibility of results for argon was improved when the gas was passed into the cavity through a calcium purifier. The neon used was of research grade purity with impurity content less than 1 part in 10^4.

Results obtained for the ratio $\Phi(F)/\Phi(0)$ of the probability of positronium formation in a field F, to that in zero field, are shown for neon and argon in Fig. 26.26, as functions of the ratio F/p of field strength to gas pressure. Absolute values of $\Phi(0)$ were measured also, as 0.55 ± 0.06 and 0.36 ± 0.06 for neon and argon, respectively, at 1·2 atm.

Marder *et al.* also obtained positive results for helium at quite low values of F/p (10 V cm⁻¹ atm⁻¹) but later experiments† have failed to confirm this. It will be seen that the general shape of the curves is similar to that expected from the foregoing theory and, indeed, by appropriate choice

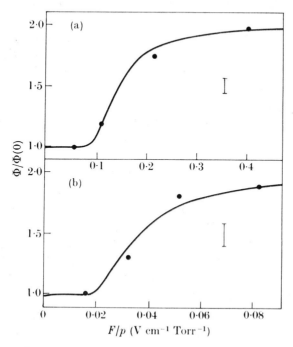

FIG. 26.26. Comparison of observed and calculated variation of Ps yield with F/p (a) for neon, (b) for argon, calculated by Teutsch and Hughes, with $Z_a = 10$, $Q_d = 0\cdot12\pi a_0^2$ for neon, and $Z_a = 18$, $Q_d = 1\cdot5\pi a_0^2$ for argon.

of Q_a and Q_d near the lower end of the Öre gap, a quite good fit is obtained in each case (see Fig. 26.26). Similar results are also obtained for H_2, D_2, and N_2 except that, for the same pressure, the electric field $F_{\frac{1}{2}}$ required to produce a ratio $\Phi(F)/\Phi(0) = 1+\frac{1}{2}g$, where $1+g$ is the saturation value for strong fields, is much larger than for the rare gases. This is expected theoretically because of the larger rate of energy loss per collision at energies close to the Öre gap (see eqn (5)).

The effect of a relatively larger rate of energy loss per collision is shown by the experiments carried out in CO_2, CH_4, CCl_2F_2, and C_2H_6 which at pressures of 2–2·7 atm showed no increased positronium formation even for field strengths as high as 2200 V cm⁻¹. SF_6 on the other hand

† LEE, G. F., ORTH, P. H. R., and JONES, G., *Phys. Lett.* **28A** (1969) 674. LEUNG, C. Y. and PAUL, D. A. R., *J. Phys.* **B2** (1969) 1278.

showed a steady slow decrease at least out to fields as high as 1600 V cm^{-1}, as seen from Fig. 26.24, which was interpreted by Brandt and Feibus as described above.

In a later investigation Obenshain and Page† applied the angular correlation method, as well as the single-quantum annihilation-spectrum technique. For the former experiments they used the apparatus described on p. 3147, the angular resolution being 5 mrad.

The broad component of the angular distribution, with width around 10 mrad, arises from direct, pick-off annihilation of free positrons in collisions with atomic electrons. In the presence of an electric field the relative number of free positrons will be reduced through an increased probability of Ps formation so that the intensity in the broad component, at some angle such as 10 mrad, should fall off as the field increases. Fig. 26.27 shows observed results in a mixture of 75 per cent Ar, and 25 per cent N$_2$, at 13 atm. The intensity begins to fall suddenly as the field strength exceeds 3000 V cm^{-1}, to reach a lower saturation value at fields greater than about 15 000 V cm^{-1}. As a check that no important contribution to the coincidence rate was coming from annihilation of singlet Ps atoms possessing considerable kinetic energy, similar observations were made with the same angular resolution, but centred round an angle of 13 mrad. The same fractional reduction in intensity at saturation fields was obtained, and the shape of the curve was consistent with that found at the smaller angle. Finally, to exclude the possibility of any significant contribution from the narrow component due to annihilation of thermalized singlet Ps, observations were made with a magnetic field present of sufficient strength to quench out the triplet Ps atoms with $m = 0$. No effect of the field on the curve shown in Fig. 26.27 was found.

From observed values of B_s (the ratio of the saturated intensity at large fields to that when no field is present), measured in the wings of the distribution curve, the fraction f_s of all positrons forming positronium in the presence of a strong electric field may be obtained. Thus,

$$(1-f_s) = B_s(1-f_0),$$

where f_0 is the fraction in zero field, measured as described on p. 3140. For pure Ar, pure N$_2$, and a mixture of 75 per cent Ar with 25 per cent N$_2$, Obenshain and Page found f_s to be 0·87, 0·86, and 0·84, respectively. The ratio of field strength $F_{\frac{1}{2}}$ to gas pressure p for which $B = \frac{1}{2}(1+B_s)$ was found for the respective gases to be 180, 440, and 300 V cm^{-1} atm^{-1}.

† OBENSHAIN, F. E. and PAGE, L. A., *Phys. Rev.* **125** (1962) 573.

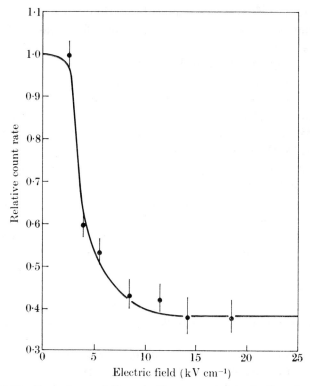

Fig. 26.27. Dependence of the coincidence count rate, in the wing of the angular distribution ($\theta = 10$ mrad), on electric field strength, for two-quantum annihilation of positrons in a mixture at 13 atm of 0·75 Ar with 0·25 N_2, as observed by Obenshain and Page.

These are to be compared with 150 for Ar and 400 for N_2, observed by Marder et al. For O_2, which is of special interest because of its strong quenching powers for orthoPs, Obenshain and Page found, from observations at 5 mrad, that $F_{\frac{1}{2}}/p$ is 300 V cm^{-1} atm^{-1} for this gas, and that f_s is between 0·8 and 0·9.

Observations in SF_6, using both the angular correlation and annihilation spectrum methods, confirmed the initial decrease of Φ as the field increases (see p. 3163 and Fig. 26.24) and in addition, because much larger fields could be applied, it was possible to pass through this region to that in which Φ increases with F to saturation. $F_{\frac{1}{2}}/p$ was found to be as high as 10 000 V cm^{-1} atm^{-1}, while f_s is about 0·75.

7.4. Summary

Summarizing, it appears that, from observation of the electric field strength at which the yield function begins to rise appreciably above the value in zero field, quite good information can be obtained about the diffusion cross-section at the positronium formation threshold, at least for monatomic gases. For polyatomic gases an effective energy loss cross-section μQ_d at the threshold is obtained where μ is the mean fractional energy loss per collision. From the shape of the sigmoid curve representing the rise of the yield function to saturation as the field strength increases, the ratio Q_d/Q_a is obtained unless Q_f is so small that the approximations of eqn (35) or (37) apply, in which case the shape is determined by Q_f/Q_a near the threshold.

If measurements can be made of the rate of the initial decrease of fractional yield as the field increases, $(Q_f Q_d)^{\frac{1}{2}}$ can be obtained, again at energies near the threshold. For monatomic and simpler polyatomic gases the field at which the function W begins to rise appreciably above the zero field value is so small that the minimum in the yield curve (see Fig. 26.25) is too small to permit usefully accurate measurements of the initial slope but for complex molecules such as SF_6 this restriction does not apply.

Information derived from measurements of the positronium yield function as a function of electric field strength is discussed in § 14.

8. Summary of experimental possibilities for studying reactions of positrons and of positronium in gases

Essentially there are five techniques which are available for providing information about reactions involving positrons or Ps atoms:

(a) Measurement of annihilation rate as a function of time by delayed coincidence measurements.

(b) Measurement of the rate of two-quantum events under different conditions.

(c) Measurement of the rate of three-quantum events by observation of triple coincidences.

(d) Observation of the 'valley' to 'peak' ratio in the energy spectrum of single annihilation quanta.

(e) Angular correlation measurements for quanta emitted in two-quantum decay events.

Measurements of this kind may be carried out in gases of different composition, pressure, and temperature and in the presence of electric and/or magnetic fields. Advantage may also be taken of the fact that nitric oxide is a very efficient quencher for orthoPs and that the free positron annihilation rate is very fast in freon and very slow in SF_6, relative to other substances.

From experiments using these techniques information may be obtained about the following processes:
  (i) The rate of annihilation of free positrons in gases.
  (ii) The probability that positrons will be annihilated while bound as Ps.
  (iii) The probability of quenching of orthoPs.

Study of the process (i) involves measurements of type (*a*) and provides, more or less directly, the number of annihilation electrons per atom as a function of the mean energy of the positrons involved. This mean energy may be changed by observing at different temperatures and with different electric fields applied.

Decay processes involving Ps may be distinguished in measurements of type (*a*) in terms of the relative proportion of slow decays (10^{-7} s) and the dependence of the decay rates on gas pressure, in measurements of type (*c*) in which the rate of orthoPs decay is directly observed, in (*d*) which employs quantum energy to distinguish three-quantum events, and in (*e*) in which the Ps decay events are concentrated in the sharp peak and those involving free positron decay in the wings of the observed angular correlation between the gamma rays. In addition use may be made of magnetic quenching which selectively affects the slow decaying population of orthoPs. Once Ps processes may be distinguished the probability of Ps formation may be investigated as a function of the variables concerned, absolute values being conveniently obtained by making use of magnetic quenching as described on p. 3150. For processes under (iii) it is necessary to distinguish specifically orthoPs, and again this may be done using (*a*), (*c*), and (*d*). By working in the presence of a magnetic field sufficiently strong to quench completely the ^3S ($m = 0$) state it is possible also to use (*e*) by observing the effect of the field on the height of the sharp peak.

From observations of the annihilation rate of free positrons in thermal equilibrium in a particular atomic gas as a function of the ratio F/p of electric field to gas pressure, information about the momentum-transfer and annihilation cross-section, as a function of positron velocity, can be derived by a trial and error analysis. This may be facilitated if information is available about the temperature variation of the annihilation rate, and supplemented from analysis of the 'shoulder' region in the observed variation of annihilation rate with time. Any guidance from theory is also valuable. In a polyatomic gas or gas mixture the analysis is more complicated because of the importance of inelastic collisions in which excitation of molecular vibration and rotation occur.

Analysis of observations of the rate of positronium formation in atomic gases in an electric field gives information about the momentum-transfer and annihilation cross-sections near the lower energy end of the Öre gap and, if an initial decrease of formation fraction with increasing field is observed, about a mean cross-section for positronium formation over the energy range of the gap. Observations under (ii) also yield information on the effectiveness of different substances in inhibiting positronium formation. This includes polyatomic gases which accelerate the rate of energy loss of positrons through inelastic collisions. Information may then be obtained about the mean fractional energy loss per collision with such molecules.

Observations of the quenching of orthoPs together with those on the inhibition of Ps formation provide the basis for development of the chemistry of Ps in a remarkably comprehensive way. This work is not confined to studies in the gas phase and indeed has proved especially fruitful for solid and liquid phase Ps chemistry.†

9. Some historical notes

The possibility that transitions between bound states of an electron–positron pair may be of significance in stellar spectra was discussed by Mohorovičić‡ as early as 1934, only two years after the discovery of positrons. Eleven years later, Ruark§ suggested the now universally accepted designation—positronium. Detailed theoretical analysis of the properties of such systems then followed.‖

The first experimental evidence of the existence of Ps came from the observation in 1949 by Shearer and Deutsch†† of a decay period for positrons in Ar, N_2, and CH_4 as long as 10^{-7} s, comparable with that predicted theoretically for orthoPs, and independent of the gas pressure. They used the delayed-coincidence technique, described on p. 3133, with equipment very similar to that used in later experiments on orthoPs quenching by Heymann, Osmon, Veit, and Williams which are described on p. 3137. Further confirmation came from use by Deutsch‡‡ in 1951 of the technique for measuring the energy spectrum of single annihilation

---

† GOLDANSKI, V. *Atomic Energy Review*, Vol. VI, p. 1. (International Atomic Energy Agency, Vienna, 1968).
‡ MOHOROVIČIĆ, S., *Astr. Nachr.* **253** (1934) 94.
§ RUARK, A. E., *Phys. Rev.* **68** (1945) 278.
‖ WHEELER, J. A., *Ann. N.Y. Acad. Sci.* **48** (1946) 221; PIRENNE, J., *Archs. Sci. phys. nat.* **29** (1947) 121, 207; BERESTETZKI, V. B., and LANDAU, L. D., *Zh. eksp. teor. Fiz. SSSR* **19** (1949) 673; BERESTETZKI, V. B., ibid. 1130; FERRELL, R. A., *Phys. Rev.* **84** (1951) 858.   †† SHEARER, J. W. and DEUTSCH, M., ibid. **76** (1949) 462.
‡‡ DEUTSCH, M., ibid. **82** (1951) 455; **83** (1951) 866.

quanta (see p. 3137). He observed such spectra in a pure rare gas and then with a small admixture of NO and interpreted the results in the now generally accepted way in terms of the quenching of orthoPs.

Further confirmation of the production of Ps came from the observation of annihilation quanta in triple-coincidence by de Benedetti and Siegel† in 1952. The historical development of the study of the effect of electric and magnetic fields has already been outlined in §§ 6 and 7.

10. The theory of the collisions of slow positrons with atoms
10.1. *Introduction*

In Chapter 8 we discussed the theory of the collisions of slow electrons with atoms. It was shown that the scattering, which is entirely elastic at the energies concerned, is mainly determined by three effective interactions—the mean static field of the undisturbed atom, the non-local interaction arising from electron exchange, and a distortion potential arising from the distortion of the atom by the incident electron. A major component of the latter potential is a dipole polarization term. For collisions with many atoms, including hydrogen and the rare gas atoms, good results are obtained if these effects are taken into account by the polarized orbital method (Chap. 8, § 2.4). If we now consider how these ideas might be extended to positrons with energies below the threshold for Ps formation we note that, while an exchange interaction no longer arises, we must allow for the possibility that a significant non-local term will be introduced through virtual positronium formation. Also, in the positron case the mean static field, which is now repulsive, is of opposite sign to the distortion field which is attractive as for electrons. Because these two interactions will act in opposite ways we must expect a greater sensitivity in the positron case to any inaccuracy in the approximations used, so the theoretical problem is likely to be a more challenging one than for electrons.

A further novel feature in the positron case is the fact that, in addition to the momentum-transfer cross-section Q_d it is also necessary to calculate theoretically the annihilation cross-section Q_a or effective number of annihilation electrons Z_a, which depend on the wave-function of the system when the positron is in close interaction with the electrons. This means that, to give good results for Q_a as well as Q_d, it is necessary to use a wave-function which is a good approximation under these conditions as well as at the larger positron–atom separations which are important for determining Q_d.

† DE BENEDETTI, S. and SIEGEL, R., ibid. **85** (1952) 371.

If $r_1, r_2, ..., r_Z$ denote the coordinates of the Z atomic electrons and r that of the positron, relative to the atomic nucleus, the wave-function describing the collision will have the asymptotic form, for large r,

$$\Psi(r_1, r_2, ..., r_Z; r) \sim \{e^{i k.r} + f(\theta) r^{-1} e^{ikr}\} \psi(r_1, r_2, ..., r_Z), \quad (41)$$

where ψ is the wave-function of the ground state of the atom and k is the wave-vector of the motion of the incident positron when at a large distance from the atom. The momentum-transfer cross-section is given by

$$Q_d = 2\pi \int_0^\pi (1-\cos\theta)|f(\theta)|^2 \sin\theta \, d\theta. \quad (42)$$

As usual, we have

$$f(\theta) = \frac{1}{2ik} \sum (2l+1)(e^{2i\eta_l}-1) P_l(\cos\theta), \quad (43)$$

where η_l is the real phase shift for scattering of positrons with angular momentum $\{l(l+1)\}^{\frac{1}{2}}\hbar$. At very low energies η_0 is alone important but as the energy increases higher order phase shifts make significant and eventually dominating contributions (see Chap. 6, § 3.4). For positron–atom collisions η_0 is quite small at low energies because the mean static field and polarization fields are of opposite sign. This increases the relative importance of the higher order phase shifts.

The effective number of annihilation electrons is given by

$$Z_a = \int |\Psi(r_1, r_2, ..., r_Z; r)|^2 \sum_{i=1}^Z \delta(r_i - r) \, dr_1 ... dr_Z \, dr. \quad (44)$$

It is also possible to measure the angular correlation of two-quantum annihilation gamma rays. This effectively gives the distribution $P(q_3)$ of a single component of the momentum of an annihilating electron–positron pair (see p. 3145). In terms of the wave-function Ψ

$$P(q_3) = \iint S(q) \, dq_1 \, dq_2, \quad (45\text{a})$$

where

$$S(q) = \sum_i \int \left| \int \delta(r_i - r) e^{iq.r_i} \Psi(r_1, ..., r_Z; r) \, dr \, dr_i \right|^2 \prod_{s \neq i} dr_s. \quad (45\text{b})$$

This provides a further test of the accuracy of the approximate wave-function.

Although experiments which yield information on the cross-sections for positron collisions with hydrogen atoms are still beyond the range of present technique, we begin by discussing in some detail the theory of these collisions. Their theoretical simplicity makes it possible to determine a very good approximation to Ψ, by variational methods, using elaborate trial functions. This may then be used as a basis for testing the usefulness of approximations of sufficient simplicity to be applicable to collisions with more complex atoms.

10.2. *Collisions with hydrogen atoms*

We discuss first the calculation of the zero order phase shift (η_0) which is the most important at positron energies of less than a few eV, as may be seen by comparison of Table 26.1 with Fig. 26.30. Results obtained by approximations which have been applied to electron scattering are first compared with the reliable variational values and found to be less satisfactory. Improved methods are then discussed. The calculation of η_1 and η_2 is then considered on similar lines.

10.2.1. *The zero-order phase shift.* The elaborate trial function used by Schwartz for calculating η_0 for collisions of slow electrons with hydrogen atoms by the Kohn variational method (see Chap. 8, § 2.6) may be applied at once to positron collisions. Results for η_0 obtained in this way by Schwartz[†] are given in Table 26.1 and will be used as a basis against which to compare results from other approximations.

An approximate wave-function incorporating the three main contributions which we expect will arise in slow positron collisions, can be written

$$\Psi_t^*(\mathbf{r}_1, \mathbf{r}) = \{\psi(\mathbf{r}_1) + \phi(\mathbf{r}_1, \mathbf{r})\} F(\mathbf{r}) + \omega(|\mathbf{r}-\mathbf{r}_1|) G\{\tfrac{1}{2}|\mathbf{r}_1+\mathbf{r}_2|\}, \quad (46)$$

where ψ is the ground-state wave-function of the hydrogen atom and $\omega(|\mathbf{r}-\mathbf{r}_1|)$ is that of positronium. $\phi(\mathbf{r}_1, \mathbf{r})$ represents the disturbance of ψ due to the positron and satisfies the condition

$$\int \psi(\mathbf{r}_1)\phi(\mathbf{r}_1, \mathbf{r}) \, d\mathbf{r}_1 = 0. \quad (47)$$

At low energies of impact both F and G may be taken as spherically symmetrical and written

$$F(r) = r^{-1} f_0(r), \qquad G(\rho) = \rho^{-1} g_0(\rho). \quad (48)$$

f_0 and g_0 must then have the asymptotic form

$$f_0(r) \sim k^{-1} \sin kr + \{(e^{2i\eta_0}-1)/2ik\}\exp(ikr), \quad (49\,\text{a})$$

$$g_0(\rho) \sim c\exp(-\mu\rho), \quad (49\,\text{b})$$

where $\mu^2 = -2k^2 + a_0^{-2}$.

If we follow the same prescription as that of Temkin and Lamkin for scattering of slow electrons (Chap. 8, § 2.4), $\phi(\mathbf{r}_1, \mathbf{r})$ is obtained as follows. The positron is supposed to be moving so slowly that the atomic electron can be regarded as moving under the combined influence of the proton

[†] SCHWARTZ, C., *Phys. Rev.* **124** (1961) 1468.

and positron, supposed fixed at **r**. The perturbation is then expanded in a multipole series for $r > r_1$ and the dipole term alone included. This gives

$$\phi(\mathbf{r}_1, \mathbf{r}) = -(a/r)^{-2}\left(\frac{r_1}{a_0} + \frac{1}{2}\frac{r_1^2}{a_0}\right)\cos\vartheta\,\psi(r_1), \qquad (50)$$

where ϑ is the angle between $\mathbf{r}_1$ and $\mathbf{r}$. For $r < r_1$ it is assumed that $\phi = 0$. Having obtained ϕ, f_0 and g_0 are obtained by requiring

$$\int \psi(\mathbf{r}_1) H \Psi_t\,d\mathbf{r}_1 = 0, \qquad (51\,\mathrm{a})$$

$$\int \omega(|\mathbf{r}_1 - \mathbf{r}|) H \Psi_t\,d(\mathbf{r}_1 - \mathbf{r}) = 0, \qquad (51\,\mathrm{b})$$

where H is the Hamiltonian for the complete system.

At this stage the following remarks may be made. First, we draw attention to the determination of ϕ, when $r < r_1$, i.e. when the electron is further from the proton than is the positron. Although it is clearly incorrect to regard the positron as moving slowly compared with the electron under these circumstances, it is a very arbitrary assumption to take ϕ as zero. This does not seem to be serious for electron collisions but it does not follow that it will also be unimportant for positrons. The second point is that eqns (51) do not result from a variational condition which leads to an upper bound for the scattering length. Results of calculations carried out with the above prescription are given in Table 26.1. To obtain an idea of the relative importance of the three main interactions, results are given for the four cases,

(a) mean static field only included ($\phi = 0$, $G = 0$),
(b) mean static field plus virtual positronium but no distortion ($\phi = 0$, $G \neq 0$),
(c) mean static field plus distortion, but no virtual positronium ($G = 0$, $\phi \neq 0$),
(d) all terms included ($\phi \neq 0$, $G \neq 0$).

For case (a) the scattering length a is > 0 because the mean static field is repulsive. It remains > 0 though smaller in case (b), showing that the effective non-local interaction due to virtual Ps formation, while attractive, is not strong enough to overcome the repulsive mean static field. Inclusion of distortion, even without virtual Ps, decisively changes the sign of a although its magnitude is considerably smaller than the accurate value given from the Schwartz variational function. If now the Temkin–Lamkin prescription is followed with full inclusion of virtual Ps, as in case (d), the magnitude obtained for the negative value of a considerably exceeds that given by Schwartz. Since this latter is a lower bound it is clear that (d) overestimates the effective attraction. The same general features are present in the comparison of zero-order phase shifts at finite positron energies. It seems then that it is not sufficient for the positron case merely to follow the same procedure as proved effective for electrons.

An important step forward was taken by Drachman† in 1965. He approached the problem of determining the distortion function $\phi(\mathbf{r}_1, \mathbf{r})$ in eqn (46) in a somewhat different way, although still regarding the perturbation due to the positron as adiabatic. Under these conditions we have, without further approximation,

$$\left\{\frac{\hbar^2}{2m}\nabla_1^2 + \frac{e^2}{r_1} - V(\mathbf{r},\mathbf{r}_1)\right\}(\psi+\phi) = -(E_0+\epsilon(\mathbf{r}))(\psi+\phi), \quad (52)$$

where E_0 is the energy of the ground state of hydrogen and ϵ the additional energy due to the perturbation $V = e^2/|\mathbf{r}_1-\mathbf{r}|$ by the positron when its coordinate is $\mathbf{r}$. To calculate ϕ to the second order, in terms of this perturbation, we neglect terms involving the products $\epsilon\phi$, $V\phi$, and take

$$\epsilon(\mathbf{r}) = \langle V(\mathbf{r},\mathbf{r}_1)\rangle = \int V(\mathbf{r},\mathbf{r}_1)\psi^2(\mathbf{r}_1)\,\mathrm{d}\mathbf{r}_1. \quad (53)$$

Since

$$\left(\frac{\hbar^2}{2m}\nabla_1^2 + \frac{e^2}{r_1} + E_0\right)\psi = 0, \quad (54)$$

we now have

$$\left(\frac{\hbar^2}{2m}\nabla_1^2 + E_0 + \frac{e^2}{r_1}\right)\phi = \{V(\mathbf{r},\mathbf{r}_1) - \langle V\rangle\}\psi. \quad (55)$$

Writing $\phi = \chi\psi$, and using eqn (54) once more, gives

$$\psi\nabla_1^2\chi + 2\nabla\psi \cdot \nabla_1\chi = \{V(\mathbf{r},\mathbf{r}_1) - \langle V\rangle\}\psi. \quad (56)$$

This equation has been solved by Dalgarno and Lynn‡ who found it to be separable in spheroidal coordinates. The proper solution so obtained is determined to within an arbitrary additive function of $\mathbf{r}$. This may be chosen so as to make

$$\int |\psi|^2 \chi\,\mathrm{d}\mathbf{r}_1 = 0,$$

and hence to satisfy the requirement (47).

The resulting form for ϕ differs from that used by Temkin and Lamkin in two essential ways. It includes to the same order contributions from terms of all polarity, including a monopole term which is important at small distances. Further, it is taken to be defined at all positron distances r from eqn (55). When this form for ϕ is substituted in eqn (46), and

† DRACHMAN, F. J., Phys. Rev. A138 (1965) 1582.
‡ DALGARNO, A. and LYNN, N., Proc. phys. Soc. A70 (1957) 223.

eqn (51 a) applied, we find that the effective distortion potential acting on the function $F(\mathbf{r})$ is given, in atomic units, by

$$V_\mathrm{d} = r^{-2}[5-(4r^2+8r+10)\mathrm{e}^{-2r}+(4r^3+7r^2+8r+5)\mathrm{e}^{-4r}-$$
$$-2(r+1)^2(\mathrm{e}^{-2r}+\mathrm{e}^{-4r})\{Ei(2r)-2\ln(2\gamma r)\}-$$
$$-2Ei(-2r)\{(r-1)^2\mathrm{e}^{2r}+(r^2+2r-3)+4(r+1)\mathrm{e}^{-2r}\}]. \quad (57)$$

For large r, $V_\mathrm{d} \sim -\tfrac{9}{2}r^{-4}$, the asymptotic form of the dipole polarization term which alone is included in the prescription of Temkin and Lamkin. On the other hand $V_\mathrm{d} \to -1$ for small r.

It seems clear that the most serious failing of the form for ϕ introduced by Drachman is the large contribution of the monopole term at small positron distances, where the whole adiabatic treatment is suspect. This is probably responsible for the fact that, even with no inclusion of virtual Ps, the magnitude of the scattering length obtained is considerably larger than the upper bound given by Schwartz (see Table 26.1). To improve this situation, Drachman isolated the contribution from the monopole term so that a chosen fraction of it could be subtracted from the full distortion potential of eqn (57). To do this the functions $\chi(\mathbf{r},\mathbf{r}_1)$ and $V(\mathbf{r},\mathbf{r}_1)$ are expanded in Legendre polynomials in the form

$$\chi(r,r_1,\vartheta) = \sum_{s=0}^{\infty} \chi_s(r,r_1)P_s(\cos\vartheta),$$

$$V(r,r_1,\vartheta) = \sum_{s=0}^{\infty} V_s(r,r_1)P_s(\cos\vartheta). \quad (58)$$

The monopole term $\chi_0(r,r_1)$ satisfies the equation

$$\psi\nabla_1^2\chi_0+2\nabla\psi\cdot\nabla_1\chi_0 = \{V_0(\mathbf{r},\mathbf{r}_1)-\langle V_0(\mathbf{r},\mathbf{r}_1)\rangle\}\psi. \quad (59)$$

By solving this equation for χ_0 we may determine the contribution $V_{\mathrm{d}0}$ to the effective distortion potential. Drachman then considers the scattering determined by the mean static potential together with distortion potentials

$$V_\mathrm{d}+(\gamma-1)V_{\mathrm{d}0}, \quad (60)$$

for various values of γ between 0 and 1. $\gamma = 0$ corresponds to complete elimination of the monopole component, whilst $\gamma = 1$ includes it fully. It was then found that a very close fit to the scattering length and phase shifts given by Schwartz could be obtained by taking $\gamma = 0\cdot1$ (see Fig. 26.28) involving almost complete monopole elimination. In fact taking $\gamma = 0$ gave results which were quite close.

It seemed then, that with Drachman's choice for ϕ, together with nearly complete suppression of the monopole term, a good approximation could be obtained to the correct scattering length and phase shifts, without needing to include explicitly any virtual Ps in the collision wave-function. The empirical procedure used to remove the major part of the short-range monopole term is unsatisfactory however, and Drachman† proceeded still further to a stage in which a variational method

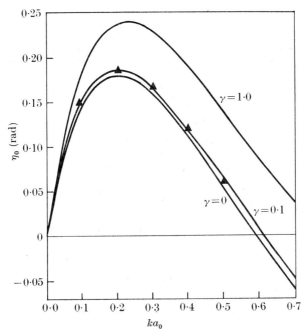

FIG. 26.28. Zero order phase shifts (η_0) for scattering of positrons by hydrogen atoms. ▲ calculated by Schwartz by a many-parameter variational method, ——— calculated by the adiabatic method of Drachman, for different values of the monopole suppression parameter γ, as indicated.

was used to determine the degree of suppression. The trial function to be used with the Kohn variational method was chosen to be of the form

$$\Psi_t = \psi(\mathbf{r}_1)\{F(\mathbf{r}) + S(\mathbf{r})\chi(\mathbf{r}, \mathbf{r}_1)\}, \tag{61}$$

where χ is given from eqn (56). This involves two undetermined functions $F(\mathbf{r})$ and $S(\mathbf{r})$. To satisfy the Kohn variational principle we must have, in terms of the bracket notation,

$$\langle L(\mathbf{r}_1) \rangle = \int |\psi(\mathbf{r}_1)|^2 L(\mathbf{r}_1) \, d\mathbf{r}_1,$$

$$\langle (H-E)(F+S\chi) \rangle = 0, \tag{62a}$$

$$\langle \chi(H-E)(F+S\chi) \rangle = 0. \tag{62b}$$

† DRACHMAN, R. J., *Phys. Rev.* **173** (1968) 190.

To satisfy these conditions F and S must be proper solutions of the coupled equations

$$(\nabla^2 + k^2 - U_0)F = U_1 S, \qquad (63\,\text{a})$$

$$\left\{N(\nabla^2 + k^2) + U_1 - U_2 - W - U_N \frac{\text{d}}{\text{d}r}\right\}S = U_1 F, \qquad (63\,\text{b})$$

where

$$U_p(r) = \langle \chi^{p-1} U \rangle, \qquad W(r) = -\langle \chi \nabla_r^2 \chi \rangle, \qquad N(r) = \langle \chi^2 \rangle,$$

$$U_N(r) = -\text{d}N/\text{d}r, \qquad U = (2me^2/\hbar^2)\left(\frac{1}{r} - \frac{1}{|\mathbf{r}-\mathbf{r}_1|}\right). \qquad (64)$$

Thus $\hbar^2 U_0/2me^2$ is the mean static potential $\langle V_0 \rangle$.

The functions F and S may be expanded in harmonics in the form

$$F(\mathbf{r}) = r^{-1} \sum f_l(r) P_l(\cos\theta), \qquad (65\,\text{a})$$

$$S(\mathbf{r}) = r^{-1} \sum s_l(r) P_l(\cos\theta), \qquad (65\,\text{b})$$

so that f_l is coupled only to s_l. We then obtain

$$\frac{\text{d}^2 f_l}{\text{d}r^2} + \{k^2 - l(l+1)r^{-2} - U_0(r)\}f_l = U_1 s_l, \qquad (66\,\text{a})$$

$$\frac{\text{d}^2 s_l}{\text{d}r^2} + \{k^2 - l(l+1)r^{-2} + N^{-1}(U_1 - U_2 - W + U_N r^{-1})\}s_l - N^{-1} U_N \frac{\text{d}s_l}{\text{d}r}$$
$$= N^{-1} U_1 f_l, \qquad (66\,\text{b})$$

where

$$f_l \sim k^{-1} \sin(kr - \tfrac{1}{2}l\pi + \eta_l), \qquad (67\,\text{a})$$

$$s_l(0) = 0, \quad \text{and} \quad \int_0^\infty s_l^2 \,\text{d}r \text{ is finite}. \qquad (67\,\text{b})$$

Drachman solved the eqns (66) for $l = 0$ and $l = 1$, for a number of values of k, to obtain the phase shifts η_0 and η_1. His solutions for $k = 0$, $l = 0$ are illustrated in Fig. 26.29, in which $r^{-1}f_0$ and s_0/f_0 are plotted as functions of r. The normalization is such that, as $r \to \infty$,

$$r^{-1}f_0 \sim 1 - a/r, \qquad s_0 \sim f_0,$$

a being the scattering length. s_0/f_0 is a measure of the extent to which the adiabatic modification of the atomic wave-function is suppressed at small r, and it will be seen that it falls well below unity as r falls below $4a_0$. An unexpected feature, however, is the increase above unity to a maximum of about 1·15 near $r = 6a_0$.

The phase shifts η_0 are compared with those calculated by other methods in Table 26.1.

It will be seen from columns (G) and (V), of Table 26.1, that the magnitude of the scattering length and the phase shifts are appreciably smaller than the corresponding values obtained by the elaborate variational

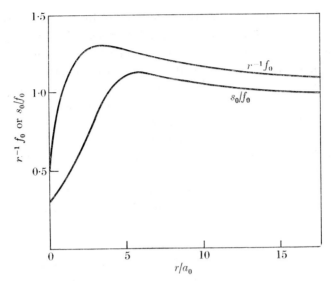

Fig. 26.29. Functions appearing in Drachman's variational treatment, in the zero energy limit for positron–hydrogen scattering.

TABLE 26.1

Phase shift η_0 (rad) for elastic scattering of positrons by hydrogen atoms

Wave number k (atomic units)	(A)	(B)	(C)	(D)	(V)	(E)	(F)	(G)	(H)
0	(0·582)	(−1·267)	(0·170)	(−3·06)	(−2·10)	(−2·07)	(−2·54)	(−1·85)	
0·1	−0·058	0·080	—	0·226	0·152			0·128	−0·005
0·2	−0·114	0·085	−0·046	0·293	0·188			0·158	−0·043
0·3	−0·168	0·053	—	0·275	0·168			0·135	−0·093
0·4	−0·218	0·005	−0·134	0·223	0·120			0·089	−0·147
0·5	−0·264	−0·048	−0·156	0·159	0·062			0·034	−0·199
0·6	−0·304	−0·099	−0·238	−0·095	0·007			−0·022	−0·246
0·7	−0·340	—	−0·286	—	−0·054			−0·074	—

(A) mean static field only,† (B) polarized orbital method with no virtual Ps,† (C) static field+virtual Ps,† (D) polarized orbital+virtual Ps,† (V) full variational method (Schwartz),‡ (E) Drachman adiabatic with full monopole suppression,§ (F) Drachman adiabatic with no monopole suppression,§ (G) Drachman, variational,‖ (H) truncated (1s–2s–2p) eigenfunction method.†† (A), (C), (V), (G), and (H) all provide lower bounds to the scattering length and to the phase shifts. Values in brackets for $k = 0$ are scattering lengths in atomic units.

† CODY, W. J., LAWSON, J., MASSEY, H. S. W., and SMITH, K., *Proc. R. Soc.* A**278** (1964) 479. ‡ loc. cit., p. 3171. § loc. cit., p. 3173. ‖ loc. cit., p. 3175.
†† BURKE, P. G. and SMITH, K., *Rev. mod. Phys.* **34** (1962) 458.

method of Schwartz. In fact the values given by the non-variational, adiabatic method of Drachman, with full monopole suppression, are more accurate—as may be seen from comparison of column (E) with column (V). However, because of its semi-empirical character one cannot assume that for another atom, such as helium, it will be so satisfactory, and it cannot even be relied upon to produce upper bounds to the phase shifts. In column (H) of Table 26.1 we include results obtained using the truncated eigenfunction expansion method (Chap. 8, § 2.2.3), in which 1s, 2s, and 2p bound states were included. The results are quite unsatisfactory.

10.2.2. *The higher order phase shifts.* In Fig. 26.30 we illustrate the phase shifts η_1 and η_2 calculated by the following approximations:

(V) Full variational method (η_1 only),†
(E) Drachman, adiabatic with full monopole suppression,‡
(F) Drachman, adiabatic with no monopole suppression,‡
(G) Drachman, variational,§
(J) Estimated by Kleinman, Hahn, and Spruch.‖

The methods used for obtaining results for (E), (F), and (G) have already been described. The full variational method (V) is an extension of the elaborate trial function of Schwartz but has not been applied to η_2. Kleinman, Hahn, and Spruch‖ obtained estimates of η_2 which should be reliable. They used trial functions in the form of truncated eigenfunction expansions. By studying systematically the dependence of the phase shifts on the number of excited atomic states of given angular momentum quantum number (l_1) included in the expansion, and also on the coupling between states of different l_1, they were able to extrapolate to the case in which all states were included. This gives very good results for η_1 by comparison with the results of the full variational method and would be expected to be of comparable accuracy for η_2.

Just as for the electron case (Chap. 8, § 2.7), the importance of atom distortion is enhanced for these higher order phase shifts. For the same reason the contribution from a monopole component of this distortion becomes less important as l increases, and there is little difference between the results obtained by (E), (F), and (G), which are, however, less satisfactory than for $l = 0$. The reason for this is not clear but may be due to increased importance of virtual Ps formation.

† ARMSTEAD, R. L., Thesis, University of California, Berkeley (1964).
‡ loc. cit., p. 3173.   § loc. cit., p. 3175.
‖ KLEINMAN, C. J., HAHN, Y., and SPRUCH, L., *Phys. Rev.* **A140** (1965) 413.

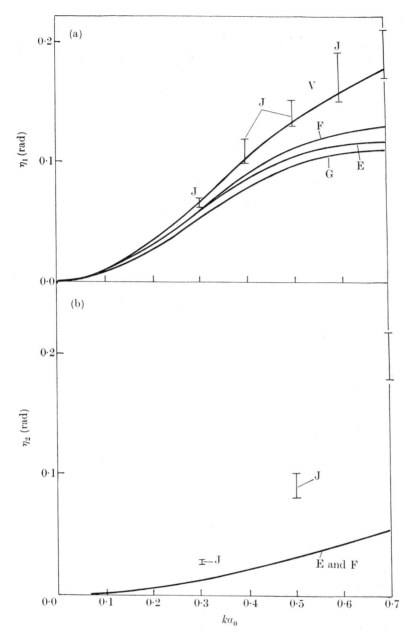

FIG. 26.30. (a) First order phase shifts (η_1) for scattering of positrons by hydrogen atoms, calculated by the following approximations, as indicated. Full variational method (V). Drachman's adiabatic method, with full monopole suppression (E), with no monopole suppression (F), Drachman's variational method (G), and estimated by Kleinman et al. (J). (b) Second order phase shifts (η_2) for scattering of positrons by hydrogen atoms calculated by the approximations (E), (F), and (J), as in (a).

10.2.3. *Summarizing remarks.* Judged from the point of view of effectiveness in determining the scattering phase shifts, of the relatively simple approximate methods which may be applied at least to collisions with helium atoms, Drachman's adiabatic method (E) with full monopole suppression gives the best results, particularly at low energies for which the η_0 phase shifts alone are important. On the other hand this method is semi-empirical and its applicability to other atoms, such as helium, cannot be guaranteed. It also has the disadvantage from this point of view of not yielding upper bounds. The variational method (G) of Drachman gives results for η_l with $l \geqslant 1$ which do not differ much from the adiabatic method, but for η_0 is less accurate for hydrogen. It does, however, yield upper bounds.

By no means does it follow that the effectiveness of a particular approximation in determining phase shifts, and hence the momentum-transfer cross-section, will also apply to the calculation of the effective number Z_a of annihilation electrons, and of the angular correlation between the directions of emission of two-quanta annihilation gamma rays. Comparison of the different approximations in these respects has not been carried out for hydrogen.

Humberston and Wallace† have examined the rate of convergence of values of Z_a calculated in the low velocity limit using elaborate trial variational functions of the form (in atomic units)

$$\Psi_t = (2\pi)^{-1}\Bigg[\bigg\{1 - \frac{a}{r}(1-\mathrm{e}^{-\delta r}) + \frac{b_1}{r^2}(1-\mathrm{e}^{-\delta r})^2 +$$

$$+ \frac{b_2}{r^2}(1-\mathrm{e}^{-\delta r})^3(r_1 + \tfrac{1}{2}r_1^2)\Big(\frac{\mathbf{r}.\mathbf{r}_1}{rr_1}\Big)\bigg\}\mathrm{e}^{-r_1} +$$

$$+ \exp\{-\alpha(r+r_1) - \gamma|\mathbf{r}-\mathbf{r}_1|\}\sum_i c_i r^{l_i}r_1^{m_i}|\mathbf{r}-\mathbf{r}_1|^{n_i} +$$

$$+ \exp\{-\beta(r_1 + |\mathbf{r}-\mathbf{r}_1|)\}\sum_i d_i r^{l_i}r_1^{m_i}|\mathbf{r}-\mathbf{r}_1|^{n_i}\Bigg], \tag{68}$$

a is the 'scattering length' while terms involving b_1 and b_2 represent the effect of dipole distortion of the atom. In the adiabatic approximation $b_1 = -2 \cdot 25$ and $b_2 = 1$. The summation over i includes all terms such that $l_i + m_i + n_i = N$ where l_i, m_i, n_i, and N are non-negative integers. The Schwartz function is obtained using the adiabatic values of b_1 and b_2, and taking $\gamma = d_i = 0$. If the number of linear parameters in the Schwartz function is s that in the more elaborate function (68) will be $2s+2$ as in it b_1 and b_2 are treated as variable parameters.

In Table 26.2 the values obtained for the scattering length a, and for the number of annihilation electrons Z_a, are given for different values of N both for the Schwartz function, and for the more elaborate function (68), in which there are three non-linear parameters. It will be seen that, although the simpler Schwartz function

† HUMBERSTON, J. W. and WALLACE, J. B. G., *J. Phys.* B **5** (1972) 1138.

gives quite rapid convergence for a, it is considerably less effective for Z_a for which the more elaborate function gives more rapid convergence. This shows clearly how difficult accurate calculation of Z_a for more complicated atoms will be. The large value for Z_a is associated with the relatively large polarizability of atomic hydrogen.

TABLE 26.2

Values of the scattering length a and the number Z_a of annihilation electrons obtained for e^+–H collisions, obtained using trial functions with different numbers of adjustable linear parameters

Schwartz function				
($\gamma = d_i = 0$ in eqn (68))				
number of linear parameters	4	10	20	35
a (in a_0)	$-0{\cdot}68$	$-1{\cdot}73$	$-2{\cdot}02$	$-2{\cdot}09$
Z_a	1·37	5·15	7·51	8·48
Complete function (eqn (68))				
number of linear parameters	10	22	42	72
a (in a_0)	$-1{\cdot}927$	$-2{\cdot}073$	$-2{\cdot}099$	$-2{\cdot}102$
Z_a	7·83	8·12	8·77	8·83

10.3. Collisions with helium atoms

Although it is still possible to carry out detailed calculations for helium, the presence of the second electron adds sufficient complication to reduce the reliability of the results as compared with hydrogen. No elaborate variational calculation corresponding to that of Schwartz has yet been carried out. It would be considerably more complex than for hydrogen, especially because the ground state wave-function of the helium atom is not known accurately. It is, nevertheless, a practical possibility for η_0. Other approximate methods such as the polarized orbital method with (D), and without (B), inclusion of virtual Ps formation, the adiabatic method of Drachman† with full monopole suppression (E), or the variational method (G) of Drachman have been applied‡ but, in all cases, the reliability is less than for hydrogen because of inaccurate knowledge of the ground state wave-function. Indeed, in the application of the polarized orbital method with, and without, virtual Ps formation and of the adiabatic method of Drachman, the simplest form of approximate wave-function for the ^1S state of helium

$$\psi_0(r_1, r_2) = (Z^3/\pi a_0^3)\exp\{-Z(r_1+r_2)/a_0\} \tag{69}$$

has been used. The best approximation to the energy of the state is given by the well known Hylleraas value $Z = 27/16 = 1{\cdot}687$. However, this

† DRACHMAN, R. J., *Phys. Rev.* **144** (1966) 25. ‡ Ibid. **173** (1968) 190.

does not give the correct polarizability of the atom which requires $Z = 1·599$.

10.3.1. *The zero-order phase shift.* Kraidy and Fraser† calculated η_0 for helium by the polarized orbital method using eqn (69) with $Z = 27/16$, and their results are shown in Fig. 26.31, together with those

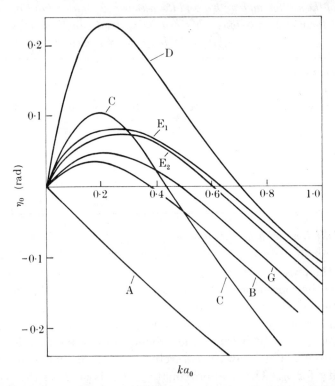

Fig. 26.31. Zero-order phase shifts (η_0) for scattering of positrons by helium atoms calculated by the following approximations, as indicated. Scattering by mean static field only (A), by mean static field and polarization (B), by mean static field and virtual Ps effects (C), by mean static field, polarization, and virtual Ps effects (D), Drachman's adiabatic method with full monopole suppression $Z = 27/16$ and normalized polarizability (E_1), and with $Z = 1·599$ (E_2), and Drachman's variational method (G).

obtained for scattering by the mean static field alone, again using eqn (69). The assumption of the simple form (69) makes the extension of Drachman's variational method (G) to helium relatively simple. However, it must be remembered that the resulting phase shifts are no longer strictly upper bounds because eqn (69) is not an exact wave-

† KRAIDY, M. and FRASER, P. A., *Abstract of 5th international conference on physics of electronic and atomic collisions*, Leningrad (1967), p. 110.

function. Results obtained by Drachman† using eqn (69), with $Z = 1\cdot599$, are shown in Fig. 26.31. In extending his adiabatic method (E), with full monopole suppression, to helium, Drachman used a Hartree self-consistent field function to evaluate the mean static field $\langle V_0 \rangle$ but retained the simpler form (69) for the calculation of the atomic distortion terms. The results are shown in Fig. 26.31. Two curves are given which do not differ very greatly. In one (E_1), Z has been taken as $27/16$ so the asymptotic polarization potential is only $0\cdot806$ of the observed value. This was renormalized by multiplying the distortion function ϕ by $1\cdot240$. For the second curve (E_2), calculated using the smaller value of Z which gives the correct polarizability the derived function ϕ was used without modification. Arguing by analogy with the hydrogen case we would take these latter results to be the most accurate, as far as the phase shifts (η_0) are concerned. However, the situation clearly differs quite considerably from that for hydrogen. Thus the effect of virtual Ps formation is much more marked and the peak positive phase shift given by approximation (C) is actually higher than that from (E). In hydrogen the effective attraction due to virtual Ps is not on its own sufficient at any energy to overcome the repulsive effect of the mean static field.

Judging by the hydrogen case, the polarized orbital method (D), including virtual Ps, is likely to overestimate, and Drachman's variational method (G), to underestimate, η_0. Reference to Fig. 26.31 shows that method (D) does give much larger positive phase shifts than any of the other methods. There seems no doubt that the situation would be much clarified if an elaborate and reliable variational calculation could be carried out.

10.3.2. *The higher order phase shifts.* Fig. 26.32 shows the phase shifts η_1 and η_2, calculated by approximations similar to those used for η_0. By analogy with hydrogen we expect the correct phase shifts to be somewhat larger. For positive energies less than about 3 eV all phase shifts η_l, with $l > 0$, are probably given with sufficient accuracy by the formula (see eqn (54 c), Chap. 6)

$$\eta_l = \pi\alpha k^2/(2l-1)(2l+1)(2l+3), \tag{70}$$

where α is the polarizability.

10.3.3. *The momentum-transfer cross-section.* Fig. 26.33 shows the momentum-transfer cross-section Q_d calculated using the various approximations for the phase shifts as indicated. Because η_0 is quite

---

† DRACHMAN, R. J., *Phys. Rev.* **173** (1968) 190.

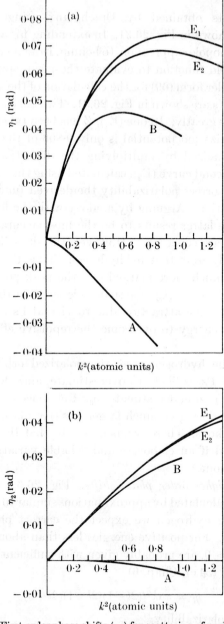

FIG. 26.32. (a) First order phase shifts (η_1) for scattering of positrons by helium atoms calculated by the following approximations, as indicated. Scattering by mean static field only (A), mean static field and polarization (B), Drachman's adiabatic method with full monopole suppression, $Z = 27/16$, and renormalized polarizability (E_1), and with $Z = 1\cdot599$ (E_2). (b) Second order phase shifts (η_2) for scattering of positrons by helium atoms calculated by the approximations as indicated in (a).

small over the energy range of interest, due to partial cancellation of effects due to attractive and repulsive interactions, an important contribution comes from η_1 at quite low energies, much lower than for electron collisions (see Chap. 8, Table 8.6). For energies less than 3 eV

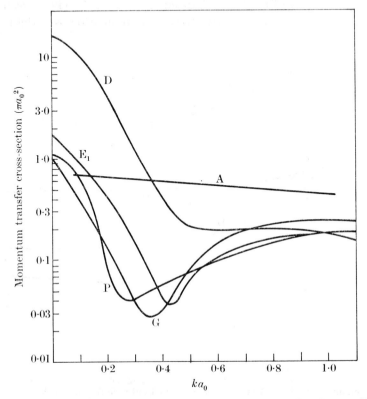

FIG. 26.33. Momentum-transfer cross-sections for collisions of positrons with helium atoms, (A) scattering by mean static field only, (D) including mean static field, polarization, and virtual Ps effects, (E_1) adiabatic method with full monopole suppression, $Z = 27/16$, and renormalized polarizability, (G) Drachman's variational method for η_0 with phase shift η_1, as in adiabatic method with full monopole suppression and $Z = 27/16$, (P) semi-empirical form used by Leung and Paul.

no appreciable uncertainty is introduced from η_1 because of the applicability of eqn (70). However, when the energy exceeds 3 eV, the relatively larger uncertainty in knowledge of η_1 is the most serious source of error.

We include also the semi-empirical curve used by Leung and Paul[†] in the analysis of their experimental results for the time spectrum of delayed annihilation events in helium described in § 11.1. Discussion

[†] LEUNG, C. Y. and PAUL, D. A. L., *J. Phys.* B2 (1969) 1278.

of the comparison between theory and experiment will be deferred to that section.

10.3.4. *The effective number Z_a of annihilation electrons.* Fig. 26.34 shows the values of Z_a, calculated from eqn (44) using the various approximate wave-functions, Ψ, discussed above. It will be seen that Z_a is considerably more sensitive to the approximation used than is Q_d.

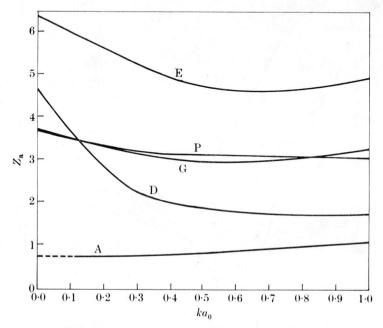

FIG. 26.34. Effective number (Z_a) of annihilation electrons in helium, (A) scattering by mean static field only, (D) including mean static field, polarization, and virtual Ps effects, (E) adiabatic method with full monopole suppression, (G) Drachman's variational method, (P) semi-empirical form used by Leung and Paul.

A remarkable result is the large value given by Drachman's adiabatic method (E) with full monopole suppression. As for Q_d we include the semi-empirical curve used by Leung and Paul† in the analysis of their experimental results.

Discussion of the comparison between theory and experiment is given in § 11.1.

10.3.5. *The angular correlation between two-photon annihilation quanta.* Drachman‡ has calculated the angular correlation function $P(q_3)$ from eqn (45), using the wave-function, Ψ, obtained by his adiabatic method

---

† loc. cit., p. 3185. ‡ DRACHMAN, R. J., *Phys. Rev.* **179** (1969) 237.

(E), with full monopole suppression, and by his variational method (G). The results are shown in Fig. 26.45 in connection with the experimental data discussed in § 11.1.

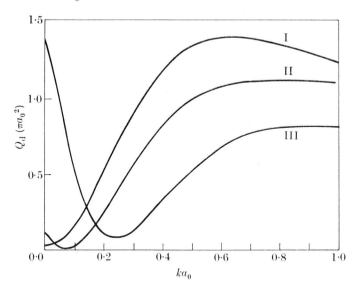

Fig. 26.35. Momentum-transfer cross-sections for collisions of positrons with neon atoms: I calculated by Massey et al., including $2p \to d$ virtual transitions only, II calculated by Montgomery and La Bahn, including $2p \to d$ virtual transitions only, III calculated by Montgomery and La Bahn, including $2p \to d$ and $2s \to p$ virtual transitions, but normalizing to the observed polarizability.

10.4. Collisions with neon and argon atoms

As described in Chapter 8, § 7.1.2, the polarized orbital method as developed by Temkin and Lamkin has been applied with considerable success by Thompson, to the scattering of slow electrons by neon and argon atoms. For neon, the contribution to the polarizability α from the $2p \to d$ virtual transitions only was taken into account. Using Hartree–Fock wave-functions, this leads to a polarizability of $2 \cdot 2 a_0^3$, as compared with an observed value of $2 \cdot 7 a_0^3$.[†] Massey, Lawson, and Thompson[‡] using the same atom distortion, applied the method to the scattering of slow positrons, obtaining the momentum-transfer cross-sections shown in Fig. 26.35. Montgomery and La Bahn[§] extended this

[†] CUTHBERTSON, C. and CUTHBERTSON, M., *Proc. R. Soc.* A **135** (1932) 40.
[‡] MASSEY, H. S. W., LAWSON, J., and THOMPSON, D. G., *Quantum theory of atoms, molecules, and the solid state.* (Edited by Löwdin), Academic Press, New York (1966), p. 203.
[§] MONTGOMERY, R. E. and LA BAHN, R. W., *Can. J. Phys.* **48** (1970) 1288.

work by taking into account the contribution to α from the $2s \to p$ virtual transitions. They obtained $1\cdot34a_0^3$ for this, giving a total value for α larger than the experimental value. To allow for this, they normalized the effective polarization interaction so that it had the correct asymptotic form. In addition, they included in the atomic distortion term, ϕ, the perturbed function for all values of the positron distance r,

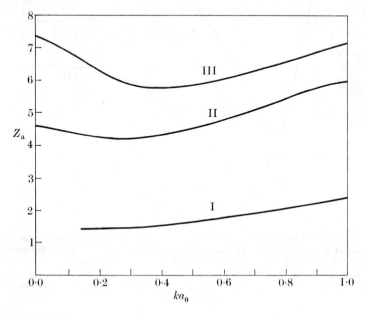

Fig. 26.36. Effective number (Z_a) of annihilation electrons in neon: I calculated by Massey et al., ignoring the atomic distortion term in the wave-function, II calculated as in II of Fig. 26.35, III calculated as in III of Fig. 26.35.

in contrast to the Temkin–Lamkin prescription, according to which ϕ was taken as zero when $r < r_1$, where r_1 is the electron coordinate concerned. These results are also shown in Fig. 26.35. To illustrate the effect of the different prescriptions for ϕ, when $r < r_1$, results obtained by Montgomery and La Bahn when $2p \to d$ virtual transitions alone are included are also shown. Fig. 26.36 shows corresponding results for Z_a calculated from eqn (44). The importance of the inclusion of the atomic distortion term (ϕ) in the wave-function, Ψ, may be seen from curve I of Fig. 26.36, calculated ignoring ϕ.

For argon, Massey et al.† used Thompson's calculations for electrons in which only $3p \to d$ virtual transitions were used in calculating α.

† loc. cit., p. 3187.

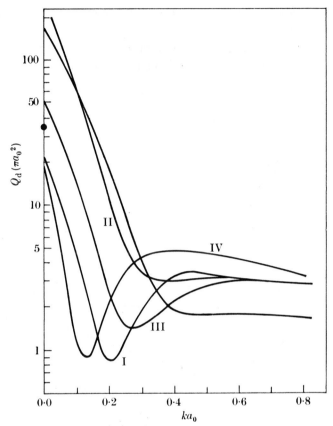

Fig. 26.37. Momentum-transfer cross-sections for collisions of positrons with argon atoms: I calculated by Massey et al., including $3p \to d$ virtual transitions only, II calculated by Montgomery and La Bahn, including $3p \to d$ virtual transitions only, III calculated by Montgomery and La Bahn, including $3p \to d$ virtual transitions only, but normalizing to the observed polarizability, IV calculated by Montgomery and La Bahn including $3p \to d$, $3p \to s$, and $3s \to p$ virtual transitions, ● derived by Orth and Jones from their observations (see § 11.2).

This gives a polarizability of $14 \cdot 2a_0^3$, somewhat larger than the observed value $11a_0^3$.[†] Their results for the momentum-transfer cross-section for positron collisions are shown in Fig. 26.37. In the same figure results obtained by Montgomery and La Bahn,[‡] using the same procedure as in their calculations for neon are also shown. Fig. 26.38 shows the corresponding results for Z_a. These theoretical values are discussed in relation to the rather extensive observations of positron annihilation in argon in § 11.2.

[†] CUTHBERTSON, C. and CUTHBERTSON, M., loc. cit., p. 3187.
[‡] loc. cit., p. 3187.

10.5. *Theoretical evaluation of cross-sections above the* Ps *formation threshold*

The problem of calculating cross-sections for processes occurring at positron energies above the threshold for Ps formation is even more

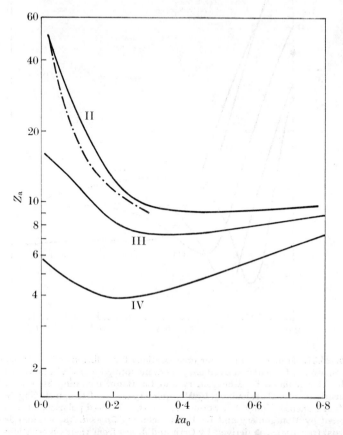

FIG. 26.38. Effective number (Z_a) of annihilation electrons in argon, II, III, and IV calculated as detailed in Fig. 26.37, —·—·— derived from observations of Miller, Orth, and Jones (see § 11.2).

difficult than at lower energies, because it is to be expected that there will be close coupling between purely elastic processes and those in which real Ps formation occurs. It is not surprising that no reliable calculations either of elastic scattering, or Ps formation, cross-sections are available. Bransden and Jundi† have carried out close coupling calculations in which the trial wave-function is of the form of eqn (46), but with the function G now having the asymptotic form of an outgoing spherical

† BRANSDEN, B. H. and JUNDI, Z., *Proc. phys. Soc.* **92** (1967) 880.

wave. In one set of calculations the distortion function (ϕ) was neglected, but in a second set they included not only the dipole distortion of the hydrogen wave-function, but also that of the Ps by the Coulomb field of the proton. It was found that the inclusion of these distortion terms had a profound effect on the calculated cross-sections both for elastic scattering and Ps formation. The great sensitivity of the results to these terms is an indication of the difficulty of obtaining reliable values.

11. Analysis and discussion of experimental results on delayed annihilation of free positrons in gases

11.1. *Helium*

Experimental data are available on the variation of the annihilation rate with time of positrons in helium with, and without, electric fields present. Information is also available about the effects of electric fields on the probability of positronium formation. In all these experiments there is a special need to ensure a very high degree of purity in the helium used, and some of the earlier work suffered because of inadequacy in this respect.

In the most recent experiments, at the time of writing, special precautions were taken. Leung and Paul†, working with helium pressures up to 20 atm, used Matheson ultra-high purity helium containing, in parts per million, 1·5 N_2, 0·3 O_2, 0·1 H_2, 2 H_2O, and 3·1 Ne. It was further purified by passage through a zeolite molecular sieve filter held at 77 °K during the gas filling, so that the only remaining impurity was estimated to be about 2 parts per million of Ne. Lee, Orth, and Jones‡ also used helium of ultra-high purity supplied by Matheson of Canada, and fitted a hot titanium purifier to the pressure chamber to reduce the impurity content further.

Fig. 26.39 (a) shows results obtained by Lee, Orth, and Jones‡ for the delayed coincidence rate up to 500 ns, at 300 °K in the absence of an electric field. It will be seen that there is some evidence of a shoulder. This is a little more clearly seen in the results obtained at a considerably lower temperature (77 °K) by Leung and Paul, shown in Fig. 26.39 (b). The variation of the mean value of Z_a with the ratio F/N of field strength to gas concentration when the free positrons have attained equilibrium with the gas, derived from analysis of the data obtained in these two sets of experiments, is shown in Fig. 26.40. There is very little difference between the results, which apply to different gas temperatures. At

† loc. cit., p. 3163.
‡ LEE, G. F., ORTH, P. H. R., and JONES, G., *Phys. Lett.* **28A** (1969) 674.

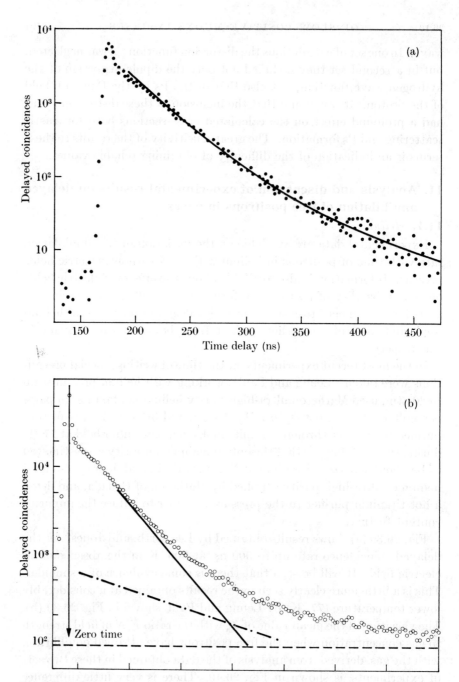

FIG. 26.39. Delayed-coincidence annihilation spectra for positrons in helium, (a) observed by Lee, Orth, and Jones, for a helium atom concentration of $1 \cdot 18 \times 10^{21}$ cm^{-3} and 300 °K, (b) observed by Leung and Paul for a helium atom concentration of $1 \cdot 08 \times 10^{21}$ cm^{-3} and 77 °K. Analysis into the free positron (———), triplet Ps (—·—·—·), components and background (– – –), is shown.

300 °K, Lee et al. find the mean value of Z_a in thermal equilibrium in zero field to be $3\cdot628\pm0\cdot041$ while Leung and Paul find $3\cdot680\pm0\cdot025$ at 77 °K. It seems from these results that Z_a does not vary much with positron velocity over the range covered. Leung and Paul analysed their data in the shoulder region to obtain a mean value of Z_a as a function of delay time. Some of their results are shown in Fig. 26.41.

In none of the experiments was any evidence obtained of an increase in Ps formation with electric field strength, even when the full precautions described above were taken to ensure gas purity. Lee et al. did find that, if the titanium purifier in their arrangement was inactive, an increase of Ps formation was found for F/p of about 10^{-2} V cm^{-1} torr^{-1}, ($F/N = 2\cdot4\times10^{-19}$ V cm^2) which is in rough agreement with the threshold value found by Marder et al. in their experiments in helium (see p. 3163). It seems that, with pure helium, the threshold value of F/p is greater than 5×10^{-2} V cm^{-1} torr^{-1} ($F/N = 1\cdot2\times10^{-18}$ V cm^2).

To relate these observations to theory we note first that, since it is apparent that Z_a varies little with positron velocity, the observed values of the mean of Z_a at 77 and 300 °K, which do not differ much, should be comparable with that calculated for positrons of nearly zero energy. Reference to Fig. 26.34 shows that there is good agreement with that calculated using Drachman's variational method (G). Other approximate methods gave much less satisfactory results. As remarked in § 10.3.4 the results are very sensitive to the approximation used.

Extension of the comparison to higher relative velocities of collision is less direct. For the analysis of their equilibrium rate at 300 °K Lee et al., following the procedure outlined in § 7, derived the velocity distribution function for positrons at different F/N, using, for the momentum-transfer cross-section Q_d, that calculated by Drachman using the adiabatic method (E) with full monopole suppression. By combining these distribution functions with the values of Z_a as a function of v, calculated by Drachman from the variational method (G) (see Fig. 26.40 (a)), quite good agreement is obtained with the observed mean values out to the highest values of F/N in the experiments. There are nevertheless some indications of discrepancies, which are difficult to identify because the observed results are still not accurate enough. The calculated values are not very sensitive to Q_d because of the small dependence of Z_a on v. The fact that Lee et al. used Q_d calculated from a different approximation is not very serious, and there is little doubt that the values of Z_a given from Drachman's variational method (G) are close to the true values.

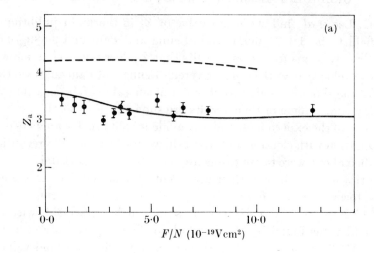

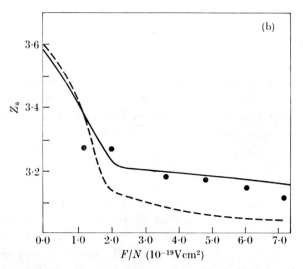

FIG. 26.40. Variation of the mean value of Z_a for positrons in helium as a function of the ratio F/N of field strength to atom concentration:
(a) at 300 °K ● observed by Lee, Orth, and Jones,
—— calculated using Q_d as in Fig. 26.33 (E_1), and Z_a as in Fig. 26.34 (G),
- - - calculated using Q_d as in Fig. 26.33 (D) and Z_a as in Fig. 26.34 (D);
(b) at 77 °K ● observed by Leung and Paul,
—— calculated using Q_d as in Fig. 26.33 (P), and Z_a as in Fig. 26.34 (P),
- - - calculated using Q_d as in Fig. 26.33 (P) and Z_a as in Fig. 26.34 (G).

Leung and Paul carried out a more elaborate analysis of their data in which they took account of the shoulder, as well as the equilibrium, region. For this purpose the momentum-transfer cross-section is

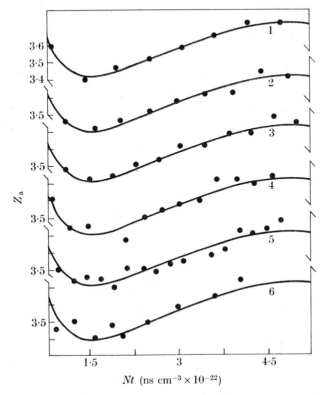

Fig. 26.41. Variation of Z_a with time t for positrons in the 'shoulder' region of the delayed-coincidence annihilation spectrum at 350 °K, in helium, at different atom concentrations N, derived by Leung and Paul from analysis of their observations. Curves 1 to 6 correspond to concentrations N, cm^{-3}, from 70 to 20 L, in steps of 1·0 L where L is Loschmidt's number ($2\cdot7 \times 10^{19}$). The solid curves are of an average shape reproduced for each set of data, the rise on the left-hand side being instrumental.

required over a wider velocity range. In their first analysis† they took as a flexible form, involving adjustable parameters,

$$Q_d = A\exp(-Bv^2) + C[1+\exp\{D(v_1-v)\}]^{-1}. \tag{71}$$

A and B were chosen to give agreement with Q_d, calculated by Drachman's variational method (G), at low velocities. At higher velocities, the parameters C and v_1 were chosen so that Q_d had a velocity dependence

† Reported in a preprint by Leung, C. Y. and Paul, D. A. L.

similar to that calculated from the adiabatic method (E) with full monopole suppression. D was then chosen, in conjunction with a certain assumed variation of Z_a with v so as to give agreement with the dependence of mean Z_a on time derived from analysis of the shoulder region (see Fig. 26.41).

Z_a was assumed to have the form
$$Z_a = a\exp(-bv)+c, \qquad (72)$$
again involving three parameters which were chosen, in conjunction with Q_d given as described above, to give the best fit with all the data. The set of values given in Table 26.3 was found to give good agreement. The corresponding forms for Q_d and $Z_a(v)$ are included in Figs. 26.33 and

TABLE 26.3

Values of parameters in semi-empirical forms of Q_d and Z_a which give a good fit to Leung and Paul's data for helium

$A = 1\cdot046\pi a_0^2$	$B = 70\cdot2$ rydberg^{-1}	$C = 0\cdot22\pi a_0^2$
$D = 4\cdot4$ rydberg$^{\frac{1}{2}}$	$v_1 = 0\cdot65$ rydberg$^{\frac{1}{2}}$	
$a = 0\cdot56$	$b = 5\cdot15$ rydberg$^{-\frac{1}{2}}$	$c = 3\cdot1$

Figs. 26.34, respectively. Fig. 26.40 (b) shows the comparison between calculated and observed values for the mean value of Z_a as a function of F/N, together with values obtained using Z_a as calculated from Drachman's variational method. The semi-empirical values represent a significant improvement at the higher F/N. Satisfactory agreement was obtained with the observed results shown in Fig. 26.41 for the variation of the mean Z_a with time, in the shoulder region, but it must be remembered that the analysis is less definite in this region because of uncertainty about the velocity distribution of positrons entering the region.

Finally, the semi-empirical form for Q_d is compatible with the failure to observe increased Ps formation even for F/N as high as $1\cdot2\times10^{-18}$ V cm^2. According to the analysis on p. 3160 this requires that at the Ps formation threshold $Q_d > 0\cdot12\pi a_0^2$. As realized by Leung and Paul this analysis is not unambiguous particularly because of its dependence on the less reliable analysis of data in the shoulder region. Nevertheless, it indicates quite clearly that the theoretical values of Z_a given by Drachman's variational method cannot be far wrong.

We must now call attention to the interesting results first obtained by Roellig and Kelly,[†] who observed the delayed annihilation of positrons

[†] ROELLIG, L. O. and KELLY, T. M., *Phys. Rev. Lett.* **15** (1965) 746.

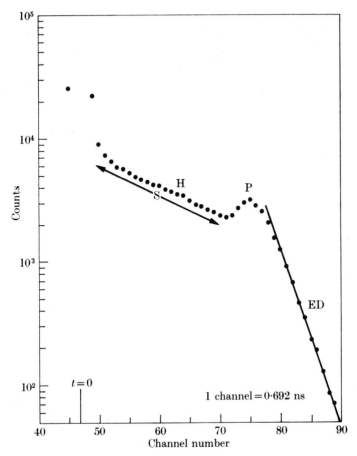

Fig. 26.42. Delayed annihilation spectrum of positrons in helium at a temperature of 5·50 °K and density of 0·023 gm cm⁻³, observed by Canter.

in dense helium gas at 4·2 °K, as well as at 77 °K. At the former temperature they found that, at densities greater than 0·012 gm cm⁻³, the variation of annihilation rate with time exhibits a new characteristic feature. Further investigations† have provided a fairly complete picture of the variation of the delayed annihilation spectrum with density and temperature.

Fig. 26.42 shows a typical observed spectrum at a temperature of 5·50 °K and density of 0·023 gm cm⁻³. The prompt peak is followed by a broad shoulder region (S) with a slight hump at H, much the same as at 300 °K. However, the shoulder is followed by a peak (P) and thence

† CANTER, K. and ROELLIG, L. O., ibid. **25** (1970) 328; CANTER, K., Thesis. Wayne State University, Detroit (1970).

by a rapid exponential decay rate (ED). At any particular density, ρ (> 0.012 gm cm^{-3}), the peak P is only present at temperatures below a particular value $T_c(\rho)$ which increases as the density increases. Fig. 26.43 shows how the final equilibrium decay rate varies with temperature at these densities. It will be seen that at sufficiently high temperatures the decay rates tend to constant values proportional to the density.

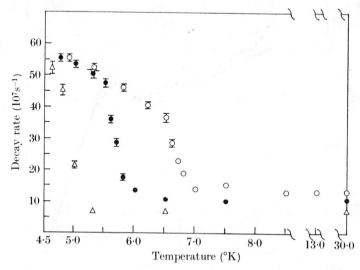

FIG. 26.43. Equilibrium decay rates, observed by Canter, of low energy positrons in helium as a function of temperature at three different densities in gm cm^{-3} as follows:
△ 0.016, ● 0.023, ○ 0.030.

Further, the constant value is the same, within experimental error, as that observed at 300 °K. On the other hand as the temperature falls below a certain value, $\simeq T_c(\rho)$, the decay rate increases abruptly and then more slowly so that at the lowest temperature of observation, 4.6 °K, it is much the same for all the densities concerned.

For a particular density (ρ) comparison of the spectra at two temperatures, one below and one above $T_c(\rho)$, shows that the two are almost coincident over the shoulder region which extends over a time t_s given by

$$\rho t_s = 2500 \times 10^{-4} \text{ gm cm}^{-3} \text{ ns}.$$

After this time the spectrum at $T > T_c$ shows the same equilibrium exponential decay as at 300 °K while for $T < T_c$ the peak appears, followed by the much faster equilibrium decay. The fact that the peak is absent at 8 °K at all densities investigated, suggests that the fast decay will not occur when the mean energy of the positron is as high as that for thermal equilibrium at this temperature. It would then follow that, in the interval before the peak occurs, the positrons must have slowed down to an energy less than 0·001 eV, the mean thermal energy at 8 °K. Most of the time, t_s, is spent in the shoulder region in reducing the positron energy from the inelastic threshold, which can only occur through elastic collisions. To place a lower limit on the mean momentum-transfer cross-section $\bar{Q}_d$ we have that the rate of change of the kinetic energy E of the positron under these conditions is given by

$$\frac{dE}{dt} = (2m/M)En\bar{Q}_d v, \qquad (73)$$

where v is the positron velocity, m and M are the respective masses of a positron and a gas atom, and n is the concentration of gas atoms. Integrating this equation we have, for the time (τ) to reduce the positron energy from E_i to E_f,

$$E_f^{-\frac{1}{2}} - E_i^{-\frac{1}{2}} = (2m)^{\frac{1}{2}}\bar{Q}_d n\tau/M. \qquad (74)$$

Taking E_f as 0·001 eV, we may neglect $E_i^{-\frac{1}{2}}$ and we then find, using eqn (74), that $\bar{Q}_d \simeq \pi a_0^2$. In default of a detailed theoretical interpretation of the origin of the peak, the estimated lower limit for $\bar{Q}_d$ remains doubtful, but it is not inconsistent with theoretically expected values (see Fig. 26.33). Considerations of this kind were first pointed out by Tao and Kelly,[†] who showed that, if the same assumptions are made about Q_d and Z_a as those of Lee et al.,[‡] namely Q_d given from the adiabatic method (E) with complete monopole suppression, and Z_a from Drachman's variational method, the shape of the delayed annihilation-time plot, observed by Roellig and Kelly, is quite well reproduced if the phenomenon responsible for the post-shoulder peak provides a very strong sink for positrons at an energy of 0·0030 eV. This is shown in Fig. 26.44. Once again a small uncertainty is introduced from ignorance of the initial velocity distribution of the positrons on entry into the shoulder region but this is probably not serious.

It was suggested by Roellig[§] in 1965 that the fast decay arises from the formation of clusters of helium atoms round the positrons, thereby markedly increasing the electron density in their neighbourhood. While cluster formation is an attractive hypothesis to explain the low temperature behaviour much further work needs to be carried out to check whether it is indeed correct.

[†] TAO, S. J. and KELLY, T. M., *Phys. Rev.* **185** (1969) 135.
[‡] loc. cit., p. 3191.
[§] ROELLIG, L. O., *Positron annihilation* (ed. Stewart, A. T., and Roellig, L. O.). Academic Press, New York (1967), p. 127.

A further test of theoretical approximations may be made from the observations of the angular correlation of the gamma rays resulting from two-photon annihilation of positrons in liquid helium.† Drachman‡ has applied eqn (46) to calculate the correlation function $P(\theta)$ defined in eqn (45). Fig. 26.45 shows the very good agreement between the observed

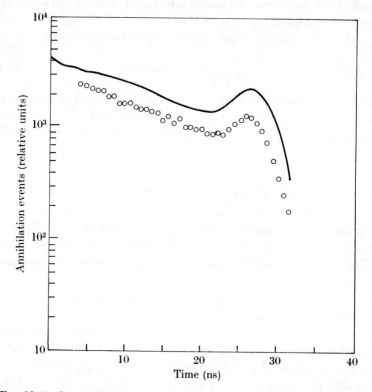

FIG. 26.44. Comparison of observed and calculated time spectra for positrons in helium, at a density of 0·0168 gm cm⁻³ at 4·2 °K, ○ observed by Roellig and Kelly, —— calculated by Tao and Kelly using Q_d as in Fig. 26.33 (E_1) and Z_a as in Fig. 26.34 (G).

function and that calculated, using the wave-function Ψ, determined by the variational method (G). The agreement is not so good if Ψ is obtained from the adiabatic method with full monopole suppression. For further comparison we also show the half-width of the distribution expected if the relative velocity spread leading to $P(\theta)$ is supposed to be entirely due to the motion of the helium atomic electrons.

† BRISCOE, C. V., CHOI, S-I., and STEWART, A. T., *Phys. Rev. Lett.* **20** (1968) 493.
‡ DRACHMAN, R. J., *Phys. Rev.* **179** (1969) 237.

To summarize the present position it seems that the momentum-transfer cross-section given from the adiabatic method (E) with complete monopole suppression is the best approximation of those we discussed

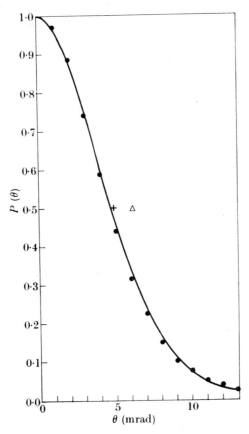

FIG. 26.45. Angular correlation function for two-quantum annihilation of zero-energy positrons in helium:
- ●   observed by Briscoe, Choi, and Stewart,
- ——— calculated by Drachman, using his variational method (G),
- +   half-width calculated by the adiabatic method with full monopole suppression (E),
- △   half-width assuming the distribution is due only to the motion of the helium atomic electrons.

in § 10.3. However, this same approximation does give considerably too large a value of $Z_a(v)$. On the other hand Drachman's variational method (G) gives quite good results for $Z_a(v)$. There is need for further theoretical work to remove this inconsistency. It is also important to understand the nature of the cluster process.

11.2. Argon and neon

The most extensive investigations of annihilation in argon have been those of Jones and his collaborators.† They have not only observed the annihilation rate as a function of time at room temperature in the absence of an electric field, but have obtained corresponding data over

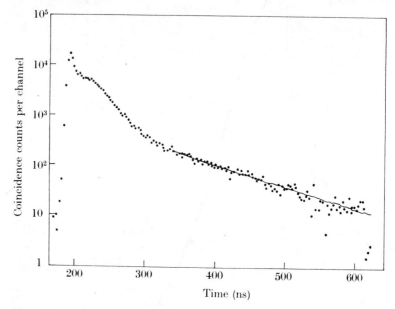

FIG. 26.46. Delayed-coincidence annihilation-time spectrum in argon gas at 300 °K and a density of 1.66×10^{-2} gm cm^{-3} in zero electric field, observed by Orth and Jones.

a range of temperatures and also for a wide range of F/p (the ratio of electric field strength to gas pressure) at room temperature. The apparatus which they used has been described in § 5.1. Fig. 26.46 shows a typical set of observations of annihilation rate as a function of time at room temperature and in zero electric field. The prompt peak and the shoulder are clearly seen. On the logarithmic plot the linear region immediately following the shoulder is due to annihilation of thermalized free positrons while at much greater times the plot takes the form of a straight line of much smaller slope arising from orthoPs annihilation. From analysis of plots of this sort, taken at different gas pressures, into its separate components, as discussed on p. 3135, the rate of annihilation

† ORTH, P. H. R. and JONES, G., Phys. Rev. **183** (1969) 7; FALK, W. R., ORTH, P. H. R., and JONES, G., Phys. Rev. Lett. **14** (1965) 447; MILLER, D. B., ORTH, P. H. R., and JONES, G., Phys. Lett. **27**A (1968) 649.

of thermalized positrons which gives the best fit to the observations is given by
$$\lambda_a/N = 5\cdot 77 \pm 0\cdot 043 - (0\cdot 041 \pm 0\cdot 10)N, \qquad (75)$$
where N is the ratio of the number of gas atoms per cm³ to Loschmidt's number. If, however, the term proportional to N in eqn (75) is required

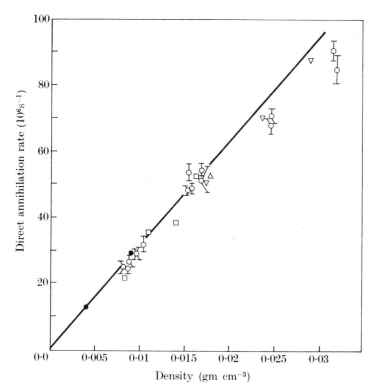

Fig. 26.47. Variation of the direct annihilation rate of positrons in argon with gas density as observed by different experimenters: ▽ Tao, Bell, and Green; ● Paul; □ Falk and Jones; △ Paul and St. Pierre; ○ Orth and Jones.

to be exactly zero, the best fit, which is somewhat less satisfactory, is
$$\lambda_a/N = 5\cdot 27 \pm 0\cdot 043. \qquad (76)$$
Assuming that the limit, for vanishing N, of λ_a/N lies between 5·76 and 5·27 the effective number of annihilation electrons Z_a per argon atom, averaged over the thermal velocity distribution of the positrons, is $27\cdot 3 \pm 1\cdot 3$. The time-resolution of the counting equipment was not great enough to provide observations of high accuracy at high gas pressures to confirm or otherwise the existence of a density dependent term in Z_a.

Fig. 26.47 compares the results obtained by Orth and Jones for λ_a,

as a function of gas pressure, with those obtained in earlier investigations.†
The agreement is good and not inconsistent with a small departure from
linearity at high pressures.

The extent of the shoulder region was found to be $340N^{-1}$ ns and its
reproducibility, from run to run, was taken to be a valuable confirmation
of the continuing purity of the gas sample. Also the fact that it is larger
than values found in earlier investigations—the nearest are those of Paul

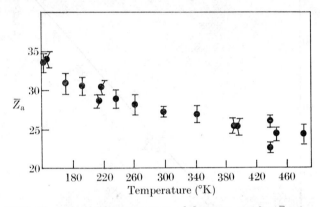

FIG. 26.48. Variation with temperature of the mean number $\bar{Z}_a$ of annihilation
electrons in argon as observed by Miller, Orth, and Jones.

who obtained $190–300N$ ns—indicates that at least the gas purity was
higher than had been attained previously. Miller, Orth, and Jones‡
measured λ_a/N over a temperature range from 140 to 480 °K and concentrations (N) between 8 and 10·5. The argon gas of 99·999 per cent
purity (supplied by Matheson of Canada Ltd.) was used, and the continuing purity was checked by continuously monitoring the shoulder
width and the orthoPs lifetime. Fig. 26.48 shows Z_a, averaged over the
Maxwellian distribution of velocities, as a function of temperature.

To derive Z_a as a function of positron velocity from these observations
two empirical expressions representing the variation of the observed $\bar{Z}_a$
with temperature were used, namely,

$$\bar{Z}_a = 138T^{-0·285} \tag{77}$$

and
$$\bar{Z}_a = 60·0 - 2·89T^{\frac{1}{2}} + 0·056T. \tag{78}$$

Using the procedure described on p. 3136, $Z_a(v)$ was then derived with
the results shown in Fig. 26.49 which indicate a rapid rise at low positron

† TAO, S. J., BELL, J., and GREEN, J. H., Proc. phys. Soc. 83 (1964) 453; PAUL,
D. A. L., ibid. 84 (1964) 563; FALK, W. R. and JONES, G., Am. J. Phys. 42 (1964) 1751;
PAUL, D. A. L. and ST. PIERRE, L., Phys. Rev. Lett. 11 (1968) 493.
‡ MILLER, D. B., ORTH, P. H. R., and JONES, G., Phys. Lett. 27 (1968) 649.

velocities. Comparison with calculated results, obtained by different approximations, in Fig. 26.38, does not reveal any very close agreement with any calculated set of values. The best of the latter are those obtained using the polarized orbital method in which only $3p \to d$ transitions were used in calculating the atomic distortion.

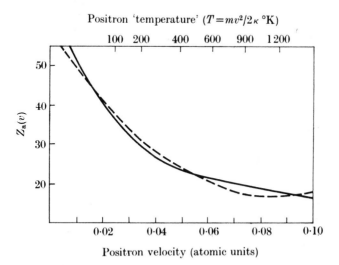

FIG. 26.49. Variation of Z_a with positron velocity in argon derived from the observations of $\bar{Z}_a$ as a function of temperature shown in Fig. 26.48, —— derived assuming Z_a is given by eqn (77), --- derived assuming Z_a is given by eqn (78).

Fig. 26.50 shows the variation of $\bar{Z}_a$ with F/N, the ratio of electric field strength to atom concentration, observed by Orth and Jones and also in earlier work by Falk.† The agreement between the two sets of results is good and it is clear that Z_a must vary quite strongly with positron velocity above the mean thermal velocity at room temperature. At low fields these results give

$$\frac{\Delta(\lambda_a/N)}{\Delta(F/N)^2} = -3\cdot 70 \times 10^3. \tag{79}$$

Combining this with the observed variation of λ_a with temperature at 300 °K, when $F = 0$, gives

$$\frac{\partial \lambda_a}{\partial(F^2)} \Big/ \frac{\partial \lambda_a}{\partial T} \simeq 0\cdot 65 \text{ °K V}^{-2} \text{ cm}^2. \tag{80}$$

Using eqn (25) we then find, at room temperature,

$$Q_d \simeq 33\pi a_0^2. \tag{81}$$

† FALK, W. R., Ph.D. Thesis. University of British Columbia (1965).

Referring to Fig. 26.37, we see that this (admittedly very rough) value falls between that given from the polarized orbital calculations of Montgomery and La Bahn,† and of Massey, Lawson, and Thompson (curves III and I)‡ in each of which the $3p \to d$ virtual transition alone was used in calculating the atomic distortion.

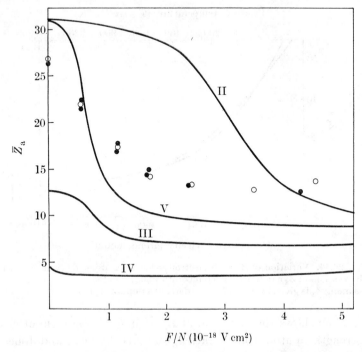

Fig. 26.50. Variation of $\bar{Z}_a$ with F/N for positrons in argon at 300 °K. Observed: ● by Falk; ○ by Orth and Jones. —— calculated curves. II, III, and IV with Q_d as in Fig. 26.37 and Z_a as in Fig. 26.38, (V) with Q_d as in III of Fig. 26.37 and Z_a as in II of Fig. 26.38.

The comparatively good results obtained using these calculations are not so apparent when they are applied to obtain Z_a as a function of F/N at 300 °K. This may be seen by reference to Fig. 26.50. Once again, it is clear that the problem of calculating Z_a, as well as Q_d, is more severe than for the corresponding problem of calculating Q_d for electron collisions. The fact that Z_a depends on the collision wave-function where the positron and electron coordinates are small, and not merely on the asymptotic form of the function, aggravates the difficulty. It is of interest to note, again from Fig. 26.50, that considerable improvement

† loc. cit., p. 3187.   ‡ loc. cit., p. 3187.

is obtained if $\bar{Z}_a$ is calculated using Q_d calculated with the polarizability normalized to the observed value, and Z_a without such normalization. This may be due to the fact that the wave-function, calculated with the normalization, underestimates the effective attraction on the positron at the close distances which are important in determining Z_a. Omitting the normalization, so that the atomic distortion is increased, may partially correct for this.

It is clear that much still remains to be done before the analysis of the major data is complete. Orth and Jones applied an empirical procedure in which they treated the positron–argon collisions as one-body processes, with an interaction given by the Hartree–Fock potential plus an empirical polarization potential

$$V_p = -\tfrac{1}{2}\alpha(r^2+r_0^2)^{-2}, \tag{82 a}$$

with α the observed polarizability, and r_0 an adjustable parameter. In this treatment, Z_a is calculated without inclusion of an atom distortion term in the wave-function, Ψ. It is not surprising, therefore, that if r_0^2 is chosen to have the value which gives the best fit for low-energy electron scattering ($r_0^2 = 2\cdot 5a_0^2$), Z_a is greatly underestimated. The best fit was obtained with $r_0^2 = 0\cdot 62a_0^2$, but even then was not very satisfactory. They then tried in place of eqn (82 a)

$$V_p = -\tfrac{1}{2}\alpha r^{-4}[1-\exp\{-(r/r_0)^8\}], \tag{82 b}$$

but again did not obtain any better fit.

No analysis of the shoulder region in argon is available at the time of writing.

The experimental data for argon which we have been discussing have all been obtained at pressures of less than 10 atm so that, even at the lowest temperatures concerned (140 °K, see Fig. 26.48) the gas density was less than $1\cdot 6 \times 10^{-2}$ gm cm^{-3}. Observations at higher gas densities at the lower temperatures have been carried out by Canter[†] to determine whether effects appear in argon comparable with those at densities above $1\cdot 2 \times 10^{-2}$ gm cm^{-3} and temperatures below 8 °K in helium. Fig. 26.51 shows his results for a density of 0·0356 gm cm^{-3} and it is clear that, at temperatures below 180 °K, at this density, the same effects appear, namely the shoulder is followed by a peak and then a fast equilibrium decay rate. Unlike helium, however, the width of the shoulder region does not remain constant as the temperature falls.

Fig. 26.52 shows the observed behaviour of the mean number of annihilation electrons $\bar{Z}_a$ as a function of temperature for four gas

† Thesis. Wayne State University, Detroit (1970).

densities, including that used by Miller, Orth, and Jones in obtaining the results shown in Fig. 26.48. It will be seen that, above 180 °K, $\bar{Z}_a$ is independent of temperature, but at lower temperatures increases quite rapidly as the temperature falls. The onset of this rise is delayed

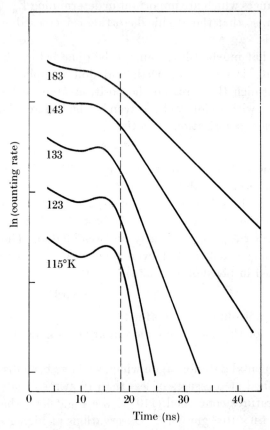

FIG. 26.51. Delayed annihilation spectra of positrons in argon at a density of 0·0356 g cm^{-3}, and different temperatures, as indicated, in °K. The curves are smoothly fitted to the observed results.

as the density falls and Miller, Orth, and Jones did not find any such rise in their experiments down to 140 °K.

It is not certain that the phenomena involved at the higher densities are essentially the same in nature as observed in helium. Canter† has produced arguments which show that it is not possible to interpret the argon high-density results by assuming an appropriate variation of Z_a with positron velocity, in single collisions. No sign of these effects has

† Thesis. Wayne State University, Detroit (1970).

been found in neon. Thus Fig. 26.53 shows observed equilibrium decay rates as a function of density at different temperatures. No departure from linearity is seen up to densities as high as 0·9 g cm^{-3}, and there is no dependence on temperature from 34 to 300 °K.

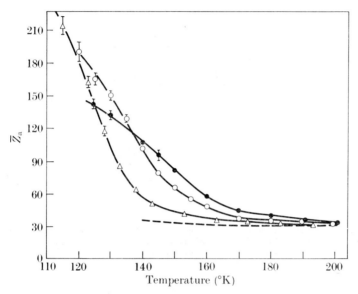

Fig. 26.52. Mean number $\bar{Z}_a$ of annihilation electrons per argon atom as a function of temperature for different densities in 10^{-2} g cm^{-3} as follows:
● 7·26, ○ 5·41, △ 3·56—(observed by Canter);
--- 1·4 to 1·8 (observed by Miller, Orth, and Jones (see Fig. 26.48)).

We show in Fig. 26.54, Z_a calculated as a function of F/N for room temperature, using the approximations discussed on p. 3187. The observed value, for zero electric field, derived from the results shown in Fig. 26.53, agrees quite well with that calculated in which both $2p \to d$ and $2s \to p$ virtual transitions are taken into account in determining the atom distortion, but the resulting polarizability is normalized to agree with the observed value. Whether this good agreement will persist when observed data are available for finite F/p remains to be seen.

11.3. *Other gases*

Results for other gases are limited to observations of Z_a at room temperature for thermalized positrons in zero electric field. For Kr, Falk and Jones[†] find $\bar{Z}_a = 60·5$, while for N_2, Tao, Bell, and Green[‡] find $\bar{Z}_a = 27·2$.

[†] FALK, W. R. and JONES, G., *Can. J. Phys.* **42** (1964) 1751.
[‡] TAO, S. J., BELL, J., and GREEN, J. H., *Proc. phys. Soc.* **83** (1964) 453.

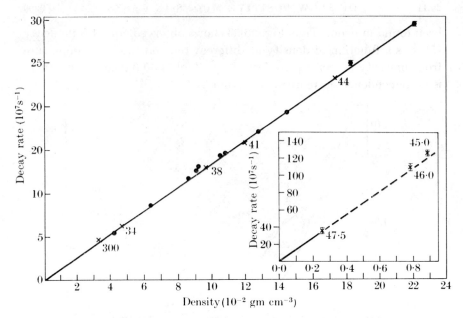

FIG. 26.53. Equilibrium decay rates (observed by Canter) of low-energy positrons in neon, as a function of density at various temperatures, ● observations at 77 °K; × observations at other temperatures, as indicated.

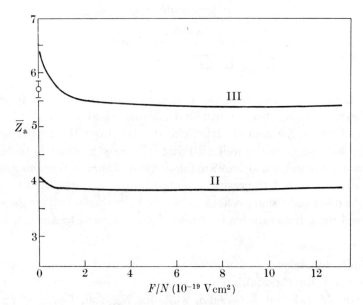

FIG. 26.54. Variation of $\bar{Z}_a$ with F/N for positrons in neon.
— calculated curves II and III with Q_d as in Fig. 26.35 and Z_a as in Fig. 26.36,
⌷ observed by Roellig.

26.11 COLLISIONS OF SLOW POSITIVE MUONS IN GASES

Analysis of data obtained in molecular gases, or in mixtures of such gases with rare gases, is complicated by the occurrence of collisions in which positrons lose energy by excitation of molecular vibration and rotation. These collisions modify the equilibrium velocity distribution

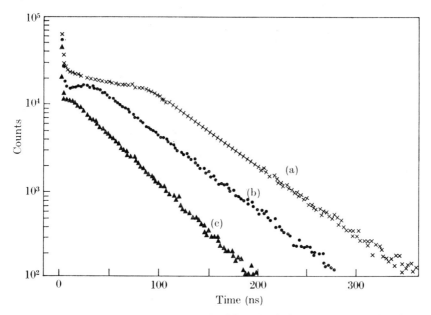

FIG. 26.55. Free positron component of the delayed coincidence spectrum of positrons in argon containing partial pressures of nitrogen. (a) pure argon at 3000 torr, (b) argon with 120 torr of N_2 added, (c) argon with 746 torr of N_2 added.

of the positrons at any particular value of F/p, and hence change $\bar{Z}_a$ from the value it would have if elastic collisions alone were important. Furthermore, they reduce the time taken for the positrons to be slowed down to thermal energies. This will appear in the usual delayed-coincidence annihilation-time spectra as a reduction in the shoulder region. Thus we have already pointed out (see p. 3136) the importance of removal of molecular impurities for accurate observation of this region in rare gases and particularly helium.

Paul and Leung† have carried out observations in mixtures of argon with nitrogen, and with methane, specifically to investigate the effect on the shoulder region. Fig. 26.55 shows results which they obtained for Ar–N_2 mixtures in which the shoulder region is seen to be progressively reduced by addition of higher partial pressures of N_2. To analyse these

† PAUL, D. A. L. and LEUNG, C. Y., Can. J. Phys. **46** (1968) 2779.

results a little further, Paul and Leung plotted the inverse shoulder width as a function of the N_2 pressure, obtaining the results shown in Fig. 26.56. The criterion adopted in determining the shoulder breadth, which was somewhat arbitrary, was as follows. If $\bar{Z}_a^0$ is the steady mean value when the positrons are thermalized, the temporal range of the

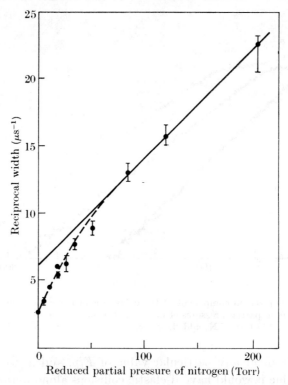

FIG. 26.56. Inverse 'shoulder' widths for nitrogen–argon mixtures reduced to s.t.p. for the argon component.

shoulder was taken to extend out until the observed value of $\bar{Z}_a$ approached to within $0 \cdot 1 \, \Delta Z_a$, where ΔZ_a is the difference between Z_a^0 and the minimum value $Z_a^{\min}$ over the shoulder range for pure argon. It will be seen from Fig. 26.56, in which the times are normalized to correspond to argon at s.t.p., that the curve becomes linear at N_2 pressures above 100 torr. This suggests that, at such pressures, the slowing-down time is almost exclusively determined by the N_2. If this is so, we obtain for the slowing-down time in N_2

$$\tau = 16 \text{ ns at s.t.p.}$$

If this were entirely due to elastic collisions the mean value of Q_d

would be as large as $5\pi a_0^2$, considerably higher than for argon. Instead we must suppose that the mean fraction μ of energy lost by a positron in a collision with N_2 is greater than $2m/M$, where m, M are the respective masses of a positron and a molecule. For 0·4 eV electrons in N_2, $\mu = 15 \times 2m/M$ (see Chap. 11, Table 11.1).

Similar observations were made for methane giving

$$\tau = 0{\cdot}25 \text{ ns at s.t.p.},$$

which is very much smaller than for N_2.

12. The theory of the collisions of slow Ps atoms with other atoms

12.1. *Introduction*

In § 4.2 we discussed the processes which could lead to quenching of orthoPs atoms in a gas—exchange quenching, pick-off quenching, and quenching due to chemical reactions and to radiative or three-body capture. Of these we may use collisions with the simplest atom, hydrogen, to discuss, theoretically, an example of exchange quenching, even though it is one which is at present beyond the reach of experiment. Pick-off quenching by helium can be discussed in some detail, theoretically, and here we have a direct check with experiment. It is difficult, on the other hand, to deal in any detail with a useful theoretical model of quenching due to a chemical reaction. Our theoretical discussion will be confined, therefore, to the collisions of Ps with H and He atoms. The presence of the extra electron so complicates the problem as to make it impracticable to attempt a theory of any more complicated case. It is of interest to note one or two special general features of Ps–atom collisions. Of three of the main contributory interactions to positron–atom scattering, the mean static field, the polarization of the atom, and the non-local interaction due to virtual Ps formation, the first two are absent for the Ps case, and in place of the third we have the non-local interaction due to electron exchange between the Ps and the struck atom. Nevertheless, this is a very strong interaction and cannot be treated by perturbation methods. Although there is no equivalent of atomic polarization, a van der Waals interaction will be present at long range.

12.2. *Collision with hydrogen atoms—exchange quenching*

As in other two-electron collision problems, we must distinguish two cases depending on whether the electron spins are parallel or antiparallel. We thus have two wave-functions $\Psi^\pm(\mathbf{r}_1, \mathbf{r}_2, \mathbf{r})$, corresponding to these respective cases, where $\mathbf{r}_1$ and $\mathbf{r}_2$ are the coordinates of the two electrons and **r** that of the positron, relative to the proton. Ignoring any distortion of the wave-functions of the atom and of the Ps, such as would give rise to a van der Waals interaction, we write as trial forms for $\Psi^\pm$,

$$\Psi_t^\pm = \psi_0(r_1)\omega_0(\rho_2)F^\pm(\boldsymbol{\sigma}_2) \pm \psi_0(r_2)\omega_0(\rho_1)F^\pm(\boldsymbol{\sigma}_1), \tag{83 a}$$

where
$$\boldsymbol{\rho}_{12} = \mathbf{r}_{12}-\mathbf{r}, \qquad \boldsymbol{\sigma}_{12} = \tfrac{1}{2}(\mathbf{r}_{12}+\mathbf{r}). \tag{83 b}$$

ψ_0, ω_0 are the ground state wave-functions of the H and Ps atoms. $F^\pm$ are proper functions which have the asymptotic form

$$F^\pm(\boldsymbol{\sigma}) \sim e^{i\mathbf{k}\cdot\boldsymbol{\sigma}} + f\left(\frac{\mathbf{k}\cdot\boldsymbol{\sigma}}{k\sigma}\right)\sigma^{-1}e^{ik\sigma}, \tag{84}$$

where the wave number (k) is given in terms of the kinetic energy E of the Ps atom by
$$k^2 = 4mE/\hbar^2,$$
m being the electron mass.

As usual for low-energy collisions, we need consider only the zero order term in the expansion of $F^\pm(\boldsymbol{\sigma})$ in spherical harmonics of $(\mathbf{k}\cdot\boldsymbol{\sigma})/k\sigma$. Then if we take

$$\sigma F^\pm(\sigma) = G^\pm(\sigma)$$
$$\sim k^{-1}\sin k\sigma + \alpha^\pm \exp(ik\sigma), \tag{85}$$

the momentum-transfer cross-section Q_d is given by

$$Q_d = \frac{4\pi}{k^2}\{\tfrac{1}{4}|\alpha^+|^2 + \tfrac{3}{4}|\alpha^-|^2\}. \tag{86}$$

We are concerned, also, with the cross-section for electron exchange which can lead to quenching of the orthoPs. This will be given by (see Chap. 8, § 3 and Chap. 18, § 8.2)

$$Q_{\text{ex}} = \frac{\pi}{k^2}|\alpha^+ - \alpha^-|^2. \tag{87}$$

To evaluate Q_d and Q_{ex} the trial functions (83 a) are substituted in the Kohn variation formula (Chap. 8, p. 502), and equations for $G^\pm(\sigma)$ are obtained, of the form

$$\left(\frac{d^2}{d\sigma_1^2}+k^2\right)G^\pm(\sigma_1) = \pm\int N(\sigma_1,\sigma_2)G^\pm(\sigma_2)\,d\sigma_2. \tag{88}$$

Although these are single integro-differential equations, the kernel N is quite complex to evaluate, because the problem involves three light particles.

Fraser[†] has solved the equations (88) numerically and finds that, in the low energy limit

$$Q_d = 1{\cdot}70\times 10^{-14}\text{ cm}^2, \qquad Q_{\text{ex}} = 0{\cdot}30\times 10^{-14}\text{ cm}^2,$$

† FRASER, P. A., *Proc. phys. Soc.* **78** (1961) 329.

while at a Ps energy of 0·5 eV,
$$Q_d = 0\cdot 31\times 10^{-14}\text{ cm}^2, \qquad Q_{ex} = 0\cdot 045\times 10^{-14}\text{ cm}^2.$$

It will be seen that the quenching cross-section, even at 0·5 eV, is quite large, larger than the greatest observed, that for NO (see Table 26.5). Since the exchange interaction is attractive and likely to be strong for collisions with most atoms, it is clear that, when exchange can produce quenching, the effective quenching cross-section will be difficult to predict as it may have any value between zero and infinity depending on the details of the interaction.

12.3. *Collisions with helium atoms—pick-off quenching*

In this case, there is no possibility of quenching through electron exchange. For this to occur the helium would have to be in a triplet state, the lowest of which requires 19·77 eV excitation energy. However, both the momentum-transfer cross-section and that for pick-off quenching may be calculated approximately, and both may be compared with observed values.

We have now, in place of eqn (83 a) just one trial function which, for collisions of orthoPs with helium takes the form

$$\Psi'_t = 3^{-\frac{1}{2}} \sum_{1,2,3} F(\boldsymbol{\sigma}_1)\omega_0(\boldsymbol{\rho}_1)\psi(r_2,r_3)\chi^1_1(s,s_1)\chi^0_0(s_2,s_3), \qquad (89)$$

where χ^1_1 is the triplet spin wave-function for the orthoPs and χ^0_0 for the singlet function for the helium atom, s, s_1, s_2, s_3, being the spin coordinates of the positron and of the electrons, respectively.

To calculate the pick-off cross-section we first determine the probability P_1 that, with a trial function of the form (89), the positron and electron are in a singlet state with respect to each other. This is given by

$$\phi(\mathbf{r},\mathbf{r}_1;\mathbf{r}_2,s_2;\mathbf{r}_3,s_3) = \left|\sum_{s,s_1}|\chi^0_0(s,s_1)|^2\Psi'_t\right|^2. \qquad (90)$$

The effective number of annihilation electrons is then given by

$$Z_a = 3\sum_{s_2,s_3}\int |\phi(\mathbf{r},\mathbf{r}_1;\mathbf{r}_2,s_2;\mathbf{r}_3,s_3)|^2 \, d\mathbf{r}_1 d\mathbf{r}_2 d\mathbf{r}_3. \qquad (91)$$

Fraser[†] obtained the integro–differential equation which results when eqn (89) is substituted as a trial function in the Kohn variational method. Substituting the simple Hylleraas form (eqn (69)), with $Z = 27/16$ for the helium wave-function, he solved the equation to obtain the spherically symmetrical solutions. This work was extended by Fraser and Kraidy,[‡] who obtained solutions for states of relative motion with angular momentum numbers 0, 1, 2, and also calculated Z_a from eqn (91).

[†] Ibid. **79** (1962) 721.   [‡] FRASER, P. A. and KRAIDY, M., ibid. **89** (1966) 533.

Somewhat later, Barker and Bransden† repeated these calculations and also took into account the distortion of the Ps wave-function by the long-range van der Waals interaction with the helium atom. A Ps atom is highly polarizable (the polarizability is 8 times that of an H atom), so the van der Waals force is large. Following a procedure similar to that of the polarized orbital method for slow electron and positron scattering (see Chap. 8, § 2.4 and Chap. 26, § 10.1), Barker and Bransden replaced the undisturbed Ps wave-function $\omega(\rho)$ in eqn (89) by

$$\omega_p = \omega(\rho)\{1+A(\sigma)C(\rho, r_2, r_3)/\sigma^3\}, \tag{92}$$

where C/σ^3 is the asymptotic form, for large σ, of the dipole–dipole interaction between the Ps and He atoms. $A(\sigma)$ was determined by a variational method which gives for the van der Waals interaction $19\cdot3/\sigma^6$ in atomic units. Table 26.4 gives the results obtained for the phase shifts η_0, η_1, and η_2, as well as for Q_d and Z_a. It will be seen that the effect of distortion of the Ps by the dipole–dipole interaction, though significant, is not very great. Just as for collisions of atoms with free positrons, Z_a depends most strongly on a different volume of configuration space than does Q_d and it is to be expected that it will be easier to obtain a good approximation to Q_d than to Z_a.

Drachman and Houston‡ have investigated the importance of inter-particle correlation terms by representing the non-local exchange interaction by an equivalent local interaction. This so simplifies the problem as to make practicable the use of a variational method involving a large number of linear parameters associated with products of powers of the distances between the different particles. Thus the Ps electron may be regarded as distinguishable from the two helium electrons, and its interaction with that atom represented as $Ae^{-\alpha r}$, where r is its distance from the helium nucleus. No other interactions between the Ps and helium atoms are included, and the parameters A and α were chosen so as to give the same values of a and Z_a at zero impact energy, as calculated by Fraser and Kraidy, when a trial function including no correlation terms was used. Having obtained A and α, a and Z_a were recalculated using an elaborate trial function of the form, in the notation of eqn (89),

$$\Psi_t = \omega(\rho)\psi(r_2, r_3)\sigma^{-1}\{a(1-e^{-\delta\sigma})-\sigma\}+e^{-\delta\sigma}e^{-\beta\rho}e^{-\gamma r}\sum_{i=1}^{N} c_i \sigma^{l_i}\rho^{m_i}r^{n_i}, \tag{93}$$

δ, β, and γ being non-linear parameters, and the c_i linear parameters. The suffix 1 distinguishing the Ps electron has been dropped. The correlation terms allow for distortion of the Ps, but not of the helium atom. It was found that the scattering length changed from $1\cdot448$ to $1\cdot389a_0$ as N was increased from 4 to 35. Z_a changed only from $0\cdot090$ to $0\cdot098$, indicating good convergence and a considerable increase above the values obtained without allowance for correlation.

† BARKER, M. I. and BRANSDEN, B. H., *J. Phys.* B **1** (1968) 1109.
‡ DRACHMAN, F. J. and HOUSTON, S. K., ibid. **3** (1970) 1657.

TABLE 26.4

Calculated phase shifts η_0, η_1, and η_2, momentum-transfer cross-sections Q_d, and effective number of annihilation electrons Z_a for Ps collisions in He

k (atomic units)	Phase shifts (radians)						Q_d (πa_0^2)		Z_a	
	η_0		η_1		η_2					
	(1)	(2)	(1)	(2)	(1)	(2)	(1)	(2)	(1)	(2)
0	(1·80)	(2·05)	—	—	—	—	13·05	16·9	0·035	0·044$_5$
0·01	−0·017	−0·021	—	—	—	—	13·05	16·9	0·035	0·046
0·05	−0·091	−0·103	−0·0$_3$3	−0·0$_3$3	—	—	12·9	16·8	0·037	0·048
0·07	−0·127	−0·143	−0·0$_3$7	−0·0$_3$8	—	—	12·8	16·4	0·038	0·048$_5$
0·10	−0·181	−0·203	−0·0$_2$1	−0·0$_2$2	—	—	12·6	15·8	0·040	0·050
0·15	−0·270	−0·299	−0·0$_2$68	−0·0$_2$73	—	—	12·0	14·7	0·044	0·054
0·20	−0·359	−0·391	−0·015$_5$	−0·017	−0·0$_3$1	−0·0$_3$1	11·4	13·4	0·050	0·059
0·40	−0·701	−0·716	−0·097	−0·103	−0·0$_3$3	−0·0$_3$3	8·5	8·7	0·080	0·089
0·60	−1·011	−0·987	−0·236	−0·246	−0·0$_2$7	−0·0$_2$8	6·3$_5$	6·0$_5$	0·119	0·126
					−0·039	−0·049				

(1) Calculated without inclusion of distortion of the Ps wave-function.
(2) Calculated allowing for distortion of the Ps wave-function by the dipole–dipole interaction.
Bracketed values for η_0 are the zero energy scattering lengths.

13. Discussion of experimental results in the quenching of orthoPs atoms in gases

13.1. *Results at room temperature*

We have described techniques for measuring the quenching rates of orthoPs in gases, in §§ 5.1 and 5.2. Table 26.5 summarizes the results obtained at room temperature for a number of gases. The method used is indicated with the references.

It will be seen that, apart from the special cases of NO and O_2, the mean quenching cross-sections are of order 10^{-21} cm^2, corresponding to a mean number of annihilation electrons smaller than unity. For NO the cross-section is as high as $1 \cdot 4 \times 10^{-17}$ cm^2 which is not unexpected because exchange-quenching is clearly operative. The situation is somewhat less clear for O_2 for which the cross-section is about 10^{-19} cm^2. This is much too large to result from pick-off quenching but is much smaller than atomic dimensions. The question has therefore been raised as to whether it is too small to result from exchange-quenching. However, such quenching will certainly occur for O_2 and, according to quantum theory, may have any value depending on details of the interaction—the equivalent of a Ramsauer–Townsend effect occurring is quite possible, and may explain the relatively small cross-section. Some support for the effectiveness of O_2 in reorienting the spin of orthoPs was provided in early experiments[†] in which it was found that, in the presence of an admixture of oxygen, a magnetic field quenches more than a third of the three-quantum decay. However, doubt is thrown on the interpretation, because it was found that NO was less effective in this regard. In this connection it was found,[†] in the same experiment, that, at partial pressures of O_2 in argon less than 10^{-3} atm, O_2 quenched as effectively as NO but at higher pressures was 100 times less effective. It was suggested[‡] that this large quenching is due to excitation of the $a^1\Delta_g$ state (see Vol. 2, Fig. 13.69) at 1·6 eV, by unthermalized orthoPs.

The situation for helium is less clear as the spread of the observed data is considerable. This is due to the low quenching rates which have to be measured. It is notable that reasonably good agreement is obtained between the result of Duff and Heymann on the one hand, who used freon to quench the free positron component (see p. 3135), and Canter, who worked with pure helium as described on p. 3197. Their values are

[†] GITTELMAN, B., DULIT, E. P., and DEUTSCH, M., *Bull. Am. phys. Soc.* **1** (1956) 69.
[‡] DEUTSCH, M. and BERKO, S., *Alpha-, beta-* and *gamma-ray spectroscopy* (edited by K. Siegbahn), vol. 2. North Holland Publishing Co., Amsterdam. 1965, pp. 1583–98.

TABLE 26.5

Observed rates of quenching of orthoPs atoms by various gases at 300 °K

Gas	He	H_2	Ne	Ar	N_2	CO_2	NO	O_2
Quenching rate (10^6 atm^{-1} sec^{-1})	$0.087^2 \pm 0.007$ $0.095^7 \pm 0.006$ $0.133^9 \pm 0.012$	0.164^7 ± 0.003	0.222^8 ± 0.006	$0.248^{1,2} \pm 0.03$ $0.249^{3,4} \pm 0.03$ $0.264^5 \pm 0.3$	0.18^6 $0.206^3 \pm 0.08$ $0.221^4 \pm 0.08$	0.55^1	2500^1	18^1 20^4
Mean number of annihilation electrons	$0.178^1 \pm 0.010$ $0.126^7 \pm 0.008$ $0.180^9 \pm 0.016$	0.211^7 ± 0.004	0.312^8 ± 0.09	$0.339^{1,2} \pm 0.007$ $0.340^{3,4} \pm 0.007$ $0.36^5 \pm 0.05$	0.24^6 $0.28^3 \pm 0.02$ $0.30^4 \pm 0.02$	0.77^1		
Mean quenching cross-section (10^{-21} cm^2)	0.490^1 0.524^7 0.750^9	0.84^7	1.30^8	$1.41^{1,2}$ $1.41^{3,4}$ 1.50	1.00^6 1.12^3 1.25^4	3.2^1	$14\,000^1$	100^1 114^2

1. HEYMANN, F. F., OSMON, P. E., VEITT, J. J., and WILLIAMS, W. F., *Proc. phys. Soc.* **78** (1961) 1038. γ-ray spectrum.
2. DUFF, B. G. and HEYMANN, F. F., *Proc. R. Soc.* **A270** (1962) 517. Pressure variation of long-lived annihilation component.
3. CELITANS, G. J. and GREEN, J. H., *Proc. phys. Soc.* **83** (1964) 823. Three-photon coincidences.
4. CELITANS, G. J., TAO, S. J., and GREEN, J. H., *Proc. phys. Soc.* **83** (1964) 833. Pressure variation of long-lived annihilation component.
5. JONES, G. and ORTH, P. H. R. Private communication to Fraser, P. Pressure variation of long-lived annihilation component.
6. TAO, S. J., BELL, J., and GREEN, J. H., *Proc. phys. Soc.* **83** (1964) 453.
7. CANTER, K., Thesis. Wayne State University, Detroit (1970). Pressure variation of long-lived annihilation component.
8. GOLDANSKII, V. I., quoted in *Atomic Energy Review*. International Atomic Energy Agency, Vienna (1968), by HOGG, B. G., LAIDLAW, G. M., GOLDANSKII, V. I., and SHANTAROVICH, V. P., **6** (1968) 149.
9. BEERS, R. H. and HUGHES, V. W., *Bull. Am. phys. Soc.* **13** (1968) 633.

also not far from that calculated by Drachman and Houston in their semi-empirical theory. Too much weight should not be given to this agreement until the experimental situation is more definite, although some support is given by the fact that the scattering length for Ps–He deduced from low temperature observations of Ps quenching, as described in § 13.2, is quite close to the prediction of the semi-empirical theory.

No quantitative values are available for quenching rates for halogens in the gas phase although these rates are known to be large, the corresponding cross-sections being of the same order as for NO. In these cases the large values are almost certainly due to the occurrence of reactions such as
$$\mathrm{Ps} + \mathrm{I}_2 \to \mathrm{PsI} + \mathrm{I}. \tag{94}$$
The positron will be exposed to a greater electron density in the compound PsI and will therefore annihilate quicker. When it is remembered that, chemically, Ps can be expected to behave in many ways like a light H atom the occurrence of reactions such as (94) is not unexpected. A considerable amount of information about quenching, and inhibition of Ps formation, in condensed systems, is available and although the discussion of reactions in such systems is outside the scope of this book we later summarize (§ 14) some of the methods which may be used to identify the nature of the phenomena concerned, and make detailed consideration of one or two cases. Meanwhile we proceed to describe the experimental results on the quenching of Ps at low temperatures which has yielded some very interesting information about bubble formation.

13.2. Results at low temperatures—bubble formation

Whereas a low-energy free positron is attracted to an atom through long range polarization, so that the scattering length is negative, the opposite applies to a Ps atom. This may be seen from Table 26.4, the scattering length for Ps–He collisions being positive. If the enhanced rate of annihilation observed with slow positrons in helium at low temperatures arises from cluster formation round an otherwise free positron we might expect that, under similar conditions, a Ps atom may dig a cavity for itself in the gas, and so lower the pick-off annihilation rate. We shall now describe some results which suggest that this does in fact occur.

In Fig. 26.57 we show the annihilation rate in H_2, as a function of gas density ρ, at 300 °K and at 77·3 °K, observed by Canter.† At the

† loc. cit., p. 3207.

higher temperature the rate closely follows the linear law

$$\Lambda = \lambda + \nu\rho,$$

where λ is the annihilation rate for free orthoPs ($0.68\pm0.03\times 10^7$ s^{-1}) and ν is a constant ($2.01\pm0.04\times 10^9$ cm^3 s^{-1} g^{-1}). This also holds at 77·3 °K, for densities less than 8×10^{-3} gm cm^{-3}, but at higher densities ν decreases with increasing ρ.

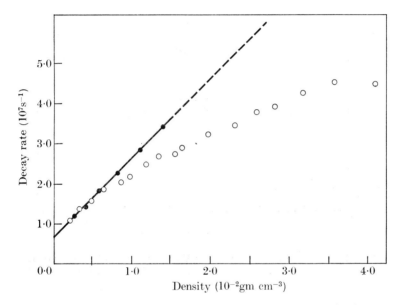

Fig. 26.57. Decay rates for orthoPs in H_2 as a function of gas density at two temperatures. ○ 77·3 °K; ● 300 °K, as observed by Canter.

For helium, less detailed data are available at any temperature below 300 °K but Fig. 26.58 shows clear evidence of similar effects. Thus, at 30 °K, the annihilation rate is not far from that at 300 °K for densities below 2.5×10^{-2} gm cm^{-3}, but is well below at 3×10^{-2} gm cm^{-3}. At a much lower temperature of 4·8 °K the observed rate is well below that at 300 °K, even at densities of 1.5×10^{-2} gm cm^{-3}.

For neon and argon the situation is less clear cut, as may be seen from Figs. 26.59 and 26.60, though there seems to be evidence of a decreased annihilation rate for neon at densities greater than 0·12 gm cm^{-3}, and temperatures of 41 °K. Otherwise the accuracy of measurement is not high enough to be sure that any density effect is present.

To analyse the matter further consider an orthoPs atom within a spherical cavity of radius R_0. Let **r**, **R** be the relative, and centre of mass coordinates, of the orthoPs, the latter being measured with respect to the centre of the cavity as origin. We assume for simplicity that the effect of the cavity may be represented by a potential well $V(R)$, which affects R only, and does not modify the internal relative motion of the

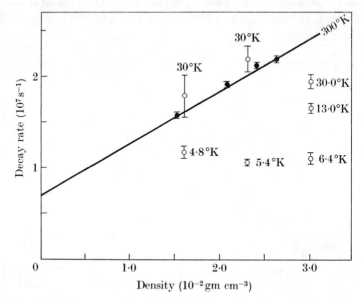

FIG. 26.58. Decay rates for orthoPs in He as a function of gas density at different temperatures: ○ as indicated; ——— at 300 °K, ● experimental points.

electron and positron in the orthoPs. It will further be assumed that the well is only deep enough to include one bound state for motion of the centre of mass of the orthoPs. Taking

$$V = 0, \quad R < R_0,$$
$$= V_0, \quad R > R_0, \qquad (95)$$

we obtain for the wave-function for the orthoPs,

$$\psi(\mathbf{r}_+, \mathbf{r}_-) = \omega(r) F(R), \qquad (96)$$

where $\mathbf{r}_+$, $\mathbf{r}_-$ are the respective coordinates of the positron and electron, $\omega(r)$ is the internal wave-function, and $F(R)$ is given by

$$F(R) = A R^{-1} \sin \alpha R, \qquad R < R_0, \qquad (97\text{ a})$$
$$= B R^{-1} \exp(-\beta R), \qquad R > R_0 \qquad (97\text{ b})$$

where, in terms of the energy E_0 of the bound state,

$$\alpha^2 = 4m E_0/\hbar^2, \qquad \beta^2 = 4m(V_0 - E_0)/\hbar^2, \qquad (98)$$

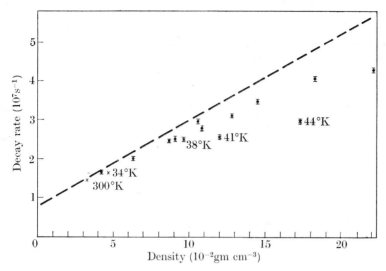

FIG. 26.59. Decay rates for orthoPs in Ne as a function of gas density at different temperatures:

— — — 300 °K (Goldanskii),
● 77·3 °K (Canter),
× as indicated (Canter).

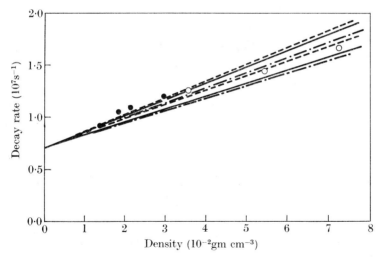

FIG. 26.60. Decay rates for orthoPs in Ar as a function of gas density at different temperatures:

Between the lines – – – – observed by Tao
Between the lines ——— observed by Orth and Jones
Between the lines – · – · – observed by Goldanskii
● observed by Canter
○ observed by Canter at 100–200 °K.

} 300 °K

and A and B are such as to normalize $F(R)$ to unity and provide continuity of F and dF/dR, at $R = R_0$. m is the mass of electron or positron.

The probability that the positron be found within a distance between r_+ and $r_+ + dr_+$ from the centre of the cavity is given by

$$P(r_+)\,dr_+ = r_+^2 \int_0^\pi \int_0^{2\pi} \left\{ \int |\psi(\mathbf{r}_+, \mathbf{r}_-)|^2\,d\mathbf{r}_- \right\} \sin\theta_+\,d\theta_+\,d\phi_+\,dr_+. \qquad (99)$$

We may, therefore, write for the probability P_{out} of finding the positron outside the cavity and so in the normal gas of density ρ,

$$P_{\text{out}} = \int_{R_0}^\infty P(r_+)\,dr_+. \qquad (100)$$

To a good approximation eqn (100) gives

$$P_{\text{out}} = \sin^2\xi (1 - \xi \cot\xi)^{-1}, \qquad (101)$$

$$E_0 = V_0 \sin^2\xi, \qquad (102)$$

where

$$\xi/\sin\xi = (\alpha^2 + \beta^2)^{\frac{1}{2}} R_0. \qquad (103)$$

The pick-off quenching rate is then given by

$$\nu P_{\text{out}}\, \rho. \qquad (104)$$

If the cavity is not entirely empty but contains gas at density $\rho_i < \rho$ then eqn (104) becomes

$$\nu \{P_{\text{out}}\, \rho + (1 - P_{\text{out}})\rho_i\}. \qquad (105)$$

We next consider the determination of the equilibrium cavity radius R_0, which may be done in terms of the Helmholtz free-energy A of the system. Writing this in the form

$$A = A_g + A_c, \qquad (106)$$

where subscripts g and c refer to the gas and the cavity, respectively, we have at temperature T,

$$A_c = -\kappa T \ln Z_c, \qquad (107)$$

where Z_c is the partition function for the cavity. This is given by

$$(2\pi m^* \kappa T / h^2)^{\frac{3}{2}} \mathscr{V}_c e^{-E_0/\kappa T}, \qquad (108)$$

where m^* is the hydrodynamic mass of the cavity and $\mathscr{V}_c$ is its volume. All we need to know about m^* is that it is proportional to $\mathscr{V}_c$ and hence to R_0^3.

For equilibrium

$$\frac{\partial A}{\partial \mathscr{V}_c} = 0, \qquad (109)$$

so that, using eqns (106), (107), and (108),

$$\frac{\partial A_g}{\partial \mathscr{V}_c} = -\frac{1}{4\pi R_0^2} \frac{\partial E_0}{\partial R_0} + \frac{9\kappa T}{8\pi R_0^3}. \qquad (110)$$

But, since the total volume of the gas–cavity system is constant,

$$\frac{\partial A_g}{\partial \mathscr{V}_c} = -\frac{\partial A_c}{\partial \mathscr{V}_c} = p, \tag{111}$$

where p is the pressure on the cavity due to the gas atoms. Eqn (110) then becomes

$$p = -\frac{1}{4\pi R_0^2}\frac{\partial E_0}{\partial R_0} + \frac{9\kappa T}{8\pi R_0^3}. \tag{112}$$

In terms of the macroscopic gas pressure p_0

$$p = p_0(1 - \rho_i/\rho). \tag{113}$$

It only remains to estimate ρ_i/ρ. The effect of finite ρ_i is to reduce the potential well depth to $V_0(1-\rho_i/\rho)$. This would result in a decrease of the zero-point energy E_0 by an amount ΔE_0 equal, to first order in ρ_i/ρ, to $V_0 \rho_i/\rho$. ΔE_0 is the amount of energy which must be imparted to the orthoPs atom in order that a density ρ_i of gas should enter the cavity. ρ_i/ρ can then be regarded as the probability that a gas atom will have an energy greater than ΔE_0. This may readily be calculated from the Maxwell distribution function. At low temperatures ρ_i/ρ will usually be negligible. Neglecting ρ_i/ρ we have from eqns (102) and (103),

$$p_0 = \frac{1}{2\pi}V_0^{5/2}(4m/\hbar^2)^{3/2}\sin^4\xi \cos\xi/\xi^2(\xi \cot\xi - 1). \tag{114}$$

This equation only has a solution if

$$p_0/V_0^{5/2} \geqslant 4\cdot 8 \times 10^3 \text{ atm eV}^{-5/2}, \tag{115}$$

a result which may be used to obtain an upper limit on V_0, when no cavity effects are observed at a pressure p_0. For given V_0, R_0, and hence P_{out} may now be calculated. As a simple approximation to V_0 we may use an optical model in which the refractive index of the gas for orthoPs is calculated in terms of the scattering length a. This gives

$$V_0 = \pi a \hbar^2 n/m, \tag{116}$$

where n is the concentration of gas atoms.

We are now in a position to apply this model to analyse the data for hydrogen and helium. By trial and error, the well depth V_0 and the cavity radius R_0 may be adjusted to give agreement with the experimental data. Fig. 26.61 shows V_0 as a function of gas density derived from the H_2 data, the cavity radii being given at the extreme points. It will be seen that these results do not agree with eqn (116) as V_0 is not proportional to the density. For helium the results, shown in Fig. 26.62, are more promising. At low temperatures and large cavity radii, the

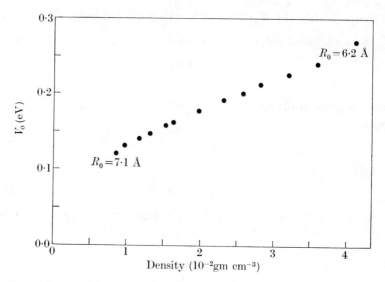

Fig. 26.61. Derived barrier heights V_0 against adiabatic penetration of orthoPs into gaseous H_2. The equilibrium cavity radius R_0 is indicated at the extreme points.

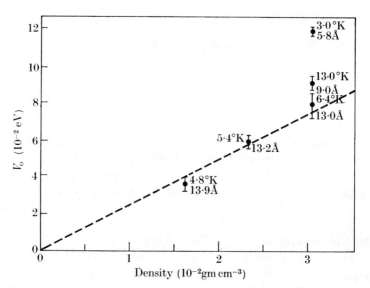

Fig. 26.62. Derived barrier heights V_0 against adiabatic penetration of orthoPs into gaseous He. The equilibrium cavity radius R_0, and the temperature, are indicated at each point.

derived values of V_0 are proportional to the gas density and give
$$a = 1 \cdot 45 \pm 0 \cdot 08 a_0.$$
This is quite close to that given by the semi-empirical theory of Drachman and Houston (see p. 3216) and, although too much weight should not be given to the closeness of the agreement, it does appear that the cavity theory is on the right lines.

It seems possible that the simple theory of the cavity we have sketched out is applicable at large cavity radii but fails at smaller radii. This is not surprising in view of the assumptions made, especially that of assuming no distortion of the internal wave-function of the orthoPs atom.

14. Positronium chemistry

Relatively little has been done in investigating chemical reactions involving Ps atoms in the gas phase but a great deal of work has been devoted to the study of reactions in liquids and gases. Although this falls outside the scope of the present book it is of such interest and importance, and is so closely related to many of the subjects discussed in the preceding sections, that we shall summarize some of the methods used and illustrate these by describing in some detail a particular case. For further details the reader is referred to the book on positronium chemistry by Green and Lee,† and the review article by Goldanskii.‡

A particular chemical substance may act as an inhibitor of Ps formation or as a quencher through formation of a Ps compound. If it is an inhibitor, it will reduce the magnitude of the long-lived component of the annihilation spectrum corresponding to orthoPs decay, but not change the long lifetime. A chemical quencher will both reduce the magnitude and the long lifetime. This will be true provided the quenching reaction occurs with thermal Ps. If, however, the reaction occurs only with superthermal Ps the long lifetime component will be reduced in magnitude, but that which remains will decay at the normal rate. It is, therefore, difficult to distinguish these possibilities.§ Further information about the nature of the effects produced by a particular substance may be obtained from angular correlation measurements of the two-quantum decay photons.

By way of illustration we now describe briefly experiments which have been carried out on the effect of I_2, and of CCl_4, on Ps inhibition

† GREEN, J. and LEE, J., *Positronium chemistry*. Academic Press, New York (1964).
‡ GOLDANSKII, V. I., *Atomic Energy Review* **6** (1968) 3.
§ McGERVEY, J. D., *Positron annihilation* (edited by Stewart and Roellig). Academic Press, New York (1967), p. 143.

and quenching. In 1958 Hatcher and Millett† observed the quenching of orthoPs decay in solutions of iodine in normal heptane by observing the rate of decrease with iodine pressure of the long lifetime component in the annihilation spectrum. Heptane was chosen as the solute because

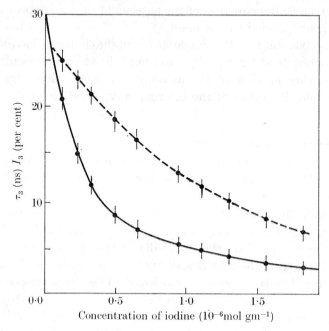

FIG. 26.63. Lifetime (τ_3) and intensity (I_3) of the long-lived component of the annihilation spectrum as functions of the concentration of I_2 in a silica gel.
———— lifetime τ_3 – – – intensity I_3.

it is known not to interact with I_2, and the annihilation spectrum in the pure liquid has a sufficiently strong long-lived component. A quenching cross-section of 10^{-17} cm² was measured. More than ten years later Chuang and Tao‡ studied the effect of I_2 and of CCl_4 absorbed in silica gels. CCl_4 was known,§ in solution, to reduce the intensity of the long-lived component without changing its lifetime. Chuang and Tao were especially concerned with establishing whether or not the CCl_4 acts as an inhibitor alone.

The lifetime spectrum in silica gels exhibits two long lifetime components τ_2 and $\tau_3 \simeq 1\cdot8$ ns and 32 ns, respectively, with relative intensity 3 and 27 per cent. τ_2 and τ_3 may be attributed respectively to the

† HATCHER, C. R. and MILLETT, W. E., *Phys. Rev.* **112** (1958) 1924.
‡ CHUANG, S. Y. and TAO, S. J., *J. chem. Phys.* **52** (1970) 749.
§ ORMROD, J. H. and HOGG, B. G., ibid. **34** (1961) 624.

orthoPs annihilating inside the silica particles and in the pores of the gel. OrthoPs atoms within the pores will move under the attraction of a van der Waals force and the magnitude of this force will determine their rate of annihilation. If now a chemical quenching agent is adsorbed on the inner surface of the pores a chemical attractive force will arise

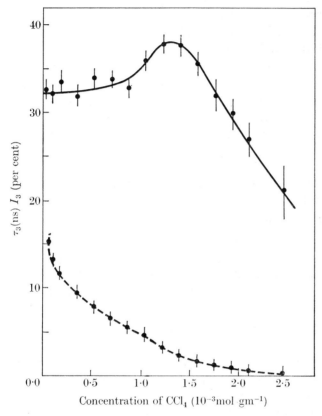

FIG. 26.64. Lifetime (τ_3) and intensity (I_3) of the long-lived component of the annihilation spectrum, as functions of the concentration of CCl_4 in a silica gel.
——— lifetime τ_3 – – – intensity I_3.

between the orthoPs and the pore surface, so that the lifetime τ_3 should be reduced. Figs. 26.63 and 26.64 show the results obtained with adsorbed I_2 and CCl_4, respectively. In each case τ_3 and I_3, the intensity of the long-lived component, are shown as functions of the concentration of the adsorbed material. The fact that τ_3 and I_3 fall rapidly with iodine concentration confirms that I_2 is a good chemical quencher of orthoPs. From the data, the rate constant for the quenching reaction may be derived, and was found to be 1.66×10^{14} gm mol^{-1} s^{-1}, when the iodine concentration is given in moles/gm of gel. If ρ is the density of the gel in gm cm^{-3}, the quenching cross-section comes out to be $1.5\rho \times 10^{-17}$ cm^2 which is comparable with the value measured in the earlier experiments.

For CCl$_4$ the picture is quite different. τ_3 at first changes very little with increasing CCl$_4$ concentration, whilst I_3 falls quite rapidly. However, as the concentration increases still further τ_3 rises to a peak of 38 ns, after which it falls rapidly. If it is assumed that the sudden increase is due to formation of a monolayer of CCl$_4$ the presence of the peak indicates that the attractive interaction between the monolayer and orthoPs, within the pores, is weaker than for the pure gel. This suggests that the CCl$_4$ is acting purely as an inhibitor, and not as a quencher, of superthermal orthoPs. Decrease of τ_3 at high concentrations may be interpreted as due to decrease of the free volume within the gel.

15. The passage of positive muons through matter and muonium formation

15.1. *Introductory*

The study of the interaction of slow positively charged mesons in matter is of interest both for its own sake and also because of the similarity of the phenomena involved to those involved in the interaction of slow protons. Studies of this kind have so far been limited almost entirely to the positive muon, μ^+, a particle which behaves essentially like a heavy positron, with mass 206·76 times that of the positron. Such muons passing through matter may capture electrons, to form neutral entities resembling atomic hydrogen, known as muonium. These would clearly be expected to have many analogous properties also to positronium.

The possibility of studying the processes of the capture and loss of electrons by μ^+, and the chemical reactions involving muonium, is provided by the existence of the intrinsic spin, $\tfrac{1}{2}\hbar$, and intrinsic magnetic moment, $-\tfrac{1}{2}g_\mu(m/m_\mu)\mu_0$ of the muon (where m_μ = muon mass, m = electron mass, $\mu_0 = e\hbar/2mc$, the Bohr magneton, and g_μ is the Landé factor of the muon); by the fact that μ^+, produced by π^+ decay in the process

$$\pi^+ \to \mu^+ + \nu \quad (\nu = \text{neutrino})$$

are completely polarized in the pion rest-frame in a direction opposite to the muon direction; and by the non-conservation of parity in the μ^+ decay as a result of which the decay positrons emitted in the decay process $\mu^+ \to e^+ + \nu + \bar{\nu}$ are emitted asymmetrically relative to a plane perpendicular to the muon spin. The polarization of the muon beam can be measured by studying the angular distribution of the positrons,

$$I(\theta) = I_0(1 - a \cos \theta), \qquad (117)$$

where θ is the emission angle of the positron relative to the direction of the muon momentum. For a weak interaction of the form $V-A$, characteristic of the μ^+ decay process, where V, A are respectively the vector and axial vector parts of the interaction, $a = \frac{1}{3}$ for a fully polarized beam so that for a muon beam of longitudinal polarization P, $a = \frac{1}{3}P$.

15.2. Muonium

Changes in the μ^+ spin polarization can be interpreted in terms of muonium formation. The properties of muonium have been reviewed by Hughes.† They are summarized in this section. The symbol Mu refers to the muonium atom which can also be written symbolically (μ^+e^-).

15.2.1. *Energy levels and fine structure splitting of muonium.* Since the charge and spin of the positive muon and proton are the same the energy levels of Mu are very similar to those of atomic hydrogen. The Schrödinger energy levels are given by

$$E(n) = -R_{\mathrm{Mu}}\, hc/n^2, \quad n \text{ integral}, \qquad (118)$$

where the Rydberg constant R_{Mu}, for Mu, differs from that for atomic H only due to the small difference between the reduced masses of the two systems, and is given by

$$R_{\mathrm{Mu}} = R_\infty \Big/ \left(1 + \frac{m}{m_\mu}\right) = 109209 \cdot 12 \pm 0 \cdot 02 \text{ cm}^{-1},$$

about 0·5 per cent smaller than for H. Here R_∞ is Rydberg's constant for infinite mass. The fine structure splittings in terms of the fine structure constant, α, are, ignoring radiative corrections,

$$\Delta E(n, j = l \pm \tfrac{1}{2}) = -\frac{R_{\mathrm{Mu}}\, ch\alpha^2}{n^3}\left(\frac{1}{j+\tfrac{1}{2}} - \frac{3}{4n}\right),$$

the same as for atomic hydrogen except for the modified Rydberg constant.

15.2.2. *Hyperfine splitting of muonium.* In the ground ($n = 1$) state the hyperfine structure gives rise to two levels of angular momentum $F\hbar$ and energy

$$E(F) = E(1) + \frac{\hbar\omega}{4} \quad \text{for } F = 1,$$

$$E(F) = E(1) - \frac{3\hbar\omega}{4} \quad \text{for } F = 0, \qquad (119)$$

† Hughes, V. W., *Ann. Rev. nucl. Sci.* **16** (1966) 445.

where $E(1)$ is given by eqn (118) and ω, the angular frequency corresponding to the hyperfine splitting, is given by Dirac's theory, with radiative corrections calculated from quantum electrodynamics by

$$\omega = 2\pi\left(\frac{16}{3}\alpha^2 c R_\infty \frac{\mu_\mu}{\mu_0}\right)\left(1+\frac{m}{m_\mu}\right)^{-3}\left(1+\frac{3}{2}\alpha^2\right)\left(1+\frac{\alpha}{2\pi}-0\cdot 328\frac{\alpha^2}{\pi^2}\right)\times$$

$$\times\left(1+\frac{\alpha}{2\pi}+0\cdot 75\frac{\alpha^2}{\pi^2}\right)(1-1\cdot 81\alpha^2)\left(1-\frac{3\alpha}{\pi}\cdot\frac{m}{m_\mu}\ln\frac{m}{m_\mu}\right), \quad (120)$$

where $\mu_\mu (= (m/m_\mu)\mu_0)$ is the muon magneton. The factors in parentheses in eqn (120) are, respectively, the Fermi value of the hyperfine splitting, a reduced mass correction, the Breit relativistic correction, half the g-values (gyromagnetic ratios) for the electron and muon, respectively (i.e. the ratios of the magnetic moments in terms of the Bohr and muon magnetons, respectively, to the angular momentum in units of $\hbar$), a second order radiative correction, and a relativistic recoil factor. Numerically $\Delta\nu = \omega/2\pi \simeq 4463$ Mc/s, corresponding to a relaxation time $\omega^{-1} \simeq 3\cdot 6\times 10^{-11}$ s.

In the presence of a magnetic field H the energies of the four hyperfine structure states ($F = 1$, $M_F = \pm 1, 0$; $F = 0$, $M = 0$) are given by†

$$E(F, M_F) = E(1) - \frac{\hbar\omega}{4} - \mu_0 g_\mu \frac{m}{m_\mu} H M_F \pm \frac{\hbar\omega}{2}[1+2M_F x+x^2]^{\frac{1}{2}}, \quad (121)$$

where the $+$ and $-$ signs are to be taken for the $F = 1$ and $F = 0$ states, respectively, and

$$x = \omega_L^{(\text{Mu})}/\omega \simeq 2(1-m/m_\mu)\mu_0 H/\hbar\omega \simeq H/1580, \quad (122)$$

if H is expressed in gauss, $\omega_L^{(\text{Mu})}$ being the Larmor angular frequency of Mu. In the limit of strong fields where the electron and muon spins couple independently to the applied magnetic field (projection quantum numbers M_s, M_μ respectively) the four hyperfine states go over directly to

$(F, M_F) \equiv (1, 1) \rightarrow (M_s, M_\mu) \equiv (\tfrac{1}{2}, \tfrac{1}{2})$ (a),
$\phantom{(F, M_F) \equiv{}}(1, 0) \rightarrow \phantom{(M_s, M_\mu) \equiv{}} (\tfrac{1}{2}, -\tfrac{1}{2})$ (b),
$\phantom{(F, M_F) \equiv{}}(1, -1) \rightarrow \phantom{(M_s, M_\mu) \equiv{}} (-\tfrac{1}{2}, -\tfrac{1}{2})$ (c),
$\phantom{(F, M_F) \equiv{}}(0, 0) \rightarrow \phantom{(M_s, M_\mu) \equiv{}} (-\tfrac{1}{2}, \tfrac{1}{2})$ (d),

so that it is legitimate to label these states as $(a), (b), (c), (d)$, respectively. Fig. 26.65 shows the hyperfine splitting given by eqn (121), as a function of x. In the presence of a u.h.f. field of the resonance frequency magnetic

† BREIT, G. and RABI, I. I., *Phys. Rev.* **38** (1931) 2082.

resonance transitions would be expected corresponding to spin flip of the muon (i.e. $a \to b$) and of the electron (i.e. $a \to d$). For weak fields the angular frequency of this latter transition tends to ω as $H \to 0$.

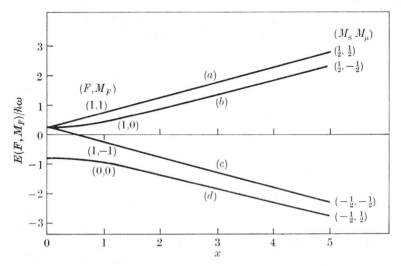

Fig. 26.65. Energy of the four hyperfine states (a), (b), (c), (d) in a magnetic field H. The energy $E(F, M_F)$ expressed in units of the hyperfine splitting, $\hbar \omega$ ($\omega = 4463$ Mc s^{-1}), is plotted as a function of $x = H/1580$ (H in gauss).

15.3. *Magnetic moment of muonium*

The magnetic moment of Mu in the state $(F, M_F) \equiv (1, 1)$ is given by $-\mu_e + \mu_\mu$, neglecting effects arising from its magnetic interaction with an external magnetic field. In the presence of a weak magnetic field perpendicular to the direction of spin the magnetic moment will precess with the Larmor frequency,

$$\nu_L^{(Mu)} (= \omega_L^{(Mu)}/2\pi) \simeq 1{\cdot}40 H \text{ Mc s}^{-1} \tag{123}$$

(H expressed in gauss).

A free muon will also precess in a transverse magnetic field but in this case

$$\nu_L^\mu = \frac{\mu_\mu H}{\hbar/2} \simeq \frac{\nu_L^{(Mu)}}{103} = \frac{(\mu_\mu - \mu_e)H}{\hbar} \simeq 13{\cdot}5H \text{ kc s}^{-1}. \tag{124}$$

Observation of the muon precession rate in a transverse field should therefore enable free muons to be distinguished from muons bound in Mu.

15.4. *Depolarization of muons by muonium formation*

Suppose the muons are initially polarized along the axis of quantization corresponding to the spin state $M_\mu = \frac{1}{2}$. In Mu the possible hyperfine states are then the states labelled (a), (b), (d) above with

$(F, M_F) \equiv (1,1), (1,0),$ and $(0,0),$ respectively,

and in the absence of an external magnetic field these states will be populated in the proportions $\frac{1}{2}, \frac{1}{4}, \frac{1}{4}$. At formation, the two states $(b), (d)$, will have a definite magnetic moment of magnitude $|\mu_\mu + \mu_e|$, if the muon has a definite spin polarization at the instant when muonium formation occurs. Depolarization of the muon is brought about by the action of the magnetic field, H_s at the position of the muon, arising from the spin s of the electron,

$$H_s = (32\pi mc/e\hbar)|\psi(0)|^2 s, \qquad (125)$$

where $\psi(0)$ is the electron wave-function at the position of the muon. The action of this field causes a spin flip of the muon, accompanied by a recoil flip of the electron spin, so that angular momentum is conserved. Eventually in the absence of an external magnetic field, the muonium in these states will contain equal populations with the muon spin parallel, and antiparallel, to the direction of quantization. The relaxation time, t_R, required to attain this equilibrium is given approximately by $t_R \simeq \omega^{-1} \simeq 3.6 \times 10^{-11}$ s. When equilibrium has been attained the Mu formed, in state (a), will still of course maintain its polarization, whilst the polarization of the μ^+, in states (b) and (d), is destroyed, so that the overall μ^+ polarization has been reduced to one-half its initial value. In the presence of an external magnetic field the final polarization of the muon in muonium will be greater than one-half the initial value, since the field inhibits the spin flip of the electron and thence also of the μ^+.

Simple perturbation theory considerations indicate that the decrease (ΔP) in polarization, in time Δt, is not linear in Δt, but in the absence of an external magnetic field is given by

$$\Delta P = \frac{1}{2}\left\{1 - \frac{\sin\{\omega(\Delta t)\}}{\omega(\Delta t)}\right\} \simeq \frac{\omega^2(\Delta t)^2}{12}, \quad \text{where } \Delta t \text{ is small.} \quad (126)$$

In the light of this discussion one can distinguish three types of Mu, classified on the basis of its spin states, namely

(i) Triplet muonium, Mu^T, in the state (a), $(F, M_F) \equiv (1,1)$, which does not undergo depolarization as a result of spin flip produced by the interaction of the electron and muon magnetic fields.

(ii) Depolarized muonium, Mu^D, in the states $(b), (d), (F, M_F) \equiv (1,0)$ or $(0,0)$, after a time long compared with the relaxation time t_R so that in zero or weak magnetic fields the muon polarization has disappeared due to spin flip.

(iii) Muonium formed in states (b), (d) at times short compared with t_R so that the muon retains its polarization. This is referred to as singlet muonium and written Mu^s despite the fact that it may involve a mixture of $F = 0$ and $F = 1$ states.

It is seen that in the absence of an external field the spin flip effects discussed above lead eventually to equal concentrations of Mu^T and Mu^D.

15.5 *Muonium formation in presence of longitudinal magnetic field*

In the presence of a magnetic field in the direction of the muon spin, the muonium spin eigenfunctions can be expressed in terms of the spin eigenfunctions $\alpha_\mu, \beta_\mu, \alpha_e, \beta_e$ for the muon and electron respectively,† namely,

$$\begin{aligned}
\chi_{11} &= \alpha_e \alpha_\mu & (a), \\
\chi_{10} &= C\alpha_e \beta_\mu + S\beta_e \alpha_\mu & (b), \\
\chi_{1-1} &= \beta_e \beta_\mu & (c), \\
\chi_{00} &= C\beta_e \alpha_\mu - S\alpha_e \beta_\mu & (d),
\end{aligned} \quad (127\,\text{a})$$

where

$$S = \frac{1}{\sqrt{2}}\left[1 - \frac{x}{\sqrt{(1+x^2)}}\right]^{\frac{1}{2}}, \quad C = \frac{1}{\sqrt{2}}\left[1 + \frac{x}{\sqrt{(1+x^2)}}\right]^{\frac{1}{2}}, \quad (127\,\text{b})$$

and x is given by eqn (122). Initially only the states $\alpha_e \alpha_\mu$, $\beta_e \alpha_\mu$ are present, and with equal populations. The latter can be written in terms of the eigenfunctions χ_{10}, χ_{00},

$$\beta_e \alpha_\mu = S\chi_{10} + C\chi_{00},$$

so that the relative populations of states (a), (b), (c), (d) are, respectively, $\tfrac{1}{2}, \tfrac{1}{2}S^2, 0, \tfrac{1}{2}C^2$. Finally, after spin relaxation, when equilibrium has been attained, i.e. $t \gg \sqrt{\{\omega^2 + (\omega_L^e)^2\}}$, where $\omega_L^e = \dfrac{\mu_0 H}{\hbar/2}$ is the Larmor precession angular frequency of the electron, the muon polarization is

$$P = \tfrac{1}{2} + \tfrac{1}{2}(S^4 + C^4) - S^2 C^2$$

$$= \frac{1}{2}\left(1 + \frac{x^2}{1+x^2}\right), \quad (128)$$

i.e. $P > \tfrac{1}{2}$ if $H > 0$. In particular, for strong fields, where $\omega_L^e \gg \omega$, corresponding to $H \gg 1580$ G ($x \gg 1$), depolarization is effectively inhibited since electron spin flip cannot occur.

† BREIT, G. and HUGHES, V. W., *Phys. Rev.* **106** (1957) 1293.

16. Direct experimental identification of muonium formation in gases

Two different experimental methods have been employed to establish Mu production directly. In one method the Larmor precession frequency characteristic of the magnetic moment of Mu is observed. In the other u.h.f. magnetic resonance transitions characteristic of the hyperfine structure of Mu are detected.

Even before Mu production had been detected directly its formation had been inferred and used to account for the observed depolarization of thermalized μ^+. These observations were made using condensed moderating materials where the phenomena are more complex but their experimental study is easier. The work of Hughes and his colleagues on phenomena associated with Mu formation and μ^+ depolarization in gaseous argon, has been of outstanding importance in elucidating the phenomena involved. In line with the philosophy of the present work these experiments are described first, although many of the basic experimental methods were developed in the course of the study of μ^+ depolarization in condensed materials. The phenomena observed in condensed materials and their interpretation in terms of the formation and collision processes of Mu are discussed briefly later.

16.1. *Direct observation of muonium precession frequency*

The direct detection of the Mu precession frequency (§ 15.3) is difficult on account of the large magnetic moment of Mu, close to that of the electron. In order to obtain a precession frequency in a convenient experimental range of a few Mc/s, very small magnetic fields need to be employed.

Hughes *et al.*† first observed the precession of Mu, formed by slowing down μ^+ in argon, at a pressure of 50 atm. The precession frequency was about 5 Mc/s in a magnetic field of 3·91 gauss. The experiment was repeated using magnetic fields of 4·34, 4·50, and 4·96 gauss.‡

Fig. 26.66 shows the arrangement used. A mixed beam of μ^+ and π^+ of momentum 190 MeV/c was used, the π^+ stopping in the carbon absorber, and the μ^+ in the gas target containing argon at a pressure of 50 atm. 1–5 were scintillation counters. The coincidence signal $12\overline{3}$ indicated the stopping of a μ^+ in the target. Decay positrons were observed by means of a delayed coincidence 45 during a gated time

† HUGHES, V. W., MCCOLM, D. W., ZIOCK, K., and PREPOST, R., *Phys. Rev. Lett.* 5 (1960) 63.
‡ HUGHES, V. W., MCCOLM, D. W., ZIOCK, K., and PREPOST, R., *Phys. Rev.* A1 (1970) 595.

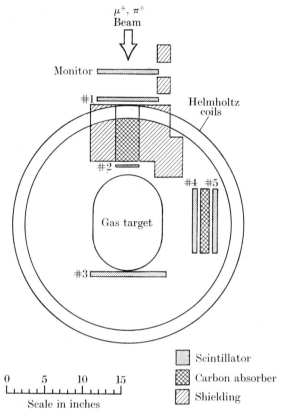

Fig. 26.66. Experimental arrangement used by Hughes and co-workers to observe the formation of Mu in gaseous argon.

interval of 1·8 μs, delayed by 0·2 μs after the signal 123. A magnetic field of 3 to 5 G was used corresponding to a Mu precession frequency between 4·2 Mc/s and 7·0 Mc/s. Its magnitude was measured to an accuracy of 1 per cent with an electron resonance magnetometer using DPH free radicals. The time distribution of the decay positrons was measured by means of a time to pulse-height converter and pulse-height analyser, with overall time resolution and accuracy of about 20 ns. The argon gas was purified by recirculation over titanium heated to about 773 °K.

The time distribution of the decay positrons was fitted to the form

$$y(t_i) = \exp(-t_i/\tau)[C + A\exp(-t_i/\tau')\sin\{2\pi f(t_i+t_0)\}], \quad (129)$$

where τ is the muon mean lifetime, τ' a parameter introduced to allow for line broadening due to processes that gradually remove the Mu, t_0 is an instrumental constant, and f the trial value for the precession frequency.

Fig. 26.67 shows the results of a least-squares analysis of their results. The solid curve (Fig. 26.67 (a)) represents the percentage amplitude of A in eqn (129), as a function of the trial value of the precession frequency. The error bar corresponds to an error of one standard deviation. The

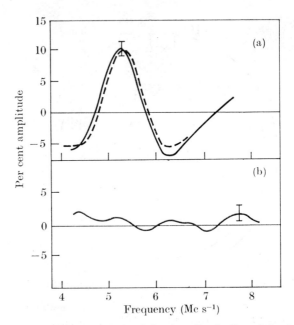

FIG. 26.67. (a) Frequency analysis of the time distribution of the decay positrons from Mu, fitted to eqn (129) (Hughes et al.). τ' ($= 0\cdot 5$ μs) is a parameter introduced into the frequency analysis to allow for line broadening. (b) Similar frequency analysis for an incident beam of unpolarized μ^+.

dashed curve is a theoretical line shape centred at the Mu precession frequency predicted from the measured value of the magnetic field. For comparison Fig. 26.67 (b) shows the same analysis applied to a similar experiment using unpolarized muons. The experiment provides a direct confirmation of Mu production. It was not possible to obtain an accurate estimate of the proportion of μ^+ that form Mu, but the observed amplitude of precession was consistent with a value between $\tfrac{1}{2}$ and 1 for this proportion.

16.2. Observation of microwave resonance spectrum of muonium

Confirmation of the existence of Mu has also been provided by the observation by Hughes and his colleagues[†] of the microwave resonance

[†] BAILEY, J. M., CLELAND, W. E., HUGHES, V. W., PREPOST, R., and ZIOCK, K., Phys. Rev. A3 (1971) 871.

spectrum for μ^+ stopped in argon at frequencies very close to those expected for free muonium. In the apparatus used in an accurate measurement of the Mu microwave spectrum by Cleland et al.,† shown

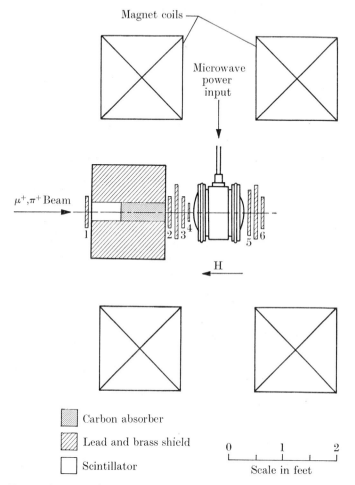

Fig. 26.68. Experimental arrangement used by Cleland et al. to study microwave resonance spectrum of Mu.

in Fig. 26.68, the target was a thin-walled, high Q microwave cavity filled with argon gas at pressures in the range 10–60 atm, and contained in a stainless steel pressure tank. The argon gas was continuously purified by recirculation over titanium, at a temperature of about 1023 °K. The external longitudinal magnetic field provided by the

† CLELAND, W. E., BAILEY, J. M., ECKHAUSE, M., HUGHES, V. W., PREPOST, R., ROTHBERG, J. E., and MOBLEY, R. M., *Phys. Rev.* A5 (1972) 2338

Helmholtz coils was homogeneous over the target volume to within 2 parts in 10^4 in the high field case, and was calibrated by means of a proton NMR probe. The microwave cavity operated in the TM_{110} mode with the axial direction coincident with that of the static magnetic field. It was fed by a 1 kW klystron amplifier, and frequency stability better than 1 part in 10^6 and power stability to within 2 per cent were achieved. The coincidence $123\overline{45}$ of the coincidence counters indicated the stopping of a positive muon in the target while the delayed coincidence $\overline{3}56$ indicated the emission of a decay positron in the forward direction. Positrons were observed over a decay time of 4 μs. The ratio $(1+\eta)$ of the number of positrons emitted in the forward direction with the microwave field on and off, respectively, was measured as a function of the static magnetic field, H. At resonance this ratio should pass through a maximum.

Writing $a_1(t)$, $a_2(t)$ for the state amplitudes of the two hyperfine states (1, 2) concerned, $\hbar\omega_H$ for the energy difference between the states corresponding to the static magnetic field H, ω_c for the angular frequency of the u.h.f. field in the cavity, $\hbar b e^{-i\omega_c t}$ for the matrix element linking states 1 and 2 of the interaction between the Mu atom and the u.h.f. field (so that b^2 is proportional to the intensity of this field), and $\gamma\,(=\tau^{-1})$ for the muon decay rate, standard perturbation theory leads to the equations

$$\dot{a}_1 = -ia_2 b \exp(i(\omega_c - \omega_H)t) - \tfrac{1}{2}\gamma a_1,$$
$$\dot{a}_2 = -ia_1 b^* \exp(-i(\omega_c - \omega_H)t) - \tfrac{1}{2}\gamma a_2. \qquad (130)$$

Supposing that initially the Mu is in state 1, so that $a_1 = 1$, $a_2 = 0$, then at time t,

$$|a_2|^2 = e^{-\gamma t}\frac{|2b|^2}{|2b|^2 + (\omega_H - \omega_c)^2}\sin^2\{\tfrac{1}{2}t[|2b|^2 + (\omega_H - \omega_c)^2]^{\frac{1}{2}}\} \qquad (131)$$

gives the probability that the Mu is in state 2. The probability that the muon will decay from the state 2 is then equal to

$$\int_0^\infty \gamma |a_2|^2\,dt = \frac{4|b|^2}{4|b|^2 + \gamma^2 + (\omega_c - \omega_H)^2}. \qquad (132)$$

The experimentally determined quantity (η) should be proportional to expression (132).

The width at half maximum is

$$2\Gamma = 2\{4|b|^2 + \gamma^2\}^{\frac{1}{2}}, \qquad (133)$$

and the signal strength, S, at resonance is

$$S = \frac{4|b|^2}{4|b|^2 + \gamma^2}. \qquad (134)$$

The actual width contains a contribution from the power-broadening term $4|b|^2$ which is proportional to the intensity of the u.h.f. field.

16.2.1. *Observation at high magnetic fields.* The high field experiments were carried out using a magnetic field of around 5400 G. Measurements were carried out for argon gas pressures in the range from 9·5 to 63 atm and a microwave power in the cavity of 24 W. In each case the microwave frequency was held constant and the static magnetic field varied over a range of about 100 G on either side of the resonance value. The resonance observed in the high field case corresponds to a transition between states (a) and (b).

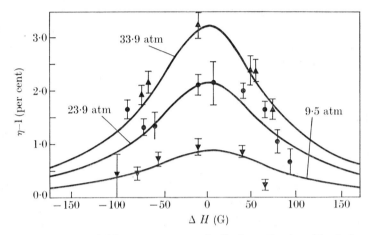

FIG. 26.69. Strong field resonance curves for the hyperfine transition between states (a) and (b) (Fig. 26.65) for three different pressures of argon as indicated. The signal is shown as a function of static magnetic field variation about a central value of approximately 5400 G. The microwave frequency was maintained fixed around a value of approximately 1840 Mc s^{-1}.

Fig. 26.69 shows three resonance curves obtained for argon pressures of 9·5, 23·9, and 33·9 atm, respectively. In these, the quantity $\eta-1$ is expressed as per cent. The shapes of the resonance curves shown in the figure represent least-squares fits to the experimental points, using the theoretical line shapes given by eqn (132). In making these fits account was taken of

(a) the variation of the magnetic field H over the target,
(b) the variation of the microwave field over the cavity,
(c) the variation of effective solid angle subtended by the positron counters for different points of the target,
(d) the spatial distribution of muons stopping in the cavity, and
(e) the angular distribution of the decay positrons.

Power broadening effects increased the line width by a factor of about four over that to be expected from the natural line width.

The measurements made by the Yale group of Hughes have been repeated by the Chicago group of Telegdi† using a method essentially the same in principle but incorporating a number of innovations to improve the accuracy. For example the uncertainty arising from the variation of the magnetic field over the target was greatly reduced by working at a magnetic field of magnitude 11 327 gauss (the 'magic field') for which $\partial \omega_{ab}/\partial H = 0$, where $\hbar\omega_{ab}$ is the difference in energy of states (a), (b), given by eqn (121). Muons stopping in the walls of the resonator were excluded using proportional counters located within the resonator. The variation of the microwave field over the cavity was reduced by operating in the TM_{210} mode. In addition the distribution in time of the decay positron pulses relative to the time of entry of the μ^+ was measured so that it was possible to follow the r.f. induced change in μ^+ polarization with time instead of only the net change integrated over the observation time. This enabled a more accurate determination of ω_{ab} from the data. From the measured value of $\hbar\omega_{ab}$ $\{= E(1,1)-E(1,0)\}$ the hyperfine interval of Mu, $\Delta\nu$ ($= \omega/2\pi$), could then be calculated from eqn (121).

16.2.2. *Observation at low magnetic fields.* Hughes and his colleagues have also observed microwave magnetic resonance in Mu with a very weak magnetic field H of 0·01 gauss.‡ This measurement provides practically a direct measurement of the hyperfine splitting frequency ω ($= 2\pi\nu$) without requiring an accurate knowledge of the gyromagnetic ratios as in the high field experiment. This is a very difficult experiment owing to the small magnetic field and resonance signal strength. The cavity operated in the TM_{220} mode and a typical resonance signal, obtained by varying the microwave frequency is shown in Fig. 26.70 for the $(a$–$d)$ ($\Delta F = -1$) transition. The time distribution of the positron decays was also measured in the later experiments and enabled a better value of the Mu hyperfine splitting to be derived.

The low field method as used by Hughes and colleagues also has the disadvantage that the line width is much wider than for the high field method (e.g. 1400 kHz for 90 per cent saturation, compared with 520 kHz). This is related to the fact that the $F = 1$ level contains three sub-levels which contribute their natural widths to the total line width.

---

† DE VOE, R., MCINTYRE, P. M., MAGNON, A., STOWELL, D. Y., SWANSON, R. A., and TELEGDI, V. L., *Phys. Rev. Lett.* **25** (1970) 1779; EHRLICH, R. D., HOFER, H., MAGNON, A., STOWELL, D. Y., SWANSON, R. A., and TELEGDI, V. L., *Phys. Rev.* **A5** (1972) 2357.

‡ THOMPSON, P. A., AMATO, J. J., CRANE, P., HUGHES, V. W., MOBLEY, R. M., ZU PUTLITZ, G., and ROTHBERG, J. E., *Phys. Rev. Lett.* **22** (1969) 163; see also CRANE, T. CASPERSON, D., CRANE, P., EGAN, P., HUGHES, V. W., STAMBAUGH, R., THOMPSON, P. A., and ZU PUTLITZ, G., ibid. **27** (1971) 474.

Telegdi and his colleagues† have adapted an ingenious resonance method developed by Ramsey‡ in atomic beam studies to a precision measurement of the Mu hyperfine interval in weak fields (<0·010 gauss).

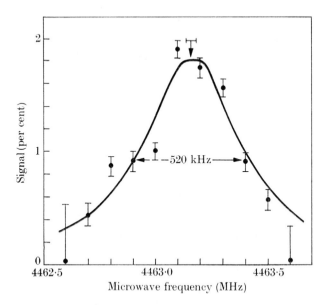

FIG. 26.70. Low-field resonance curve for hyperfine transitions between states (a) and (d) of Mu (Fig. 26.65) showing the resonance signal as a function of microwave frequency in a static magnetic field of 0·010 gauss for μ^+ stopping in argon gas at 7600 torr pressure. The solid curve is a least-squares fit by a Lorentzian with the centre at 4463·159±0·016 MHz (Crane et al.).

Using this method the width of the resonance signal can be made even narrower than the natural line width (140 kHz). Muons with initial polarization, $P_z(0) = 1$, are formed in the $F = 1$, $M = 1$ state in a very weak static field ($\leqslant 0·01$ gauss). A microwave field $\mathbf{B}(\omega')$ oscillating in the x direction with frequency ω' ($\simeq \omega$, the Mu hyperfine splitting frequency) induces transitions to the $F = 0$, $M = 0$ state and from it back to both $|M| = 1$ states. $P_z(t)$ will vary through 0 to -1, then back to $+1$ and so on.

Suppose $P_z(\tau) = 0$, so that the $M = \pm 1$ states are equally populated. The polarization vector, $\mathbf{P}$, is then in the (x, y) plane. If the microwave field $\mathbf{B}(\omega')$ is switched off at time τ, $\mathbf{P}$ continues to oscillate in the (x, y) plane with frequency ω since the system is described by a coherent

† FAVART, D., MCINTYRE, P. M., STOWELL, D. Y., TELEGDI, V. L., and DE VOE, R., Phys. Rev. Lett. 27 (1971) 1336.
‡ RAMSEY, N. F., Molecular beams (Clarendon Press, Oxford (1956), p. 124.

superposition of $F = 1$ and $F = 0$ states. If the microwave field is switched on again at time T later the rotation of **P** into the $-z$ direction to give $P_z = -1$ will continue if and only if $\mathbf{B}(\omega')$ and **P** are in the same phase relationship relative to each other at time $T+\tau$ as they were at time τ. This clearly requires the same resonance condition, $\Delta \ (\equiv \omega'-\omega) = 0$, but the width of the resonance is now determined primarily by the time T for which **B** was switched off.

One can then write for the signal, S,

$$S(\Delta) = \frac{N_{\text{off}}-N_{\text{on}}}{N_{\text{off}}+N_{\text{on}}} = \frac{A[P_z(0)-P_z(T+2\tau)]}{2P_z(0)},$$

where the N are the decay positron rates for $t > T+2\tau$ with r.f. on and off respectively, and A is the effective muon decay asymmetry for the experimental set-up.

Instead of counting with r.f. on and off Favart et al. shifted the phase of the second pulse alternatively by $\pm\tfrac{1}{2}\pi$ so that the signal observed is

$$S_\delta(\Delta) = \{N(\tfrac{1}{2}\pi)-N(-\tfrac{1}{2}\pi)\}/\{N(\tfrac{1}{2}\pi)+N(-\tfrac{1}{2}\pi)\}.$$

They derived $\qquad S_\delta(\Delta) = A\,G^2(\tau)\sin\Delta(T+\tau),$

where $G(\tau)$ depends on Δ, on the matrix element for the hyperfine transition, and on the microwave field strength.

$S_\delta(\Delta)$ passes through zero at resonance ($\Delta = 0$) and has full line-width $\Gamma = \tfrac{1}{3}\pi(T+\tau)$, i.e. $\Gamma \simeq 50$ kHz for $T+\tau \simeq 3\cdot4$ s, an improvement of nearly a factor of 30 over the direct method. Fig. 26.71 shows the signal obtained by Favart et al. in Kr at a pressure of 2742 torr. A curve of the form $a\sin\Delta(T+\tau)$, with $T+\tau = 3\cdot4$ µs, is fitted to the data points.

16.2.3. *Pressure shift in the microwave resonance spectrum.* The theory of the pressure shift of spectral lines has been discussed earlier (Vol. 1, Chap. 5). It arises in the present context from the distortion of the wave-functions of Mu by neighbouring argon atoms.

A correct estimate of the pressure shift is very important in using measurements of the microwave resonance spectrum to obtain accurate values of the hyperfine interval, $\omega/2\pi$, of Mu from which an accurate value of the fine structure constant, α, can be derived. Earlier work used a linear extrapolation of the measured values of $\omega/2\pi$ to zero pressure, but an indication of the necessity of including a quadratic term in the pressure dependence was given by Ehrlich et al.[†] Crane et al.[‡] fitted

---

[†] EHRLICH, R. D., HOFER, H., MAGNON, A., STOWELL, D., SWANSON, R. A., and TELEGDI, V. L., *Phys. Rev. Lett.* **23** (1969) 513.

[‡] loc. cit., p. 3242.

their low field data on ω to the quadratic function

$$\omega(p) = \omega(0)\{1+ap+bp^2\}. \tag{135}$$

The quadratic term could evidently arise from the effect of three-body collisions involving two gas atoms and the muonium atom.

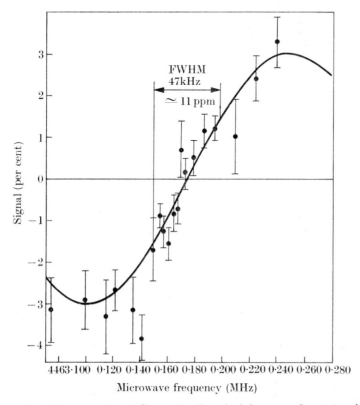

FIG. 26.71. Resonance signal, S_δ, as a function of r.f. frequency for μ^+ stopping in krypton gas at 2742 torr pressure. The observations are fitted to the curve $a \sin \Delta(T+\tau)$ with $T+\tau = 3 \cdot 4\ \mu\text{s}$ ($\Delta = \omega-\omega_0$). With the procedure employed the signal passes through zero at the position of the Ramsey resonance (Favart et al.).

Fig. 26.72 shows the variation of the hyperfine separation, $\omega/2\pi$, in argon gas at various pressures measured by Cleland et al.† Best fits to linear and quadratic pressure variation laws are shown. From the measurements of Cleland et al. it would be difficult to establish unambiguously the need for a quadratic term. Also shown on the same figure, however, are points obtained by de Voe et al.‡ using the high 'magic'

† loc. cit. ‡ loc. cit.

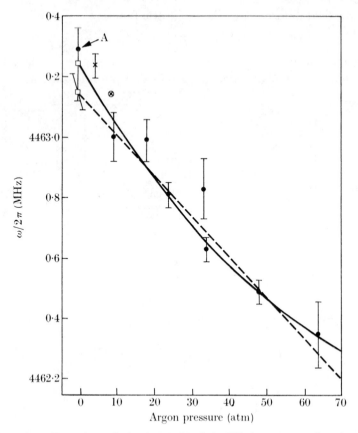

FIG. 26.72. Hyperfine splitting interval ($\omega/2\pi$) of Mu in argon as a function of the gas pressure.
● Cleland et al. (strong field); ✻ de Voe et al. (strong 'magic' field); ⊗ Favart et al. (weak-field Ramsey resonance). The dashed and full lines are best fits to the data of Cleland et al. only. The 'best' extrapolation of all the data is indicated by A.

field method, and by Favart et al.† using the low-field Ramsey resonance method. The errors of the various measurements are indicated. That for the Ramsey method is smaller by an order of magnitude than for the other measurements so that the error can scarcely be seen on the scale of the plotted points. When the points of Ehrlich et al.† and Favart et al.‡ are added to those plotted by Cleland et al.§ the evidence for the quadratic term in the variation becomes stronger.

The pressure shift of the hyperfine structure interval has been measured for atomic hydrogen in argon and krypton (see Chap. 5). No

† loc. cit.   ‡ loc. cit.   § loc. cit.

evidence for a quadratic term in the pressure dependence has been found, i.e. in eqn (135) $b = 0$. For the coefficient a, the fractional pressure shift, values of $a = -(4\cdot78\pm0\cdot03)\times10^{-9}\,\text{torr}^{-1}$ have been found for H in Ar,† and $a = -(10\cdot4\pm0\cdot2)\times10^{-9}\,\text{torr}^{-1}$ for H in Kr.‡ The values of a for Mu in these gases would be expected to be almost the same since the only difference expected is that due to the small difference in reduced mass of the electron. Ehrlich et al. have therefore fitted all the data on the pressure shift of the hyperfine structure of Mu in these two gases to the quadratic form of eqn (135), with a constrained to have the value given above for H.

The hfs interval obtained

$$\frac{\omega}{2\pi} = 4463\cdot288\pm0\cdot008 \text{ MHz from the Ar data}$$
$$= 4463\cdot279\pm0\cdot012 \text{ MHz from the Kr data}.$$

The coefficient b of the quadratic term in eqn (135) is

$$b = (8\cdot3\pm2\cdot8)\times10^{-15}\,\text{torr}^{-2} \text{ for Ar}$$

and

$$= (7\cdot3\pm2\cdot0)\times10^{-15}\,\text{torr}^{-2} \text{ for Kr}.$$

Comparison of the absolute magnitude of ω for Mu and H gives for the ratio μ_μ/μ_p of the magnetic moment of μ and the proton, the value

$$\mu_\mu/\mu_p = 3\cdot183349\pm0\cdot0_4 15,$$

giving for the fine structure constant α, derived from eqn (128),

$$\alpha^{-1} = 137\cdot03646\pm0\cdot0_3 36, \S$$

slightly higher than, but consistent with, the 'best' value obtained from other measurements,

$$\alpha^{-1} = 137\cdot03602\pm0\cdot0_3 21.\|$$

Detailed comparison of the calculated pressure shifts of atomic hydrogen in various buffer gases with experimentally observed results reveals a marked discrepancy. For example, Clarke†† calculated for the pressure shift coefficient of atomic hydrogen in helium the value $+1\cdot73\times10^{-9}\,\text{torr}^{-1}$ at 273 °K, compared with the experimental value $+3\cdot7\times10^{-9}\,\text{torr}^{-1}$,‡‡ reflecting the inadequacy of the perturbed wavefunctions employed.

Ray and Kaufman§§ have calculated the magnitude of three-body

---

† BROWN, R. A. and PIPKIN, F. M., Phys. Rev. 174 (1968) 48.
‡ ENSBERG, E. S. and MORGAN, C. L., Phys. Lett. 28A (1968) 106.
§ JARECKI, J. and HERMAN, R. M., Phys. Rev. Lett. 28 (1972) 199.
‖ TAYLOR, B. N., PARKER, W. H., and LANGENBERG, D. N., Rev. mod. Phys. 41 (1969) 375.   †† CLARKE, G. A., J. chem. Phys. 36 (1962) 2211.
‡‡ ANDERSON, L. W., PIPKIN, F. M., and BAIRD, J. C., Phys. Rev. Lett. 4 (1960) 69; Phys. Rev. 120 (1960) 1279; 122 (1961) 1962.
§§ RAY, S. and KAUFMAN, S. L., Phys. Rev. Lett. 29 (1971) 895.

effects for Mu in Ar and have been able to account for a quadratic dependence of the hyperfine pressure shift of about the observed magnitude.

17. Muonium formation in condensed materials

Direct observation of Mu formation in condensed materials through observation of μ^+ spin precession at a rate characteristic of Mu is comparatively recent. There is, however, a large earlier literature in which the formation of Mu in condensed materials was assumed, in order to explain the observed depolarization effects on μ^+ stopping in various materials. Although we are mainly concerned with the interactions of μ^+ in gases in this section, we summarize, for completeness, the evidence for Mu formation in condensed materials.

17.1. *Direct evidence for muonium formation in condensed materials*

Triplet Mu formation in condensed materials (crystalline and fused quartz, solid CO_2, and ice) has been observed by Myasishcheva et al.[†] Their apparatus,[‡] which was also used in their studies of chemical reactions involving Mu (see § 19), is shown in Fig. 26.73. The target containers in their experiments were hollow cylindrical duralumin vessels 100 mm diameter, of wall thickness 0·7 mm, and approximately 1 litre in volume. The substances studied were solids and liquids (usually transparent) of high degree of chemical purity, this being monitored, where appropriate, by measuring the refractive index, by means of a precision refractometer, at a temperature held constant to within 0·05 °K. The deviations of the measured refractive indices from published data were less than 0·00010. After filling, the targets were carefully de-aerated and evacuated. A μ^+ beam of momentum 155 MeV/c was extracted with the aid of a system of focusing lenses and deflecting magnets, and entered the apparatus through the hole in the Pb collimator. The thickness of the Cu block was adjusted so that the μ^+ stopped in the target. Since the targets had the same dimensions for all materials the corresponding thicknesses in g cm^{-2} were not identical, and corrections were made for these differences by estimating the ionization losses of the decay posi-

[†] MYASISHCHEVA, G. G., OBUKHOV, YU. V., ROGANOV, V. S., and FIRSOV, V. G., *Zh. eksp. teor. Fiz.* **53** (1967) 451 (English translation, *Soviet Physics, JETP* **26** (1968) 298); ibid. **56** (1969) 1199 (English translation *Soviet Physics, JETP* **29** (1970) 645).

[‡] BABAYEV, A. I., BALATS, M. YA., MYASISHCHEVA, G. G., OBUKHOV, YU. V., ROGANOV, V. S., and FIRSOV, V. G., *Zh. eksp. teor. Fiz.* **50** (1966) 877 (English translation, *Soviet Physics, JETP* **23** (1966) 583); see also MINAĬCHEV, E. V., MYASISHCHEVA, C. G., OBUKHOV, YU. V., ROGANOV, V. S., SAVIL'EV, G. I., and FIRSOV, V. G., ibid. **58** (1970) 1586 (English translation, *Soviet Physics, JETP* **31** (1970) 849); ADRIANOV, D. G., MYASISHCHEVA, G. G., OBUKHOV, YU. V., ROGANOV, V. S., FIRSOV, V. G., and FISTUL, V. I., ibid. **56** (1969) 1195 (English translation, *Soviet Physics, JETP* **29** (1969) 643).

trons. The coincidence counter signal $12\bar{3}$ denoted a stopping μ^+ while the delayed coincidence $\bar{2}34$ signalled a decay positron emission. The delay time was measured by means of a time to pulse-height converter, and a 100 channel pulse-height analyser. Channel widths of 73·8, 18·9, and 4·9 ns were used in different experiments. The resolution of the electronics as a whole was 2·5 ns, but usually decays that occurred less than 40 ns after μ^+ stopping were not observed, although in some experiments this starting-time delay was reduced to 10–20 ns.

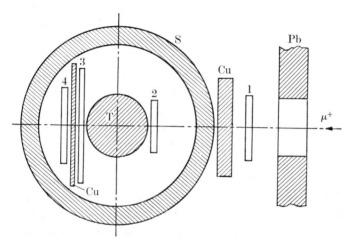

Fig. 26.73. Apparatus used by Myasishcheva, Babaev, and colleagues to observe Mu formation in various substances and to study Mu chemistry. S, solenoid; T, target; 1, 2, 3, 4, counters.

The transverse magnetic field producing the μ^+ precession was produced by a watercooled solenoid of 30 cm diameter and 60 cm height, the maximum magnetic field being 300 G with a homogeneity better than 0·5 per cent over the target. Three pairs of compensating coils, not shown in the diagram, compensated for external magnetic fields. A magnetic field of 7·2 G was used for the direct observation of Mu precession.

The positron counting rate was fitted to the expression

$$N(t) = N_0 e^{-t/\tau}\{1 - a\cos(\omega_L^{Mu} t + \delta)\} + B, \tag{136}$$

B being the background, and δ the initial phase of the curve. Mu formation is only likely to be detectable, with available electronic techniques, in chemically inert substances, since if it enters into chemical combination the covalent chemical bond pairs the electrons of Mu and of the radical with which it combines so that the observed muon spin precession frequency will be close to that of a free μ^+. We return to the discussion of chemical processes involving Mu in § 19.

Fig. 26.74 shows the modulation, observed by Myasishcheva et al., at the Mu precession frequency, of the counting rate of positrons as a function of delay-time after stopping of the μ^+ in (a) fused quartz at 303 °K, (b) fused quartz at 77 °K, (c) crystalline quartz at 293 °K, and (d) ice at 77 °K. The amplitude of the modulation provides a measure of the Mu concentration. The amplitude is seen to be damped with a damping time of the order of 0·5–1 μs. In order to obtain an estimate of the initial Mu concentration the modulation amplitude was extrapolated back to zero time, to give values of the absolute asymmetry amplitude, a.

Experiments were also carried out to study the precession in a magnetic field of 50 G, at a frequency corresponding to that for precession of the free μ^+. The observed asymmetry coefficients, a_μ, a_{Mu}, corresponding to precession at the free μ^+ and Mu Larmor frequencies, respectively, were compared with the value $a_\mu = 0.280 \pm 0.006$ obtained for μ^+ stopping in bromoform ($CHBr_3$). This value is very close to that expected from the estimated polarization of the μ^+ beam, and indicates that very little depolarization takes place in $CHBr_3$, which can therefore be used as a standard material for depolarization measurements. Control experiments also indicated that target thickness corrections were small. Table 26.6 gives the results obtained by Myasishcheva et al. It is seen that no evidence was obtained for Mu formation in powdered SiO_2, powdered Al_2O_3, liquid N_2, fused B_2O_3, benzene, polyethylene, and paraffin. For those materials for which Mu precession was observed Table 26.6 gives also the mean damping time, t_{Mu} for the disappearance of Mu.

For quartz (fused or crystalline), the Mu polarization (P) relative to that for μ^+ in bromoform $= \dfrac{a_{\text{Mu (fused quartz)}}}{a_{\mu^+(\text{CHBr}_3)}} \simeq 0.5$. The discussion of § 15.4 indicates this to be the value expected, if all the μ^+ form Mu. A similar conclusion for μ^+ stopping in fused quartz was reached by Gurevich et al.†

The results given in Table 26.6 indicate that, in certain insulators, the μ^+ polarization persists after thermalization. For a given material, the polarization does not depend strongly on the temperature but, as

---

† GUREVICH, I. I., MAKAR'INA, L. A., MEL'ESHKO, E. A., NIKOL'SKII, B. A., ROGANOV, V. S., SELIVANOV, V. I., and SOKOLOV, B. V., Zh. eksp. teor. Fiz. **54** (1968) 432 (English translation, Soviet Physics, JETP **27** (1968) 235).

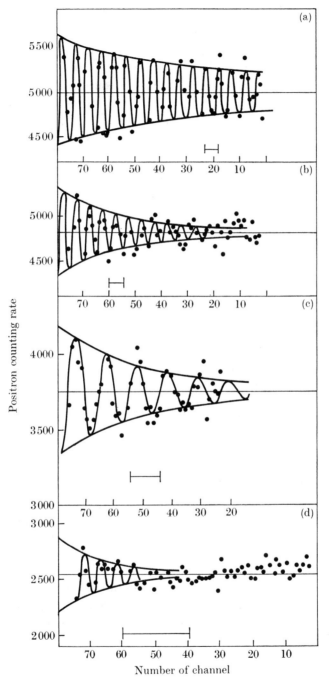

FIG. 26.74. Modulation of the positron counting rate in the experiments of Myasishcheva et al. at the Mu precession frequency for polarized μ^+ stopping in
(a) fused quartz at 303 °K,
(b) fused quartz at 77 °K,
(c) crystalline quartz at 293 °K,
(d) ice at 77 °K.
In (a), (b), and (c) the magnetic field was 7·2 G, in (d) it was 29 G. ——— indicates 10^{-7} s, in each case.

shown for fused and powdered quartz, may depend strongly on the physical state of the substance. On the other hand, the damping time t_{Mu}, corresponding to the disappearance of polarized Mu in fused SiO_2, is strongly temperature dependent, being considerably smaller at the

TABLE 26.6

Substance	Temperature (°C)	Muonium asymmetry coefficient a_{Mu}	Muonium damping time t_{Mu} (µs)	Free muon asymmetry coefficient a_μ
Fused SiO_2	+ 30°	0·161±0·012	1·3±0·2	0·050±0·006
	−196°	0·148±0·014	0·5±0·1	0·048±0·007
Crystalline SiO_2	+ 30°	0·167±0·013	0·4±0·1	0·056±0·006
Powdered SiO_2	+ 30°	0·02 ±0·02	—	0·110±0·005
Powdered Al_2O_3	+ 30°	0·04 ±0·05	—	0·111±0·007
Fused B_2O_3	+ 30°	0·02 ±0·03	—	0·127±0·006
	−196°	0·01 ±0·03	—	—
Solid CO_2	− 78°	0·070±0·015	0·4±0·1	0·038±0·013
	−196°	0·075±0·015	0·6±0·1	—
Solid H_2O	−196°	0·16	0·08	0·066±0·004
Liquid N_2	−196°	0·01 ±0·02	—	0·037±0·006
Benzene	−196°	0·02 ±0·03	—	—
Polyethylene	−196°	0·01 ±0·02	—	—
Paraffin	−196°	0·01 ±0·02	—	—

lower temperature. The damping may suggest in some cases the involvement of the Mu in a chemical reaction that proceeds over a time interval of some tens or hundred of nanoseconds. Cases where Mu was not observed may simply represent cases where t_{Mu} was very small so that no Mu precession signal could be obtained with the time-resolution of the apparatus.

μ^+ precession, at a frequency close to the free muon precession frequency, is also observed in those materials for which Mu has been detected. This would be expected if Mu eventually enters into chemical combination as discussed above (see also § 19).

Gurevich et al.† observed beats in the modulated precession of Mu in fused quartz. With a transverse field of 95 gauss a Mu precession frequency of ∼ 140 MHz was observed, modulated by a beat frequency of ∼ 4 MHz. They interpreted the beats as arising from interference between the precession frequencies of different hyperfine states of Mu,

† GUREVICH, I. I., IVANTER, I. G., MAKARIYNA, L. A., MEL'ESHKO, E. A., NIKOL'SKY, B. A., ROGANOV, V. S., SELIVANOV, V. I., SMILGA, V. P., SOKOLOV, B. V., SHESTAKOV, V. D., and JAKOVLEVA, I. V., Phys. Lett. 29B (1969) 387.

the coefficient a in eqn (136) being given by

$$a = a_0 \cos \Omega t$$

with $\qquad \Omega = (\omega_L^{Mu})^2/\omega,$

where ω is the hyperfine interval angular frequency.

A beat phenomenon of this kind was used by Schenck and Crowe[†] to explain the apparent slow muon depolarization for muons stopping in a single crystal of gypsum ($CaSO_4 \cdot 2H_2O$). A μ^+ replacing a proton in the crystal lattice possesses four resonance frequencies in an external magnetic field. In a field of 4500 gauss the beat frequency corresponding to interference between the main resonances is $\sim 0 \cdot 1$ MHz, so that only one quarter-cycle of the beat is seen before the μ^+ decay erases the signal. The apparent damping in this case is therefore to be related to this beat phenomenon rather than to chemical combination. A close correlation was found with the spin relaxation time for proton n.m.r. in gypsum.

17.2. *Indirect evidence of muonium production from the study of the spin depolarization of μ^+ stopped in various materials*

Measurements of spin depolarization for μ^+ stopped in various materials were interpreted in terms of Mu formation long before its production had been directly observed in the experiments described above. Most of these experiments have been carried out using condensed materials. Two types of experimental arrangement have been used, employing, respectively, a longitudinal magnetic field parallel to the direction of spin polarization, and a transverse magnetic field in which muons precess. The former arrangement permits a study of the muon polarization irrespective of whether it is free, associated with an electron in Mu, or bound in a molecule. In principle, the second arrangement enables a distinction to be drawn between μ^+ in these different situations.

17.3. *Measurement of μ^+ polarization in a longitudinal magnetic field*

The simplest type of arrangement for studying the polarization of muons, at the time of their decay, is one in which the positron emission rate into a small range of solid angle, about the forward direction, is compared with that into a similar range of solid angle about the backward direction. Typical of the experimental set-up is that of Gorodetzky et al.[‡] shown in Fig. 26.75. With this arrangement it was also possible to study the effect on the depolarization of a longitudinal magnetic field.

[†] SCHENCK, A. and CROWE, K. M., *Phys. Rev. Lett.* **26** (1971) 57.
[‡] GORODETZKY, S., MULLER, TH., PORT, M., and ZICHICHI, A., *Phys. Lett.* **2** (1962) 133.

The brass collimators, A, defined the μ^+ beam which then passed through the scintillation counters 1 and 2, separated by the Perspex absorber, B. The μ^+ were brought to rest in F, after being degraded in the carbon absorber, C. The seven scintillation counters served to distinguish different types of event. A stopping μ^+ was defined by the prompt coincidence $1234\bar{5}\bar{6}$, and forward and backward decay positrons

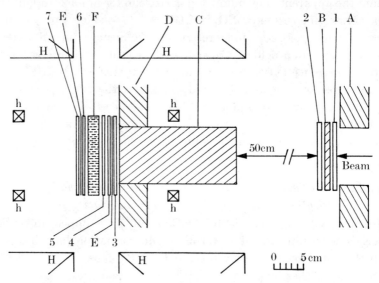

FIG. 26.75. Apparatus of Gorodetzky and co-workers for studying μ^+ depolarization in a longitudinal magnetic field. (A) Brass collimator. (B) Perspex absorber. (C) Carbon absorber. (D) Collimator. (E) Copper sheet. (F) Target. (H) Helmholtz coils.

by the delayed coincidences $\overline{45}67$ and $\overline{67}34$, respectively. The μ^+ polarization at decay was then measured by the ratio of positron emission rate in the backward and forward directions. The small Helmholtz coils, h, were used to cancel the strong vertical cyclotron field so that measurements of the polarization could be made at zero field. The large Helmholtz coils, H, enabled the measurement of the polarization for different longitudinal fields. The thickness of the carbon absorber, C, was varied to ensure that for each target the centres of the stopping region coincided with the centre of the target.

In fact, a measurement of the electron emission rate in the forward direction alone provides a sensitive monitor of μ^+ polarization, since the effect of depolarization will increase the electron emission rate. This method is indeed used in the microwave resonance measurements on Mu (§ 16.2.1).

17.4. *Measurement of μ^+ polarization in a transverse magnetic field*

This method has already been described in § 17.1 in relation to the detection of Mu formation in condensed materials. Values of P_{rel}, the polarization relative to some standard material, in which the depolarizing effects are believed to be negligible, are estimated from $P_{\text{rel}} = a/a_{\text{s}}$, where a_{s} is the value of a observed for the standard material (carbon, bromoform, anthracene, etc.).

The transverse field method possesses the advantage over the longitudinal field method that, since the observed precession frequency is sensitive to the state of chemical combination of Mu, it permits (at least in principle) fast chemical reactions of Mu to be followed. Most of the work has, however, been concerned with measuring the modulation amplitude corresponding to precession of free μ^+, and this gives information only on the polarization of free μ^+.

17.5. *Results of muon depolarization measurements*

Table 26.7 gives typical results of depolarization measurements of positive muons stopped in various materials in the absence of magnetic field and in various transverse and longitudinal magnetic fields. The figures given are for P_{rel}, the polarization relative to some standard material such as carbon or bromoform, in which polarization is believed not to occur. In assessing these values, account has to be taken of the experimental method employed. The measurements using a zero or longitudinal magnetic field refer to the overall μ^+ polarization, whether free or bound in muonium. The measurements using a transverse magnetic field show separately the polarization of free μ^+ and of μ^+ bound in muonium, so that the overall μ^+ polarization is the sum of these two polarizations.

It is seen that for μ^+ stopping in insulators, or in semi-conductors, the values of P_{rel} are small (in the absence of a magnetic field), while large values of P_{rel} are observed for μ^+ decaying in conducting materials. It is further observed that P_{rel} is strongly dependent on the longitudinal magnetic field in the case of μ^+ stopping in an insulator or semiconductor, and at a sufficiently high field the value of P_{rel} is restored, practically, to that observed for μ^+ stopping in conductors.

17.6. *Measurement of spin relaxation time for μ^+ stopping in a condensed medium*

In § 17.1 we have seen that the signal corresponding to Mu precession, in some materials, is attenuated with a characteristic time of the order

TABLE 26.7

Residual polarization of μ^+ stopping in various materials

Material	Overall P_{rel} in longitudinal field $H_\parallel$							P_{rel} for component that precesses at free μ^+ precession frequency in transverse field $H_\perp$				P_{rel} for component that precesses at Mu precession frequency	
	$H_\parallel$ (G)							$H_\perp$ (G)				$H_\perp$ (G)	
	0	50	200	500	1000	3000	5000	7·5	50	800	3500	7·5	
C (graphite)	0·74± 0·04†	0·81± 0·07†	0·85± 0·08†	0·91± 0·09†					1·00± 0·04§	1·01± 0·03§	0·95± 0·03§	—	
C (diamond)									0·20± 0·04§				
Al									0·91± 0·05§				
Be									0·97± 0·06§				
Si									1·10± 0·05§				
Mg									1·04± 0·04§				
S	0± 0·05†	—	0·15± 0·08†	0·17± 0·08†	0·16± 0·08†	0·21± 0·07†	0·33± 0·06†		0·06± 0·04§	0·26± 0·04††	0·22± 0·04††		
water	0·80± 0·08†	—	0·83± 0·08†	0·79± 0·08†	0·90± 0·08†	1·05± 0·08†	1·05± 0·08†		0·62± 0·06§		0·64± 0·05††		
ice								0·24± 0·02‡‡				0·57‡‡	

Solid CO$_2$												0.26±0.06‡‡
Plastic scintillator	0.04±0.06†	—	0.35±0.07†	0.74±0.07†	0.85±0.08†	0.90±0.08†	0.97±0.08†	0.14±0.05‡‡		0.27±0.04††	0.21±0.04††	
Photographic emulsion	0.30±0.05†	0.31±0.05†	0.32±0.05†	0.35±0.05†	0.39±0.07†	0.56±0.08†	0.65±0.10†		0.20∥	0.26±0.04††	0.23±0.04††	
									0.38±0.04§	0.37±0.05††	0.45±0.06††	
Fused SiO$_2$								0.18±0.02‡‡	0.17±0.04§			0.58±0.04‡‡

† GORODETSKY, S., MULLER, TH., PORT, M., and ZICHICHI, A., *Phys. Lett.* **2** (1962) 133.

‡ ALI-ZADE, S. A., GUREVICH, I. I., and NIKOLSKII, B. A., *Zh. eksp. teor. Fiz.* **40** (1961) 452 (English translation, *Soviet Physics, JETP* **13** (1961) 313).

§ SWANSON, R. A., *Phys. Rev.* **112** (1958) 580.

∥ BUHLER, A., MASSAM, T., MULLER, TH., SCHNEEGANS, H., and ZICHICHI, A., *Nuovo Cim.* **39** (1965) 812.

†† GUREVICH, I. I., MAKAR'INA, L. A., MEL'ESHKO, E. A., NIKOL'SKII, B. A., ROGANOV, V. S., SELIVANOV, V. I., and SOKOLOV, B. V., *Zh. eksp. teor. Fiz.* **54** (1968) 432 (English translation, *Soviet Physics, JETP* **27** (1968) 235).

‡‡ MYASISHCHEVA, G. G., OBUKHOV, YU. V., ROGANOV, V. S., and FIRSOV, V. G., *Zh. eksp. teor. Fiz.* **53** (1967) 451 (English translation, *Soviet Physics, JETP* **26** (1968) 298).

of some tenths of a microsecond. In a similar way Gurevich et al.† have looked for overall μ^+ spin polarization, as a function of time, in the interval 0–150 ns after thermalization of the muon in condensed materials. In their experiment a time to pulse-height converter and 100-channel pulse-height analyser were used to study the emission rate of decay

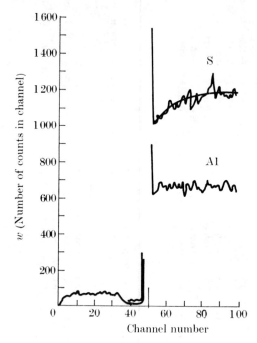

FIG. 26.76. Time dependence of the number of decay positrons emitted in the backward direction for μ^+ stopped in S and Al, with zero magnetic field. Each channel of the pulse-height analyser corresponds to 2·3 ns. Channel 51 corresponds to zero time. Channels 1–50 represent random background.

positrons over this time interval, in the backward direction. The emission rate in this direction would be expected to increase as the μ^+ becomes depolarized. Fig. 26.76 shows the time dependence of this emission rate in the case $H = 0$, for Al and S targets. Each channel of the pulse-height converter corresponds to an interval of 2·3 ns. The electronics was so arranged that the positron emission signal is recorded in channels 51–100,

† GUREVICH, I. I., MAKAR'INA, L. A., MEL'ESHKO, E. A., NIKOL'SKII, B. A., ROGANOV, V. S., SELIVANOV, V. I., and SOKOLOV, B. V., *Zh. eksp. teor. Fiz.* **54** (1968) 432 (English translation, *Soviet Physics, JETP* **27** (1968) 235); GUREVICH, I. I., IVANTER, I. G., MEL'ESHKO, E. A., NIKOL'SKII, B. A., ROGANOV, V. S., SELIVANOV, V. I., SMILGA, V. P., SOKOLOV, B. V., and SHESTAKOV, V. D., ibid. **60** (1971) 471.

channel 51 representing zero time. Channels 1–50 record the random background. An increase of the emission rate over a time interval of the order of 33 ns was observed for S but none for Al, or any of the thirteen other materials studied. Evidently some depolarization mechanism with a lifetime of some tens of nanoseconds is present for μ^+ stopped in S, but all such depolarization has ceased in times of the order of a few nanoseconds for the other materials. The full line in the figure shows the growth of signal fitted to a variation of the form of eqn (136) (with a negative).

17.7. *Interpretation of the observations on muon spin depolarization*

The results of the experiments described in the previous section indicate that, in many cases, positive muons which are strongly polarized at production have, to a large extent, lost their polarization when they decay after being reduced to rest. Further, the degree of polarization retained at the instant of decay is markedly affected by the presence of a magnetic field in the direction of polarization. Estimates have been made† of the amount of depolarization expected during the moderation process in which the muons lose their kinetic energy in collision with the electrons of the moderating material. The loss of polarization from this cause is expected to be small, and quite unable to account for the depolarization observed.

The possibility of Mu formation suggests a mechanism for depolarization since, as pointed out in § 15.4, muons which capture electrons with antiparallel spin will eventually lose their polarization, and Mu formation with a high probability has been observed in some materials (§ 16). This interpretation of the loss of μ^+ polarization is supported also by the observation that μ^+ brought to rest in conducting materials maintain their polarization (§ 17.3). The formation of triplet positronium is inhibited in conductors.‡ This is attributed to the screening effect of the conduction electrons as a result of which any system of positron and electron could, at best, be very weakly bound and readily ionized, so that its existence is transient. A similar effect would be expected for Mu formation. A longitudinal magnetic field would be expected to inhibit the depolarization of the muon in singlet Mu, exactly as observed. Estimates of the magnitude of the depolarization expected as a result of Mu formation show that the moderation time in condensed materials, or indeed in all materials except gases at low pressure, is too short to

---

† MÜHLSCHLEGEL, B. and KOPPE, H., *Z. Phys.* **150** (1958) 474.
‡ FERRELL, R. A., *Rev. mod. Phys.* **28** (1956) 308.

account for the amount of depolarization observed. The maximum depolarization effect of a single process of muonium formation reduces the polarization to one-half of its initial value, since half the μ^+ which form triplet muonium ($F = 1$, $M_F = 1$) are not subject to depolarization effects. The $\mathrm{Mu^T}$ may eventually be ionized, and the polarization of the μ^+ released can be reduced to $\tfrac{1}{2}$ by a subsequent act of Mu formation. In this way, after n cycles of muonium formation and reionization, the overall μ^+ polarization could, in principle, be reduced to 2^{-n} of its initial value.

As pointed out in § 15.4, however, $\mathrm{Mu^s}$ becomes depolarized only after a time considerably longer than t_R ($= 3\cdot6 \times 10^{-11}$ s). The time between consecutive cycles of Mu formation and ionization can be estimated from the cross-section Q_{ex} for muonium formation as a result of electron capture from the moderating medium. Consider the case of μ^+ in He. The charge exchange cross-section, Q_{ex}, for protons in He, passes through a maximum of $\sim 3 \times 10^{-16}$ cm^2 for a proton energy of 27 keV, corresponding to a velocity of 2×10^8 cm s^{-1} (see Chap. 24, § 2.2.4), while the cross-section for re-ionization of the hydrogen at this energy, Q_i, is about one-tenth of this. These correspond to mean free paths in helium at a pressure of 50 atm of 2×10^{-6} cm and 2×10^{-5} cm, respectively. One would expect these same mean free paths to apply, approximately, for Mu formation and re-ionization, at an energy of 3 keV, corresponding to the same velocity of 2×10^8 cm s^{-1}. A μ^+ of this energy is, therefore, expected to capture an electron to form Mu in a flight-time of 10^{-14} s, and after a further time of 10^{-13} s the Mu will be re-ionized to give a free muon. The rate of energy loss of a μ^+ of energy 3 keV is approximately 1200 MeV gm^{-1} cm^{-2} so that in 10^{-14} s, in He at this pressure, it will lose approximately 27 eV. A further loss of about 12 eV occurs each time an electron is extracted from the He to form Mu, so that after about 40 cycles of electron capture and loss, the energy of the μ^+ will be reduced below 1 eV. The change of polarization (ΔP) during each cycle, given by eqn (126) is approximately $1\cdot5 \times 10^{-6}$, so that the overall depolarization during moderation is negligible. As the velocity of the meson is reduced further, however, Q_i decreases more rapidly than Q_{ex}. At an energy of 100 eV the Mu lifetime before re-ionization would be expected to be about 10^{-11} s, with $\Delta P \simeq 0\cdot015$, still very small. In a condensed medium the moderation time, and thence the depolarization during moderation due to Mu formation, will be much smaller. Most of the depolarization must therefore occur after thermalization of the Mu.

Hughes et al.† have estimated the cross-section for Mu formation by charge-exchange in He ($\mu^+ + \text{He} \to \text{He}^+ + \text{Mu}$), using the adiabatic approximation at muon energies below 50 eV, an impact parameter method at energies up to 10 keV and Born's approximation at higher energies. They estimated a maximum cross-section of $\sim 2 \times 10^{-16}$ cm² at 2 keV. Belkič and Janev have made similar calculations for Mu formation in He and atomic H using a Coulomb–Born approximation‡ method in the high-energy region and an adiabatic approximation§ in the low-energy region. The results of these calculations would not appreciably alter the above conclusions about the change of polarization of the μ^+ during moderation.

The experiments of Gurevich et al. (§ 17.6) have shown that μ^+ spin relaxation processes had ceased in a time interval less than 10^{-8} s for all the condensed materials they investigated, with the exception of sulphur, for which the relaxation time was of the order of 3×10^{-8} s. The failure of Myasishcheva et al.§ to observe free Mu in many condensed materials (see Table 26.6) suggests also that in most cases it is no longer present at times of 10^{-7}–10^{-6} s after μ^+ production. It is assumed that the disappearance of Mu may be related to its entering eventually into chemical combination. In a Mu compound the electron spins would be expected to be paired so that the depolarizing magnetic field at the position of the μ^+ would be very small. Practically, spin depolarization effects would be expected to cease, therefore, with the entry of Mu into chemical combination. It has been suggested by Breskman and Kanofsky‖ that chemical combination should produce a chemical shift in the μ^+ precession frequency so that accurate measurement of this frequency should provide a means of studying chemical reactions involving μ^+.

Nosov and Yakovleva†† and Ivanter and Smilga‡‡ have given a phenomenological theory of spin depolarization of μ^+ in thermalized Mu. They postulate some interaction between the electron in the Mu and the surrounding medium, as a result of which its spin is reversed. The characteristic time for this process is α^{-1}. The magnetic interaction between the electron and μ^+, in the Mu, then induces a spin flip of the latter in a time of the order ω_0^{-1}. These effects continue for a time T,

---

† HUGHES, V. W., MCCOLM, D. W., ZIOCK, K., and PREPOST, R., *Phys. Rev.* A1 (1970) 595.   ‡ BELKIČ, DZ. S. and JANEV, R. K. *J., Phys.* B5 (1972) L237.
§ In course of publication.
‖ BRESKMAN, D. and KANOFSKY, A., *Nuovo Cim.* 68B (1970) 147.
†† NOSOV, V. G. and YAKOVLEVA, I. V., *Nucl. Phys.* 68 (1965) 609.
‡‡ IVANTER, I. G. and SMILGA, V. P., *Zh. eksp. teor. Fiz.* 54 (1968) 559 (English translation, *Soviet Physics, JETP* 27 (1968) 301); ibid 60 (1971) 1985; IVANTER, I.G., ibid. 56 (1969) 1419. See also SAMOILOV, V. M., ibid 58 (1970) 2202.

the mean time before chemical combination of the Mu takes place, after which they suppose that depolarization ceases.

The spin state of the Mu interacting with a medium in thermal equilibrium can be described by a density matrix ρ which obeys the Wangsness–Bloch equation.† Nosov and Yakovleva obtained approximate solutions of this equation in two limiting cases, while Ivanter and Smilga obtained an exact solution for the residual polarization $(P(H))$ in a longitudinal magnetic field (H)

$$P(H) = 1 - \frac{(1+2\alpha T)}{2[1+x^2+\alpha T]+2(1+2\alpha T)^2/\omega^2 T^2}. \qquad (137)$$

For a given value of T, P actually passes through a minimum value for

$$\alpha_{\min} = -\frac{1}{2T} + \frac{\omega}{2}(x^2+\tfrac{1}{2})^{\frac{1}{2}}. \qquad (138)$$

From eqn (137) it is seen that the condition for a large measure of depolarization, in the case $H = 0$, is $1 \ll \alpha T \ll \omega T$.

Ivanter and Smilga have also calculated the effect of a transverse magnetic field on the μ^+ polarization. Fig. 26.77 shows their results for P, as a function of α/ω, for a number of different values of ωT for both transverse fields $(H_\perp)$ and longitudinal fields $(H_\parallel)$. It is seen that $P(\alpha/\omega)$ depends significantly on the magnetic field only at comparatively small values of α/ω. The large observed dependence of $P(\alpha/\omega)$ on $H_\parallel$ suggests $\alpha/\omega < 1$ for such materials. On the other hand, the observed insensitivity of P to $H_\perp$‡ suggests $\alpha/\omega > 0.1$. It appears, therefore, that for the materials investigated α/ω lies between 0.1 and 1, i.e. in the vicinity of $\alpha_{\min}$, given by eqn (138). The strong dependence of $P(\alpha/\omega)$ on $H_\parallel$, together with its very weak dependence on $H_\perp$ restricts acceptable values of α/ω to a rather small range. The behaviour of a number of stopping materials could be understood assuming $\alpha/\omega \simeq 0.2$ and $\omega T \simeq 20$.

17.8. *Physical processes involved in μ^+ depolarization*

The theory given in the last section did not involve any assumption about the physical nature of the interaction leading to the spin flip of the electron in Mu. Some random spin flip process occurring during an average time of the order of a few tenths of a nanosecond is needed.

The most obvious process is electron exchange with the medium in the process

$$\mathrm{Mu^{T,S}} + \mathrm{X} \to \mathrm{Mu^{S,T}} + \mathrm{X}, \qquad (139)$$

† WANGSNESS, R. K. and BLOCH, F., *Phys. Rev.* **89** (1953) 728.
‡ GUREVICH, I. I. *et al.*, loc. cit., p. 3258.

where X is a molecule present in the medium. Such a process is possible if X is a paramagnetic molecule containing unpaired electron spins so that the exchange of electrons with opposite spin is energetically possible.

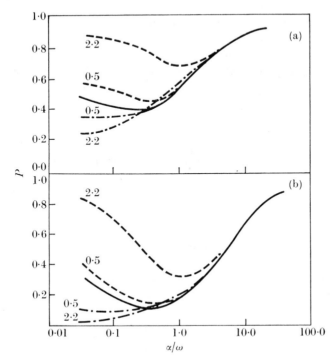

FIG. 26.77. Calculations of Ivanter and Smilga of the residual polarization P of μ^+ stopping in a medium. Depolarization is supposed to arise from an interaction of the electrons with the medium producing an electron spin flip at a rate α. The Mu interacts chemically with the medium with a mean life T. P is shown as a function of α/ω ($\omega^{-1} = 3 \cdot 57 \times 10^{-11}$ s).

(a) $T = 1 \cdot 43 \times 10^{-10}$ s ($\omega T = 4$),
    $x = 0 \cdot 5$ and $2 \cdot 2$, as indicated,
    ------ longitudinal magnetic field,
    —·—·— transverse magnetic field,
    ——— zero magnetic field.

(b) $T = 7 \cdot 15 \times 10^{-10}$ s ($\omega T = 20$),
    $x = 0 \cdot 5$ and $2 \cdot 2$, as indicated,
    as in (a).

Most of the materials for which μ^+ depolarization is observed are not paramagnetic, but they may contain paramagnetic molecular impurities. Process (139) is also possible if X is a free electron, but, for most of the media in which large depolarizations have been observed, the density of free electrons is too small to produce the observed effect. In some materials, electron spin exchange between Mu and molecules present in the moderating medium leading to the transition $\text{Mu}^T \rightarrow \text{Mu}^S$ is possible, but an activation energy is involved. Since, for only a portion of the collision of Mu with molecules of the medium will the electron spin exchange be energetically possible, the existence of the activation energy will markedly reduce the effective cross-section for the process. For materials such as the rare gases, where electron spin exchange collisions require the excitation of an electron into an outer shell, the activation energy needed would of course be very large. However, the activation energy required to excite higher energy states within the same electronic configuration is much smaller. Another possible depolarizing mechanism is Mu⁻ ion formation (attachment energy 0·7 eV). Subsequent loss of one of the electrons could then leave the muonium in the Mu^S state even though it was in the Mu^T state before negative ion formation. Such a mechanism would require the presence in the medium of molecules with ionization energy less than 0·7 eV. Other possible depolarization mechanisms, discussed by Nosov and Yakovleva,[†] include spin–orbit interaction in Mu and the interaction of Mu with the crystalline field, in the case of a solid medium, but neither process appears able to account for much depolarization.

Therefore, although the observed depolarization of μ^+ brought to rest in non-conducting condensed media can be discussed phenomenologically, the details of the actual physical processes involved still remain somewhat obscure.

18. Measurement of electron spin-exchange cross-sections for muonium

Direct evidence has been obtained that, in the case of μ^+ depolarization in argon, the physical process involved is electron spin exchange between Mu and impurities present in the argon. It has indeed been possible to estimate the cross-sections for Mu electron spin-exchange collisions with a number of materials. The process is closely similar to the electron exchange process for Ps (§§ 4.2–12). Both the microwave resonance method (§ 16.2), and the μ^+ polarization method in a longitudinal magnetic field (§ 17.3), have been applied for μ^+ stopping in gaseous argon containing known concentrations of various impurities to obtain these cross-sections.

18.1. *Effect of impurities on the microwave resonance spectrum of muonium*

Mobley et al.[‡] have studied the effect of various impurities present in the argon on the amplitude of the u.h.f. resonance signal due to

[†] loc. cit., p. 3261.
[‡] MOBLEY, R. M., BAILEY, J. M., CLELAND, W. E., HUGHES, V. W., and ROTHBERG, J. E., *J. chem. Phys.* **44** (1966) 4354.

stopping μ^+. The experiments were carried out under high field conditions ($H = 5200$ G) (§ 16.2), using impurities of O_2, C_2H_4, NO, and H_2. Fig. 26.78 shows the dependence of the amplitude of the Mu resonance transition signal on the impurity molecule concentration. It is seen that very small concentrations of the paramagnetic molecular impurities, O_2

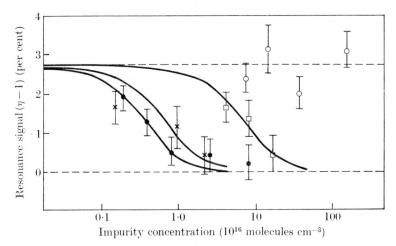

FIG. 26.78. Signal at resonance corresponding to the hyperfine (a) → (b) transition of Mu in a strong magnetic field (5200 G). For μ^+ stopping in argon containing various impurities (●, O_2; □, C_2H_4; ×, NO; ○, H_2) as a function of impurity concentration.

[The fitted curves are of the form
$$S = S_0\{4|b|^2 + (\gamma + \lambda_R)^2\}^{-1},$$
where γ is the mean decay rate of the μ^+ and λ_R the electron exchange collision rate between Mu and the impurity molecules.]

and NO, are sufficient to destroy the Mu signal. Larger concentrations of the unsaturated hydrocarbon molecule, C_2H_4, produce the same effect while the signal appears to be unaffected by the presence of quite large concentrations of H_2. The effect is attributed to electron exchange collisions between Mu^T and the impurity molecules, as a result of which the electron in the muonium may be replaced by one with opposite spin component, thus leading to the replacement of Mu^T by Mu^S, or in the nomenclature of states introduced in § 15.22, in the presence of a magnetic field, the replacement of state (a) by state (d). Since the observed u.h.f. resonance spectrum arises from the transition (a) → (b), electron exchange collisions of this kind lead to the disappearance of the resonance transition.

The possible occurrence of electron spin exchange collisions leading to the disappearance of hyperfine structure state (a) of the Mu requires

the modification of eqn (132) of § 16.2, as a result of which γ is replaced by $(\gamma+\lambda_R)$, where
$$\lambda_R = n\bar{v}Q_R, \tag{140}$$
where n is the impurity concentration, $\bar{v}$ the mean relative velocity of Mu and impurity atoms, and Q_R the electron exchange cross-section that leads to the disappearance of the state (a) of Mu. The signal strength S at resonance (eqn (134)) then becomes replaced by
$$S = \frac{4|b|^2}{4|b|^2+(\gamma+\lambda_R)^2}. \tag{141}$$
The solid curves to which the results of Fig. 26.78 are fitted are of this form, the value of λ_R being adjusted to give the best fit. From these values of λ_R the electron exchange cross-section Q_R shown in Table 26.8 have been deduced from eqn (140).

Since C_2H_4 is not a paramagnetic molecule there are no unpaired electron spins in its structure and, in this case, an electron exchange process would require an activation energy of some eV, so that the effective cross-section would be expected to be much smaller. On the other hand, the disappearance of free Mu, in this case, could be associated with its capture to form the radical C_2H_4Mu. We return to the discussion of reactions of this type in § 19.

18.2. *Effect of impurities on μ^+ spin relaxation time in argon*

Mobley et al.† have measured the overall spin polarization in a longitudinal magnetic field as a function of depolarization time for μ^+ stopping in argon. For pure argon very little depolarization is observed, but Mobley et al. studied the effect of adding small amounts of various impurities. They used an arrangement of the type illustrated in Fig. 26.75 (§ 17.3), a precision digital time analyser developed by Meyer et al.,‡ to obtain the asymmetry of positron emission in the range of depolarization times 0·2–5 μs. Fig. 26.79 shows the results they obtained for the depolarization $(1-P)$ as a function of the time, and for argon containing NO concentrations of 5 and 15 parts per million with a longitudinal magnetic field of 5200 G.

The curves are fitted to a time dependence of the form
$$P(t) = P_0 e^{-\lambda_D t}. \tag{142}$$
Since it is believed that almost all μ^+ stopping in argon form Mu, these

† MOBLEY, R. M., BAILEY, J. M., CLELAND, W. E., HUGHES, V. W., and ROTHBERG, J. E., *J. chem. Phys.* **44** (1966) 4354; MOBLEY, R. M., AMATO, J. J., HUGHES, V. W., ROTHBERG, J. E., and THOMPSON, P. A., ibid. **47** (1967) 3074.

‡ MEYER, S. L., ANDERSON, E. W., BLESER, E., LEDERMAN, L. M., ROSEN, J. L., ROTHBERG, J., and WANG, I-T., *Phys. Rev.* **132** (1963) 2693.

measurements must refer to depolarization of μ^+ in Mu. The cross-section Q_D for the depolarization process involved can be determined from
$$\lambda_D = n\bar{v}Q_D, \qquad (143)$$
where n is the impurity concentration, and $\bar{v}$ the mean relative velocity of the Mu and impurity atoms. Values of λ_D were obtained for different

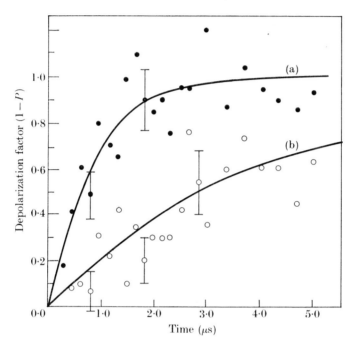

FIG. 26.79. Depolarization $(1-P)$ plotted against time for μ^+ stopping in argon containing NO impurities at pressure of approximately 50 atm. in longitudinal magnetic field of 5200 G.
(a) NO concentration 15 parts per million,
(b) NO concentration 5 parts per million.
The solid curves assume a μ^+ polarization variation of the form $P = P_0 e^{-\lambda_D t}$.

values of the longitudinal magnetic field, and these are shown in Fig. 26.80. For the paramagnetic molecules O_2, NO, λ_D was found to increase markedly as the magnetic field decreased. For the non-paramagnetic molecules C_2H_4 and NO_2 the variation of λ_D with H was more complicated.

These experiments measure the overall μ^+ polarization whether free, or bound with an electron to form Mu. However, since the interpretation of the results of the experiment on μ^+ stopping in argon are consistent with 100 per cent Mu formation, the dominant process producing μ^+ spin relaxation is once again considered to be electron exchange between the molecular impurities and Mu. One cannot, however, immediately identify the depolarization cross-section Q_D with the electron spin exchange cross-section (Q_{ex}) for the following reason. In the asymmetric decay experiment, an interaction between a Mu atom and an impurity molecule does not destroy the polarization of the Mu. In the zero

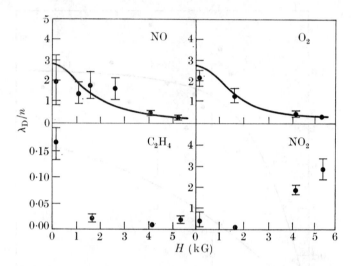

FIG. 26.80. Observed values of depolarization rate per impurity molecule (λ_D/n) as a function of longitudinal magnetic field (H) obtained from observations of the quenching of μ^+ polarization in argon containing impurities of NO, O_2, C_2H_4, NO_2. The solid curves for NO and O_2 give a fit of the results to the expected variation $(1+x^2)^{-1}$ with x (eqn (146)).

magnetic field case it reduces the polarization by one-half but this reduction in polarization is much smaller for a strong magnetic field. The depolarization cross-section is a function of x ($= H/1580$), $Q_D(x)$, and we have

$$Q_D(0) = \tfrac{1}{2} Q_{ex}. \tag{144}$$

In the u.h.f. resonance experiment, however, the electron exchange process depopulates the state (a), and thus completely suppresses the signal corresponding to the resonance transition $(a) \to (b)$, so that

$$Q_R = Q_{ex}. \tag{145}$$

The variation of $Q_D(x)$ with x is given by the theory of Ivanter and Smilga (§ 17.7). In eqn (137), write $\omega T \gg 1$, $\alpha T < 1+x^2$, so that

$$P(x) = \left(1 - \frac{1}{2(1+x^2)}\right)\left(1 + \frac{\alpha T}{1+x^2}\right), \tag{146}$$

and
$$Q_D(x) \propto \frac{1}{\alpha T} \frac{dP}{P} \left(\simeq \frac{1}{1+x^2} \right)$$

so that
$$Q_D(x) = \frac{Q_D(0)}{1+x^2}. \tag{147}$$

The results given in Fig. 26.80 for NO and O_2 are fitted to a variation of the form of eqn (147). The corresponding values of Q_{ex} are given in Table 26.8. For NO, which has $S = \frac{1}{2}$, eqn (144) can be used but for O_2, with $S = 1$, a more detailed analysis of the change of polarization of the μ^+ in Mu, following electron capture, gives a further factor of 27/32 in estimating Q_{ex} from $Q_D(0)$.

TABLE 26.8

Electron spin exchange cross-section of Mu *and* H ($\times 10^{-16}$ cm^2)

Molecule		O_2	NO
Mu	Resonance method (Q_R)	5·4±2·5	3·2±1·5
	Depolarization method	5·9±0·6	7·1±1·0
H		21±2·1	25±2·5

The agreement with values of Q_R from the Mu u.h.f. resonance experiment is satisfactory for O_2 but less so for NO. It suggests that paramagnetic impurities, also, may be an important factor in giving rise to muon spin depolarization in condensed materials.

For comparison Table 26.8 gives the electron spin exchange cross-sections for atomic H in NO and O_2.† The cross-sections in these cases are three or four times as large as those deduced for Mu in these gases. Mobley *et al.* have pointed out that this could be understood since, for Mu and H atoms of the same relative kinetic energy, fewer partial waves will be important for Mu collisions owing to its smaller mass.

19. Muonium chemistry

As pointed out already, the chemical similarity between Mu and H suggests the use of the former to study the high-speed chemical reactions of H. The most extensive studies so far reported are those of Firsov and his collaborators‡ studying the polarization of μ^+ brought to rest in a

† BERG, H. C., *Phys. Rev.* A**137** (1965) 1621.
‡ FIRSOV, V. G. and BYAKOV, V. M., *Zh. eksp. teor. Fiz.* **47** (1964) 1074 (English translation, *Soviet Physics, JETP* **20** (1965) 719); FIRSOV, V. G., *Zh. eksp. teor. Fiz.* **48** (1965) 1179 (English translation, *Soviet Physics, JETP* **21** (1965) 786); BABAYEV, A. I., BALATS, M. YA., MYASISHCHEVA, G. C., OBUKHOV, YU. V., ROGANOV, V. S., and FIRSOV, V. G., *Zh. eksp. teor. Fiz.* **50** (1966) 877 (English translation, *Soviet Physics, JETP* **23** (1966) 583).

large number of materials in the presence of transverse magnetic fields of different strengths. They used the apparatus already described in § 17.1. They consider the reactions that may occur when μ^+ are brought to rest in a hydrogenous compound of molecular structure RH, where R is any radical that combines with hydrogen. The reactions they considered were

$$\text{Mu} + \text{RH} \rightarrow \text{R} + \text{MuH} \quad \text{(substitution reaction)}, \quad (148)$$

$$\text{Mu} + \text{RH} \rightarrow \text{RMuH} \quad \text{(capture reaction)}. \quad (149)$$

Thermalized Mu in the three states Mu^T, Mu^S, and Mu^D will be present, the μ^+ being polarized in Mu^T and Mu^S and depolarized in Mu^D. Depolarization was supposed to arise from electron exchange collisions with the RH molecules in the process

$$\text{Mu}^{T,S} + \text{RH} \rightarrow \text{Mu}^{S,T} + \text{RH}. \quad (150)$$

Since chemically pure materials were used, it was not necessary to take account of the presence of paramagnetic molecular impurities. Nor was account taken of the possibility of electron exchange in collisions of the radical RMuH leading to further μ^+ depolarization. It was assumed that depolarization could take place in free Mu, through the process

$$\text{Mu}^S \rightarrow \text{Mu}^D. \quad (151)$$

The rate coefficients (number of collisions/s/unit concentration) for reactions (148) and (149) are written k_s, k_c, respectively. That for reaction (150), k_e, is related to the electron exchange cross-section, Q_{ex}, by

$$k_e = \eta \bar{v} Q_{\text{ex}} N_0, \quad (152)$$

where $\bar{v}$ is the mean relative velocity of the two particles concerned in the electron exchange process and N_0 the number of molecules/cm³/unit concentration, of the molecule RH. The quantity η (< 1) allows for the activation energy of process (150). If there is no activation energy, $\eta = 1$. The mean rate of process (151) was approximated by

$$\lambda_d = 2\omega/\pi = 1\cdot 78 \times 10^{10} \text{ s}^{-1}$$

obtained from the equation

$$\int_0^\infty \sin \omega t \, \mathrm{d}t = \int_0^\infty e^{-\lambda_d t} \, \mathrm{d}t \quad (153)$$

(compare eqn (126)). All processes take place in competition with μ^+ decay at the rate $\lambda_0 = 4\cdot 52 \times 10^5$ s^{-1}.

In equilibrium the production and destruction rates of the various products are equal so that, writing [RH] for the concentration of the molecule RH, etc., we have

$$k_s[\text{RH}][\text{Mu}] = \lambda_0[\text{MuH}], \tag{154}$$

$$k_c[\text{RH}][\text{Mu}] = \lambda_0[\text{RMuH}], \tag{155}$$

$$\lambda_d[\text{Mu}^S] = \lambda_0[\text{Mu}^D]. \tag{156}$$

In eqns (154) and (155), [Mu] refers to the total concentration $[\text{Mu}^S]+[\text{Mu}^T]$ of muonium for which the μ^+ remains polarized and [MuH], [RMuH] to the similar concentrations of molecules containing a polarized μ^+. [Mu^D] on the other hand refers to the total concentration of muonium containing a depolarized μ^+ whether free or bound in a muonic molecule. Further, since Mu^T, Mu^S have equal production rates their destruction rates must also be equal and this leads to

$$\frac{[\text{Mu}^T]}{[\text{Mu}^S]} = 1 + \frac{\lambda_d}{(k_s+k_c+2k_e)[\text{RH}]+\lambda_0} = \gamma-1, \text{ say}. \tag{157}$$

In the absence of external fields the value of the polarization expected P_{rel} (i.e. the ratio of the experimental asymmetry coefficient to the asymmetry coefficient observed in the standard medium for which depolarization is absent) is equal to the ratio

$$\frac{[\text{Mu}^S]+[\text{Mu}^T]+[\text{MuH}]+[\text{RMuH}]}{[\text{Mu}^S]+[\text{Mu}^T]+[\text{MuH}]+[\text{RMuH}]+[\text{Mu}^D]} = P_{\text{rel}} \tag{158}$$

or, using eqn (157),

$$P_{\text{rel}} = \frac{\lambda_0+(k_s+k_c)[\text{RH}]}{\lambda_0+(k_s+k_c)[\text{RH}]+\lambda_d/\gamma}. \tag{159}$$

Generally, in the presence of a transverse magnetic field, two precession frequencies (ν_μ, ν_{Mu}) corresponding to free μ^+ and to Mu precession, respectively, can be expected, with different values of P_{rel}, determined from the relative amplitude of the modulation of the signal due to the precession.

The substitution reaction (148) leads to the formation of the stable molecule MuH with paired electrons in the form of either orthoMuH (parallel μ^+ and p spin) or paraMuH. Since for paraMuH the resultant spin is 0, no precession will be observed. The remaining half of the MuH molecules in the ortho-state will be expected to precess about the field direction at a rate $0.657\nu_\mu$ corresponding to a resultant magnetic moment of 1·314 muon magnetons and spin $\hbar$. At the same time precession at the frequency ν_{Mu} is expected giving a value of P_{rel} proportional to the concentration of polarized free Mu (i.e. proportional to

$$[\text{Mu}^T]+[\text{Mu}^S]).$$

This is the state of affairs for small external fields. At fields above 20 G the coupling between Mu and H is broken, and the μ^+ bound in MuH will precess with the normal μ^+ precession frequency.

The radical RMuH formed in the capture reaction (149) has one unpaired electron so that its resultant magnetic moment is close to that of the electron (or Mu). So long as the applied magnetic field is insufficient to uncouple the μ^+ spin from the rest of the molecule, precession will

TABLE 26.9

Values of P_{rel} expected in various magnetic fields H

Magnetic field (G)	Precession frequency	$\{\lambda_0+(k_c+k_s)[RH]+\lambda_d/\gamma\}P_{rel}$
$H = 0$	(longitudinal field apps. of Fig. 26.75)	$\lambda_0+(k_c+k_s)[RH]$
$H < 20$	ν_{Mu}	$\lambda_0+k_c[RH]$
	$0.657\nu_\mu$	$\tfrac{1}{2}k_s[RH]$
$20 < H < 200$	ν_{Mu}	$\lambda_0+k_c[RH]$
$20 < H < 200$	ν_μ	$k_s[RH]$
$200 < H < 6000$	ν_{Mu}	λ_0
$200 < H < 1600$	ν_μ	$(k_c+k_s)[RH]$
$H > 1600$	ν_{Mu}	decoupled
$H > 1600$	ν_μ	$\lambda_0+(k_c+k_s)[RH]$

occur at a frequency close to that of Mu. Decoupling is expected for magnetic fields above about 200 G, much smaller than that required for decoupling of the μ^+ and electron spins in Mu ($\simeq 1580$ G) because the mean separation between the μ^+ and the uncoupled electron in RMuH is expected to be much larger than in Mu. Above 200 G, therefore, the μ^+ bound in RMuH would also be expected to precess with the μ^+ precession frequency. Above about 1600 G, the μ^+ and electron in Mu are decoupled, no precession would be observed at the Mu precession frequency, and the value of P_{rel}, derived from measurements of the precession signal at the μ^+ precession frequency, should be the same as P_{rel} determined at zero field in the longitudinal-field type of apparatus (Fig. 26.75).

Table 26.9 shows the values of P_{rel} expected under different magnetic field conditions for the various precession frequencies.

By carrying out experiments using different pressures ([RH]) and magnetic fields, and observing the precession at the μ^+ and Mu precession frequencies it appears possible, in principle, to determine the rate coefficients k_c, k_s, k_e, and thence the cross-sections for the processes involved. Babayev et al.† have applied the method to observe P_{rel} values corresponding to the μ^+ precession frequency for transverse fields in the range $20 < H < 200$ G, and from these have derived rate constants k_{s}, k_{c} for the chemical reactions of Mu with a number of acceptor molecules. Measurements were carried out using the method of competing

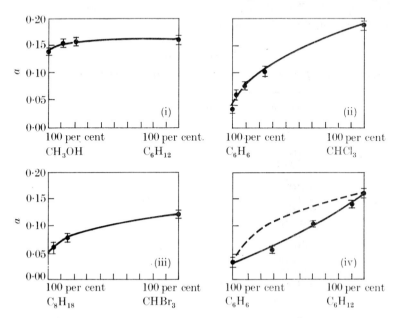

FIG. 26.81. Dependence of the asymmetry coefficient ($a = \tfrac{1}{3}P$) on the concentration of the components of the solution (in mol. per cent) in the experiments of Babayev and colleagues.

(i) CH_3OH–C_6H_{12} (ii) C_6H_6–$CHCl_3$
(iii) C_8H_{18}–$CHBr_3$ (iv) C_6H_6–C_6H_{12}.

acceptors in which asymmetry coefficients (and thence P_{rel}) were determined in binary mixtures of compounds, with both of which Mu undergoes chemical reactions. Generally, the asymmetry at the μ^+ resonance frequency with magnetic field conditions $20 < H < 200$ was observed. Measurements with a single substance would enable only the k_{s} rate to be determined, assuming k_{c} to be negligible. By measuring P_{rel} for a range of different concentrations it is possible to derive both k_{s} and k_{c}. For example, Fig. 26.81 shows the observed values of the asymmetry parameter a at the μ^+ resonance frequency for binary mixtures of different relative concentrations. If only k_{s} is taken into account a variation of the form given by the broken line would be expected. The

† loc. cit., p. 3269.

observed form of variation with concentration was consistent with finite k_c and, in fact, as seen from Table 26.10, $k_c > k_s$. Table 26.10 gives the cross-sections Q_s, Q_c for the reactions (148) and (149) respectively, derived from their rate coefficient measurements. These assumed a value of the mean velocity $\bar{v}$ corresponding to a temperature of 288 °K, and put $\eta = 1$. The statistical errors of these determinations were usually between 20 and 30 per cent.

TABLE 26.10

Cross-sections for substitution (Q_s) and capture (Q_c) reactions of Mu

Substance	Cross-sections ($\times 10^{-17}$ cm²)	
	Q_s	Q_c
C_6H_{12}	0·27	—
CH_3OH	0·061	—
C_8H_{18}	0·25	—
$CHCl_3$	1·1	—
C_6H_6	0·08	0·29
$CHBr_3$	3·6	—
DPPH	25	—
C_6H_{10}	0·22	0·10
C_6H_5Cl	0·17	0·30
C_6H_5Br	0·39	0·30
$C_2H_4Cl_2$	0·16	—
CH_2I_2	0·21	—

It is of interest to compare the values of Q_s and Q_c for C_6H_6 with the measurements of Melville and Robb[†] of the total reaction rate of H with C_6H_6. They obtained for the rate, a value of 6×10^8 litre mole⁻¹ s⁻¹. Allowing for the difference in mass between Mu and H, the corresponding quantity for Mu reactions with C_6H_6 derived from the cross-section of Table 26.10 is 5.5×10^8 litre mole⁻¹ s⁻¹. Babayev *et al.*, also, have used these methods for studying the chemical reactions of Mu with inorganic ions of variable valence in aqueous solutions.

Hughes[‡] and his colleagues have concentrated on the study of Mu reactions with molecules in the gaseous phase, present as impurities in the pressurized argon in which the μ^+ stop. They have used both the method of quenching of Mu microwave resonance spectrum (§ 16.2), and that of spin relaxation of the μ^+ in Mu (§ 18.2). The capture of Mu by a molecular impurity will lead to the quenching of the resonance

[†] MELVILLE, H. W. and ROBB, J. C., *Proc. R. Soc.* A**202** (1950) 181.
[‡] MOBLEY, R. M., BAILEY, J. M., CLELAND, W. E., HUGHES, V. W., and ROTHBERG, J. E., *J. chem. Phys.* **44** (1966) 4354.

spectrum, and using this method the cross-section for the formation of C_2H_4Mu, when known amounts of C_2H_4 are introduced into Ar as an impurity, has been estimated as $(2 \cdot 9 \pm 1 \cdot 6) \times 10^{-17}$ cm².

Spin relaxation measurements of the μ^+ in Mu formed when μ^+ are stopped in argon containing various impurities have also led to estimates for the following depolarization cross-sections in a longitudinal magnetic field of strength 100 G,

C_2H_6, $(2 \pm 1) \times 10^{-18}$ cm²; CH_3Cl, $(2 \pm 0 \cdot 5) \times 10^{-19}$ cm²;

Cl_2, $1 \cdot 1 \pm 0 \cdot 9 \times 10^{-17}$ cm²; CO_2, $(4 \pm 2) \times 10^{-19}$ cm².

Muonium chemistry is clearly in its infancy but it appears to offer advantages in the study of the fast interactions of atomic hydrogen, and the possibility of disclosing features of the spin states of intermediate products that would be difficult to elucidate directly.

27

COLLISION PROCESSES INVOLVING MESIC ATOMS

1. Introduction

NEGATIVELY charged mesons can replace electrons in atoms, forming mesic atoms. The properties of the particles that can form atomic structures of this kind are summarized in Table 27.1.

For a meson moving in a circular orbit of principal quantum number n in a mesic atom of (point) nuclear charge Ze, the orbital radius r_n, and energy E_n, are given by

$$r_n = n^2\hbar^2/\mu_R e^2 Z \tag{1}$$

and $$E_n = -Ze^2/2r_n = -\mu_R C^2 Z^2 \alpha^2/2n^2, \tag{2}$$

where μ_R is the reduced mass of the meson and nucleus. The relation between E_n and r_n is independent of μ_R, and is the same as for a normal hydrogenic atom so long as the finite extension of the nuclear charge is neglected. However, the principal quantum number n of an orbit, of given radius r_n, is proportional to $\mu_R^{\frac{1}{2}}$. The radius of the lowest orbit is proportional to μ_R^{-1}, so that the meson moves very much closer to the nucleus than does the electron in the lowest state of a normal atom. It is this property that enables mesic atoms to be used as effective probes of nuclei. Mesic atoms can also be used in studies of atomic collision processes, playing a role somewhat analogous to that of tracer elements in studies of chemical reactions. Such phenomena involving mesic atoms are discussed here.

2. Processes leading to mesic atom formation

The negatively charged particles which form mesic atoms are emitted from high energy interactions or from the decay of unstable heavier particles. Three stages can be identified in the subsequent process of formation of mesic atoms, namely,

    (a) energy loss of the fast mesons chiefly due to collisions with electrons of the medium through which they are passing, until they are moving with speeds comparable with those of the atomic electrons,

(b) capture of the slow moving mesons to form mesic atoms in states of high excitation,
(c) de-excitation of the mesic atom to the ground state or to a low lying state in which nuclear interaction may take place.

TABLE 27.1

Particle†	Mass (in units of m)	Spin (in units of $\hbar$)	Magnetic moment (in units of $e\hbar/2mc$)	Mean lifetime (s)
μ^-	$m_\mu = 206 \cdot 708 \pm 0 \cdot 004$	$\tfrac{1}{2}$	$\left(\begin{array}{c}1 \cdot 00000658 \\ \pm 0 \cdot 00000031\end{array}\right)\dfrac{m}{m_\mu}$	$(2 \cdot 1983 \pm 0 \cdot 0008) \times 10^{-6}$
π^-	$m_\pi = 273 \cdot 144 \pm 0 \cdot 027$	0	0	$(2 \cdot 604 \pm 0 \cdot 007) \times 10^{-8}$
K^-	$m_K = 966 \cdot 36 \pm 0 \cdot 22$	0	0	$(1 \cdot 235 \pm 0 \cdot 004) \times 10^{-8}$
$\bar{p}$	$M = 1836 \cdot 10 \pm 0 \cdot 01$	$\tfrac{1}{2}$	$\left(\begin{array}{c}2 \cdot 78953 \\ \pm 0 \cdot 00003\end{array}\right)\dfrac{m}{M}$	—
Σ^-	$M_\Sigma = 2348 \cdot 08 \pm 0 \cdot 22$	$\tfrac{1}{2}$	$(2 \cdot 5 \pm 0 \cdot 5)\dfrac{m}{M_\Sigma}$	$(1 \cdot 64 \pm 0 \cdot 06) \times 10^{-10}$

† Atoms incorporating $\bar{p}$ and Σ^- in this way are referred to also as 'mesic' atoms for convenience. Other particles that could form such atoms are the hyperons Ξ^- and Ω^-, but no observations on such atoms have been made.

2.1. Energy loss of fast negatively-charged particles

The rate of energy loss due to collisions of fast charged particles is well given by the Bethe–Bloch formula,†

$$-\frac{dE}{dt} = \frac{4\pi e^4 NZ}{mv} \ln\left(\frac{2mv^2}{I}\right), \tag{3}$$

where v, E are, respectively, the velocity and energy of the meson, m the electron mass, N the atomic concentration, Z the atomic number, and I the mean ionization potential of the atoms of the moderating medium. For hydrogen, I has been calculated explicitly as $I = 6 \cdot 92 R\hbar \simeq 15$ eV, where R is Rydberg's constant. For heavy elements $I = KZ$, where K is nearly constant and equal to $8 \cdot 8$ eV. For lighter elements K is larger, being $11 \cdot 5$ eV for aluminium and increasing to 15 eV for hydrogen.

Eqn (3) gives a good representation of the rate of energy loss in hydrogen where the particle velocity

$$v \gg e^2/\hbar \quad (= c/137),$$

the velocity of the atomic hydrogen electron in its ground state, corresponding to kinetic energy of 2·7 keV for muons and greater, in proportion to their masses, for other particles. In other materials, too, the

† MOTT, N. F. and MASSEY, H. S. W., *The theory of atomic collisions*, 3rd edition. Clarendon Press, Oxford (1965), p. 614.

Bethe–Bloch formula applies fairly well under these conditions, since the velocities of the most loosely-bound electrons, which contribute most to the energy loss, are of the same order as those of hydrogen. The mesons, or other negatively-charged particles, which are eventually captured to form 'mesic' atoms are commonly produced with kinetic energies of many tens or even hundreds of MeV. The time taken for the slowing down of such particles from an initial kinetic energy of 100 MeV to an energy of a few keV, as calculated by the Bethe–Bloch formula, is of the order of 10^{-9} to 10^{-10} s in condensed matter, and 1000 times as long in a gas at s.t.p. Comparing these figures with the mean lifetimes given in Table 27.1, it is clear that negatively charged particles in condensed media will mostly come to rest, and form mesic atoms, rather than decay in flight. Most π and K mesons, moderated in a gaseous medium at ordinary pressure, will decay rather than form mesic atoms, but muons will predominantly form muonic atoms, even when moderated in such a medium.

For velocities of the order of $e^2/\hbar$ the Bethe–Bloch formula is no longer applicable. Wightman[†] has evaluated the full expression for the rate of energy loss in hydrogen given by Bethe,[‡] using Born's approximation. It turns out that under these conditions the energy loss to electrons is very small and it is necessary to take account of the contribution to the energy loss arising from nuclear Coulomb collisions, given by

$$-\frac{dE}{dt} = \frac{2\pi e^4 N Z^2}{Mv} \ln(\Delta E_{max}/\Delta E_{min}), \quad (4)$$

where
$$\Delta E_{max} = 2\mu_R^2 v^2/M$$

is the maximum possible energy transfer in a head-on collision of the meson with a nucleus of mass M, μ_R being the reduced mass of the meson and nucleus. ΔE_{min}, the minimum energy transfer, can be chosen to take account of screening effects. For hydrogen Wightman took $\Delta E_{min} = e^2/6a_0$ and he estimated, in this case, the retardation times required from a velocity of $7e^2/\hbar$ to $0.8e^2/\hbar$ as varying from 7.4×10^{-13} s, for μ^-, to 6.3×10^{-12} s, for the heavier $\bar{p}$, where the effects of nuclear collisions are more important.

2.2. *Capture of negative mesons*

Capture of negative mesons to form mesic atoms occurs predominantly by the Auger type process

$$(Me)^- + A \rightarrow A_\mu + e, \quad (5)$$

[†] WIGHTMAN, A. S., *Phys. Rev.* **77** (1950) 521.
[‡] BETHE, H. A., *Annln Phys.* **5** (1930) 325.

where A, A_μ represent normal and mesic atoms, respectively, and $(Me)^-$, any negatively charged meson. Some meson capture will occur while the meson is still moving with a velocity $v = e^2/\hbar$, or greater, but most capture will take place for $v \ll e^2/\hbar$.

2.2.1. *Capture of negative mesons by atomic hydrogen.* Baker† has calculated the meson capture cross-section in atomic hydrogen by process (5), using a Coulomb–Born approximation, and supposing the ratio of the masses of the meson and proton is small. The use of the Coulomb–Born approximation limits the applicability of the calculation to the region $v > e^2/\hbar$. The transition rate $\bar{\omega}$ is given by

$$\frac{2\pi}{\hbar} \rho(k) \int \psi_f^*(\mathbf{r}_1) \phi_f^*(\mathbf{r}_2) \frac{e^2}{|\mathbf{r}_1 - \mathbf{r}_2|} \psi_i(\mathbf{r}_1) \phi_i(\mathbf{r}_2) \, d\tau_1 \, d\tau_2, \qquad (6\,a)$$

where $\mathbf{r}_1$, $\mathbf{r}_2$ refer to the coordinates of the electron and meson, respectively, $\rho(k)$ is the density of available final states of the electron, and the transition is supposed to occur as a result of the perturbation due to the Coulomb interaction between the meson and the electron. $\psi_i(\mathbf{r}_1)$, the electron wave-function in its initial state, is given by

$$\psi_i(\mathbf{r}_1) = (\pi a_0^3)^{-\frac{1}{2}} \exp(-r_1/a_0), \qquad (6\,b)$$

while the meson is supposed to be captured into a hydrogen-like state $\phi_{nlm}(\mathbf{r}_2) = R_{nl}(r_2) \Theta_{lm}(\theta_2) \Phi_m(\phi_2)$, with R_{nl}, Θ_{lm}, Φ_m defined in the usual way. For the meson wave-function, $\phi_i(\mathbf{r}_2)$ in its initial state, and the electron wave-function, $\psi_f(\mathbf{r}_1)$ in its final state, Coulomb wave-functions were used. If $k\hbar$ is the momentum of the incident meson, and n the total quantum number of the capture orbit, Baker found that over a considerable range of values of n and k, the capture cross-section, $Q(n, \mu, k)$, could be represented, approximately, by

$$Q(n, \mu, k) \simeq A n^5 (\mu/m)^{-3} k^{-2} \{1 - \exp(-Bn)\} \exp\{-[(C+Dn^2)^{\frac{1}{2}} - C^{\frac{1}{2}}]\}, \qquad (7)$$

where

$A = 26 \cdot 8 + 138(\mu/m)^{-\frac{1}{2}} - 268(\mu/m)^{-1}$,
$B = 0 \cdot 123 + 2 \cdot 60(\mu/m)^{-\frac{1}{2}}$,
$C = 40 \cdot 6\{k^2(\mu/m)^{-1} a_0^2\}\{1 + 4 \cdot 44(ka_0)^2(\mu/m)^{-1}\}^{-1}$,
$D = \{[(0 \cdot 615)^2 + 27 \cdot 1(ka_0)^2(\mu/m)^{-1}]^{\frac{1}{2}} - 0 \cdot 615\}(\mu/m)^{-\frac{1}{2}}$,

and μ is the meson mass. The k^{-2} factor in eqn (7) ensures that the cross-section increases very rapidly as k decreases, while the n^5 factor shows that capture will take place into the mesic orbit of largest n value

† BAKER, G. A., *Phys. Rev.* **117** (1960) 1130.

consistent with energy being conserved, i.e. $n \simeq (\mu/m)^{\frac{1}{2}} (= n_0)$ for very slow mesons.

Fig. 27.1 shows the distribution of radial distances from the nucleus at which mesic capture occurs, according to Baker's calculations, for

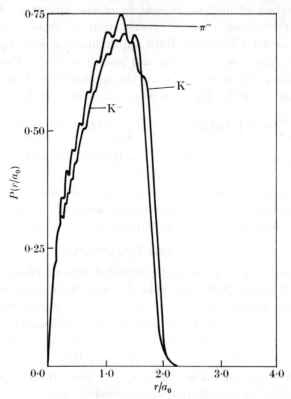

Fig. 27.1. Distribution of capture probability $P(r/a_0)$ for capture of π^- and K^- mesons of low energy on hydrogen, as calculated by Baker.

π^- and K^- mesons of very low energy, captured by hydrogen. The most probable capture distance is seen to be slightly greater than the orbital radius of the ejected electron.

The capture cross-section into an orbit of given n value was found to vary markedly with the l value of the capture orbit. This is illustrated in Fig. 27.2, which shows the proportion of the total capture cross-section corresponding to capture into states of given (n, l) values of slow K^- mesons in hydrogen. The maximum cross-section is seen to arise for $l \simeq \frac{1}{2}n$, or slightly less.

The results of Baker, based on Born's approximation, are not applicable to the most interesting and important case, where the meson velocity $v \ll e^2/\hbar$. Wightman† calculated the capture cross-section, in this case, using an adiabatic approach in which, to a first approximation, the meson and nuclei are regarded as at rest with separation R, the electrons moving in stationary orbitals around them. The

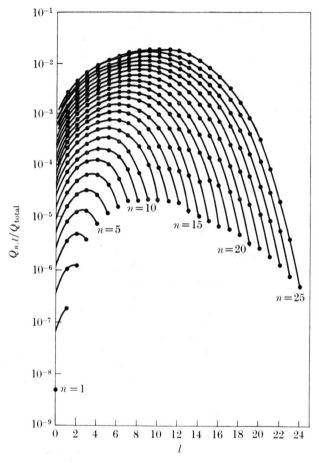

FIG. 27.2. Partial Coulomb capture cross-sections Q_{nl} for low energy K⁻ in hydrogen expressed as a fraction of the total capture cross-section calculated by Baker. The lines represent constant n values.

kinetic energy is then introduced as a perturbation (method of perturbed stationary states (see Chap. 23)). The negative meson and the proton constitute a dipole of moment eR, and Wightman examined the state of an electron moving in the field of such a dipole.‡ For sufficiently small R ($<$ a critical radius, R_0), such a dipole has no bound states.

† loc. cit., p. 3278.
‡ This problem has been examined independently by BATES, D. R., LEDSHAM, K., and STEWART, A. L., Phil. Trans. R. Soc. A246 (1953) 215. They do not give details of their results but state that they are in agreement with those of Wightman.

The wave-function (ψ) of the system is separable in prolate spheroidal coordinates, $\psi = M(\xi)N(\eta)\Phi(\phi)$ leading to the equations

$$\frac{d^2\Phi}{d\phi^2} - m^2\Phi = 0, \tag{8a}$$

$$\frac{d}{d\xi}(\xi^2-1)\frac{dM}{d\xi} + \left[B + \epsilon\xi^2 - \frac{m^2}{\xi^2-1}\right]M = 0, \quad 1 \leqslant \xi \leqslant \infty, \tag{8b}$$

$$\frac{d}{d\eta}(1-\eta^2)\frac{dN}{d\eta} + \left[-B - \epsilon\eta^2 - 2R\eta - \frac{m^2}{1-\eta^2}\right]N = 0, \quad -1 \leqslant \eta \leqslant 1, \tag{8c}$$

with $\epsilon = R^2/4(E+2/R)$, and B a dimensionless constant.

In these equations a_0 is taken as the unit of length, and the rydberg ($= e^2/2a_0$) as the unit of energy. The condition that N be quadratically integrable applied to eqn (8c) determines a denumerable set of quantities

$$B = B_{l,m}(\epsilon, R), \tag{9}$$

while a similar condition on M applied to eqn (8b) determines another set

$$B = B_{n,m}(\epsilon). \tag{10}$$

Since these two sets of B's must be the same, the equation

$$B_{n,m}(E) = B_{l,m}(E, R) \tag{11}$$

enables R to be determined for any given energy E.

Fig. 27.3 shows the results of such calculations for the total energy, $E(R)$, of a π^-H system, as a function of the separation (R) of the two particles (curve II of Fig. 27.3(a)). At a critical radius, $R_c = 0.639 a_0$, this energy equals the Coulomb energy of attraction between the proton and the negative meson. For smaller separations an electron in the ground state becomes unbound and leaves the atom. Each time this happens, the meson leaves the atom with an energy smaller than the incident energy by an amount equal to the ionization energy, $e^2/2a_0$.

Wightman estimated the rate of energy loss of the meson due to ionization, in the adiabatic approximation, by treating the motion of the meson as that of a classical particle moving in the potential $E(R)$ of Fig. 27.3(a). If p_m is the largest impact parameter for which the orbit of the meson reaches the critical radius the rate of energy loss,

$$-\frac{dE}{dt} = -v\frac{dE}{dx} = -Nv\int \Delta E \, dQ_{\Delta E} \geqslant \left(\frac{e^2}{2a_0}\right)\pi p_m^2 Nv$$

$$> \frac{e^2}{2a_0}\pi R_c^2 Nv, \tag{12}$$

where $dQ_{\Delta E}$ is the cross-section for energy loss in a small range of energy around ΔE. The meson will continue to lose energy in steps of approximately $e^2/2a_0$, in the adiabatic region, until the energy drops below $e^2/2a_0$. The next time the meson approaches within the critical distance (R_c) of the proton, so that an electron is ejected, the meson finds itself with insufficient energy to escape. If T' is the kinetic energy with which it entered its last collision it will find itself captured in a mesonic orbital of ionization energy $e^2/2a_0 - T'$, and of total quantum number n, given by

$$\frac{e^2}{2a_0} - T' = \frac{\mu e^4}{2n^2\hbar^2}. \tag{13}$$

Gershtein† pointed out that since the energy spectrum of the incident mesons

† GERSHTEIN, S. S., *Zh. eksp. teor. Fiz.* **39** (1960) 1170 (English translation, *Soviet Physics, JETP* **12** (1961) 815).

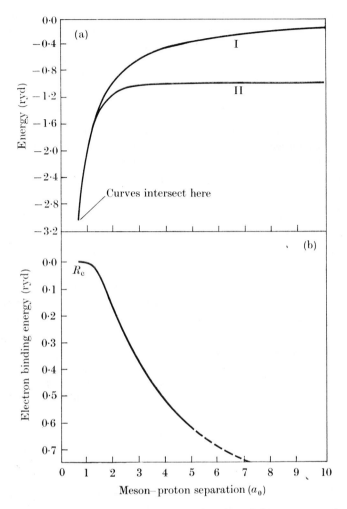

FIG. 27.3 (a) Energy of π^-H system as a function of the π^-p separation as calculated by Wightman. The system cannot bind its electron if the π^- and p get closer than $0.639a_0 \, (= R_c)$.

I Coulomb energy of attraction of H^+ and negative meson,
II Total energy $E(R)$ of π^-H system.

The two curves intersect at $R = R_c$.
(b) Binding energy of electron to π^-H system as a function of π^-p separation.

is practically constant over intervals of the order $e^2/2a_0$, and the energy losses in the adiabatic ionization region are multiples of $e^2/2a_0$, the energy distribution of the slowed-down mesons is constant in the energy interval between 0 and $e^2/2a_0$. The number of mesons captured into an orbital of total quantum number n ($> n_0 = (\mu/m)^{\frac{1}{2}}$) is proportional, therefore, to $\dfrac{d}{dn}\left(\dfrac{\mu e^4}{2n^2\hbar^2}\right)$, i.e. to n^{-3}.

Gershteĭn also pointed out that the maximum value of the azimuthal quantum number, $l_{\max}$, of the meson capture orbit is not given by the usual condition, $l_{\max} = n-1$, but is determined from the condition that the meson of energy T' reaches the critical distance R_c in the presence of the centrifugal barrier $\hbar^2 l^2/2\mu R_c^2$, so that

$$\hbar^2 l^2/2\mu R_c^2 - e^2/R_c \leqslant T' - e^2/2a_0 = -\mu e^4/2n^2\hbar^2; \qquad (14)$$

whence

$$l_{\max}^2 = \frac{2\mu R_0 e^2}{\hbar^2}\left(1 - \frac{\mu R_0 e^2}{2n^2\hbar^2}\right)$$

$$= 1\cdot 28 n_0^2 \{1 - 0\cdot 32(n_0/n)\}^2 \simeq n_0^2 \; (= \mu/m). \qquad (15)$$

A meson with a larger l value than this will not be able to approach the nucleus sufficiently closely to cause ionization of the atom.

2.2.2. *Capture of negative mesons in molecular hydrogen.* Wightman† considered the ionization processes produced by negative mesons in molecular hydrogen in order to calculate the slowing-down time in hydrogen, in the adiabatic region, before capture. He was not able to establish the existence of a region, within the hydrogen molecule, where the presence of a negative meson would 'ease out' one of the electrons in the molecule. However, he, plausibly, suggested the existence of a region, where the presence of a meson reduces the binding energy of the last electron to a value less than $0\cdot 01 e^2/2a_0$ ($= 0\cdot 135$ eV), if not zero.

The binding energy of an electron to the system $(\text{Me})^-\text{H}_2$ is the difference between the energy of $(\text{Me})^-\text{H}_2$ and that of $(\text{Me})^-\text{H}_2^+$, where $(\text{Me})^-$ could stand for any negatively charged meson. Consider the special case in which the meson approaches the H_2 molecule along the internuclear axis. Then at large separations (R) of the meson from the molecule, the binding energy of the second electron rises like

$$E(R) = e^2/R + \tfrac{1}{2}(\alpha_{\text{H}_2^+} - \alpha_{\text{H}_2})(e/R^2)^2. \qquad (16)$$

The first term represents the additional Coulomb attraction between the meson and the charged H_2^+. The second term represents the difference in energy of the two systems due to the polarization of the respective molecules by the field of the meson, α_{H_2}, $\alpha_{\text{H}_2^+}$ being the polarizability in the respective cases. The internuclear distance was taken as $1\cdot 4 a_0$, the value appropriate for H_2, since in the time of the meson collision the nuclei could not move far. When the meson reaches the position of one of the protons, its charge cancels that of the proton, and the binding energy of the $(\text{Me})^-\text{H}_2$ system becomes equivalent to that of the H^- ion for which not only the electron affinity, but also the charge distribution, is accurately known (see Chap. 15, § 5.1). This enables the slope of the $E(R)$ curve to be determined at $R = 1\cdot 4 a_0$, since this slope is the difference between the electric fields, at $R = 1\cdot 4 a_0$ of H^- and H, namely,

$$\frac{d}{dR}(E_{\text{Me}^-\text{H}_2}(R) - E_{\text{Me}^-\text{H}_2^+}(R))\}_{R=1\cdot 4 a_0} = -\frac{e^2}{(1\cdot 4 a_0)^2} \int_{R=0}^{R=1\cdot 4 a_0} \{(dZ)_{\text{H}^-} - (dZ)_{\text{H}}\}$$

$$= 0\cdot 0664 e^2/2 a_0^2, \qquad (17)$$

where $e(dZ)$ is the mean charge in dR, in the respective cases. The situation for the approach of the meson along the internuclear axis is illustrated in Fig. 27.4.

† loc. cit. p. 3278.

If the curve $E(R)$ were extrapolated linearly, from $R = 1\cdot 4a_0$ toward smaller R, it would reach zero energy at a distance of $0\cdot 4a_0$ from the nucleus, so that there would be a length $0\cdot 6a_0$ along the internuclear axis where the presence of the meson would ionize the H_2 molecule. If this result also held for the meson slightly off the nuclear axis, one might perhaps conclude that the cross-section for ionization

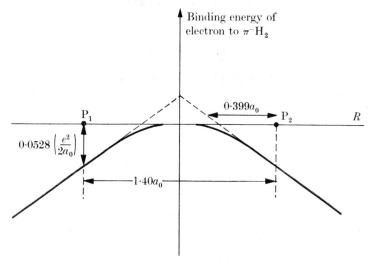

FIG. 27.4. Binding energy, $E(R)$, of electron to $\pi^- H_2$ as a function of meson position in the case when it moves along the H_2 internuclear axes. (R is the distance of the meson from the molecular centre.) The two protons (p_1, p_2) are at the H_2 equilibrium distance ($1\cdot 4a_0$). The position and slope of the binding energy curve when the π^- is at the position of one proton are determined from information on the H^- atomic ion. The dashed lines are linear extrapolations of the slope. The solid line gives the expected behaviour of the binding energy curve (after Wightman).

loss, in the adiabatic limit, should be of the same order for H_2 as for H. Even if this conclusion is not justified, however, it seems reasonable to conclude that the presence of the meson will make the binding energy of the second electron in H_2 so small that the molecule will have a large probability of being ionized by non-adiabatic processes resulting from the finite velocity of the meson. If, as Wightman conjectures, the energy separation (ΔE) of the bound state of (Me)$^-$H$_2$ from the continuum is not more than $0\cdot 01 e^2/2a_0$, the associated time ($\hbar/\Delta E$) is $4\cdot 18 \times 10^{-15}$ s. On the other hand, the collision time (a_0/v) for a meson of kinetic energy T (in eV) is approximately $4 \times 10^{-15}/T^{\frac{1}{2}}$ s for a π or μ meson so that, even though the meson is moving slowly with a kinetic energy of some tens of eV, under the conditions envisaged it still has an energy well above the adiabatic region, when ΔE is taken to be $0\cdot 01 e^2/2a_0$ rather than $e^2/2a_0$, the appropriate value when the meson is far away from the molecule.

On the basis of arguments such as these, Wightman used a model of the non-adiabatic collision process between a meson and an H_2 molecule to estimate that, in the velocity range $e^2/\hbar > v > 5 \times 10^{-4} e^2/\hbar$, the cross-section for ionization of a hydrogen molecule, by a negative meson, will be $> 0.05\pi a_0^2$, and will probably be $\simeq 0.1\pi a_0^2$. Using this cross-section he estimated the time required for moderation of the meson in the velocity range $6 \times 10^{-2}c$ to $5 \times 10^{-5}c$, in liquid H_2. The results of Wightman's estimates of the moderation time of different negatively charged particles in liquid H_2, in the various velocity regions considered, are given in Table 27.2.

TABLE 27.2

Moderation times in liquid H_2 (s)†

Particle	μ^-	π^-	K^-	$\bar{p}^-$
Moderation range				
10 MeV to $v/c = 5 \times 10^{-2}$	8.6×10^{-10}	4.8×10^{-10}	2.9×10^{-10}	2.4×10^{-10}
$v/c = 5 \times 10^{-2}$ to 6×10^{-3}	0.74×10^{-12}	0.97×10^{-12} 2.5×10^{-12}‡	3.4×10^{-12} $(12 \times 10^{-12}$‡$)$	6.3×10^{-12}
$v/c = 6 \times 10^{-3}$ to 5×10^{-5}	0.79×10^{-12}	0.74×10^{-12} 1.2×10^{-12}‡	1.5×10^{-12}	2.4×10^{-12}

† The estimates given in this table are those of Wightman, except where indicated.
‡ These estimates are those of Day, T. B. (1960), using the methods developed by Wightman (University of Maryland, Physics Dept. *Technical Report* No. 175).§ For π^- mesons Day took the boundary between the two moderation regions to be 0.1 instead of $0.006c$. For K^- mesons the figure given in brackets is total moderation time from $v = 0.05c$ to molecular capture.

2.2.3. *Capture of negative mesons by atoms other than hydrogen.* Calculations of meson capture cross-sections by the Auger type process (5), for more complex atoms, have been carried out by de Borde,∥ by Mann and Rose,†† and by Martin.‡‡ De Borde, and Mann and Rose, used Born's approximation, and de Borde showed that this gave cross-sections well above the unitarity limit. Martin used Coulomb-distorted waves for the incident meson and ejected electron, so that his calculations are similar to those of Baker§§ for hydrogen, and are applicable only when the meson velocity $v > Z_{\text{eff}} e^2/\hbar$, where Z_{eff} is the effective nuclear charge at the initial orbit of the ejected electron. He found capture to occur most probably to a mesic atom state with orbital velocity approximately equal to the incident meson velocity, and that the most tightly bound electron, consistent with energy conservation, was ejected in the process. Although some mesic capture into orbits with $n \simeq (\mu/m)^{\frac{1}{2}}$ occurs, it is much more

§ Quoted by LEON, M. and BETHE, H. A., *Phys. Rev.* **127** (1962) 636.
∥ DE BORDE, A. H., *Proc. phys. Soc.* **67** (1954) 57.
†† MANN, R. A. and ROSE, M. E., *Phys. Rev.* **121** (1961) 293.
‡‡ MARTIN, A. D., *Nuovo Cim.* **27** (1963) 1359.   §§ loc. cit., p. 3279.

probable for the meson to be moderated to lower energies so that capture occurs into mesic orbits of higher n value, with the ejection of one of the outermost electrons. For K$^-$ mesons, for example, initial capture is most probable into states with $n \simeq 100$, for Hg, and $n \sim 60$, for C, much greater than $(\mu/m)^{\frac{1}{2}}$ ($\simeq 30$). For a given n, the distribution of l values is quite similar to that obtained by Baker for hydrogen, and the capture probability passes through a broad maximum for $l \simeq \frac{1}{3}n$.

For mesons in the adiabatic region ($v < e^2/\hbar$) ionization of an atom of charge Z should be possible due to a process similar to that discussed by Wightman for hydrogen, provided the atom of charge $Z-1$ does not form a negative ion. In this latter case, no adiabatic ionization will occur, even when the position of the meson coincides with the nucleus. Non-adiabatic ionization processes will occur however, possibly with an appreciable cross-section owing to the effect of the negative meson inside the atom in reducing the ionization energy of the valence electrons. Gershtein† estimated that, in this case, the most probable l value of the captured orbit would be $(\mu/m)^{\frac{1}{2}}$, thus generally leading to the population of orbits with $l \ll n$.

In a mixture of substances containing equal numbers of atoms of different Z, Fermi and Teller predicted that the probability of capture of the meson by an atom of given Z should be proportional to Z (Fermi–Teller Z law). For mixtures of gases this law appears to hold reasonably well, but there are wide departures from it when the capture takes place in chemical compounds.‡ Ponomarev§ attributes the breakdown of the Z law to the initial capture, in some cases, of the negatively charged mesons on highly excited molecular orbitals, forming a large mesic molecule. De-excitation of the meson may then occur by transitions to meso-atomic levels about the different atoms in the molecule. This process causes a marked departure from the Z law for molecules containing H.

For capture of negatively charged particles in a chemical compound of type $Z_m H_n$ capture is assumed to occur either directly into an atomic orbital of the heavy atom or into a molecular orbital of the mesic molecule, the relative probabilities of these two capture processes being $mZ:n$, so that this assumes a kind of Z law for the initial capture process.

† loc. cit., p. 3282.
‡ BARJAL, J. S., DIAZ, J. A., KAPLAN, S. N., and PYLE, R. V., *Nuovo Cim.* **30** (1963) 711.
§ PONOMAREV, L. I., *Yadern. Fiz.* **2** (1965) 223 (English translation, *Sov. J. nucl. Phys.* **2** (1966) 160); ibid. **6** (1967) 389 (English translation, *Sov. J. nucl. Phys.* **6** (1968) 281). See also GERSHTEIN, S. S., PETRUKHIN, V. I., PONOMAREV, L. I., and PROKOSHKIN, YU. D., *Usp. Fiz. Nauk* **97** (1969) 3 (English translation, *Soviet Physics—Uspekhi* **12** (1969) 1).

Radiative transitions from the molecular orbital may take place to atomic orbitals of the hydrogen or of the heavier atom Z, the relative probabilities of these two transitions being Z^{-2} so that the probability, W, that nuclear capture of the negative particle occurs on the free proton can be written

$$W \simeq \frac{a_\text{L} n}{(mZ+n)Z^2}. \tag{18}$$

An extensive series of measurements of W for π^- capture in hydrogenic compounds has been made by Krumshtein et al.† They stopped the π^- meson in a vessel containing the compound under study, placed between two Cerenkov total-absorption spectrometers operated in coincidence. The π^- captured by a free proton was identified by the high-energy γ rays produced in the process $\pi^- + p \to \pi^0 + n \to \gamma + \gamma + n$. For $Z \leqslant 8$ they found reasonable agreement with expression (18) for W with $a_\text{L} \sim 1.3$

Van der Velde-Wilquet et al.,‡ measured the probability W for K⁻-mesons stopping in a mixture of C_2H_6 and C_3H_8, of known proportions in a heavy liquid bubble chamber. They observed the interaction K⁻ + p → Σ^- + π^+. When capture occurs at rest on a free proton, the π^+ has a unique momentum of 173 MeV/c. They found $W = 0.040 \pm 0.005$, corresponding to $a_\text{L} \sim 4.8$.

Pawlewicz et al.§ have measured W for antiprotons stopping in C_3H_8 in a bubble chamber. They looked for examples of the reaction
$$\bar{p} + p \to \begin{cases} \pi^+ + \pi^- \\ K^+ + K^- \end{cases}.$$ The observation of a collinear pair of mesons provided a signature for annihilation at rest on a free proton. They found $W = 0.11 \pm 0.03$ corresponding to $a_\text{L} \sim 13.2$.

Evidently the model of Ponomarev needs modification to allow for the dependence of W on the mass of the captured negative particle. Panofsky et al.‖ suggested that after formation of mesic hydrogen there is a high probability of transfer of the negative particle to a heavier atom in collisions taking place during the cascade time. Since the size of the neutral mesic atom is inversely proportional to the mass of the bound negative particle the probability of such a transfer process might be expected to be least for the $\bar{p}$ case and greatest for the π^- case. Writing§

† KRUMSHTEĬN, Z. V., PETRUKHIN, V. I., PONOMAREV, L. I., and PROKOSHKIN, YU. D., Zh. eksp. teor. Fiz. **54** (1968) 1690 (English translation, Soviet Physics, JETP **27** (1968) 906).

‡ VAN DER VELDE-WILQUET, C., KEYES, G. S., SACTON, J., WICKENS, J. H., DAVIS, D. H., and TOVEE, D. N., Lettere al Nuovo Cim. **5** (1972) 1099.

§ PAWLEWICZ, W. T., MURPHY, C. T., FETKOVICH, J. G., DOMBECK, T., DERRICK, M., and WANGLER, T., Phys. Rev. D2 (1970) 2538.

‖ PANOFSKY, W. K. H., AAMODT, R. L., and HADLEY, I., Phys. Rev. **81** (1951) 565.

$a_L = b_L(1-p_M)$, where p_M is the probability that a negative particle initially in an atomic orbit around a proton is transferred to an orbit around the heavier nucleus, agreement with the observed values of W can be obtained by taking $p_M = 0$ for antiprotons, 0·6 for K^-, and 0·85 for π^-.

2.2.4. *Depolarization of negative muons in the capture process.* Negative muons emitted in processes $\pi^- \to \mu^- + \bar{\nu}$, or $K^- \to \mu^- + \bar{\nu}$, are fully polarized along their direction of motion, in the rest system of the parent particle. This is necessitated by the fact that the antineutrino, emitted in association with the muon, is a particle with definite helicity. The effect of the motion of the parent particle on the initial polarization in the muon rest system has been shown normally to be small,† so that at production the μ^- polarization $P_{\mu^-} \simeq 1$. Depolarization of the muons during moderation, as a result of multiple Coulomb scattering, has been shown to be small.‡

Mann and Rose§ have investigated the polarization,

$$P = \frac{\int \phi^* \sigma_z \phi \, d\tau}{\int \phi^* \phi \, d\tau},$$

of the meson immediately after capture, where the z axis is taken as the initial direction of emission of the muon. If the capture process involves the emission of an s electron then capture into a state of angular momentum j will leave the muon with polarization

$$P_\mu = 1/2j \tag{19}$$

if the muon is still moving in the z direction at the time of capture. Mann and Rose show, however, that as a result of multiple scattering, the direction of motion of the muons will be almost random by the time capture occurs. Under these conditions, they derived the result for the polarization in an excited muonic atom orbital immediately after capture,

$$\begin{aligned} P_\mu &= \tfrac{1}{3}(j+1)/j, & j &= l+\tfrac{1}{2} \\ &= -\tfrac{1}{3}, & j &= l-\tfrac{1}{2} \end{aligned} \tag{20}$$

This result depends only on the angular part of their wave-functions, and so is not invalidated by their use of undistorted radial wave-functions. However, the overall value of P_μ, being an average over the distribution of j values of the capture orbitals, will depend to some extent on these assumptions.

† JENSEN, S. H. P. and ØVERAS, H., *K. norske Vidensk. Selsk. Forh.* **31** (1958) 34.
‡ MÜHLSCHLEGEL, B. and KOPPE, H., *Z. Phys.* **150** (1958) 474.
§ loc. cit., p. 3286.

3. De-excitation processes of mesic atoms

From the state of very high excitation in which a mesic atom is usually formed, electromagnetic transitions—either Auger or radiative—rapidly drive it towards its ground state. For particles that interact only weakly with nucleons, like μ^-, the ground ($1s$) state of the mesic atom will be reached long before nuclear capture can occur. For strongly interacting particles, like π^- or K^-, nuclear capture will commonly occur before the mesic atom has been de-excited to its ground state. In the case of mesic hydrogen atoms in a condensed medium, such as liquid hydrogen, collision processes have a strong influence on the transition rate. Such effects may also occur for other mesic atoms, but this is less certain. It is of interest to estimate the effect of such processes on the de-excitation time, since one of the few experimental quantities that can readily be measured for the capture process is the sum of the moderation time from a given initial velocity and the de-excitation time before nuclear capture. Such a measurement can therefore give some information about the cross-sections for the de-exciting collision processes.

3.1. *De-excitation of mesic hydrogen* (Me⁻p)†

The de-excitation of mesic hydrogen presents some special features, which distinguish it from the de-excitation of other mesic atoms. In most other atoms, the de-excitation in the early stages occurs predominantly through the internal Auger effect. However, the absence of orbital electrons excludes this effect for (Me⁻p). In a rarefied gas of H_2 it would be necessary to rely entirely on radiative transitions for de-excitation. Bethe and Salpeter‡ have given for the radiative transition rates, Γ_R, for transition between states of total quantum number n_i, n_f, of energy difference ΔE,

$$\Gamma_R^{n_i n_f} = \tfrac{4}{3}(\Delta E)^3 (R_{n_i}^{n_f})^2 (\mu)^{-2} \times 1{\cdot}60 \times 10^{10}\ \mathrm{s}^{-1}, \tag{21}$$

where $R_{n_i}^{n_f}$ is the radial matrix element of the dipole moment, between the initial states and final states, averaged over the l values of initial states, and summed over the possible l values of the final state. A statistical distribution was assumed for the l values of the initial state, i.e. each state was assigned the weight $2l+1$. In general, this would be a rather dubious assumption. We have seen that in the initial capture process the distribution is far from statistical. We return to the discussion of this point later.

† We use the symbol (Me⁻p) to signify any mesic hydrogen atom.
‡ BETHE, H. A. and SALPETER, E. E., *Quantum mechanics of one- and two-electron atoms*. Academic Press, New York (1957), §§ 59, 60, 63.

In most experiments, however, the (Me⁻p) atom has been formed in a condensed medium, as in the liquid hydrogen bubble chamber and, under these circumstances, radiative transitions are important only for the lowest transitions from states with $n = 4$. De-excitation takes place predominantly as a result of interaction with other molecules. There are three such processes of importance.

At very high excitation ($n \gg (\mu/m)^{\frac{1}{4}}$) a chemical dissociation process,

$$(Me^-p) + H_2 \to (Me^-H_3^+)$$
$$\to (Me^-p) + H + H, \quad (22)$$

is possible. The mesic atom loses an excitation energy $\gtrsim 4\cdot 7$ eV, in this process, and emerges from the impact with a distribution of kinetic energies around 1 eV. Leon and Bethe† estimated a cross-section

$$Q_{\text{chem}} \simeq \tfrac{1}{2}\pi a_n^2 = \tfrac{1}{2}\pi n^2 m a_0/\mu, \quad (23)$$

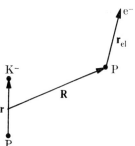

FIG. 27.5. Coordinate system for calculation of the cross-section for the external Auger process.

and considered it to be the dominant dissociation process for (Me⁻p), in liquid hydrogen, when $n > (\mu/m)^{\frac{1}{2}}$.

For $n \simeq (\mu/m)^{\frac{1}{2}}$ the second type of de-exciting collision becomes important, namely, the external Auger process

$$(Me^-p)_{n_i} + H \to (Me^-p)_{n_f} + p + e. \quad (24)$$

The mesic hydrogen atom makes a transition between the states n_i, n_f, and the energy difference is used to eject an electron from the hydrogen atom. The interaction leading to this process is illustrated in Fig. 27.5. If $\mathbf{r}$ is the meson–proton separation in the mesic atom, $\mathbf{r}_{\text{el}}$ the similar electron–proton separation in the H atom, and $\mathbf{R}$, the position vector of the H atom, relative to the centre of mass of the mesic atom, the perturbation producing the transitions is

$$H = \frac{e^2}{|\mathbf{R} - (1-\epsilon)\mathbf{r} + \mathbf{r}_{\text{el}}|} - \frac{e^2}{|\mathbf{R} + \epsilon\mathbf{r} + \mathbf{r}_{\text{el}}|}, \quad (25)$$

where $\epsilon = \mu/(\mu + m)$.

Using Born's approximation the cross-section for the process can be written (Chap. 23, § 2.2)

$$Q_{if} = \frac{2\pi}{v_i} \frac{k_f M_R}{(2\pi)^3} (4\pi)^2 \iiint |\phi_{n_f}^*(\mathbf{r})\psi_f^*(\mathbf{r}_{\text{el}})[\exp\{i(1-\epsilon)\mathbf{q}\cdot\mathbf{r}\} - \exp\{-i\epsilon\mathbf{q}\cdot\mathbf{r}\}] \times$$
$$\times \exp\{-i\mathbf{q}\cdot\mathbf{r}_{\text{el}}\}\phi_{ni}(\mathbf{r})\psi_{1s}(\mathbf{r}_{\text{el}})|^2 \frac{k^2}{q^4} dk d\omega_k d\omega_{k_f}, \quad (26)$$

where $k_i \hbar$, $k_f \hbar$ are, respectively, the initial and final relative momenta of the two atoms, $\mathbf{q}\hbar = (\mathbf{k}_i - \mathbf{k}_f)\hbar$, the momentum transfer, v_i the initial velocity of the mesic atom, k the momentum of the outgoing electron, ω_k, ω_{k_f}, respectively, the solid angles of the outgoing electron and the direction of the relative momentum of the two atoms in the final state, and M_R the reduced mass of the two atoms.

† LEON, M. and BETHE, H. A., *Phys. Rev.* **127** (1962) 636.

ψ_{1s}, ψ_f are respectively the wave-functions of the electron in its initial $1s$ state and final continuum state; ϕ_{n_i}, ϕ_{n_f}, the wave-functions of the meson in the states n_i, n_f respectively.

Conservation of energy gives

$$\frac{k^2}{m} = \frac{2\Delta E}{\hbar^2} + (k_i^2 - k_f^2)/M_R \simeq \frac{2\Delta E}{\hbar^2}, \qquad (27\text{a})$$

where

$$\Delta E = \frac{e^2}{2a_0}\left\{\frac{m^2}{\mu^2}\left(\frac{1}{n_f^2} - \frac{1}{n_i^2}\right) - 1\right\}. \qquad (27\text{b})$$

Leon and Bethe evaluated eqn (26), approximately, as

$$Q_{if} \simeq \frac{16\pi}{3v_i \mu^2}(k^2 + 1\cdot 39)^{-\frac{1}{2}}(R_{n_f}^{n_i})^2, \qquad (28)$$

where the quantities are expressed in atomic units. The average rate of Auger de-excitation is then given by

$$\Gamma_A^{n_i n_f} = NvQ_{if} = 4\cdot 3 \times 10^{15}(R_{n_f}^{n_i})^2 \mu^{-2}(2\Delta E + 1\cdot 39)^{-\frac{1}{2}}, \qquad (29)$$

where the concentration N of hydrogen atoms, in liquid hydrogen, has been taken as $4\cdot 3 \times 10^{22}$ cm^{-3}.

The third interaction process of mesic hydrogen with the surrounding molecules is of great importance for $n < (\mu/m)^{\frac{1}{2}}$. It was pointed out by Day, Snow, and Sucher† that in such a state a (Me$^-$H) atom will appear as a compact neutral object, able to approach close to the nuclei of other atoms. The electric field experienced by the mesic atoms in such close collisions gives rise to Stark mixing of states of different l, but the same n, and induces transitions between them. These transitions have the effect of increasing the population of levels of low l value from which nuclear interaction is much more likely owing to the close approach of the meson to the nucleus in such states. In this way they reduce the overall de-excitation time.

The matrix element for this Stark mixing can be written

$$V_{l-1}^l = \langle n, l-1 | \mathbf{F}_0 \cdot \mathbf{r} | n, l \rangle, \qquad (30)$$

where $\mathbf{F}_0$ is the shielded electric field produced by the hydrogen atom at the position of the mesic atom (position vector $\mathbf{r}$). Leon and Bethe wrote

$$V_{l-1}^l/\hbar = \frac{e^2}{a_0^2} Z_{\text{eff}} R_{n,l}^{n,l-1}/\mu\hbar = n(n^2 - l^2)^{\frac{1}{2}}\mu^{-1} \times 4 \times 10^{16} \text{ s}^{-1}, \qquad (31)$$

where Z_{eff} is the effective nuclear charge at the position of the mesic atom, and $R_{n,l}^{n,l-1}$ the hydrogenic radial dipole matrix element. For mesic atom kinetic energies, of the order of 1 eV ($v \simeq 10^6$ cm s^{-1}), the collision time $(a_0/v) \simeq 10^{-14}$ s, so that if eqn (31) gives the order of magnitude

† Day, T. B., Snow, C. A., and Sucher, J., *Phys. Rev. Lett.* **3** (1959) 61; *Phys. Rev.* **118** (1960) 864.

of the transition rate between states of different l, many virtual transitions back and forth between the states take place during the collision. Under such circumstances Born's approximation is inapplicable.

Leon and Bethe used a simplified impact parameter method to calculate the total probability of Stark transitions to the s state. The calculation in this case is greatly simplified by the circumstance that transitions of the type $\Delta l = 0$, $\Delta m_z = \pm 1$, which correspond to a rotation of the axis of quantization, during the collision, are not possible for s states. For the case where the mesic atoms enter the collisions with the different l values populated statistically except that the $l = 0$ state is unpopulated, Leon and Bethe derived for the mixing cross-section for population of the s state,

$$Q_s = \pi \rho_s^2 n^{-1}, \tag{32}$$

where ρ_s, the effective impact parameter, is the root of the equation

$$\frac{v\mu}{2n^2} = \frac{1}{\pi\rho} \int_{-\frac{1}{2}\pi}^{\frac{1}{2}\pi} e^{-2\rho \sec\theta}(1+2\rho \sec\theta + 2\rho^2 \sec^2\theta)\, d\theta. \tag{33}$$

For $(\pi^-\text{p})$ atoms ρ_s varied from $1\cdot 4a_0$, for $n = 2$, to $2\cdot 5a_0$, for $n = 6$, and $3\cdot 3a_0$, for $n = 15$. The values for (K^-p) atoms were about $0\cdot 4a_0$ less. In liquid H_2 expression (32) corresponds to a transition rate to the s state due to Stark effect

$$\Gamma_{\text{St}}(n) \simeq 4\cdot 9 \rho_s^2(n) \times 10^{12}\ \text{s}^{-1}. \tag{34}$$

This analysis assumes that all the states for a given n are degenerate. For strongly interacting particles, however, this degeneracy is destroyed, as a result of the interaction between the meson and the nucleus. This produces a shift of the energy of the mesonic atom states, strongly dependent on l. This energy shift is complex, the imaginary part corresponding to absorption of the meson by the nucleus.

The s wave energy-shift is given approximately[†] by

$$\delta E = -2\pi A |\psi(0)|^2 \mu^{-1} = -2A\mu^{-2}n^{-3}, \tag{35}$$

where A is the complex meson–nucleon scattering length. Only the s state is appreciably shifted in this way. In order to induce Stark transitions to this state the electric field must be strong enough to produce a Stark shift of the same order as the nuclear shift. The least impact parameter needed to achieve this is greater than ρ_s for $(\pi^-\text{p})$ for states with $n \geqslant 3$, so that this condition does not further reduce the cross-section for Stark transitions to the s state. For (K^-p) atoms, however,

[†] DE3ER, S., GOLDBERGER, M. L., BAUMANN, K., and THIRRING, W., *Phys. Rev.* **96** (1954) 774.

the impact parameter required to achieve a sufficiently strong electric field is less than ρ_s, for $n < 20$, so that the nuclear shift of the s levels reduces the cross-sections markedly. For (π^-p), with a statistical distribution of l values for states of a given n, the nuclear capture rate from the s state is

$$\Gamma_c^\pi(n) = \Gamma_s n^{-5} \simeq 1\cdot 1 \times 10^{15} n^{-5} \text{ s}^{-1}. \tag{36}$$

For $n < 6$, the time between Stark collisions ($\sim 10^{-13}$ s from eqn (34)) is long enough to give some depletion of the s state, so that the nuclear capture rate is reduced from the value given by eqn (36) to an effective value

$$\Gamma_{c,\text{eff}}^\pi(n) = \Gamma_{\text{St}}(n) n^{-2} \{1 - \exp[-\Gamma_s/n^3 \Gamma_{\text{St}}(n)]\}. \tag{37}$$

For (K$^-$p) the nuclear capture rate corresponding to eqn (36) is

$$\Gamma_c^K(n) \simeq 1\cdot 3 \times 10^{18} n^{-5} \text{ s}^{-1}, \tag{38}$$

and nuclear capture is practically complete between Stark collisions. An important amount of capture actually occurs during the collision. This is difficult to calculate, and introduces an uncertainty in the calculation of the cascade time for (K$^-$p) atoms.

For small impact parameters the Stark effect splitting may be large enough to produce appreciable mixing between states of total quantum numbers n, $n-1$, and can then contribute to the de-excitation of the mesic atom to states of lower n. This effect has not been calculated.

The fraction of mesons that undergo nuclear capture from a state n is given by

$$F(n) = \frac{\Gamma_{c,\text{eff}}(n)}{\Gamma_{c,\text{eff}}(n) + \Gamma_{\text{chem}}(n) + \Gamma_A(n) + \Gamma_R(n)}, \tag{39}$$

where $\Gamma_{\text{chem}}(n)$, $\Gamma_A(n)$, $\Gamma_R(n)$ are sums of transition rates from the nth state to all states of lower n for chemical dissociation, Auger, and radiative transitions, respectively.

Table 27.3 shows the results of the calculations of $F_\pi(n)$, $F_K(n)$. These calculations lead to the time τ of the cascade given in Table 27.4.

3.2. *Measurement of the de-excitation time of mesic hydrogen atoms*

Estimates have been made of the de-excitation times of mesic hydrogen atoms by measuring the proportion of a sample of stopping mesons or hyperons, in a hydrogen bubble chamber, that decay from rest.

For (π^-p) Fields et al.[†] studied $\pi \to \mu$ decays, in which the π^- meson, at the time of decay, was at rest or moving very slowly. They selected

[†] FIELDS, T. H., YODH, G. B., DERRICK, M., and FETKOVICH, J. G., *Phys. Rev. Lett.* **5** (1960) 69.

TABLE 27.3

Fraction of mesons in mesic hydrogen atoms that undergo nuclear capture from states of various n

	Fraction captured	
n	$(\pi^- p)$	$(K^- p)$
23	—	0·05
20	—	0·13
18	—	0·11
16	—	0·05
15	—	0·07
14	—	0·04
13	—	0·03
12	—	0·10
11	—	0·01
10	—	0·02
9	0·003	0·03
8	—	0·07
7	0·003	0·08
6	0·013	0·10
5	0·09	0·08
4	0·44	0·03
3	0·39	0·003
2	0·04	—

TABLE 27.4

Cascade times (in 10^{-12} s) of π^- and K^- mesons in liquid H_2†

(1) $(\pi^- p)$ atoms	
Molecular capture to state $n = 15$ ($\simeq (\mu/m)^{\frac{1}{2}}$)	1·2
$n = 15$ to $n = 7$	
(i) including all processes	0·8
(ii) neglecting Γ_{chem}	1·7
$n = 7$ to nuclear capture	
(i) including all processes	1·5
(ii) neglecting Stark mixing	16
(2) $(K^- p)$ atoms	
$n = 30$ ($\simeq (\mu/m)^{\frac{1}{2}}$) to nuclear capture	
(i) including Stark mixing	< 2·4
(ii) neglecting Stark mixing	$\simeq$ 30

† The estimate of the time from molecular capture till attaining the $n = 15$ state for $(\pi^- p)$ was obtained by Day (loc. cit.). All other values in this table are those calculated by Leon and Bethe.

only decays in which the μ^- meson was ejected in the backward hemisphere relative to the direction of the π^- meson track. By measuring the range of the μ^- meson, and the angle between the π and μ directions, they were able to select π^- mesons decaying from rest, or with velocity less than $v_0 \simeq 0.01c$.

If τ_π is the mean life of the π meson, n the observed number of $\pi \to \mu$ decays in the backward hemisphere, in the laboratory system, and N is the total number of π^- stopping in the chamber, then $\tau_c(v_0)$, the time from velocity v_0 to nuclear capture, is given by

$$\tau_c(v_0) = (2kn/N)\tau_\pi, \qquad (40)$$

where k, a correction factor, takes account of the fact that for pions in flight, the backward hemisphere, in the laboratory system, corresponds to less than a hemispherical solid angle in the pion rest system. The experiment has been repeated by Bierman et al.[†] and Doede et al.[‡] with better statistics, and starting with a smaller value of v_0 ($\simeq 0.006c$). Their results are shown in Table 27.5.

TABLE 27.5

Cascade times for mesons (hyperons) in liquid H_2

Stopping particle	v_0/c	Total number of stopping particles (N)	Number of events (n)	$\tau(v_0)$ ($\times 10^{-12}$ s)	Calculated time $\tau(v_0)$ ($\times 10^{-12}$ s)	
					with Stark transitions	without Stark transitions
π^-	0.006	283 000	11	2.0 ± 0.8	4.2	19.6
π^-	0.006	546 000	21	2.3 ± 0.6		
K^-	0.02		1	4	3.4	$\simeq 30$
K^-	0.02		1	4		
Σ^-	0.004	16 500	5	5	3	$\simeq 50$

Similar measurements have been made of the cascade time for (K^-p) atoms by Knop et al.[§] and by Cresti et al.[∥] They looked for the τ mode of decay of K^- mesons in liquid hydrogen. This very characteristic mode, $K^- \to \pi^-\pi^-\pi^+$, has a branching ratio of 5.57 per cent of all K^- decays. By selecting almost coplanar decays of this kind, and measuring the range and emission angles of the π mesons, it was possible to select decays when the K^- meson was moving with $v_0 < 0.02c$, or from rest.

Burnstein et al.[††] have made similar measurements for Σ^- hyperons. They selected Σ^- hyperons produced by stopping K^- mesons in the hydrogen bubble chamber, using the decay mode $\Sigma^- \to \pi^- + n$. They

[†] BIERMAN, E., TAYLOR, S., KOLLER, E. L., STAMER, P., and HUETTER, T., *Phys. Lett.* **4** (1963) 351.
[‡] DOEDE, J. H., HILDEBRAND, R. H., ISRAEL, M. H., and PYAKA, M. R., *Phys. Rev.* **129** (1963) 2808.
[§] KNOP, R., BURNSTEIN, R. A., and SNOW, C. A., *Phys. Rev. Lett.* **14** (1965) 767.
[∥] CRESTI, M., LIMENTANI, S., LORIA, A., PERUZZO, L., and SANTANGELO, R., ibid. **14** (1965) 847.
[††] BURNSTEIN, R. A., SNOW, C. A., and WHITESIDE, H., ibid. **15** (1965) 639.

selected events in which an associated n–p scatter of the neutron enabled the Σ^- momentum to be determined to within ± 5 MeV/c.

The results of all these measurements are presented in Table 27.5. For (π^-p), using the estimate of Wightman for the moderation time of 0.74×10^{-12} s from $v_0/c = 0.006$ to 5×10^{-5}, that of Day for the cascade time from molecular capture to the state $n = 15$ (1.2×10^{-12} s), and those of Leon and Bethe for the time required for cascade from $n = 15$ to nuclear capture, the following values are obtained for the time expected from a velocity $v_0/c = 0.006$ to nuclear capture:

(a) including all interaction effects of (π^-p) with H_2 4.2×10^{-12} s,

(b) neglecting chemical dissociation (process (22)) 5.1×10^{-12} s,

(c) neglecting Stark mixing 19.6×10^{-12} s,

(d) observed overall time $2.2 \pm 0.5 \times 10^{-12}$ s.

The calculated values all give a time considerably larger than that observed. The situation is worse if the moderation times given by Day are used, instead of those of Wightman. The results decisively support the theory of Day, Snow, and Sucher concerning the importance of Stark mixing in collisions between (π^-p) and H atoms. They can also be regarded as lending some support to the importance of the chemical dissociation process (22) in producing de-excitation, and the assumption of a larger cross-section than that used by Leon and Bethe (eqn (23)) would improve the agreement with the measured value. On the other hand it should be remembered that any scanning losses will reduce the estimated cascade time so that the experimental results are more likely to be low than high.

For (K^-p) atoms, where the theoretical estimates are less well founded, and the experimental observations based on poorer statistics, it can also be concluded that the results show the importance of Stark mixing collisions in liquid H_2. The same conclusion can be drawn from the results of (Σ^-p) although the theoretical estimates in this case are rudimentary (see Table 27.5).

Muonic X-rays have been observed when negative muons stop in hydrogen. Budick et al.[†] identified the Lyman alpha radiation for μ^- stopping in liquid hydrogen and showed that K_α ($2p$–$1s$) transitions accounted for a fraction 0.7 ± 0.2 of the total Lyman radiation. On the other hand Placci et al.[‡] found for gaseous hydrogen a value of 0.4 ± 0.1

[†] BUDICK, B., TORASKAR, J. R., and YAGHOOBIA, I., Phys. Lett. 34B (1971) 539.
[‡] PLACCI, A., POLACCO, E., ZAVATTINI, E., ZIOCK, K., CARBONI, C., GASTALDI, U., GORINI, G., and TORELLI, G., ibid. 32B (1970) 413.

for this ratio. This latter result is surprising since the Stark effect in hydrogen which will tend to favour transitions from higher p levels directly to the $1s$ state would be expected to be most important for liquid hydrogen. In the absence of Stark transitions the expected fraction of the total Lyman radiation in $2p$–$1s$ transitions is 0·9.

3.3. *The de-excitation time of mesic atoms in liquid helium*

De-excitation times of mesic atoms in liquid helium have been measured by the groups of Fetkovich and of Block for $(\pi^-\text{He})$†, $(K^-\text{He})$‡, and $(\Sigma^-\text{He})$§ atoms while Zaĭmidoroga *et al.*‖ have measured the de-excitation of $(\pi^-\text{He})$ in gaseous He^3 in a diffusion cloud chamber. The same method of selection of decays at very low velocity, or at rest, were used as for hydrogen. The measurements of Block based on the observation of 241 π^- decays, with a velocity $v_0 < 0·02c$, gave a moderation and cascade time of $3·2 \pm 0·2 \times 10^{-10}$ s. For $(K^-\text{He})$ the combined results of Block *et al.* and Fetkovich *et al.* give a cascade time of $(2·8 \pm 0·4) \times 10^{-10}$ s based on the observation of 79 κ^- decays at rest in the τ mode. For $(\Sigma^-\text{He})$ the preliminary measurements of Fetkovich *et al.* give a cascade time of $(2 \pm 1) \times 10^{-11}$ s. These times are two orders of magnitude larger than similar times for mesic hydrogen in liquid hydrogen.

Mesic helium atoms retain a single electron when formed, but this is soon lost in an internal Auger transition process. Thereafter they retain a single resultant positive charge. There is no process in helium analogous to the chemical dissociation process that apparently plays an important role in molecular hydrogen. The presence of the resultant positive charge prevents a close approach of mesic helium to the neighbouring helium nuclei, so that external Auger transitions and Stark mixing would be expected to be much less significant than in the liquid hydrogen case. Calculation of the influence of Stark mixing, and other polarization effects, on the cascade de-excitation time in liquid helium have been carried out by Day.†† He predicts a considerably lower de-excitation

---

† FETKOVICH, J. G., and PEWITT, F. G., *Phys. Rev. Lett.* **11** (1963) 290; BLOCK, M. M., KOPELMAN, J. B., and SUN, C. R., *Phys. Rev.* **B150** (1965) 143.

‡ BLOCK, *et al.*, loc. cit.; FETKOVICH, J. G., McKENZIE, J., RILEY, B. R., WANG, I. T., BUNNELL, K., DERRICK, M., FIELDS, T., HYMAN, L. G., and KEYES, G. J., *Phys. Rev.* **D2** (1970) 1803.

§ FETKOVICH, J. G., McKENZIE, J., RILEY, B. R., and WANG, I. T., *Bull. Am. Phys. Soc.* **15** (1970) 54.

‖ ZAĬMIDOROGA, O. A., KULYUKIN, M. M., SULYAEV, R. M., FALOMKIN, I. V., FILLIPOV, A. I., and TSUPKO-SITNIKOV, V. M., *Zh. eksp. teor. Fiz.* **51** (1967) 1646 (*Soviet Physics, JETP* **24** (1967) 1111); ZAĬMIDOROGA, O. A., SULYAEV, R. M., and TSUPKO-SITNIKOV, V. M., *Zh. eksp. teor. Fiz.* **52** (1967) 97 (*Soviet Physics, JETP* **25** (1967) 63).

†† DAY, T. B., *Nuovo Cim.* **18** (1960) 381.

time than that measured. The most extensive calculations have been made by Russell† who has been concerned particularly with the effect of trapping of the meson in a highly excited 'circular' ($l = n-1$) orbit, as suggested by Condo.‡ Electric dipole radiative transitions from such orbits are very improbable. Auger transitions giving sufficient energy to eject an electron from another helium atom need a change Δl of several units and are therefore also improbable. For π^- or K^- mesic He the trapping time in this state may be comparable with the mean lifetime against decay of the particle itself.

If, however, the observed long cascade times in liquid He are to be interpreted in terms of trapping in circular orbits of high excitation, the influence of Stark mixing must be considerably smaller than estimated by Day. Placci et al.§ have measured the intensity of K series muonic X-rays in liquid helium and have found that the K_α radiation accounts for 0.62 ± 0.08 of all the K series radiation. To explain this value they had to assume Stark effect mixing which is so strong as to lead to all the substates of a highly excited level n being statistically populated. Weak Stark effect mixing would be expected to produce a much larger proportion (> 0.9) of K_α radiation. The situation with regard to the importance of Stark transitions in the de-excitation of mesic atoms remains unclear. On the other hand Placci et al. observed X-rays from the two-photon transition $2s$–$1s$ in He and estimated that 3.4 ± 0.7 per cent of stopped μ^- were trapped in the $2s$ metastable state of (μ^-He) whence they underwent two-photon transitions to the ground state with a mean life of 1.8 ± 0.4 μs. The observation of double photon decay with this lifetime indicates that for the low-lying $2s$ metastable state there is no depopulation due to Stark effect or external Auger effect at a helium pressure of 7 atmospheres.

3.4. *De-excitation of heavier mesic atoms*

The de-excitation processes of more complex mesic atoms involve mainly internal Auger transitions for states of high n value, while the lowest energy transitions are mainly radiative. The dominant Auger transitions have $\Delta l = -1$, and Δn the minimum, consistent with the availability of sufficient energy to eject the electron. The details of the early stages of the cascade are very complex, and most detailed theoretical treatments have been concerned with the stages after the meson has

---

† RUSSELL, J. E., *Phys. Rev.* **188** (1969) 187; **A1** (1970) 721, 735, 742; **A2** (1970) 2284; *Phys. Rev. Lett.* **23** (1969) 63. ‡ CONDO, G. T., *Phys. Lett.* **9** (1964) 65.
§ PLACCI, A., POLACCO, E., ZAVATTINI, E., ZIOCK, K., CARBONI, G., GASTALDI, U., GORINI, G., NERI, G., and TORELLI, G., *Nuovo Cim.* **1A** (1971) 445.

reached an orbit with $n \simeq 15$, of radius approximately that of the K electron orbits for π and μ mesons. By this time, there have been so many transitions that features characteristic of the initial capture event, such as the initial distribution of l values, have been forgotten. Indeed, irrespective of the initial distribution, the selection rules for both Auger and radiative transitions tend to drive the meson toward circular orbits ($l = n-1$). Also, unlike the situation for mesic hydrogen atoms, collision processes do not apparently play very much part in the de-excitation process. For these reasons, although a considerable amount of experimental information is now available concerning the emission of mesic X-rays and Auger electrons, during the later stages of the cascade, these are of importance from the point of view of atomic and nuclear structure studies rather than for any light they throw on collision processes involving mesic atoms. Other phenomena on which much experimental information is now available are the dependence of the relative meson capture probability on the relative atomic numbers of the atoms in a compound and the transfer of mesons bound to hydrogen atoms in a compound to atoms of higher Z value.†

3.5. *The polarization of muons decaying at rest*

There is one type of measurement that has been made with muonic atoms that does reflect details of the initial capture process. This is the measurement of the polarization of muons, decaying after capture into muonic atoms. The depolarization of muons in the capture process as calculated by Mann and Rose‡ has been discussed in § 2.24. Several authors have discussed, also, the further depolarization occurring during the subsequent cascade process down to the 1s state of the muonic atom. This discussion depends on the detailed l distribution, in the initial capture state, and this depends, in turn, on the assumptions made in the calculation of the capture cross-section. Mann and Rose used the plane-wave Born approximation for these calculations, which is known to give poor results.§ They considered depolarization in the cascade in an atom with zero nuclear spin. In this case spin–orbit interaction plays the dominant role in producing depolarization. Depolarization is not very significant during the early stages of the cascade where most of the de-excitation is due to Auger transitions, and the level widths are larger than the fine structure splitting. For the lower levels, however, the level

---

† These features are summarized in BURHOP, E. H. S., *High energy physics* (ed. D. R. Bates), vol. 3. Academic Press, New York (1969), p. 109.
‡ loc. cit., p. 3286. § DE BORDE, A. H., *Proc. phys. Soc.* **67A** (1954) 57.

width, due mainly to radiative transitions, is small compared with the fine structure splitting, and under these circumstances appreciable depolarization will occur. The muon in muonic atoms generally spends the greater part of its time in the ground $1s$ state. Spin–orbit effects are absent in this state but appreciable depolarization arises through the interaction of the mean muon magnetic moment with the magnetic field of the electronic shells. For carbon, for which detailed calculations were carried out, Mann and Rose estimated that, as a result of the cascade, the muon depolarization would be reduced from the values given in eqn (20), at the capture stage, to a value $P = 0.133$.

In cases where nuclei have a non-zero spin, hyperfine structure effects become important in producing depolarization, when the hyperfine splitting becomes comparable with the level width. Bukhvostov and Shmushkevich† have made calculations of the depolarization for spin-$\frac{1}{2}$ nuclei, and Bukhvostov‡ for spin-1 nuclei and for nuclei with higher spins up to $\frac{11}{2}$. In these calculations the l value distribution, in the initial state, was taken to be statistical. For spin-$\frac{1}{2}$ nuclei, hyperfine structure effects may reduce the polarization to between $\frac{1}{2}$ and $\frac{1}{3}$ of the value for spin-zero nuclei. The residual muon polarization decreases with increasing nuclear spin but depends on the ratio of the hyperfine splitting to the muonic atom level width and on the sign of the nuclear magnetic moment.

Measurements of the polarization of muons that come to rest, and decay from a muonic atom orbit, were first made by Garwin et al.§ in the experiment that established parity non-conservation in muon decay. A typical arrangement due to Garwin et al.‖ is shown in Fig. 27.6. The muon beam from the cyclotron was led into a space in which a steady magnetic field H was established, and brought to rest in a target, after moderation in a copper block. The coincidence $123\overline{45}$ in the five scintillation counters signalled a stopping muon. The polarized muon precessed about the direction of the field with circular frequency ω_H. The delayed coincidence $23\overline{45}$ or $\overline{23}45$ which signalled the emission of a decay electron was, therefore, modulated with this circular frequency ω_H due to the asymmetry of the electron decay direction with respect to the spin direction. One would then expect the counting rate $N(t)$, into solid

† BUKHVOSTOV, A. P. and SHMUSHKEVICH, I. M., *Zh. eksp. teor. Fiz.* **41** (1962) 1895 (*Soviet Physics, JETP* **14** (1962) 1347).
‡ BUKHVOSTOV, A. P., *Yadern. Fiz.* **4** (1966) 83 (*Soviet J. nucl. Phys.* **4** (1967) 59); *Yadern. Fiz.* **9** (1969) 107 (*Soviet J. nucl. Phys.* **9** (1969) 65).
§ GARWIN, R. L., LEDERMAN, L. M., and WEINRICH, M., *Phys. Rev.* **105** (1957) 1415.
‖ GARWIN, R. L., HUTCHINSON, D. P., PENMAN, S., and SHAPIRO, G., *Phys. Rev.* **118** (1960) 271.

angle $d\Omega$, at elapsed time t after the muon comes to rest, to be given by
$$N(t) = (1/4\pi\tau)e^{-t/\tau}\{1+a\cos(\omega_H t - \theta)\}\,d\Omega, \tag{41}$$
where θ is the angle between the direction of the electron momentum and the initial muon spin, and a is an asymmetry factor. To record the whole distribution (41), over many cycles of the precession, would require

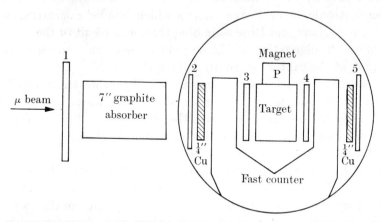

FIG. 27.6. Experimental arrangement for studying the polarization at the instant of decay of μ^- captured in muonic atoms. Muons coming to rest are identified by a $123\overline{45}$ prompt coincidence. The decay electron gives a $\overline{23}45$ delayed coincidence (forward emitted electron) or a $23\overline{45}$ delayed coincidence (backward emitted electron). P is a nuclear magnetic resonance probe.

a very large memory capacity of the pulse-height analyser and extreme linearity of the time to pulse-height converter. Therefore, a reference oscillator of fixed frequency (ω_0) was employed, and only the difference between the elapsed time, and the nearest number of cycles of the oscillator, recorded. Electrons for which the elapsed time after the muon stopped differed by a whole number of cycles of the oscillator were stored in the same channel of the pulse-height analyser. If then, $\omega_0 = \omega_H$, the contributions from each of the cycles reinforce each other, so that the number of counts in the various channels is of the form $\{1+a\cos\psi\}$, with $\psi = (\omega_0 t - 2\pi n)$, n integral. If the two frequencies differ, the count received in a given channel varies with ψ. If the gate accepting pulses is open for a time T, such that
$$T = \frac{2\pi n}{|\omega_0 - \omega_H|},$$
the signal would be the same as for unpolarized muons. As ω_H is varied around the reference frequency, ω_0, the number of counts received in the channel varies periodically with decreasing amplitude. An analysis

of the amplitude of the fluctuations then enables the asymmetry factor a of eqn (41) to be obtained.

Fig. 27.7 shows a typical distribution of the counting rate observed in a given channel as a function of ω_H as determined by the magnetizing

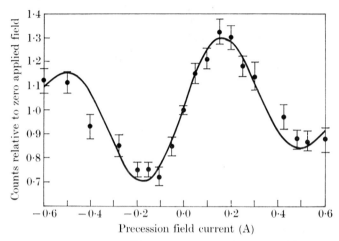

FIG. 27.7. Variation of gated $\overline{2345}$ counting rate with magnetizing current of the precession field magnet. The solid curve is computed from an assumed electron angular distribution $(1-\frac{1}{3}\cos\theta)$ with counter and gate-width resolution folded (experiment of Garwin et al.).

current of the precession magnet. The solid curve is that to be expected if the asymmetry factor $a = \frac{1}{3}$. The amplitude a, derived from such a distribution, measures the asymmetry of the emission which is expected to be related to the polarization P, at decay, by

$$a \simeq \tfrac{1}{3}P.$$

Table 27.6 given by Ignatenko et al.† lists measured values of $(-a)$ in a number of materials. The result for carbon is in good agreement with the value expected (-0.044) from the calculations of Mann and Rose, thus giving support for the theoretical value they obtained for depolarization in the capture process (eqn (20)). For each of the substances shown in the table the predominantly abundant isotope has zero nuclear spin.‡ For nuclei with non-zero spin much smaller asymmetries are observed in agreement with the calculated estimates.§

† IGNATENKO, A. E., EGOROV, L. B., KHALUPA, B., and CHULTEM, D., Zh. eksp. teor. Fiz. **35** (1958) 1131 (Soviet Physics, JETP **8** (1959) 792).

‡ See also EVSEEV, V. S., ROGANOV, V. S., CHERNAGOROVA, V. A., MYASISHCHEVA, G. G., and OBUKHOV, YU. V., Yadern. Fiz. **8** (1968) 741 (Soviet J. nucl. Phys. **8** (1969) 431).

§ BABAYEV, A. I., EVSEEV, V. S., MYASISHCHEVA, G. G., OBUKHOV, YU. V., ROGANOV, V. S., and CHERNAGOROVA, V. A., Yadern. Fiz. **10** (1969) 964 (Soviet J. nucl. Phys. **10**

3.6. *Exchange collisions of* (μ^-p) *in hydrogen*

In (μ^-p) depolarizing effects due to spin–orbit interaction would be expected during the cascade process. In the ground state, however, this effect is absent, and the absence of orbital electrons precludes muon-electron depolarization effects. It might have been expected, therefore,

TABLE 27.6

Measured asymmetry (a) in decay of stopping μ^-

Stopping substance	$-a$
C	$0 \cdot 040 \pm 0 \cdot 005$
$H_2O(O)$	$0 \cdot 043 \pm 0 \cdot 005$
Mg	$0 \cdot 058 \pm 0 \cdot 008$
S	$0 \cdot 052 \pm 0 \cdot 006$
Zn	$0 \cdot 056 \pm 0 \cdot 011$
Cd	$0 \cdot 055 \pm 0 \cdot 012$
Pb	$0 \cdot 054 \pm 0 \cdot 013$

that the polarization of μ^- in (μ^-p) atoms would be larger than for other atoms. On the contrary Ignatenko *et al.*[†] found a value $0 \cdot 029 \pm 0 \cdot 029$ (i.e. consistent with zero) for the polarization, while the measurements of Klem[‡] of asymmetry in the decay of stopped μ^-, in a hydrogen bubble chamber, indicated a polarization of $-0 \cdot 056 \pm 0 \cdot 032$. For μ^+, in contrast the observed polarization was $-0 \cdot 618 \pm 0 \cdot 132$.

The observation can be understood by noting that there are two hyperfine s states of (μ^-p) with $F = 1, 0$, respectively, the triplet state lying an amount

$$\Delta E = \frac{16\pi}{3}\beta_\mu \beta_N g_P |\psi(0)|^2 \quad (= 0 \cdot 18 \text{ eV}) \tag{42}$$

above the singlet state, where β_μ, β_N are respectively the muonic and nucleonic Bohr magnetons, $g_P = 2 \times 2 \cdot 79$ is the gyromagnetic ratio of the proton, and $\psi(0)$ the μ^- wave-function at the position of the proton. Collisions can occur between (μ^-p) and protons in which the μ^- transfers from one proton to the other. Such collisions may involve change from an $F = 1$ to an $F = 0$ (μ^-p) atom, in the exothermic process

$$(\mu^-p)_{F=1} + p \rightarrow p + (\mu^-p)_{F=0}, \tag{43}$$

if the two protons have opposite spin orientations.

(1970) 554). FAVART, D., BROUILLARD, F., GRENACS, L., IGO-KEMENES, P., LIPNIK, P., and MACQ, P. C., *Phys. Rev. Lett.* **25** (1970) 1348.
    [†] IGNATENKO, A. E., EGOROV, L. A., KHALUPA, B., and CHULTEM, D., *Zh. eksp. teor. Fiz.* **35** (1958) 894 (*Soviet Physics, JETP* **8** (1959) 621).
    [‡] KLEM, R. D., *Nuovo Cim.* **48** (1967) 743.

For mesons of low thermal velocity the reverse (endothermic) process cannot occur, so that, after a time, all the (μ^-p) atoms are in the unpolarized $F = 0$ state.

Gershteĭn† obtained for the cross-section of process (43) (see § 5, p. 3336)

$$Q_{1\to 0} = 4\pi \frac{k_0}{k} \frac{3(c_g^{(p)} - c_u^{(p)})^2}{16 + k_0^2(3c_g^{(p)} + c_u^{(p)})^2}, \tag{44}$$

where the $c_{g,u}^{(p)}$'s are the scattering lengths of the (μ^-p) atom by the proton, the suffixes referring to the cases where the muon wave-function is symmetrical with respect to interchange of the proton coordinates (g), or antisymmetrical (u). $k\hbar$, $k_0\hbar$ are, respectively, the initial and final relative momenta. If k corresponds to thermal energy,

$$k_0 \hbar \simeq \{M(\Delta E)\}^{\frac{1}{2}}, \tag{45}$$

where M is the proton mass. He estimated the values

$$c_g^{(p)} = -17\cdot 3 a_\mu; \quad c_u^{(p)} = 5\cdot 25 a_\mu,$$

where a_μ is the radius of the lowest Bohr orbit in (μ^-p).

We return to the discussion of Gershteĭn's calculation in § 5, where we shall see that the large negative value of $c_g^{(p)}$ arises from the existence of a virtual vibrational ($v = 1$) level of the (pμp) molecule with very small positive energy. Putting these values of $c_g^{(p)}$, $c_u^{(p)}$ into eqn (44) gives

$$Q_{1\to 0} \simeq 7\cdot 8 \times 10^{-19} \text{ cm}^2.$$

This gives a transition rate of $\sim 5 \times 10^9$ s^{-1} between the two states in liquid hydrogen, 10^3 times larger than the muon decay rate, so that, at the time of decay, the depolarization of (μ^-p) should be almost complete, as is observed.

Alberigi-Quaranta et al.‡ have derived values of $Q_{1\to 0}$ from studies of the diffusion of (μ^-p) atoms in gaseous hydrogen (see § 4.4.3), in good agreement with this estimate due to Gershteĭn. Gershteĭn§ has also discussed the transition between hyperfine levels in mesic deuterium atoms. Since the deuteron has unit spin the (μ^-d) hyperfine states have $F = \frac{3}{2}, \frac{1}{2}$, respectively, and the cross-section of the exchange process with spin change is

$$Q_{\frac{3}{2}\to\frac{1}{2}} = \tfrac{1}{3}\pi(c_g^{(d)} - c_u^{(d)}) k_0/k, \tag{46}$$

with the scattering lengths $c_{g,u}^{(d)}$ defined in an exactly analogous way to

---

† GERSHTEĬN, S. S., *Zh. eksp. teor. Fiz.* **34** (1958) 463 (*Soviet Physics, JETP* **7** (1958) 318); *Zh. eksp. teor. Fiz.* **36** (1959) 1309 (*Soviet Physics, JETP* **9** (1959) 927).

‡ ALBERIGI-QUARANTA, A., BERTIN, A., MATONE, G., PALMONARI, F., PLACCI, A., DALPIAZ, P., TORELLI, G., and ZAVATTINI, E., *Nuovo Cim.* **47B** (1967) 72.

§ GERSHTEĬN, S. S., *Zh. eksp. teor. Fiz.* **40** (1961) 698 (*Soviet Physics, JETP* **13** (1961) 488).

the (μ^-p), p collision case. Now, however,

$$c_g^{(d)} \simeq 6\cdot 67 a_\mu; \qquad c_u^{(d)} \simeq 5\cdot 73 a_\mu.$$

There is no resonance effect, and $c_g^{(d)}$, $c_u^{(d)}$ are of the same sign, and nearly equal, giving

$$Q_{\frac{3}{2}\to\frac{1}{2}} \simeq 0\cdot 3\pi a_\mu^2 k_0/k. \qquad (47)$$

The hyperfine separation between the two states, $\Delta E^{(D)}$, is 0·046 eV and if k corresponds to thermal energy

$$k_0 \hbar \simeq \{2M(\Delta E^D)\}^{\frac{1}{2}}, \qquad (48)$$

eqn (47) then gives, for the transition rate between the two hyperfine states in liquid deuterium, 6×10^6 s^{-1}, very much smaller than for the (μ^-p) case. Doede† has estimated this transition rate as lying between $(7\cdot 9\pm 2\cdot 0)\times 10^5$ s^{-1} and $(1\cdot 35\pm 0\cdot 5)\times 10^6$ s^{-1}, from his studies of μ-catalysed fusion in a deuterium bubble chamber. On the other hand Klem‡ also observed practically complete depolarization of μ^- in liquid deuterium, a result difficult to understand in terms of (μ^-d) exchange collisions.

4. Cross-sections and collision rates of (μ^-p) and (μ^-d) atoms obtained from the muon-catalysed fusion process, and related phenomena

4.1. *The observation of muon-catalysed fusion*

Most information concerning collision processes of muonic hydrogen has come from the investigation of the nuclear fusion processes of protons and deuterons which can be catalysed by muons.

The phenomenon of muon-catalysed nuclear fusion was first observed by Alvarez *et al.*§ when studying negative muons stopping, in a hydrogen bubble chamber. They observed a considerable number of events in which the μ^- meson was apparently re-emitted from the muonic atom or molecule in which it was bound. In each case the re-emitted particle produced a typical muon track of the unique range 1·7 cm, corresponding to a particle re-emitted with an energy of 5·3 MeV. They found that the proportion of stopping tracks showing the phenomenon depended on the deuterium content of the liquid hydrogen. In a sample of nearly 6000 muons stopping in liquid hydrogen containing 0·3 per cent deuterium, they observed 139 examples of μ^- re-emission, and in three of these cases the μ^- after coming to rest was re-emitted a second time with the same

† DOEDE, J. H., *Phys. Rev.* **132** (1963) 1782.
‡ KLEM, R. D., loc. cit.
§ ALVAREZ, L. W., BRADNER, H., CRAWFORD, F. S., CRAWFORD, J. A., FALK-VAIRANT, P., GOOD, N. L., GOW, J. D., ROSENFELD, A. H., SOLMITZ, F. T., STEVENSON, M. L., TICHO, H. K., and TRIPP, R. D., *Phys. Rev.* **105** (1957) 1127.

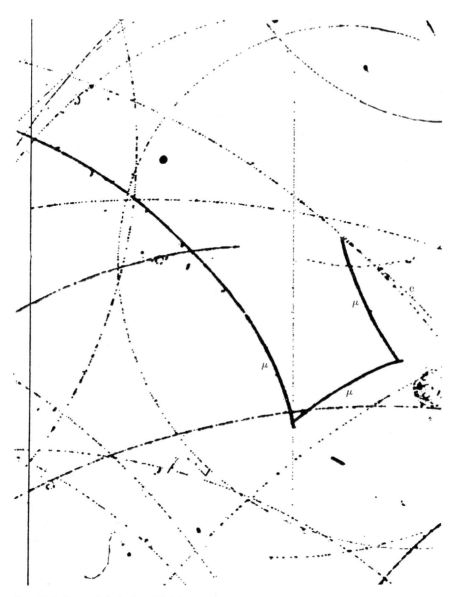

FIG. 27.8. Re-emission of μ^- following catalysis of p–d interaction with muonic deuterium in liquid hydrogen observed by Doede. The picture shows an example of two successive fusion events leading to μ^- re-emission. Note the gaps between the stopping μ^- and the rejuvenated muons in each case. (Photograph by courtesy of Professor R. H. Hildebrand and Dr. J. H. Doede.)

energy of 5·3 MeV. A beautiful example of this type of event is shown in Fig. 27.8, which was obtained in a later investigation by Doede.† Alvarez et al. noted that the energy of 5·3 MeV is close to that of the γ-ray produced in the nuclear radiative capture process

$$p+d \to He^3+\gamma, \qquad (49)$$

and he proposed an interpretation in terms of the nuclear reaction (49), between the p and d in the muonic molecular ion $(p\mu^-d)^+$, followed by the internal conversion of the photon with ejection of the muon. In such a molecule the separation between the nuclei is $\sim 5 \times 10^{-11}$ cm, smaller than that for normal H_2^+ by a factor $\sim (m/\mu)$. The internuclear separation is sufficiently small to give a finite probability of penetration of the potential barrier separating them, thus permitting the fusion process (49) to take place. The observation of single and double μ^- re-emission suggests a rate for the fusion process comparable with, or greater than, the muon decay rate $(4 \cdot 55 \times 10^5 \text{ s}^{-1})$.

Therefore, the major process involved can be written

$$(p\mu^-d) \to He^3 + \mu^-. \qquad (50)$$

A process such as this had been suggested by Frank‡ long before the observations of Alvarez et al., and detailed estimates of the rates of the various processes involved in the formation of the ion and the subsequent nuclear reaction had been made by Zel'dovich.§ Support for this interpretation of the observations was provided by an experiment of Ashmore et al.,‖ who observed γ-ray photons of energy in the range 3–7·5 MeV emitted from a liquid hydrogen target in which μ^- mesons were stopped. These were attributed to the alternative process

$$(p\mu^-d) \to He^3 + \mu^- + \gamma. \qquad (51)$$

Fig. 27.9 shows the experimental arrangement employed. Muons stopping in the hydrogen target were detected by the coincidence $123\bar{4}$ in the scintillation counters indicated. Gamma rays emitted subsequently were observed by means of the delayed coincidence $\bar{5}6$, where 6 was a NaI counter which also provided an approximate estimate of the γ-ray energy.

4.2. *Experimental study of the process*

4.2.1. *The bubble chamber method.* The most important measurement is that of the conversion muon yield Y_μ, i.e. the proportion of stopping

---

† DOEDE, J. H., *Phys. Rev.* **132** (1963) 1782.
‡ FRANK, F. C., *Nature, Lond.* **160** (1947) 525.
§ ZEL'DOVICH, YA. B., *Dokl. Akad. Nauk SSSR* **95** (1954) 493.
‖ ASHMORE, A., NORDHAGEN, B., STRAUCH, K., and TOWNES, B. M., *Proc. phys. Soc.* **71** (1958) 161.

μ^- leading to re-emission. Studies can be made of Y_μ for different deuterium concentrations. The effect of other impurities on Y_μ can also be studied.

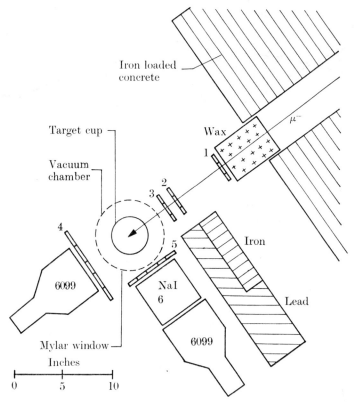

Fig. 27.9. Apparatus used by Ashmore et al. (loc. cit.) for studying photon emission following muonic catalysis of the p–d nuclear interaction.

In their original experiments Alvarez et al.† also found two examples of the muon catalysed d–d reaction, namely,

$$(d\mu^- d) \to t + p + \mu^-. \qquad (52)$$

This process has been studied directly in the deuterium bubble chamber, and the yield measured.‡

In this case the μ^- is not emitted with sufficient energy to enable it to produce a visible track in the bubble chamber, since no γ-ray internal conversion process is involved. The reaction can be detected by the

† loc. cit.
‡ FETKOVICH, J. C., FIELDS, T. H., and McILWAIN, R. L., Phys. Rev. **118** (1960) 319; DOEDE, J. H., loc. cit.

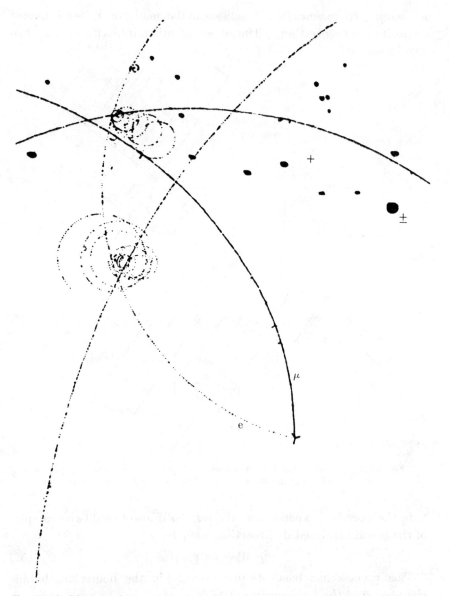

Fig. 27.10. Photograph of μ^- catalysed fusion in a deuterium bubble chamber obtained by Doede. Three successive fusions of the type $(d\mu d)^+ \to p+t+\mu^-$ are shown. The three short proton tracks are seen as well as the electron from the μ^- when it finally decays. The μ^- does not receive any appreciable energy in fusion processes of this type.

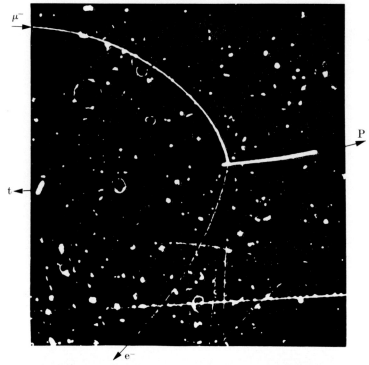

Fig. 27.11. Muonic catalysis of the d–d interaction in a diffusion chamber containing deuterium in the process $(d\mu^-d) \to t+p+\mu^-$. (Photograph by courtesy of Professor Dzhelepov.)

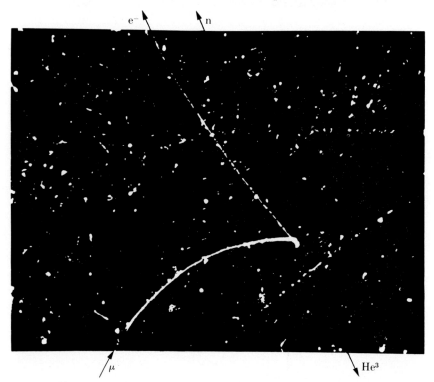

Fig. 27.12. Example of process $(d\mu^-d) \to He^3+n+\mu^-$ in a diffusion chamber containing deuterium. The neutron track is not visible (by courtesy of Professor Dzhelepov).

observation of the track due to the proton of energy 3·04 MeV (range ∼ 0·8 mm) together with the electron track from the decay of the μ^-, both coming from the end point of the primary μ^- track. The triton has too little energy to produce a visible track in the bubble chamber. The μ^- may be recycled several times, giving rise to a catalysis process each time before it decays. Fig. 27.10 shows an example obtained by Doede of three successive fusions involving process (52).

The alternative process

$$(d\mu^-d) \to He^3 + n + \mu^- \qquad (53)$$

would also be expected with equal probability but the kinetic energy of He^3 is too small to give a visible track in the deuterium bubble chamber.

A careful study of the details of the events in a hydrogen bubble chamber in which the muon was re-emitted disclosed a distinct gap of mean length 0·6 mm between the end of the primary muon track and the beginning of the track of the re-emitted muon. The mean gap length depends on the deuterium concentration and for a concentration of 1 per cent, or more, in liquid hydrogen, no gap was observed.† The distribution of gap sizes in liquid hydrogen has been studied by Schiff.‡ These gaps represent the migration distance of (μ^-d) atoms before formation of the molecule $(p\mu^-d)$. We shall see that from such measurements scattering cross-sections for low energy (μ^-d) atoms in hydrogen can be derived.

4.2.2. *The diffusion chamber method.* Muon catalysed fusion processes both in hydrogen and in deuterium have been studied in the diffusion chamber by Dzhelepov et al.§ The process (53) has been observed by Dzhelepov et al.,‖ since the track of the He^3 is long enough to be visible in the deuterium diffusion chamber. Figs. 27.11 and 27.12 show examples of the reactions (52) and (53) in the diffusion chamber. The recoil t and He^3 tracks are clearly seen. Owing to the smaller density, the proportion of muons that become bound in $(d\mu^-d)$ or $(p\mu^-d)$ molecules is much smaller, so that the yield is expected to be smaller than in liquid hydrogen or liquid deuterium. Allowance is made for this difference in quoting reaction rates derived from such experiments (see, e.g. Table 27.7). It is customary to express them as the rates that would be observed if the

---

† ALVAREZ et al., loc. cit., p. 3306.
‡ SCHIFF, M., *Nuovo Cim.* **22** (1961) 66.
§ DZHELEPOV, V. P., ERMOLOV, P. F., MOSKALEV, V. I., and FIL'CHENKOV, V. V., *Zh. eksp. teor. Fiz.* **50** (1966) 1235 (*Soviet Physics, JETP* **23** (1966) 820).
‖ DZHELEPOV, V. P., ERMOLOV, P. F., KATYSHEV, YU. V., MOSKALEV, V. I., FIL'CHENKOV, V. V., and FRIML, M., *Zh. eksp. teor. Fiz.* **46** (1964) 2042 (*Soviet Physics, JETP* **19** (1964) 1376); *Nuovo Cim.* **33** (1964) 40.

density were increased to that of liquid hydrogen (or deuterium), supposing the rate to scale linearly with the density.

The distribution of gap sizes corresponding to the migration of (μ^-d) atoms has also been studied in the diffusion chamber by Dzhelepov et al.†

4.2.3. *Study of γ-ray emission in muon-catalysed fusion.* The advantage of studying the emission of gamma rays from the muon-catalysed fusion process using counter methods is that, in addition to the γ-ray yield per stopped muon (Y_γ) it is possible to study, also, the time distribution of the γ-rays. Earlier investigations‡ used a time-to-pulse-height converter to obtain the time between the stopping of the μ^- and the emission of the γ-ray. Bleser et al.§ used a digitron device developed by Swanson‖ in which time intervals were measured by scaling the number of sine waves from a CW crystal-controlled oscillator, gated by start and stop pulses produced by the stopping of the μ^- and the emission of the γ-ray, respectively. In this way, an effective time-resolution of 0·03 μs was obtained.

The γ-ray yield, and emission time distribution, were measured for various deuterium concentrations and the effect of introducing various concentrations of neon was also studied by Schiff,†† Bleser et al.,§ and Conforto et al.‡‡ Such measurements yield information about formation rates of the μ-mesic molecules (pμ^-p), (pμ^-d), (dμ^-d), as well as transfer rates of the muon from (μ^-p) or (μ^-d) to (μ^-Ne) in a process of type (66 a) (see §§ 4.3.3, 4.4).

4.3. *Derivation of cross-sections for basic processes important in muon catalysed fusion*

4.3.1. *Description of the basic processes.* We suppose the hydrogen contains a small fractional concentration, C_D, of deuterium, and there are a total of N atoms cm^{-3}. The (μ^-p) mesic atom, formed when the muon comes to rest, is a compact neutral structure, radius $\sim 2·5 \times 10^{-11}$ cm, and able to get very close to the nuclei of the hydrogen atoms, with

---

† DZHELEPOV, V. P., ERMOLOV, P. F., MOSKALEV, V. I., FIL'CHENKOV, V. V., and FRIML, M., *Zh. eksp. teor. Fiz.* **47** (1964) 1234 (*Soviet Physics, JETP* **20** (1965) 841).

‡ BLESER, E. J., LEDERMAN, L. M., ROSEN, J. L., ROTHBERG, J. E., and ZAVATTINI, E., *Phys. Rev. Lett.* **8** (1962) 128; CONFORTO, G., FOCARDI, S., RUBBIA, C., and ZAVATTINI, E., ibid. **9** (1962) 22.

§ BLESER, E. J., ANDERSON, E. W., LEDERMAN, L. M., MEYER, S. L., ROSEN, J. L., ROTHBERG, J. E., and WANG, I. T., *Phys. Rev.* **132** (1963) 2679.

‖ SWANSON, R. A., *Rev. scient. Instrum.* **31** (1960) 149.

†† loc. cit., p. 3310.

‡‡ CONFORTO, G., RUBBIA, C., and ZAVATTINI, E., *Nuovo Cim.* **33** (1964) 1002.

which it can interact to form the mumolecular ion $(p\mu^-p)^+$ in the process
$$(\mu^-p)+H \to (p\mu^-p)^+ + e. \tag{54}$$

We write for the rate of this process $C_H \nu_{pp}^{(M)}$, where $\nu_{pp}^{(M)}$ is connected with the cross-section $Q_{pp}^{(M)}$ by $\nu_{pp}^{(M)} = Q_{pp}^{(M)} N v$, v being the velocity of the (μ^-p), and C_H the fractional concentration of hydrogen atoms, so that in the absence of other impurities $C_H = 1 - C_D$. Since He^2 has no bound state this molecule does not lead to nuclear fusion. The μ^- is trapped until it decays or, in a small proportion of cases, undergoes capture by one of the protons in the process
$$p + \mu^- \to n + \nu. \tag{55}$$

The (μ^-p) may, however, undergo a charge exchange collision in the process
$$(\mu^-p) + D \to (\mu^-d) + H. \tag{56}$$

We write for the rate of this process $C_D \nu_{pd}^{(ex)}$. This rate is evidently considerably greater than that of process (54), even for very small deuterium concentrations, so that a significant proportion of the μ^- form (μ^-d). Owing to the larger reduced mass of the muon in (μ^-d), than in (μ^-p), the binding energy of the $1s$ state of (μ^-d) (2664 eV) is larger than that of (μ^-p) (2530 eV), so that the process (56) is strongly exothermic and the (μ^-d) atom is emitted from the reaction with a kinetic energy of about 43 eV. The (μ^-d) atom diffuses through the stopping medium until it is again reduced to thermal energy, thus giving rise to the observed gaps between the stopping and re-emitted muons. The (μ^-d) atom can then interact with a hydrogen atom in the process
$$(\mu^-d) + H \to (p\mu^-d)^+ + e. \tag{57}$$

We write $C_H \nu_{pd}^{(M)}$ for the rate of this process. Nuclear fusion can then occur either through γ-ray emission in the process
$$(p\mu^-d) \to (\mu^-He^3) + \gamma,$$
at an intrinsic rate ν_γ, or, through μ^- re-emission in the process
$$(p\mu^-d) \to He^3 + \mu^-,$$
at an intrinsic rate ν_μ.

All these reactions take place in competition with the muon decay at an intrinsic rate ν_0 ($= 4\cdot 55 \times 10^5$ s^{-1}). Fig. 27.13 (a) illustrates these processes and shows their corresponding rates. The processes occurring when μ^- are stopped in liquid deuterium are similarly illustrated in Fig. 27.13 (b).

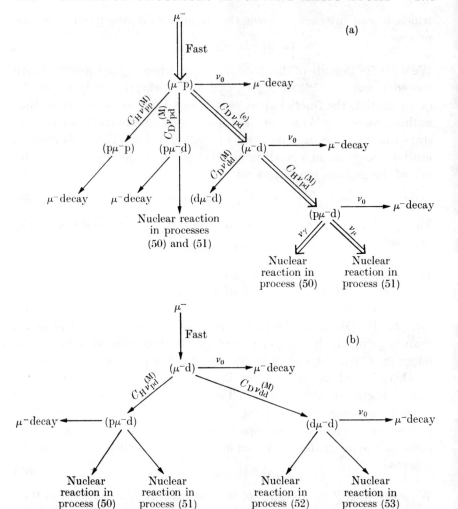

FIG. 27.13. Illustrating the competing processes taking place following μ^- capture in liquid hydrogen and liquid deuterium.

(a) Capture in liquid hydrogen with concentration C_H of hydrogen and containing a small concentration C_D of deuterium. The major process chain leading to μ^--catalysed nuclear fusion is shown by the double arrow. The other chain leading to nuclear fusion is much less important since $\nu_{pd}^{(M)} \ll \nu_{pd}^{(e)}$.

(b) Capture in liquid deuterium of concentration C_D mixed with a small concentration C_H of hydrogen.

4.3.2. *The yield of γ-rays and re-emitted muons.* The proportion Y_μ of stopping μ^- leading to re-emission in process (50) can be written (assuming $C_d \ll 1$)

$$Y_\mu = \left(\frac{C_D \nu_{pd}^{(ex)}}{C_D \nu_{pd}^{(ex)} + \nu_{pp}^{(M)} + \nu_0}\right)\left(\frac{\nu_{pd}^{(M)}}{\nu_{pd}^{(M)} + \nu_0}\right)\left(\frac{\nu_\mu}{\nu_\mu + \nu_\gamma + \nu_0}\right), \qquad (58)$$

while the γ-ray yield is
$$Y_\gamma = \frac{\nu_\gamma}{\nu_\mu} Y_\mu. \tag{59}$$

Eqn (58) can be written
$$\frac{1}{Y_\mu} = A + \frac{B}{C_\mathrm{D}}, \tag{60}$$

with
$$\frac{B}{A} = \frac{\nu_{\mathrm{pp}}^{(\mathrm{M})} + \nu_0}{\nu_{\mathrm{pd}}^{(\mathrm{ex})}}. \tag{61}$$

Measurements of Schiff† of the conversion muon yield at different deuterium concentrations gave
$$\frac{\nu_{\mathrm{pp}}^{(\mathrm{M})} + \nu_0}{\nu_{\mathrm{pd}}^{(\mathrm{ex})}} = 1\cdot 12 \left.\begin{matrix}+0\cdot 77\\-0\cdot 46\end{matrix}\right\} \times 10^{-4}. \tag{62}$$

The small value of B/A implies that Y_μ varies appreciably with C_D only at very small concentrations. The value of Y_μ saturates at $0\cdot 0264 \pm 0\cdot 0035$, and this value is already reached at a deuterium concentration of 0·3 per cent.

For the dependence of the γ-ray yield (Y_γ) on deuterium concentration, Fig. 27.14 shows a plot of $10^6/n_\gamma$ against $1/C_\mathrm{D}$, obtained by Bleser et al.‡ where n_γ is the observed γ-ray flux. A least squares fit to this curve (shown in the figure) gave
$$B/A = \frac{\nu_{\mathrm{pp}}^{(\mathrm{M})} + \nu_0}{\nu_{\mathrm{pd}}^{(\mathrm{ex})}} = 1\cdot 59 \pm 0\cdot 05 \times 10^{-4}, \tag{63}$$

which is not inconsistent with eqn (62).

4.3.3. *Time distribution of γ-ray emission.* The differential equations giving the rate of γ-ray emission at time t after the muon is stopped, are

$$\frac{\mathrm{d}n_\gamma}{\mathrm{d}t} = \nu_\gamma N_{(\mathrm{p}\mu\mathrm{d})}, \tag{64a}$$

$$\frac{\mathrm{d}N_{(\mathrm{p}\mu\mathrm{d})}}{\mathrm{d}t} = -(\nu_\gamma + \nu_\mu + \nu_0) N_{(\mathrm{p}\mu\mathrm{d})} + \nu_{\mathrm{pd}}^{(\mathrm{M})} N_{(\mu\mathrm{d})}, \tag{64b}$$

$$\frac{\mathrm{d}N_{(\mu\mathrm{d})}}{\mathrm{d}t} = -(\nu_{\mathrm{pd}}^{(\mathrm{M})} + \nu_0) N_{(\mu\mathrm{d})} + C_\mathrm{D} \nu_{\mathrm{pd}}^{(\mathrm{ex})} N_{(\mu\mathrm{p})}, \tag{64c}$$

$$\frac{\mathrm{d}N_{(\mu\mathrm{p})}}{\mathrm{d}t} = -(C_\mathrm{D} \nu_{\mathrm{pd}}^{(\mathrm{ex})} + \nu_{\mathrm{pp}}^{(\mathrm{M})} + \nu_0) N_{(\mu\mathrm{p})} + \nu_\mu N_{(\mathrm{p}\mu\mathrm{d})}, \tag{64d}$$

where $\mathrm{d}n_\gamma/\mathrm{d}t$ is the rate of γ-ray emission, and $N_{(\mathrm{p}\mu\mathrm{d})}$, $N_{(\mu\mathrm{d})}$, and $N_{(\mu\mathrm{p})}$ are, respectively, the concentrations of (pμd) molecules, (μd) atoms, and

† SCHIFF, M., *Nuovo Cim.* **22** (1961) 66.   ‡ loc. cit. p. 3312.

(μp) atoms, at time t.† The solution of these equations for the case of saturation ($C_D \nu_{pd}^{(ex)} \gg \nu_{pp}^{(M)}, \nu_{pd}^{(M)}, \nu_\gamma, \nu_\mu$) gives a simple parent–daughter decay relationship,

$$\frac{dn_\gamma}{dt} = \frac{\nu_\gamma \nu_{pd}^{(M)}}{\nu_\gamma + \nu_\mu - \nu_{pd}^{(M)}} \exp(-\nu_0 t)[\exp(-\nu_{pd}^{(M)} t) - \exp\{-(\nu_\gamma + \nu_\mu)t\}]. \quad (65)$$

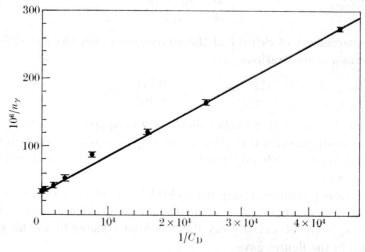

FIG. 27.14. The reciprocal of the observed γ-ray flux, n_γ as a function of the reciprocal of the deuterium concentration (C_D), in the muonic catalysis of the pd interaction. (Bleser et al., loc. cit., p. 3312.)

Fig. 27.15 shows a typical time distribution of γ-ray emission, obtained by Bleser et al., for a deuterium concentration of 62 parts per million. From an analysis of distributions of this kind the rates of the various processes involved can be obtained.

A fit of time distributions, such as that of Fig. 27.15, to the difference of two exponentials, enables the two exponents in the square bracket of eqn (65) to be obtained, but there is an ambiguity concerning the assignment of the exponents to $\nu_{pd}^{(M)}$ and $(\nu_\gamma + \nu_\mu)$, since it is not known which of these quantities is the larger. To overcome this difficulty, an impurity such as Ne is added to the hydrogen at small fractional concentration C_Z. Some of the μ^- in the (μ-d) atoms then become attached to the impurity in the process

$$(\mu^- d) + Z \rightarrow (\mu^- Z) + d, \quad (66\,a)$$

† The last term on the right-hand side of eqn (64 d) allows for the effect of the re-emitted μ^-. In the case of γ-ray emission, however, the μ^- remains also, although moving with little kinetic energy. It is assumed that the chance of its remaining trapped to the He³ nucleus after fusion is close to 1, so that recycling is unlikely in this case.

and the number of $(p\mu^-d)$ molecules formed, and thence the rate of the fusion process, is reduced. Process (66 a) is highly exothermic so that the reverse process cannot occur. If ν_{dZ} is the rate of process (66 a) for $C_Z = 1$, the rate of γ-ray emission at saturation (eqn (65)) has to be modified by replacing $\nu_{pd}^{(M)}$ by $(\nu_{pd}^{(M)} + C_Z \nu_{dZ})$, while $(\nu_\gamma + \nu_\mu)$ is unaltered.

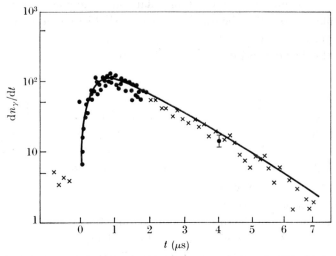

FIG. 27.15. Time distribution of fusion γ-ray emission (dn_γ/dt) obtained by Bleser et al. (loc. cit., p. 3312) for a deuterium concentration of 62 parts per million. The dots are obtained from the number of counts in a 30 ns interval, the crosses in a 150 ns interval. The fitted curve is obtained using $\nu_{pd}^{(M)}$, ν_f^{pd} of Table 27.7.

The quantities $\nu_{pd}^{(M)}$, and $(\nu_\gamma + \nu_\mu)$, can then be identified in the time distribution obtained without impurity, while the distribution obtained in the presence of impurity enables the rate ν_{dZ}, also, to be derived. We return to the discussion of the effect of impurities in experiments of this type in § 4.4.

From studies of Y_γ as a function of C_D, as illustrated in Fig. 27.14, the ratio $\nu_{pp}^{(M)}/\nu_{pd}^{(ex)}$ can be obtained. These rates can be obtained separately from analysis of the γ-ray time distribution for deuterium concentrations (C_D) well below saturation. Eqn (65) then takes the more complicated form

$$\frac{dn_\gamma}{dt} = \nu_\gamma \nu_{pd}^{(M)} C_D \nu_{pd}^{(ex)} \exp(-\nu_0 t) \times$$
$$\times \left[\frac{\exp(-\alpha t)}{(\alpha - \nu_{pd}^{(M)})(\alpha - \nu_\gamma - \nu_\mu)} + \frac{\exp(-\nu_{pd}^{(M)} t)}{(\alpha - \nu_{pd}^{(M)})(\nu_\gamma + \nu_\mu - \nu_{pd}^{(M)})} + \right.$$
$$\left. + \frac{\exp(-\nu_\gamma - \nu_\mu)t}{(\nu_\gamma + \nu_\mu - \nu_{pd}^{(M)})(\alpha - \nu_\gamma - \nu_\mu)} \right],$$

where $\alpha = C_\text{d}\,\nu_\text{pd}^\text{(ex)} + \nu_\text{pp}^\text{(M)}$, and analysis of the rate of rise of γ-emission in the time distribution, for very small C_D, enables $\nu_\text{pp}^\text{(M)}$, and thence $\nu_\text{pd}^\text{(ex)}$, to be obtained separately.

4.3.4. *Influence of hyperfine structure effects on fusion rates.* The expressions (58), (59), and (64) ignore effects due to hyperfine structure of the $(\mu^-\text{d})$ atom and $(\text{p}\mu^-\text{d})^+$ molecular ion. The spin-$\tfrac{1}{2}$ state, of the $(\mu^-\text{d})$ atom, lies 0·046 eV below the $\tfrac{3}{2}$ state, so that irreversible transitions to the spin-$\tfrac{1}{2}$ state can occur in exchange collisions between the $(\mu^-\text{d})$ atom and other deuterium atoms, in the process

$$(\mu^-\text{d})_{F=\tfrac{3}{2}} + \text{d} \to \text{d} + (\mu^-\text{d})_{F=\tfrac{1}{2}}. \qquad (66\,\text{b})$$

This effect has been discussed in § 3.6, where it has been pointed out that the cross-section for process (66 b) is much smaller than for the corresponding process (43) for $(\mu^-\text{p})$ atoms in hydrogen. At the small deuterium concentrations at which μ-catalysed fusion in liquid hydrogen has been observed, the transition rate for process (66 b) is too small to affect appreciably the relative populations of the $\tfrac{3}{2}$ and $\tfrac{1}{2}$ states before the formation of $(\text{p}\mu^-\text{d})$ molecules.

Wolfenstein† and Gershteĭn‡ pointed out, however, that at larger concentrations, C_D, this effect will lead to a changing population in the $(\mu^-\text{d})$ hyperfine structure states, the rate of change increasing markedly as C_D increases.

This, in turn, affects the population of the states of the $(\text{p}\mu^-\text{d})^+$ ion. It has, in fact, four spin states with total spin $F = 2, 1, 1, 0$. The nuclear fusion rate depends on the total spin F. As discussed later, the fusion reaction (49) takes place, predominantly, through a magnetic dipole transition, in which the proton and deuteron undergo an $S_{\tfrac{1}{2}} \to S_{\tfrac{1}{2}}$ transition to the ground state of He^3. The $F = 2$ state of $(\text{p}\mu^-\text{d})^+$, however, can be built up only from (pd) states of spin $I = \tfrac{3}{2}$, and thence does not contribute to the fusion process.

The $F = 0$ state of $(\text{p}\mu^-\text{d})$ can be built only from the spin $I = \tfrac{1}{2}$ state of (pd). For the higher energy $F = 1$ state of $(\text{p}\mu^-\text{d})$, the (pd) spin is $\tfrac{1}{2}$ for 86 per cent of the time, while for the lower energy $F = 1$ state, the (pd) spin is $\tfrac{1}{2}$ for 14 per cent of the time. Further, if the $(\text{p}\mu^-\text{d})$ molecules are formed from $(\mu^-\text{d})$ atoms in process (57), with the $(\mu^-\text{d})$ atoms distributed statistically between the $F = \tfrac{3}{2}$ and $F = \tfrac{1}{2}$ states, the resulting population of the four hyperfine states of the $(\text{p}\mu^-\text{d})$ molecule

---

† WOLFENSTEIN, L., *Proceedings of the conference on high energy physics at Rochester* (1960). University of Rochester Press, Rochester, New York, p. 529.

‡ GERSHTEĬN, S. S., *Zh. eksp. teor. Fiz.* **40** (1961) 698 (English translation, *Soviet Physics, JETP* **13** (1961) 488).

would be statistical, i.e. $5:3:3:1$. The full analysis involves a more complicated set of equations than eqns (64), with separate differential equations for the concentrations in the various hyperfine states of $(\mu^-\text{p})$, $(\mu^-\text{d})$, and $(\text{p}\mu^-\text{d})$. The factors $\dfrac{\nu_i}{\nu_0+\nu_\gamma+\nu_\mu}$ $(i=\gamma,\mu)$, in eqns (58) and (59), have also to be replaced by

$$0\cdot 25\frac{0\cdot 14\nu_i}{0\cdot 14(\nu_\gamma+\nu_\mu)+\nu_0}+0\cdot 25\frac{0\cdot 86\nu_i}{0\cdot 86(\nu_\gamma+\nu_\mu)+\nu_0}+0\cdot 083\frac{\nu_i}{\nu_\gamma+\nu_\mu+\nu_0}, \quad (67)$$

to allow for the contributions of the various hyperfine states to the fusion process.

If, as a result of process (66 b), the $(\text{p}\mu^-\text{d})$ molecules are formed only from $(\mu^-\text{d})$ atoms in the $F=\tfrac{1}{2}$ state, the relative populations of the four hyperfine states of $(\text{p}\mu^-\text{d})$ are $0:0\cdot 39:0\cdot 36:0\cdot 25$, and the factor $\dfrac{\nu_i}{\nu_0+\nu_\gamma+\nu_\mu}$ in eqns (58) and (59) would need to be replaced by

$$0\cdot 39\frac{0\cdot 14\nu_i}{0\cdot 14(\nu_\gamma+\nu_\mu)+\nu_0}+0\cdot 36\frac{0\cdot 86\nu_i}{0\cdot 86(\nu_\gamma+\nu_\mu)+\nu_0}+0\cdot 25\frac{\nu_i}{\nu_\gamma+\nu_\mu+\nu_0}. \quad (68)$$

Gershteĭn[†] solved the differential equations for the distribution between different hyperfine states for different fractional deuterium concentrations (C_D) allowing for $(\tfrac{3}{2}\to\tfrac{1}{2})$ transitions of $(\mu^-\text{d})$ at a rate given by eqn (47). On the basis of Gershteĭn's estimates Y_γ would be expected to rise by 18 per cent, when C_D increases from 0·72 per cent to 25 per cent. Bleser et al.[‡] observed such an increased yield of 17 ± 1 per cent in their experiments. The observations of Doede,[§] in a deuterium bubble chamber containing only a small fractional concentration of hydrogen, were also consistent with $(\text{p}\mu^-\text{d})$ molecule formation exclusively from the $(\mu^-\text{d})$, $F=\tfrac{1}{2}$ state.

These hyperfine structure effects are taken into account in deriving the 'best fit' curve to the observations of Bleser et al. in Figs. 27.14 and 27.15 and in the estimation of the rates of the various processes in liquid hydrogen given in Table 27.7.

A marked discrepancy is evident between the deuterium bubble chamber results of Doede, and the diffusion chamber results of Dzhelepov et al. with the reaction rates scaled linearly to liquid deuterium densities. The rate of formation $(\nu_{\text{dd}}^{(\text{M})})$ of $(\text{d}\mu^-\text{d})$ molecules in collisions of $(\mu^-\text{d})$,

---

[†] GERSHTEĬN, S. S., Zh. eksp. teor. Fiz. **40** (1961) 698 (English translation, Soviet Physics, JETP **13** (1961) 488).
[‡] BLESER, E. J., ANDERSON, E. W., LEDERMAN, L. M., MEYER, J. L., ROSEN, J. L., ROTHBERG, J. E., and WANG, I. T., Phys. Rev. **132** (1963) 2679.
[§] loc. cit., p. 3308.

TABLE 27.7

Reaction rates and yields in liquid hydrogen and liquid deuterium derived from muonic catalysis studies

(1) Liquid hydrogen (normal deuterium concentration)			
$\nu_{pd}^{(ex)}$	$\begin{cases} 1\cdot 43(\pm 0\cdot 13)\times 10^{10}\ s^{-1} \\ 1\cdot 81(\pm 0\cdot 13)\times 10^{10}\ ,, \end{cases}$	(1) (1) and (5)	(a)
$\nu_{pp}^{(M)}$	$\begin{cases} 1\cdot 89(\pm 0\cdot 20)\times 10^{6}\ ,, \\ 2\cdot 55(\pm 0\cdot 18)\times 10^{6}\ ,, \end{cases}$	(1) (5)	
$\nu_{pd}^{(M)}$	$\begin{cases} 5\cdot 8(\pm 0\cdot 3)\times 10^{6}\ ,, \\ 6\cdot 82(\pm 0\cdot 25\times 10^{6})\ ,, \end{cases}$	(1) (5)	
(c) $\nu_{f}^{pd}\ (=\nu_{\gamma}+\nu_{\mu})$	$0\cdot 305(\pm 0\cdot 010)\times 10^{6}\ s^{-1}$	(1)	
(c) $Y_{\gamma}^{(p)}$ (sat.)	$0\cdot 140 \pm 0\cdot 023$	(1)	
(c) $Y_{\mu}^{(p)}$ (sat.)	$0\cdot 0264 \pm 0\cdot 0035$	(3)	
(2) Liquid deuterium			
$\nu_{dd}^{(M)}$	$\begin{cases} 3\cdot 6(\pm 1\cdot 3)\times 10^{5}\ s^{-1} \\ 3\cdot 5(\pm 0\cdot 7)\times 10^{6}\ ,, \end{cases}$	(4) (2)	(b)
(c) ν_{f}^{dd}	$\begin{cases} 2\cdot 5(\pm 0\cdot 4)\times 10^{5}\ ,, \\ 1\cdot 13(\pm 0\cdot 06)\times 10^{5}\ s^{-1} \end{cases}$	(4) (2)	(b)
(c) $Y_{f}^{(d)}$	$0\cdot 159 \pm 0\cdot 006$ $0\cdot 18 \pm 0\cdot 04$	(4) (2)	(b)

(a) Using $(\nu_{pp}^{(M)}+\nu_0)/\nu_{pd}^{(ex)}$ obtained by Bleser et al. (1) and $\nu_{pp}^{(M)}$ determined by Conforto et al. (5)
(b) Converting diffusion chamber results to liquid deuterium density.
(c) $Y_{\gamma}^{(p)}$ is γ-ray yield per stopping μ^- in liquid hydrogen.
  $Y_{\mu}^{(p)}$ is yield of re-emitted μ^- per stopping μ^- in liquid hydrogen.
  $Y_{f}^{(d)}$ is total yield of fusion reactions per μ^- stopping in liquid deuterium.
  ν_{f}^{pd} is fusion rate for (pμ⁻d) molecule.
  ν_{f}^{dd} is fusion rate for (dμ⁻d) molecule.
(For footnote references (1)–(5) see p. 3321.)

in deuterium comes out nearly ten times as great in the diffusion chamber, compared with the bubble chamber results. Dzhelepov et al. have speculated that the difference might be due to a resonance in the cross-section for (dμ⁻d) molecule formation in the process

$$(\mu^- d) + D \rightarrow (d\mu^- d)^+ + e,$$

at an energy around 0·04 eV, the mean energy of (μ⁻d) atoms found in gaseous deuterium at room temperature. By contrast, the mean energy of (μ⁻d) atoms in liquid deuterium of 0·0025 eV would not exhibit such an effect.

Corresponding to the reaction rates of Table 27.7, Table 27.8 gives the cross-sections of the various processes involved. These are calculated assuming an atomic concentration in liquid hydrogen, or deuterium, of $N = 3\cdot 5 \times 10^{22}$ cm⁻³, and a mean relative velocity of (μ⁻p), (μ⁻d) atoms,

TABLE 27.8

Cross-sections important in the interpretation of the μ^- catalysed fusion process

Process	Cross-section (cm²)		
	Experimental	Theoretical	
			$vQ_{\rm pd}^{\rm (ex)}$
$(\mu^-{\rm p})+{\rm D} \to (\mu^-{\rm d})+{\rm H}$	$8\cdot18(\pm0\cdot75)\times10^{-18}(1)$	$6\cdot6\times10^{-18}(6)$	$3\cdot3\times10^{-13}$ cm³ s⁻¹
$(Q_{\rm pd}^{\rm (ex)})$	$10\cdot35(\pm0\cdot95)\times10^{-18}(1)$ and (5)	$6\cdot84\times10^{-18}(7)$	$3\cdot42\times10^{-13}$ cm³ s⁻¹
		$8\cdot68\times10^{-18}(8)$	$4\cdot34\times10^{-13}$ cm³ s⁻¹
	$4\cdot8\pm0\cdot8\times10^{-18}(9)$	$7\cdot8\times10^{-18}(10)$	$3\cdot9\times10^{-13}$ cm³ s⁻¹
$(\mu^-{\rm p})+{\rm H} \to ({\rm p}\mu^-{\rm p})^++{\rm e}$	$1\cdot08(\pm0\cdot12)\times10^{-21}(1)$	$1\cdot85\times10^{-21}(6)$	(a)
$(Q_{\rm pp}^{\rm (M)})$	$1\cdot46(\pm0\cdot10)\times10^{-21}(5)$	$1\cdot90\times10^{-21}(7)$	
$(\mu^-{\rm p})+{\rm D} \to ({\rm p}\mu^-{\rm d})^++{\rm e}$	$3\cdot31(\pm0\cdot15)\times10^{-21}(1)$	$1\cdot43\times10^{-21}(6)$	(a)
$(Q_{\rm pd}^{\rm (M)})$	$4\cdot90(\pm0\cdot18)\times10^{-21}(5)$	$4\cdot45\times10^{-22}(7)$	
$(\mu^{-1}{\rm d})+{\rm D} \to ({\rm d}\mu^-{\rm d})^++{\rm e}$	$2\cdot05(\pm0\cdot70)\times10^{-22}(4)$	$1\cdot70\times10^{-23}(6)$	
$(Q_{\rm dd}^{\rm (M)})$	$2\cdot00(\pm0\cdot40)\times10^{-21}(2)$	$3\cdot73\times10^{-24}(7)$	

(a) These values are half as large as those given in ref. (6) following Zel'dovich and Gershtein (loc. cit.) who pointed out that the values obtained by Cohen et al. were too large by a factor of 2 owing to an error of √2 in the normalization of the continuous spectrum wave-functions.

(1) BLESER, E. J., ANDERSON, E. W., LEDERMAN, L. M., MEYER, J. L., ROSEN, J. L., ROTHBERG. J. E., and WANG, I. T., *Phys. Rev.* **132** (1963) 2679.
(2) DZHELEPOV, V. P., FRIML, M., GERSHTEIN, S. S., KATYSHEV, Y. V., MOSKALEV, V. I., and ERMOLOV, P. F., *Proc. Int. Conf. high Energy Phys.* CERN (1962) (J. Prentki, ed.), p. 484.
(3) SCHIFF, M., *Nuovo Cim.* **22** (1961) 66.
(4) DOEDE, J. H., *Phys. Rev.* **132** (1963) 1782.
(5) CONFORTO, G., RUBBIA, C., ZAVATTINI, E., and FOCARDO, S., *Nuovo Cim.* **33** (1964) 1001.
(6) COHEN, S., JUDD, D. L., and RIDDELL, R. J., *Phys. Rev.* **119** (1960) 397.
(7) BELYAEV, V. B., GERSHTEĬN, S. S., ZAKHAR'EV, B. N., and LOMNEV, S. P., *Zh. eksp. teor. Fiz.* **37** (1959) 1652 (English translation, *Soviet Physics, JETP* **10** (1960) 1171).
(8) SHIMIZU, M., MIZUNO, Y., and IZUYAMA, *Prog. theor. Phys., Osaka* **20** (1958) 777.
(9) BERTIN, A., BRUNO, M., VITALE, A., PLACCI, A., and ZAVATTINI, E., *Lettere al Nuovo Cim.* **4** (1972) 449
(10) MATVEENKO, A. V., and PONOMAREV, L. I., *Zh. eksp. teor. Fiz.* **59** (1970) 1593 (English translation, *Soviet Physics, JETP* **32** (1971) 871).

relative to the protons or deuterons in the liquid, of $v = 5\times10^4$ cm s⁻¹, corresponding to a temperature of $0\cdot0025$ eV. The experimental results are compared with the values calculated using the theoretical methods described in § 5.

4.3.5. *Influence of muon recycling and trapping on fusion rate.* In the catalysis of the d–d reaction (processes 52, 53) the muon is left at the end of the process with very little kinetic energy. There is a possibility that it will be trapped to form the mesic atom (μ^-He^3) in the case of process (53), and (μ^-H^3) in the case of process (52). If the μ^- is not trapped in this way it may eventually be captured by a deuteron to form (μ^-d), and the cycle will be repeated. Let T be the probability of trapping, then allowing for recycling the yield, $Y_f^{(d)}$ of fusion reactions can be written (Doede)†

$$Y_f^{(d)} = \frac{C_D \nu_{dd}^{(M)} \nu_{df}}{(\nu_0+\nu_{df})[\nu_0+C_H \nu_{pd}^{(M)}+C_D\{1-(1-T)\nu_{df}/(\nu_0+\nu_{df})\}\nu_{dd}^{(M)}]}, \quad (69)$$

where ν_{df} is the rate of nuclear fusion in a (dμ^-d) meso-molecule. This expression gives the total number of fusion reactions per stopped μ^-. To obtain the proportion of stopped μ^- that give rise to a fusion reaction, Y_F, the last term in the square bracket in the denominator of eqn (69) would need to be replaced by $C_D \nu_{dd}^{(M)}$.

In the deuterium bubble chamber only reaction (52) leaves a visible signature in the form of a proton of energy 3·04 MeV. Doede counted the numbers of stopping μ^- that lead to the emission of one, two, or three such protons, N_{1f}, N_{2f}, N_{3f}. Then

$$N_F = \tfrac{1}{2}Y_F = \frac{(N_{1f}/\epsilon_{1f}+N_{2f}/\epsilon_{2f}+N_{3f}/\epsilon_{3f})}{N_s}, \quad (70)$$

where N_s is the number of stopping μ^-, ϵ_{1f}, ϵ_{2f}, ϵ_{3f} are the observational efficiencies for seeing a single proton, two protons, or three protons, of energy 3·04 MeV, while the factor $\tfrac{1}{2}$ arises because the reaction (53) is expected to occur at an equal rate to reaction (52). This equality of the two reaction rates was, in fact, observed by Dzhelepov et al.‡ in their studies of this process in the diffusion chamber. Then the trapping probability T can be determined from

$$N_{2f}/\epsilon_{2f} N_s = (1-T)N_F^2. \quad (71)$$

In this way Doede estimated a trapping probability $T = 0{\cdot}31$.

Estimates of T have been made theoretically by Jackson§ and by Judd.∥ They estimated the probability of trapping of the μ^- into the 1s state about the He^3 nucleus, in reaction (53), as 15 per cent, and about the triton, in reaction (52), as 2 per cent. The higher value of T observed suggests that trapping occurs predominantly into states of higher

† loc. cit., p. 3308.
‡ Dzhelepov, V. P., Filchenkov, V. V., Friml, M., Katyshev, Y. V., Moskalev, V. I., and Ermolov, P. F., *Nuovo Cim.* **33** (1964) 40; Dzhelepov, V. P., Ermolov, P. F., Katyshev, Y. V., Moskalev, V. I., Filchenkov, V. V., and Friml, M., *Zh. eksp. teor. Fiz.* **46** (1964) 2042 (English translation, *Soviet Physics, JETP* **20** (1965) 841).
§ Jackson, J. D., *Phys. Rev.* **106** (1957) 330.
∥ Quoted by Doede (loc. cit., p. 3308).

excitation. The total number of fusion reactions per stopping μ^-, $Y_f^{(d)}$, of eqn (69), was also estimated by Doede, from these measurements. In terms of the observed number of visible fusions (process 52) he derives the expression

$$Y_f^{(d)} = \frac{2N_{1f}/\epsilon_{1f}+4N_{2f}/\epsilon_{2f}+6N_{3f}/\epsilon_{3f}}{N_s}. \tag{72}$$

The value of $Y_f^{(d)}$ obtained in this way, by Doede, is given in Table 27.7.

4.4. *Impurity method of studying rates for collision processes of* (μp) *and* (μd) *atoms*

4.4.1. *Rates for muon transfer and muonic molecule formation.* Schiff† and Bleser et al.‡ showed that the addition of a small amount of neon to liquid hydrogen had a marked effect on the yield and time distribution of the emission of fusion gamma rays. This has already been referred to in § 4.3.3. A concentration of 0·02 per cent of neon is sufficient to reduce the maximum γ-ray yield by a factor of 4, and to reduce, markedly, the time over which fusion γ-rays are emitted. The differential equations (64 a–d) have to be modified by the addition of a term $-C_{\text{Ne}}\nu_{\text{d,Ne}}N_{(\mu\text{d})}$ to the right-hand side of eqn (64 c), and a term $-C_{\text{Ne}}\nu_{\text{p,Ne}}N_{(\mu\text{p})}$ to the right-hand side of eqn (64 d), where C_{Ne} is the fractional neon concentration, and the rates $\nu_{\text{d,Ne}}$ corresponding to process (66 b), for Ne, and $\nu_{\text{p,Ne}}$, corresponding to the analogous process

$$(\mu\text{p})+\text{Ne} \rightarrow (\mu\text{Ne})+\text{p}+\text{e},$$

refer to a concentration N atoms per cm³, as in the previous equations. Fig. 27.16 shows the time distribution of the fusion gamma rays observed by Bleser et al. for the case of 25 per cent deuterium and 0·02 per cent neon concentration. From such distributions Bleser et al. derived the value

$$\nu_{\text{d,Ne}} = 8\cdot1(\pm1\cdot0)\times10^{10}\text{ s}^{-1}.$$

Conforto et al.§ have carried out similar experiments but instead of observing the fusion gamma rays they have studied the time distribution dn_X/dt of neon muonic X-ray emission. This can be written

$$\frac{dn_X}{dt} = a\exp(-\nu_p t)+b\exp(-\nu_d t), \tag{73}$$

where
$$\nu_p = \nu_0+C_D\nu_{pd}^{(ex)}+\nu_{pp}^{(M)}+C_{\text{Ne}}\nu_{\text{p,Ne}}$$
and
$$\nu_d = \nu_0+\nu_{pd}^{(M)}+C_{\text{Ne}}\nu_{\text{d,Ne}}.$$

† loc. cit., p. 3315.   ‡ loc. cit., p. 3312.
§ CONFORTO, G., RUBBIA, C., ZAVATTINI, E., and FOCARDI, S., *Nuovo Cim.* **33** (1964) 1001.

From the analysis of the time distributions, observed for a series of different neon and deuterium concentrations, they were able to derive the values of $\nu_{pp}^{(M)}$, $\nu_{pd}^{(M)}$, $\nu_{p,Ne}$, $\nu_{d,Ne}$ given in Table 27.7 and Fig. 27.19. A typical distribution is shown in Fig. 27.17.

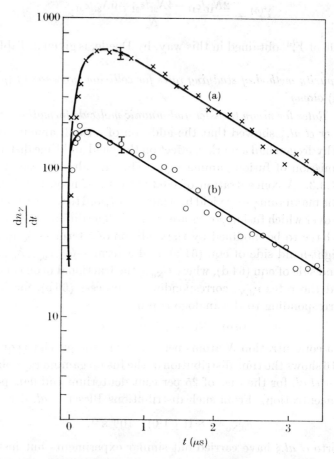

FIG. 27.16. Time distribution of fusion γ-ray emission (dn_γ/dt) for (a) mixture of 25 per cent liquid D and 75 per cent liquid H, (b) as for (a) but with 0·02 per cent Ne impurity. (Bleser et al., loc. cit., p. 3312.)

A systematic investigation of muon transfer rates from hydrogen to heavier atoms has been carried out by Alberigi-Quaranta et al.† They used gaseous hydrogen with xenon as an impurity, and measured the

† ALBERIGI-QUARANTA, A., BERTIN, A., MATONE, G., PALMONARI, F., PLACCI, A., DALPIAZ, P., TORELLI, G., and ZAVATTINI, E., Nuovo Cim. 47B (1967) 92.

27.4 COLLISION PROCESSES INVOLVING MESIC ATOMS

rate dn_X/dt of emission of xenon muonic X-rays, given by

$$\frac{dn_X}{dt} = aC_{Xe}\nu_{p,Xe}\exp(-\nu_T t), \tag{74}$$

where
$$\nu_T = \nu_0 + \rho(\nu_{pp}^{(M)} + C_{Xe}\nu_{p,Xe}).$$

Here a is a constant and ρ, the hydrogen gas density, is expressed in terms of liquid hydrogen density, so that the rates derived are directly

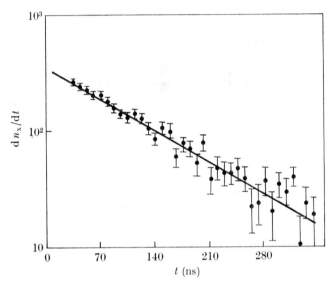

FIG. 27.17. Time distribution (dn_X/dt) of Ne muonic X-ray emission in a mixture of liquid H containing 0·46 per cent D and 0·001575 per cent Ne. (Conforto et al., loc. cit., p. 3323.)

comparable with those obtained from experiments using liquid hydrogen. By adding other rare-gas impurities it was possible to compare capture rates in Ne, A, and Kr with those in Xe. The apparatus used by Alberigi-Quaranta and his colleagues is shown in Fig. 27.18. A muon beam from the CERN synchrocyclotron came to rest in hydrogen gas at pressures up to 26 atm contained in a stainless steel vessel (b) 46 cm long and 12·5 cm diameter. The scintillation counters 1–5 defined the beam, while the target vessel itself was completely surrounded by scintillation counters 6–8 operated in anticoincidence. Xenon gas at a concentration C_{Xe}' up to 0·01 per cent was introduced into the hydrogen. (μ^-p) atoms formed when the stopped muons diffused through the gas, until the muon was transferred to Xe, to form a muonic xenon atom in an excited state.

Muonic X-rays of energy 3·8 MeV from the transition $2p \to 1s$ in muonic xenon were detected by a NaI counter (a) at the side of the target vessel. Five plates of brass (c), 0·3 cm thick and 5 cm apart, were placed inside the target vessel, in the region viewed by the NaI counter. These had the effect of reducing the size of the region in which the muons were

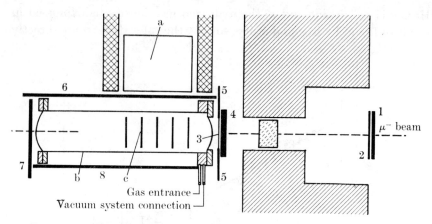

Fig. 27.18. Apparatus used by Alberigi-Quaranta et al. (loc. cit.) for studying muon transfer from hydrogen to heavier atoms. (1–8 scintillation counters. (a) NaI counter; (b) stainless steel vessel; (c) brass plates.)

stopped. The arrival of a stopping muon was signalled by the coincidence $123 4\bar{5}$. The electronics recorded the time between this signal and the observation of the (μ^-Xe) X-ray by means of the NaI counter.

The results of measurements of muon transfer rates from (μ^-p) to heavier elements, obtained in different experiments, are shown in Fig. 27.19 (a), and compared with the theoretical estimates of Gershteĭn.[†] The variation with Z is approximately linear, and the results are satisfactorily explained by the theory.

A similar kind of experimental arrangement was used by Placci et al.[‡] to study the muon transfer rates for (μ^-d) and (μ^-p) atoms to heavier atoms. However, instead of looking at the emission of muonic X-rays, they studied the differential time distribution of muon decay electrons which are similarly affected by the presence of heavy atom impurities, leading to nuclear capture of the muons in competition with their decay. The apparatus used in these experiments is shown in Fig. 27.20. The

[†] GERSHTEĬN, S. S., *Zh. eksp. teor. Fiz.* **43** (1962) 706 (English translation, *Soviet Physics, JETP* **16** (1963) 501).
[‡] PLACCI, A., ZAVATTINI, E., BERTIN, A., and VITALE, A., *Nuovo Cim.* **52**A (1967) 1274; ibid. **64**A (1969) 1053.

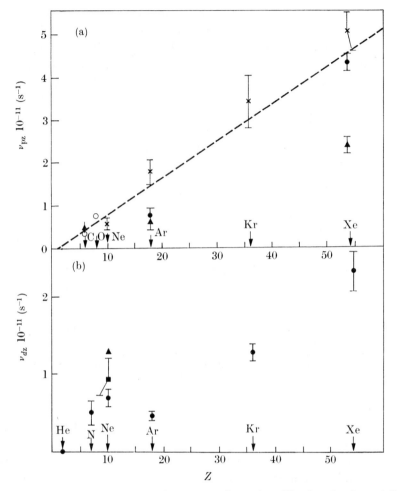

FIG. 27.19. (a) Estimated transfer rates $\nu_{\mathrm{p}Z}$ for μ^- from H to heavier element Z.
× Measurements of Alberigi-Quaranta et al. (loc. cit.).
▲ Basiladze et al.†
● Placci et al.
○ Theoretical calculations of Gershtein (loc. cit.).
(b) Estimated transfer rates $\nu_{\mathrm{d}Z}$ for μ^- from D to heavier element Z from measurements of Placci et al. (loc. cit.).

target vessel (T) contained deuterium at a pressure of 6 atm. The beam was defined by two scintillators (1, 2) and, immediately on entering the target, it passed through a proportional counter (α) consisting of two parallel planar grids, separation 7 mm, maintained at a potential

† BASILADZE, S. G., ERMOLOV, P. F., and OGANESYAN, K. O., Zh. eksp. teor. Fiz. **49** (1965) 1042 (English translation, Soviet Physics, JETP **22** (1966) 725).

difference of 14·5 kV. The size of the signal from α was used to estimate the potential range of the muon, and was accepted only if it indicated that the muon would stop within a length of 70 cm of the deuterium gas. A signal (1, 2, α, $\bar{A}_i$), where A_i ($i = 1$ to 4) indicates four scintillation counters at the sides and another (A_5) at the back of the target vessel

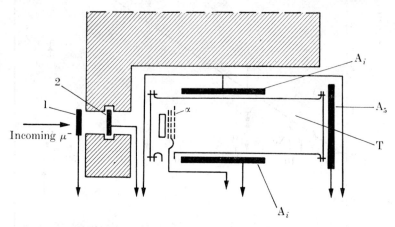

FIG. 27.20. Arrangement used by Placci et al. (loc. cit.) to study transfer rates of μ^- from (μ^-d) to heavier atoms. 1, 2 are scintillation counters defining the incident beam. A_i ($i = 1$–5) are scintillation counters arranged round the target chamber T, containing gaseous deuterium at a pressure of 6 atm. α is a parallel grid proportional counter.

opened a gate 10 μs long. During this time μ^- decay electrons were detected by the counters A_i. The differential time distribution of these electrons was measured by a pulse-height analyser and time-to-pulse-height converter, triggered by the signal for the stopping muon and the delayed pulse from the counters A_i. Fig. 27.21 shows typical time distribution spectra obtained. The distribution has the form of a combination of a number of exponential terms, the exponents of which depend on the transfer rate, ν_{dZ} (where Z stands for the heavy element). The results are shown in Fig. 27.19 (b), together with results of a few earlier measurements. They are of the same order of magnitude as the similar rates, ν_{pZ} from (μ^-p) atoms. The enormous range of transfer rates is particularly striking. The transfer rate to neon is four orders of magnitude greater than that to helium, in agreement with the theoretical predictions of Gershteĭn.†

4.4.2. *Transfer rate of negative pions from gaseous hydrogen to argon.* Relative transfer rates of π^- from pionic hydrogen to other atoms in the

† loc. cit., p. 3326.

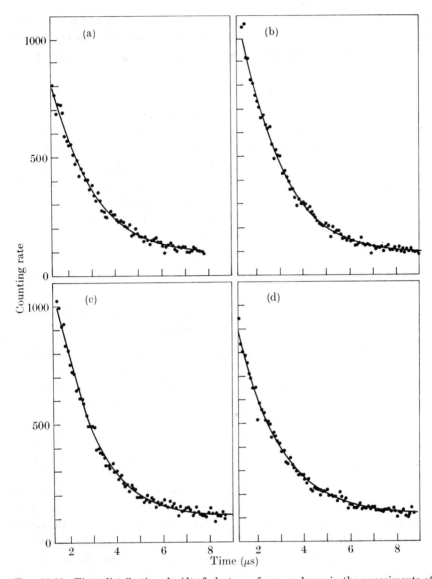

Fig. 27.21. Time distribution dn_e/dt of electrons from μ^- decay in the experiments of Placci et al. (loc. cit.) in which μ^- are stopped in deuterium gas containing impurities of (a) Ne, (b) Ar, (c) Kr, (d) Xe. (Background counts are not subtracted.)

process $(\pi^-p)+X \to (\pi X)+p$ have been measured by Petrukhin et al.[†] while Picard et al.[‡] have measured the absolute rate for the transfer to

[†] Petrukhin, V. I., Prokoshkin, Yu. D., and Suvorov, V. M., *Zh. eksp. teor. Fiz.* **55** (1968) 2173 (English translation, *Soviet Physics, JETP* **28** (1969) 1151).
[‡] Picard, J., Placci, A., Polacco, E., Zavattini, E., Carboni, G., Gastaldi, U.,

argon. In the latter work π^- mesons were stopped in a stainless steel vessel containing hydrogen at a pressure of 1 atmosphere together with a contamination of up to 20 per cent argon. Capture of π^- in the (π^-p) occurs through one of the processes

$$\pi^- + p \to n + \gamma$$
$$\to n + \pi^0 \to n + \gamma + \gamma.$$

In either case the γ-rays are very energetic (140 MeV and 70 MeV in the two processes respectively). They were identified using a coincidence counter telescope outside the pressure vessel. In (π^-Ar) nuclear capture of the pions will usually produce nuclear break-up without γ-ray emission. The proportion of stopping pions producing energetic γ-rays will then decrease as the argon contamination in the hydrogen increases. The γ-ray counting rate would be expected to vary like

$$(1 + C\tau_\pi \lambda_\pi)^{-1},$$

where C is the argon concentration, τ_π the effective average lifetime of a pion bound to a proton, and λ_π the transfer rate from (π^-p) to (π^-Ar). Fitting the observed variation to an expression of this type gave

$$\lambda_\pi = (4 \pm 1\cdot 8) \times 10^9 \text{ s}^{-1}$$

for a density of $2\cdot 5 \times 10^{19}$ atoms cm^{-3} of argon. Assuming thermal velocity this corresponds to a transfer cross-section of 6×10^{-16} cm^2. This is about 50 times larger than the transfer cross-section of muons from (μ^-p) in the 1s state to argon, suggesting that in the measurements on the (π^-p) system the transfer occurs from a state of high excitation.

4.4.3. *Scattering cross-sections of (μ^-p) and (μ^-d) atoms.* The appearance of a gap of mean length 0·6 mm between the end of a primary muon track, and the beginning of the track of the re-emitted muon (§ 4.2.1), is related to the elastic scattering cross-section of (μ^-d) atoms in hydrogen in the energy region 0–43 eV, the upper end of the range being the (μ^-d) production energy in process (56). The mean gap length decreases markedly, with increase of deuterium concentration in the hydrogen, and at a concentration of only 1 per cent deuterium the visible gap has disappeared. Evidently, this has to be interpreted as evidence that the total elastic scattering cross-section of (μ^-d) in deuterium is much larger than in hydrogen.

Dzhelepov *et al.*† observed similar gaps in a hydrogen diffusion cham-

GORINI, G., STEFANINI, G., TORELLI, G., DUCLOS, J., and MAGNON, A., *Lettere al Nuovo Cim.* **2** (1971) 957.

† DZHELEPOV, V. P., ERMOLOV, P. F., MOSKALEV, V. I., FIL'CHENKOV, V. V., and FRIML, M., *Zh. eksp. teor. Fiz.* **47** (1964) 1243 (English translation, *Soviet Physics, JETP* **20** (1965) 841).

ber containing various deuterium concentrations. In their case they looked at muons that underwent $\mu \to e$ decay, and observed a gap between the muon end point and the starting point of the decay electron. The elastic scattering cross-sections of (μ^-d) in hydrogen and deuterium can be derived from a study of the distribution of the gap sizes. Such a distribution has been measured for muons stopping in a liquid hydrogen bubble chamber by Schiff† and in the diffusion chamber filled with 23 atm of hydrogen gas by Dzhelepov and co-workers.

In the latter work, to facilitate identification of the beginning of the decay electron track, special decay tracks were selected in which the beginning of the electron track was accompanied by a small blob of droplets. These events were interpreted as those in which the μ^- underwent decay from a bound state of (μ^-O) or (μ^-C), the blob of droplets being produced by an Auger electron ejected from the muonic oxygen or carbon during the process of de-excitation. The (μ^-d) when moving slowly was assumed to have undergone an exchange reaction of type (66 a), with O or C present in the alcohol in the chamber.

Fig. 27.22 shows the gap distributions they observed for a range of deuterium concentrations, and for various concentrations of carbon and oxygen. Monte Carlo calculations to obtain the best fits to these observed distributions gave the cross-sections given in Table 27.9.

Alberigi-Quaranta et al.‡ have used a counter arrangement to study the distribution of the diffusion times of (μ^-p) and (μ^-d) atoms in hydrogen and deuterium. A layout similar to that of Fig. 27.18 was again used. In this case, however, the stainless steel target vessel contained 90 gold plates with their surfaces perpendicular to the cylinder axis, each 10 μm thick, 11 cm diameter, and spaced 1·5 mm apart. The cylinder was filled with hydrogen (or deuterium) gas up to a pressure of 26 atm. The (μ^-p) or (μ^-d) atoms formed by μ^- capture diffused in the gas until they reached a gold plate, where the μ^- was transferred to a gold atom. The muonic AuK_α X-ray, of energy 5·8 MeV, was observed in the NaI detector. The distribution of the interval between K$^-$ capture and muonic X-ray emission was measured using a method similar to that of their other experiment. This time measures the time taken for the (μ^-p) (or (μ^-d)) atom to diffuse from the μ^- stopping point to a gold plate. Fig. 27.23 shows the measured time distribution for a hydrogen pressure of 26·2 atm. In these experiments the (μ^-p) or (μ^-d) were

† loc. cit., p. 3315.
‡ ALBERIGI-QUARANTA, A., BERTIN, A., MATONE, G., PALMONARI, F., PLACCI, A., DALPIAZ, P., TORELLI, G., and ZAVATTINI, E., Nuovo Cim. 47B (1967) 72.

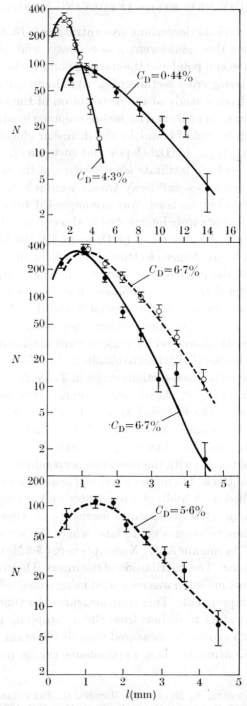

FIG. 27.22. Observed range distributions of (μ^-d) atoms in hydrogen in a diffusion chamber containing various deuterium concentrations C_D as indicated. The concentrations of O and C were varied by using as condensant —— CH_3OH and --- C_3H_7OH. The horizontal scales give the distance l between the end of the muon track and beginning of the decay electron track.

formed by direct capture of the stopping μ^- and have very much less energy than that of the (μ^-d) formed by muon exchange from (μ^-p) in the bubble chamber, or diffusion chamber, experiments. Considerations, such as those discussed in § 3.1, suggest that the initial kinetic energy of the (μ^-p) and (μ^-d) in the experiments of Alberigi-Quaranta was less

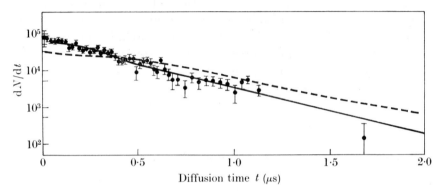

FIG. 27.23. Distribution of diffusion times of (μ^-p) atoms observed in the experiments of Alberigi-Quaranta et al. (loc. cit.). The number dN/dt per μs interval about the drift time (t) is shown.

than 1 eV. They estimated it to be 0·55 eV to give a best fit in their Monte Carlo calculation. In the experiments of Dzhelepov et al., on the other hand, the initial energy was a few eV for (μ^-p). For (μ^-d) formed in process (56) the initial energy is 43 eV (§ 4.31). Alberigi-Quaranta et al. also estimated the expected time distribution in their experiments, using a Monte Carlo method, making various assumptions about the relevant cross-section, and finding the cross-sections that gave a distribution fitting best to that observed. This best fit is the line shown in Fig. 27.23.

In making these Monte Carlo estimates account had to be taken of the hyperfine structure of (μ^-p) and (μ^-d). The (μ^-p) atom, for example, can be formed in hyperfine structure states with $F = 1$ and $F = 0$, respectively, the former having an energy $\Delta E = 0 \cdot 18$ eV above the latter, much larger than the thermal energy of a (μ^-p) atom (0·03 eV at room temperature). Therefore, so long as the (μ^-p) atom is moving with an energy above 0·18 eV, the collisions may be elastic, or may result in excitation or de-excitation between the hyperfine states in the reactions

$$(\mu^-p)_{F=1} + p \rightleftarrows p + (\mu^-p)_{F=0},$$

and the (μ^-p) atom may be regarded as a statistical mixture of the two hyperfine states. Below this energy, however, only elastic or de-exciting collisions (see eqn (43)) are possible, and almost all the atoms soon pass to the $F = 0$ state. Table 27.9, which gives the cross-sections inferred for $(\mu^- p)$ scattering in H, gives both the elastic scattering cross-section, Q_0^{el}, inferred for kinetic energies below 0·18 eV, when only the $F = 0$ state is occupied, and also the total cross-section, Q_{10} inferred for kinetic energies above 0·18 eV, when a statistical distribution between the $F = 0$ and $F = 1$ states is to be expected. It gives, also, cross-sections $Q_{1 \to 0}$, for $F = 1 \to 0$ de-exciting collisions. Gershteĭn has given expressions for these cross-sections in terms of the scattering lengths $c_g^{(p)}$, $c_u^{(p)}$, defined in § 3.6. The expression for $Q_{1 \to 0}$ is given in eqn (44). For Q_{10}, for a statistical distribution between $F = 1$ and $F = 0$ states,

$$Q_{10} = 4\pi \left(\tfrac{3}{4}(c_u^{(p)})^2 + \frac{1}{4} \frac{(c_g^{(p)})^2}{1 + k_0^2 (c_g^{(p)})^2} \right), \tag{75}$$

while, for scattering in the $F = 0$ state alone

$$Q_0^{\text{el}} = 4\pi \left(\frac{3c_u^{(p)} + c_g^{(p)}}{4} \right)^2. \tag{76}$$

Alberigi-Quaranta et al. analysed their gap distributions by assuming, for $c_u^{(p)}$, the value $5a_\mu$, which is close to that calculated theoretically by Gershteĭn,† and also by Cohen et al.‡ They then chose $c_g^{(p)}$ to give cross-sections which in the Monte Carlo calculations gave a best fit to the observed time distribution. The distribution is determined mainly by the cross-section Q_0^{el} (eqn (76)). This gives a quadratic equation to determine $c_g^{(p)}$, and this explains the ambiguity in the cross-sections given in Table 27.9. $c_u^{(p)}$ and $c_g^{(p)}$ have opposite signs, so that Q_0^{el} is much smaller than Q_{10}. The diffusion rate of $(\mu^- p)$ atoms in hydrogen gas has also been calculated by Matone.§

For deuterium the hyperfine energy difference is $\Delta E = 0{\cdot}045$ eV. For $(\mu^- d)$ atoms of energy above this, the two hyperfine states $F = \tfrac{3}{2}, \tfrac{1}{2}$ are populated statistically. The total scattering cross-section $Q_{\frac{1}{2},\frac{3}{2}}$ is given in terms of the scattering length $c_g^{(d)}$, $c_u^{(d)}$, defined in an analogous way to $c_g^{(p)}$, $c_u^{(p)}$, by

$$Q_{\frac{1}{2},\frac{3}{2}} = 4\pi \{ \tfrac{2}{3}(c_g^{(d)})^2 + \tfrac{1}{3}(c_u^{(d)})^2 \}. \tag{77}$$

For energies below ΔE, the $(\mu^- d)$ atoms quickly fall to the $F = \tfrac{1}{2}$ state,

---

† GERSHTEĬN, S. S., Zh. eksp. teor. Fiz. **34** (1958) 463 (English translation, Soviet Physics, JETP **7** (1958) 318); Zh. eksp. teor. Fiz. **36** (1959) 1309 (English translation, Soviet Physics, JETP **9** (1959) 927).
‡ COHEN, S., JUDD, D. L., and RIDDELL, R. J., Phys. Rev. **119** (1960) 397.
§ MATONE, G., Lettere al Nuovo Cim. **2** (1971) 151.

TABLE 27.9

Cross-sections and scattering lengths for (μ^-p) and (μ^-d) atoms in hydrogen and deuterium

Quantity	Experimental $(\mu^-p)+p \to (\mu^-p)+p$	Theoretical	Quantity	Experimental $(\mu^-d)+d \to (\mu^-d)+d$	Theoretical
Scattering length $c_u^{(p)}$ (unit a_μ)	5·0 (assumed)†§	5·25††, 5·0‡‡	Scattering length $c_u^{(d)}$ (unit a_μ)		5·76§§
Scattering length $c_g^{(p)}$ (unit a_μ)	$\begin{cases} -18\cdot8\pm0\cdot8\dagger \\ -11\cdot2\pm0\cdot8 \end{cases}$ (a)	$-17\cdot3\dagger\dagger,\ -11\cdot0\ddagger\ddagger$	Scattering length $c_g^{(d)}$ (unit a_μ)		6·67§§
Q_0^{el} (10^{-21} cm²)	$\begin{cases} -33\cdot0\pm2\cdot0§ \\ +\ \ 3\cdot0\pm2\cdot0 \\ \ \ \ 7\cdot6\pm0\cdot7\dagger \\ \ \ 167\pm30\ddagger \end{cases}$ (a)	$1\cdot2,\dagger\dagger\ 8\cdot2,\ddagger\ddagger\ 2\cdot5\|$	$Q_\frac{1}{2}^{el} \simeq Q_{\frac{3}{2},\frac{3}{2}}^{el}\ (=Q_{dd}^{el})$ (10^{-19} cm²)	$\begin{aligned} 0\cdot55&\pm0\cdot20\dagger \\ 4\cdot15&\pm0\cdot29\ddagger \end{aligned}$	$3\cdot3,§§\ 3\cdot5\ddagger\ddagger$
Q_{10} (10^{-19} cm²) $(=Q_{pp}^{el})$	$\begin{cases} 8\cdot8\pm0\cdot7\dagger \\ 4\cdot1\pm0\cdot5 \end{cases}$ (a)	$7\cdot9\dagger\dagger,\ 4\cdot0\ddagger\ddagger$		$(\mu^-d)+p \to (\mu^-d)+p$	
	$\begin{cases} 24\cdot0\pm5\cdot0§ \\ 1\cdot74\pm0\cdot3 \end{cases}$ (a)		Q_{dp}^{el} (10^{-21} cm²)	$\begin{aligned} 0\cdot8&+0\cdot8\ddagger \\ &-0\cdot4 \end{aligned}$	$\sim 1\ddagger\ddagger$
$Q_{1\to 0}$ (10^{-19} cm²)	$\begin{cases} 8\cdot7\pm0\cdot7\dagger \\ 4\cdot0\pm0\cdot5 \end{cases}$ (a)	$7\cdot8,\dagger\dagger\ 3\cdot9\ddagger\ddagger$		$(\mu^-d)+C,O \to (\mu^-d)+C,O$	
	$\begin{cases} 16\cdot8\pm3\cdot0§ \\ 0\cdot049\pm0\cdot013 \end{cases}$ (a)		$Q_{d C,O}^{d}$ (10^{-18} cm²)	$1\cdot2\pm0\cdot3\ddagger$	

† ALBERIGI-QUARANTA, A., BERTIN, A., MATONE, G., PALMONARI, F., PLACCI, A., DALPIAZ, P., TORELLI, G., and ZAVATTINI, E., *Nuovo Cim.* **47B** (1967) 72.
‡ DZHELEPOV, V. P., ERMOLOV, P. F., MOSKALEV, V. I., FIL'CHENKOV, V. V., and FRIML, M., *Zh. eksp. teor. Fiz.* **47** (1964) 841 (English translation, *Soviet Physics*, *JETP* **20** (1965) 841).
§ DZHELEPOV, V. P., ERMOLOV, P. F., and FIL'CHENKOV, V. V., *Zh. eksp. teor. Fiz.* **49** (1965) 393 (English translation, *Soviet Physics*, *JETP* **22** (1966) 275).
‖ MATVEENKO, A. V. and PONOMAREV, L. I., *Zh. eksp. teor. Fiz.* **59** (1970) 1593 (English translation, *Soviet Phys. JETP* **32** (1971) 871).
†† GERSHTEĬN, S. S., *Zh. eksp. teor. Fiz.* **36** (1959) 1309 (English translation, *Soviet Physics*, *JETP* **9** (1959) 927).
‡‡ COHEN, S., JUDD, D. L., and RIDDELL, R. J., *Phys. Rev.* **119** (1960) 397.
§§ ZEL'DOVICH, YA. B. and GERSHTEĬN, S. S., *Usp. fiz. Nauk* **71** (1960) 581 (*Soviet Physics–Uspekhi* **3** (1961) 593).
(a) The interpretation of the experimental distribution leads to an ambiguity in the scattering lengths and the cross-sections they determine.

and only the $F = \frac{1}{2} \to F = \frac{1}{2}$ elastic scattering is possible with cross-section

$$Q_{\frac{1}{2}}^{\text{el}} = 4\pi\left[\frac{2}{3}\left(\frac{5c_{\text{g}}^{(\text{d})}+c_{\text{u}}^{(\text{d})}}{6}\right)^2 + \frac{1}{3}\left(\frac{c_{\text{g}}^{(\text{d})}+2c_{\text{u}}^{(\text{d})}}{3}\right)^2\right]. \tag{78}$$

Theoretical calculations† suggest that $c_{\text{g}}^{(\text{d})} \simeq c_{\text{u}}^{(\text{d})}$, leading to the approximate equality of $Q_{\frac{1}{2},\frac{3}{2}}$ and $Q_{\frac{1}{2}}^{\text{el}}$ given by (77) and (78).

The cross-section for (μ^-d) scattering, given in Table 27.9, is based, therefore, on an analysis assuming a single elastic scattering cross-section. An interesting feature of these results is the small value of the cross-section Q_{d} for (μ^-d) scattering by protons, as inferred already from the remarkable sensitivity of gap length (already observed in Schiff's measurements) to the concentration of deuterium in liquid hydrogen. The differences between the cross-sections obtained by Alberigi-Quaranta et al., and of Dzhelepov et al., may be associated with the very different kinetic energies with which the (μ^-p) and (μ^-d) atoms were formed in the two experiments referred to above.

5. Theory of muonic atom collision processes

5.1. *The general theory*

The theory of the formation of (μ^-p) and (μ^-d) atoms has been given in § 2. We wish, now, to discuss their collision processes in hydrogen, deuterium, and heavier materials. Owing to the large mass of the muon, compared with that of the electron, (μ^-p) and (μ^-d) atoms form compact neutral objects of radius $2 \cdot 5 \times 10^{-11}$ cm. In nuclear collision processes, therefore, their distance of approach to the scattering nucleus will be much smaller than the mean distance of the nucleus from the surrounding electrons, so that it is legitimate to consider the collision as a three-body process involving the μ^- and the two nuclei.

Let $\mathbf{r}$ be the position vectors of the muon relative to the centre of mass of the two nuclei (1, 2), and $\mathbf{R}$ the relative vector from nucleus 1 to nucleus 2, then the Schrödinger equation for the system is

$$H\Psi(\mathbf{r},\mathbf{R}) \equiv \left\{-\frac{\hbar^2}{2m}\nabla_{\mathbf{r}}^2 - \frac{e^2}{r_1} - \frac{e^2}{r_2} + \frac{e^2}{R} - \frac{\hbar^2}{2M}\nabla_{\mathbf{R}}^2\right\}\Psi(\mathbf{r},\mathbf{R}) = E\Psi(\mathbf{r},\mathbf{R}) \tag{79}$$

where r_1, r_2 are, respectively, the distances of the μ^- from the two nuclei, and $M = \dfrac{M_1 M_2}{M_1 + M_2}$, $m = \dfrac{m_\mu(M_1 + M_2)}{M_1 + M_2 + m_\mu}$, m_μ, M_1, M_2 being, respectively, the masses of the μ^- and the two nuclei, and E the total energy. Although the ratios m_μ/M_1, m_μ/M_2 are around 0·1, or less, they are much larger than the corresponding ratios for electrons. We must expect, therefore, that the convergence of a method of obtaining solutions of eqn (79), based on the assumption that m_μ/M_1, m_μ/M_2 are

† ZEL'DOVICH, YA. B. and GERSHTEĬN, S. S., *Usp. fiz. Nauk* **71** (1960) 581 (*Soviet Physics–Uspekhi* **3** (1961) 593).

small, will be much slower than for the electron case (see Vol. 2, Chap. 12, § 1 and Chap. 13, § 2.1).

We consider, first, the calculation of the energies of the lowest states of the muonic molecules $(p\mu^- p)$, $(p\mu^- d)$, and $(d\mu^- d)$. As for the electron case, we base the trial wave-function on the $1s\sigma_g$ and $2p\sigma_u$ molecular orbitals (Chap. 13, § 2.1) for the motion of the μ^- in the field of the two nuclear centres (1, 2), at fixed separation, R. Thus we write

$$\Psi(\mathbf{r}, \mathbf{R}) = \chi_g(\mathbf{r}, \mathbf{R}) F_g(\mathbf{R}) + \chi_u(\mathbf{r}, \mathbf{R}) F_u(\mathbf{R}), \tag{80}$$

where
$$\left(-\frac{\hbar^2}{2m} \nabla_{\mathbf{r}}^2 - \frac{e^2}{r_1} - \frac{e^2}{r_2} + \frac{e^2}{R} \right) \chi_{g,u} = E_{g,u}(R) \chi_{g,u}. \tag{81}$$

As $R \to \infty$,
$$\chi_{g,u} \to 2^{-\frac{1}{2}} \{\phi(r_1) \pm \phi(r_2)\}, \tag{82}$$

where $\phi(r_1)$, $\phi(r_2)$ are $1s$ wave-functions, for motion of a μ^- in the Coulomb field of a nucleus of mass $M_1 + M_2$. The fact that the total nuclear mass appears, and not the respective masses M_1, M_2, is somewhat surprising, at first. It was not taken account of in the discussion of the theory of ordinary molecules, because the effect of finite nuclear mass on the wave-functions was neglected. For muonic molecules, however, it is more important. We shall see that, when dynamic corrections to the wave-functions of eqn (80), due to nuclear motion, are taken into account, the correct asymptotic form at large nuclear separation is regained.

We now proceed as in Vol. 3, p. 1871 for the electron case. Substituting eqn (80) in eqn (79), multiplying in turn by $\chi_g(\mathbf{r}, \mathbf{R})$, $\chi_u(\mathbf{r}, \mathbf{R})$, integrating over $\mathbf{r}$, and using eqn (81),

$$\left\{ \frac{\hbar^2}{2M} \nabla_{\mathbf{R}}^2 + E - E_g(R) \right\} F_g(\mathbf{R}) = -V_{ug}(R) F_u(\mathbf{R}) - V_{gg}(R) F_g(\mathbf{R}), \tag{83a}$$

$$\left\{ \frac{\hbar^2}{2M} \nabla_{\mathbf{R}}^2 + E - E_u(R) \right\} F_u(\mathbf{R}) = -V_{uu}(R) F_u(\mathbf{R}) - V_{gu}(R) F_g(\mathbf{R}), \tag{83b}$$

where
$$\left. \begin{array}{l} V_{jk} = \dfrac{\hbar^2}{2M} \{ 2\boldsymbol{\xi}_{jk} \cdot \nabla_{\mathbf{R}} + \nabla_{\mathbf{R}} \cdot \boldsymbol{\xi}_{jk} - \zeta_{jk} \} \\[4pt] \boldsymbol{\xi}_{jk} = \int \chi_j(\mathbf{r}, \mathbf{R}) \nabla_{\mathbf{R}} \chi_k(\mathbf{r}, \mathbf{R}) \, d\mathbf{r} = -\boldsymbol{\xi}_{kj} \\[4pt] \zeta_{jk} = \int \nabla_{\mathbf{R}} \chi_j(\mathbf{r}, \mathbf{R}) \cdot \nabla_{\mathbf{R}} \chi_k(\mathbf{r}, \mathbf{R}) \, d\mathbf{r} = \zeta_{kj} \end{array} \right\}. \tag{84}$$

The expressions (84) are the dynamical terms involving the nuclear velocities through the gradients $\nabla_{\mathbf{R}}$, and they are much more important in the muonic than the electronic case. In the hydrogen molecular ion, for example, the velocities of the nuclei are small compared with those of the electrons, and the adiabatic approximation in which the nuclei stay effectively fixed over several cycles of the electronic motion is a good approximation. The expressions (84) are negligible, and the two nuclei can be considered to move in the potentials $E_g(R)$, $E_u(R)$, in the two cases. For a muonic hydrogen ion on the other hand the nuclear velocity is much greater relative to the muon velocity, and the terms (84) are more important, modifying the effective potentials in which the nuclei move.

In the case of two like nuclei, the total Hamiltonian is symmetrical in the nuclear coordinates, so that the states of the system must be either symmetrical, or antisymmetrical, against interchange of the nuclear coordinates. Eqn (80) must then be written, for states of zero rotational quantum number L,

$$\Psi(\mathbf{r}, \mathbf{R}) = \chi_g(\mathbf{r}, \mathbf{R}) F_g(\mathbf{R})$$

for the symmetrical states, and
$$\Psi(\mathbf{r},\mathbf{R}) = \chi_u(\mathbf{r},\mathbf{R})F_u(\mathbf{R})$$
for the antisymmetrical states, and $V_{ug} = V_{gu} = 0$ so that the eqns (83) are uncoupled.

Writing $F_{g,u} = R^{-1}G_{g,u}$, eqns (83) become

$$\frac{d^2G_{g,u}}{dR^2} + \frac{2M}{\hbar^2}\{E - E_{g,u}(R) + V_{gg,uu}\}G_{g,u} = 0. \tag{85}$$

Asymptotically, the dynamical terms given by eqn (84) are related to reduced mass corrections. Cohen et al.† showed that, for $R \to \infty$,

$$\xi_{uu} = \xi_{gg} = 0, \quad \xi_{ug} \to 0, \quad \zeta_{uu}, \zeta_{gg} \to \frac{1}{2}\left(\frac{me^2}{\hbar^2}\right)^2 (f_1^2+f_2^2),$$

$$\zeta_{ug} = \zeta_{gu} \to \left(\frac{me^2}{2\hbar^2}\right)(f_1-f_2), \tag{86}$$

where $f_{1,2} = \dfrac{M_{1,2}}{M_1+M_2}$. In this limit, V_{gg} and V_{uu} tend to the constant value

$$\tfrac{1}{4}(me^2/\hbar^2)^2 \frac{\hbar^2}{2M}.$$

Also E_g, E_u tend to $me^4/2\hbar^2$ so that

$$E_g - V_{gg} \to E_u - V_{uu} = (me^4/2\hbar^2)\left(1 - \frac{1}{4}\frac{m}{M}\right)$$

$$= \frac{2m_\mu M_1}{M_1+m_\mu}\frac{e^4}{2\hbar^2} + O\left\{\left(\frac{m_\mu}{M_1}\right)^2\right\}.$$

Thus, when the dynamical corrections are included, the limiting value of the muonic binding energy, at large nuclear separation, agrees with that for a single muonic atom to terms of order $(m_\mu/M_1)^2$.

For unlike nuclei the equations corresponding to eqn (85) are coupled. Asymptotically, for $R \to \infty$, they can be uncoupled by writing

$$G_p = (G_g - G_u)/\sqrt{2}, \qquad G_d = (G_g + G_u)/\sqrt{2}, \tag{87}$$

referring to the case where nucleus 1 is a proton, and nucleus 2 a deuteron. We then find

$$\frac{d^2G_p}{dR^2} + \frac{2M}{\hbar^2}\left\{E + \left(\frac{1}{2} - \frac{2m}{9M}\right)\frac{me^2}{\hbar^2}\right\}G_p = 0,$$

$$\frac{d^2G_d}{dR^2} + \frac{2M}{\hbar^2}\left\{E + \left(\frac{1}{2} - \frac{m}{18M}\right)\frac{me^2}{\hbar^2}\right\}G_d = 0. \tag{88}$$

Asymptotically, G_p corresponds to a total wave-function in which the μ^- is centred on the proton, and G_d, the corresponding (μ^-d) case. The binding energy of the meson (corresponding to the case when the kinetic energy of relative nuclear motion is zero) is then

$$E_{\mu p} = \left(-\frac{1}{2} + \frac{2}{9}\frac{m}{M}\right)\frac{me^4}{\hbar^2}, \quad \text{for } (\mu^-p)+d,$$

and

$$E_{\mu p} = \left(-\frac{1}{2} + \frac{1}{18}\frac{m}{M}\right)\frac{me^4}{\hbar^2}, \quad \text{for } (\mu^-d)+p,$$

which, again, are the correct expressions to this order of m/M for the units used.

† loc. cit., p. 3334.

The extension of this analysis to cases in which the rotational quantum number $L \neq 0$, offers no difficulty.

5.2. States of muonic molecules

Several groups of workers† have solved eqn (81) numerically, either exactly, or by a variational method, and thence calculated the dynamical correction terms for $(p\mu^-p)$, $(d\mu^-d)$, and $(p\mu^-d)$ systems. In this way, the molecular ion potential energy curves, and energies of the bound states have been obtained. For the $(p\mu^-p)$ and $(d\mu^-d)$ systems, the potential energy functions, for even and odd states, are given by eqn (85). For the $(p\mu^-d)$ case, the wave-function $\Psi'(\mathbf{r}, \mathbf{R})$ of eqn (80) was approximated by the even term $\chi_g(\mathbf{r}, \mathbf{R})F_g(R)$, in the zero-order approximation, leading to eqn (86), for the potential energy function. The effect of the $\chi_u(\mathbf{r}, \mathbf{R})F_u(R)$ term, can then be introduced as a perturbation. The equilibrium separation between the nuclei is then approximately $2a_\mu \simeq 5 \times 10^{-11}$ cm, and the minimum in the potential energy function is approximately 580 eV. The separations between successive vibrational and rotational levels are, however, for muonic hydrogen molecules, comparable with the well depth of the potential energy function, so that muonic molecules have very few such levels, in contrast to the situation for H_2^+.

Fig. 27.24 shows potential energy curves calculated by Zel'dovich and Gershtein‡ for $(p\mu^-p)$, $(d\mu^-d)$, and $(p\mu^-d)$ muonic molecules, including dynamical corrections for different rotational quantum numbers. The positions of the vibrational levels are indicated. The potential energy curves are shown approximated by the Morse potential

$$V(R) = D(e^{-2\alpha(R-R_0)} - 2e^{-\alpha(R-R_0)}),$$

with values of the parameters D, α, R_0 given in the caption.

Table 27.10, extracted from a similar compilation by Zel'dovich and Gershtein,‡ illustrates the results of various calculations of muonic molecular energy levels. The calculations predict the presence of a virtual $L = 0$, $v = 1$ level, for the $(p\mu^-p)$ molecule, just above zero-energy, which is expected significantly to increase the probability of formation of $(p\mu^-p)$ for (μ^-p) collisions in hydrogen.

5.3. Collisions of (μ^-p) and (μ^-d) in hydrogen and deuterium

Since we are only interested in collisions with small kinetic energy we consider only the $L = 0$ case. To find a solution of eqn (85) representing

† BELYAEV, W. B., GERSHTEĬN, S. S., ZAKHER'EV, B. N., and LOMNEV, S. P., *Zh. eksp. teor. Fiz.* **37** (1959) 1652 (English translation, *Soviet Physics, JETP* **10** (1960) 1171); COHEN, S., JUDD, D. L., and RIDDELL, R. J., *Phys. Rev.* **119** (1960) 384, 397.

‡ ZEL'DOVICH, YA. B. and GERSHTEĬN, S. S., loc. cit.

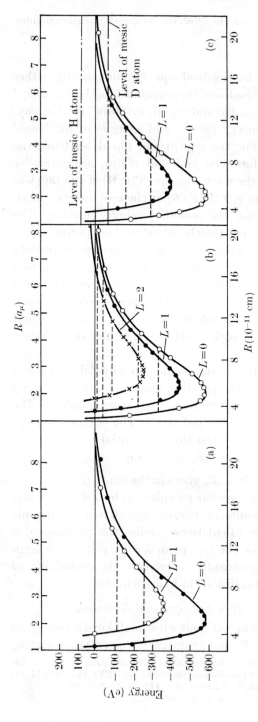

FIG. 27.24. Potential energy curves of muonic molecules showing the lowest vibrational energy levels. The points show the curves approximated by the Morse potential

$$V(R) = D(e^{-2\alpha(R-R_0)} - 2e^{-\alpha(R-R_0)})$$

with values of D, α, R_0 given (energy and length in muon atomic units).

(a) ($p\mu^-p$) for rotational states $L = 0, 1$.

L	$R_0 (a_\mu)$	$\alpha (a_\mu^{-1})$	D (eV)	symbol
0	2·15	0·67	591·6	●
1	2·62	0·69	364·5	○

(b) ($d\mu^-d$) for rotational states $L = 0, 1, 2$.

L	$R_0 (a_\mu)$	$\alpha (a_\mu^{-1})$	D (eV)	symbol
0	2·09	0·67	583·2	○
1	2·34	0·69	448·6	●
2	2·76	0·75	257·9	×

(c) ($d\mu^-p$) for rotational states $L = 0, 1$.

L	$R_0 (a_\mu)$	$\alpha (a_\mu^{-1})$	D (eV)	symbol
0	2·10	0·66	587·7	○
1	2·48	0·67	401·5	●

the scattering of (μ^-p) in deuterium we write in eqn (80) the asymptotic forms (see eqn (82))

$$\chi_g(\mathbf{r}, \mathbf{R}) \sim 2^{-\frac{1}{2}}\{\phi(\mathbf{r}_1)+\phi(\mathbf{r}_2)\},$$
$$\chi_u(\mathbf{r}, \mathbf{R}) \sim 2^{-\frac{1}{2}}\{\phi(\mathbf{r}_1)-\phi(\mathbf{r}_2)\},$$

and
$$F_g(\mathbf{R}) = 2^{-\frac{1}{2}}\{F_p(\mathbf{R})+F_d(\mathbf{R})\},$$
$$F_u(\mathbf{R}) = 2^{-\frac{1}{2}}\{F_p(\mathbf{R})-F_d(\mathbf{R})\}, \tag{89}$$

TABLE 27.10

Binding energy of muonic molecules (eV)

Rotational level	$L = 0$		$L = 1$	$L = 2$
Vibrational level	$v = 0$	$v = 1$	$v = 0$	$v = 0$
($p\mu^-p$)	252†		106†	
	241‡		93‡	
	248§			
($p\mu^-d$)	223†		95†	
	214‡		90‡	
	220‖		90‖	
($d\mu^-d$)	330†	40†	226†	88†
	322‡		223‡	82‡
	317§			

† GERSHTEĬN, S. S. (1958). Dissertation, quoted by Zel'dovich, Ya. B. and Gershtein, S. S., loc. cit.
‡ COHEN, S., JUDD, D. L., and RIDDELL, R. J. *Phys. Rev.* **119** (1960) 397.
§ KOLOS, ROOTHAAN, and SACK (preprint quoted by Zel'dovich, Ya. B., and Gershteĭn, S. S., loc. cit.).
‖ BELYAEV, V. B., GERSHTEĬN, S. S., ZAKHAR'EV, B. V., and LOMNEV, S. P., *Zh. eksp. teor. Fiz.* **37** (1959) 1652 (English translation, *Soviet Physics, JETP* **10** (1960) 1171).

where $F_p(\mathbf{R}) = R^{-1}G_p(\mathbf{R})$, $F_d(\mathbf{R}) = R^{-1}G_d(\mathbf{R})$. Hence,

$$\Psi(\mathbf{r}, \mathbf{R}) \sim \phi(\mathbf{r}_1)F_p(\mathbf{R})+\phi(\mathbf{r}_2)F_d(\mathbf{R}). \tag{90}$$

The first term then represents the state in which the μ^- is bound to the proton, and the second term that in which it is bound to the deuteron. The solution of eqn (85) representing scattering, is that which satisfies the boundary condition

$$F_p \sim \frac{\sin kR}{kR} + f\frac{e^{ikR}}{R}, \tag{91 a}$$

$$F_d \sim f'\frac{e^{ik'R}}{R}, \tag{91 b}$$

with
$$\frac{k'^2\hbar^2}{2m_d} = \frac{k^2\hbar^2}{2m_p} + E_{(\mu p)} - E_{(\mu d)}, \tag{91 c}$$

m_d, m_p being, respectively, the reduced masses of (μ^-d), (μ^-p), and $E_{(\mu p)}$, $E_{(\mu d)}$ the energies of their ground states. $(E_{(\mu p)} - E_{(\mu d)}) \simeq 134$ eV.

Unfortunately, eqn (90) does not exactly represent the scattering situation, because $\phi(\mathbf{r}_1)$, $\phi(\mathbf{r}_2)$ are wave-functions for a μ^- hydrogenic atom, with a nuclear mass equal to the sum of the proton and deuteron masses. The difference between eqn (90), and an equation that represents the scattering correctly, is small and involves terms of order m/M, but it is difficult to estimate the effect of the approximation. The most significant calculations of (μ^-p) collision processes have been made using this approximation (Belyaev et al.,[†] Zel'dovich and Gershteĭn,[‡] Shimizu et al.,[§] Cohen et al.[||]). A calculation by Burke et al.[††] using a resonating group structure method uses in place of eqn (80) the form

$$\psi(\mathbf{r}, \mathbf{R}) = \phi_p(\mathbf{r}_1) F_p(\mathbf{R}) + \phi_d(\mathbf{r}_2) F_d(\mathbf{R}), \tag{92}$$

where $\phi_p(\mathbf{r}_1)$, $\phi_d(\mathbf{r}_2)$ are the wave-functions of the 1s states of (μ^-p) and (μ^-d), respectively. This expression imposes the correct asymptotic form. For such slow particles, the neglect of polarization of the (μ^-p) and (μ^-d) atoms during collision is likely to be very serious, however, and in fact this calculation gives cross-sections much too small, and in considerably poorer agreement with the experimental results than obtained using eqn (80) with asymptotic form (90).

The substitution of eqn (89) can, in fact, be made for $\chi_g(R)$, $\chi_u(R)$, in eqns (83), and a pair of coupled differential equations obtained for $F_p(R)$ and $F_d(R)$, leading in turn to a pair of coupled ordinary differential equations for $G_p(R)$, $G_d(R)$. These lead to asymptotic forms (91) for F_p, F_d, provided $G_p(R)$, $G_d(R)$ satisfy the boundary conditions

$$\left.\begin{array}{l} G_p(0) = G_d(0) = 0 \\ G_p \sim i^l \sin kr + \alpha\, e^{ikr} \\ G_d \sim \beta\, e^{ik'r} \end{array}\right\}.$$

The method of finding scattering solutions of this form, and deducing scattering cross-sections from them is described by Mott and Massey.[‡‡]

[†] BELYAEV, V. B., GERSHTEĬN, S. S., ZAKHAR'EV, B. W., and LOMNEV, S. P., Zh. eksp. teor. Fiz. 37 (1959) 1652 (English translation, Soviet Physics, JETP 10 (1960) 1171).
[‡] ZEL'DOVICH, YA. B. and GERSHTEĬN, S. S., Usp. fiz. Nauk 71 (1960) 581 (English translation, Soviet Physics–Uspekhi 3 (1961) 593).
[§] SHIMIZU, M., MIZUNO, Y., and IZUYAMA, T., Prog. theor. Phys., Osaka 20 (1958) 777.
[||] COHEN, S., JUDD, D. L., and RIDDELL, R. J., Phys. Rev. 119 (1960) 384, 397.
[††] BURKE, P. G., HAAS, F., and PERCIVAL, I. C., Proc. phys. Soc. (London) 73 (1959) 912.
[‡‡] MOTT, N. F. and MASSEY, H. S. W., The theory of atomic collisions, 3rd edition, Clarendon Press, Oxford (1965), pp. 347–8.

The amplitudes α, β, determine, respectively, the cross-sections for the elastic scattering process

$$(\mu^- p) + d \to (\mu^- p) + d,$$

and the superelastic exchange process

$$(\mu^- p) + d \to (\mu^- d) + p.$$

The case of scattering of $(\mu^- d)$ in hydrogen is treated similarly. In this case, two regions of the incident energy have to be considered, according as

$$k^2 \hbar^2 / 2m_d < \text{ or } > E_{(\mu p)} - E_{(\mu d)}.$$

In the former case only elastic scattering is possible, namely,

$$(\mu^- d) + p \to (\mu^- d) + p,$$

and the cross-section is given in terms of a phase shift, (η_d) by

$$Q_{dp}^{el} = \frac{4\pi}{k^2} \sin^2 \eta_d. \tag{93}$$

In the latter case the inelastic exchange process

$$(\mu^- d) + p \to (\mu^- p) + d$$

can also occur, and the treatment is similar to $(\mu^- p)$ in d. For the case of like nuclei, such as the scattering of $(\mu^- p)$ in hydrogen, or $(\mu^- d)$ in deuterium, the equations for $F_g(\mathbf{R})$ and $F_u(\mathbf{R})$ are uncoupled, and writing

$$F_g(\mathbf{R}) = \frac{G_g(R)}{R}, \qquad F_u(\mathbf{R}) = \frac{G_u(R)}{R},$$

solutions can be determined for $\eta_+(R)$, $\eta_-(R)$ satisfying the boundary conditions

$$G_g(0) = G_u(0) = 0, \qquad G_g \sim \sin(kr + \eta^g), \qquad G_u \sim \sin(kr + \eta^u).$$

The elastic scattering cross-sections are then

$$\left. \begin{array}{l} Q_{pp}^{el} = 4\pi k^{-2} [\tfrac{1}{4} \sin^2 \eta_{pp}^g + \tfrac{3}{4} \sin^2 \chi_{pp}^u] \\ Q_{dd}^{el} = 4\pi k^{-2} [\tfrac{2}{3} \sin^2 \eta_{dd}^g + \tfrac{1}{3} \sin^2 \eta_{dd}^u] \end{array} \right\}, \tag{94}$$

in the two cases.

5.3.1. Scattering phases and cross-sections. Fig. 27.25 (a) shows the phase shifts, $\eta_{pp}^{g,u}$, $\eta_{dd}^{g,u}$, calculated by Cohen et al.† as a function of k (in units a_μ^{-1}), while Fig. 27.25 (b) shows η_d/k for very small values of k. A feature of the latter phase is the passage through zero for $k \sim 0.02 a_\mu^{-1}$, corresponding to an energy of 0.2 eV. This gives rise to a Ramsauer effect, and ensures that Q_{dp}^{el} is very small.

† loc. cit.

Fig. 27.26 shows the cross-sections for Q_{pp}^{el}, Q_{dd}^{el}, Q_{dp}^{el} as a function of the centre-of-mass energy, and it is seen that $Q_{dp}^{el} \ll Q_{dd}^{el}$. This small value of Q_{dp}^{el} provides the explanation of the gaps between the end point of the stopping μ^- meson, and its point of re-emission after catalysis of

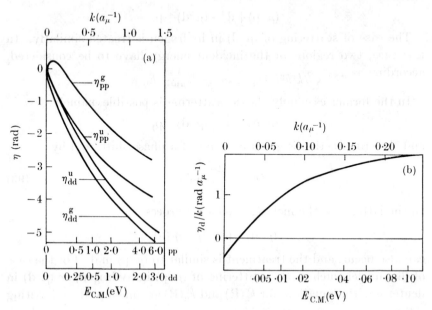

FIG. 27.25. (a) Phase shifts (η_{pp}^g, η_{pp}^u) for s wave scattering of (μ^-p) by p and η_{dd}^g, η_{dd}^u for s wave scattering of (μ^-d) by d. (b) η_d/k for scattering of (dμ) atoms by protons, illustrating the origin of the Ramsauer effect in this case (Cohen, Judd, and Riddell, loc. cit.).

the nuclear fusion process (see § 4.2.1). The value of Q_{dp}^{el} in the energy region around 1 eV, calculated by Cohen et al., is about 10^{-21} cm^2, in good agreement with the estimate made by Dzhelepov et al.† of $0.8 ^{+0.8}_{-0.4} \times 10^{-21}$ cm^2, from a study of the gap size distribution in the diffusion cloud chamber.

The values of Q_{pp}^{el} calculated in this way, are also in satisfactory agreement with the values estimated from the experiment of Alberigi-Quaranta et al.‡ as shown in Table 27.9. For Q_{dd}^{el}, however, the cross-section calculated by Cohen et al. is six times larger than the experimental estimates of Alberigi-Quaranta et al., but in closer agreement with that derived from the diffusion chamber studies of Dzhelepov et al. The value of the cross-section Q_{pd}^{el} of process (56), which plays such an

† loc. cit.   ‡ loc. cit.

important role in μ^--catalysed nuclear fusion, was calculated, in terms of the initial velocity (v) of relative motion, to be

$$3\cdot 3\times 10^{-13}/v\ \text{cm}^2,$$

by Cohen et al., about 20 per cent less than the value estimated experimentally (see Table 27.8).

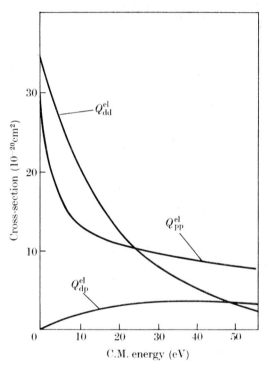

FIG. 27.26. Muonic atom elastic scattering cross-sections as a function of centre-of-mass energy.

5.3.2. *Hyperfine effects in* (μ^-p) *and* (μ^-d) *scattering*. The scattering cross-sections calculated by Cohen et al.† have ignored hyperfine effects produced by the magnetic interaction between (μ^-p) or (μ^-d) and the target nucleus. The cross-sections derived are relevant to the case of a statistical distribution, of the incident and target particles, among the possible spin states. To take account of scattering between particular hyperfine states in (μ^-p)–p collisions, say, an additional term,

$$+\tfrac{4}{3}g_i\beta_\mu\beta_N\left(\frac{\delta(\mathbf{r}_1)}{r_1^2}(\mathbf{S}\cdot\mathbf{i}_1)+\frac{\delta(\mathbf{r}_2)}{r_2^2}(\mathbf{S}\cdot\mathbf{i}_2)\right), \tag{95}$$

† loc. cit.

has to be added to the Hamiltonian (eqn (79)). In eqn (95) β_μ, β_N are, respectively, the muonic and nuclear Bohr magneton $g_i = 2 \times 2\cdot 79$, the gyromagnetic ratio of the proton, and i_1, i_2, and S the spins of the two protons and the meson, respectively.

Gershtein[†] has calculated cross-sections for (μ^-p)–p, and (μ^-d)–d scattering between particular initial and final hyperfine states. The (μ^-p) atom has hyperfine states $F = 1, 0$. We consider the collision process (see eqn (43))

$$(\mu^-p)_{F=1} + p \to p + (\mu^-p)_{F=0}.$$

The (μ^-p)–p system may have total spin values $\tfrac{3}{2}, \tfrac{1}{2}$. However, transitions to the $F = 0$ hyperfine structure state are possible only for states of total spin $\tfrac{1}{2}$. The wave-function for such a state, antisymmetric under interchange of protons, can be written

$$\psi(\mathbf{r}, \mathbf{R}) = \chi_g(\mathbf{r}, \mathbf{R}) F_g(\mathbf{R}) S^0(1, 2; \mu) + \chi_u(\mathbf{r}, \mathbf{R}) F_u(\mathbf{R}) S^1(1, 2; \mu), \tag{96}$$

where $S^0(1, 2; \mu)$, $S^1(1, 2; \mu)$ are spin functions corresponding to resultant spin 0, 1, respectively, of the particles whose indices are separated by the semicolon. The total spin of the three particles is $\tfrac{1}{2}$, and its projection on the z axis is $+\tfrac{1}{2}$. Then

$$S^0(1, 2; \mu) = \frac{1}{\sqrt{2}} [\alpha(1)\beta(2) - \beta(1)\alpha(2)]\alpha(\mu), \tag{97}$$

$$S^1(1, 2; \mu) = \frac{1}{\sqrt{6}} [\alpha(1)\beta(2) + \beta(1)\alpha(2)]\alpha(\mu) - \sqrt{\tfrac{2}{3}}\alpha(1)\alpha(2)\beta(\mu). \tag{98}$$

For a sufficiently large distance between the protons the wave-function of the system with total spin $\tfrac{1}{2}$, and spin projection $+\tfrac{1}{2}$, for the hyperfine state $F = 0$ is

$$\psi_0(\mathbf{r}, \mathbf{R}) \sim \phi(\mathbf{r}_1) F_0(\mathbf{R}) S^0(1, \mu; 2) - \phi(\mathbf{r}_2) F_0(\mathbf{R}) S^0(2, \mu; 1), \tag{99}$$

and for the $F = 1$ state,

$$\psi_1(\mathbf{r}, \mathbf{R}) \sim \phi(\mathbf{r}_1) F_1(\mathbf{R}) S^1(1, \mu; 2) - \phi(\mathbf{r}_2) F_1(\mathbf{R}) S^1(2, \mu; 1), \tag{100}$$

where, as before, $\phi(\mathbf{r}_1)$, $\phi(\mathbf{r}_2)$ are the wave-functions of the muon at the first and second protons, respectively. Using the asymptotic forms (eqn (82)) for $\chi_g(\mathbf{r}, \mathbf{R})$, $\chi_u(\mathbf{r}, \mathbf{R})$ in eqn (96), and equating (96) to a linear combination of eqn (99) and eqn (100) gives

$$\left. \begin{array}{l} F_g(\mathbf{R}) = (F_0 - \sqrt{3}\, F_1)/\sqrt{2} \\ F_u(\mathbf{R}) = (-\sqrt{3}\, F_0 - F_1)/\sqrt{2} \end{array} \right\}. \tag{101}$$

We proceed exactly as in § 5.1, substituting eqn (96) in eqn (79) with the additional term eqn (95), leading to a pair of coupled equations analogous to eqns 82 (a), (b). Then, using relations (101), these are converted into two coupled equations for F_0, F_1. Solutions of these with asymptotic form

$$\left. \begin{array}{l} F_1 \sim \dfrac{\sin kR}{kR} + f\, \dfrac{e^{ikR}}{R} \\[4pt] F_0 \sim f'\, e^{ik'R}/R \end{array} \right\}, \tag{102}$$

with $(k'^2 - k^2)\hbar^2/2M_p = \Delta E$, the hyperfine splitting, representing the scattering.

$$Q_{1 \to 1} = 4\pi |f|^2 \tag{103}$$

[†] Gershteĭn, S. S., *Zh. eksp. teor. Fiz.* **34** (1958) 463 (English translation, *Soviet Physics, JETP* **7** (1958) 318); ibid. **40** (1961) 698 (**13** (1961) 488).

is the cross-section for elastic scattering for ($F = 1$) hyperfine states while

$$Q_{1\to 0} = 4\pi \frac{k'}{k} |f'|^2 \tag{104}$$

is the cross-section for scattering with the hyperfine transition $F = 1 \to F = 0$.

The results of Gershteĭn's calculations for (μ^-p)–p and (μ^-d)–d scattering between particular hyperfine states are given in §§ 3.6, 4.4.2, and compared with the experimental results of Alberigi-Quaranta et al.† and Dzhelepov et al.‡ for Q_0^{el} ($F = 0 \to F = 0$), and $Q_{1\to 0}$, in Table 27.9. There is good agreement for $Q_{1\to 0}$ but the theoretical value for Q_0^{el} is too small by a factor of 6. On the other hand, the value for Q_0^{el} obtained by Cohen et al.§ is in good agreement with the experimental value. These calculations have been extended for mesic atom energies in the range 10^{-3} to 0.5 eV by Matveenko and Ponomarev.∥

5.3.3. *Transfer of μ^- from hydrogen to heavier elements.* In § 4.4 we have seen that, in the passage of (μ^-p) and (μ^-d) through hydrogen containing impurities of C, O, Ne or other heavier nuclei, the μ^- rapidly transfers to the heavier nucleus. Gershteĭn†† has calculated the cross-sections $Q_{p,Z}$, $Q_{d,Z}$ for such processes. The Schrödinger equation for the system is similar to eqn (79), but with $-e^2\left(\frac{1}{r_2}+\frac{1}{R}\right)$ replaced by $-Ze^2\left(\frac{1}{r_2}+\frac{1}{R}\right)$, and the theory follows along the lines of that for the transfer of the μ^- from p to d. In the present case, however, it is necessary to take account of the possibility of transfer not only to the ground state, but to several excited states of the heavier muonic atom, so that the problem involves the solution of a number of simultaneous coupled differential equations of type (83), namely,

$$\left.\begin{aligned}\left\{\frac{\hbar^2}{2M}\nabla_R^2+(E-E_0-V_{00})\right\}F_0 &= \sum_i V_{0i} F_i \\ \left\{\frac{\hbar^2}{2M}\nabla_R^2+(E-E_i-V_{ii})\right\}F_i &= V_{i0} F_0\end{aligned}\right\}, \tag{105}$$

where F_0 corresponds, for large R, to the state in which the μ^- is attached to the proton, and F_i to states where it is attached to the nucleus Z in the ith atomic state. The method of solution employed is similar to that for the transfer of excitation in the collision of atomic systems (see Vol. 3, Chap. 18, § 8.4.1). The cross-section is determined by the properties of crossing molecular terms. Consider first the potential energy term $V_{00}(R)$, corresponding at large distances to the system (μ^-p)+Z, then

$$V_{00} \simeq -\frac{9}{4}\frac{Z^2}{R^4}+O(e^{-R}), \quad R \to \infty, \tag{106a}$$

$$V_{00} \simeq \frac{Z}{R}, \quad R \to 0, \tag{106b}$$

---

† loc. cit., p. 3324.   ‡ loc. cit., p. 3321.   § loc. cit., p. 3334.
∥ Matveenko, A. V. and Ponomarev, L. J., *Zh. eksp. teor. Fiz.* **59** (1970) 1593 (English translation, *Soviet Physics, JETP* **32** (1971) 871).
†† Gershteĭn, S. S., *Zh. eksp. teor. Fiz.* **43** (1962) 706 (English translation, *Soviet Physics, JETP* **16** (1963) 501).

where we have used mesic atom units ($\hbar = m_\mu = e = a_\mu = 1$).

The terms V_{ii}, on the other hand, correspond to the motion of the atom ($\mu^- Z$) in the proton field, giving

$$V_{ii} \simeq \frac{(Z-1)}{R}, \quad \text{as } R \to \infty, \tag{107a}$$

i.e. a strong repulsion at large distances, instead of the weak attraction given by eqn (106a) for V_{00}. The potential curves, $V_{ii}(R)$, therefore cross $V_{00}(R)$.

Actually, for smaller R, eqn (107a) becomes

$$V_{ii} \simeq \frac{(Z-1)}{R} + \frac{3}{2} \frac{n(n_1 - n_2)}{ZR^2}, \tag{107b}$$

where n_1, n_2 ($n = n_1 + n_2$) are two-centre parabolic quantum numbers corresponding to the state i.

The crossing points for state i ($\equiv n_1, n_2$) are then given by

$$-\frac{1}{2}\frac{M_1}{1+M_1} + \frac{Z^2}{2n^2}\left(\frac{M_2}{1+M_2}\right) = \frac{Z-1}{R} + \frac{3n(n_1-n_2)}{2ZR^2}, \tag{108}$$

where we have ignored the R^{-4} term of eqn (106a). The terms on the left-hand side arise from the energy difference between the states 0 and i at infinite R. The situation is represented in Fig. 27.27. Following the discussion of Vol. 3, Chap. 18, § 8.4.1, in the case of a single crossing if the probability of transfer from one of the crossing terms, V_{00}, to V_{ii} is P then the transition probability during the collision is $\mathscr{P} = 2P(1-P)$. Suppose now, as in Fig. 27.27 (b), the initial term V_{00} is crossed by several terms at distances $R_1, R_2, ..., R_n$, where $R_1 > R_2 > ... > R_n$. Then, if $\mathscr{P}_i$ is the transition probability to the ith term as a result of the collision,

$$\left.\begin{aligned}
\mathscr{P}_n &= 2P_1 P_2 ... P_{n-1} P_n (1-P_n) \\
\mathscr{P}_{n-1} &= 2P_1 P_2 ... P_{n-2} P_{n-1}(1-P_{n-1})\{1 - P_n(1-P_n)\} \\
&\cdot\quad\cdot\quad\cdot\quad\cdot\quad\cdot\quad\cdot\quad\cdot\quad\cdot\quad\cdot\quad\cdot\quad\cdot\quad\cdot \\
\mathscr{P}_i &= 2P_1 P_2 ... P_{i-1} P_i(1-P_i)[1 - P_{i+1}(1-P_{i+1}) - P_{i+2}^2 P_{i+2}(1-P_{i+2})... \\
&\qquad\qquad\qquad ... P_{i+1}^2 P_{i+2}^2 ... P_{n-1}^2 P_n(1-P_n)]
\end{aligned}\right\}, \tag{109}$$

and the total transition probability to all possible states

$$\mathscr{P} = \sum_{i=1}^{n} \mathscr{P}_i = 2[P_1(1-P_1) + P_1^2 P_2(1-P_2) + ... + P_1^2 P_2^2 ... P_{n-1}^2 P_n(1-P_n)].$$

The cross-section for the transition to state i, for relative momentum $k\hbar$ is

$$Q_i = \pi k^{-2} \mathscr{P}_i. \tag{110}$$

Writing
$$P_i = e^{-\delta_i} \tag{111a}$$

we have† (Chap. 18, p. 1917)

$$\delta_i = \{2\pi |V_{i0}|^2\}/v_i |F^i - F^0|\}_{R=R_0}, \tag{111b}$$

where $F^i = -dV_{ii}/dR$, $F^0 = -dV_{00}/dR$, and v_i is the relative velocity of the nuclei at the point R_i, where the V_{ii} and V_{00} cross, involves the relative angular momentum in the collision.

Gershteĭn calculated the total transition probability $\mathscr{P}$ for carbon, and oxygen, for s state relative motion and obtained $\mathscr{P} = 0.21$ for

† STUECKELBERG, E. C. G., *Helv. phys. Acta* **5** (1932) 370.

carbon, 0·32 for oxygen. He pointed out, however, that the expression (110) cannot apply down to the smallest collision energies, where we should have $Q \propto 1/v$. By careful normalization of the solution in the

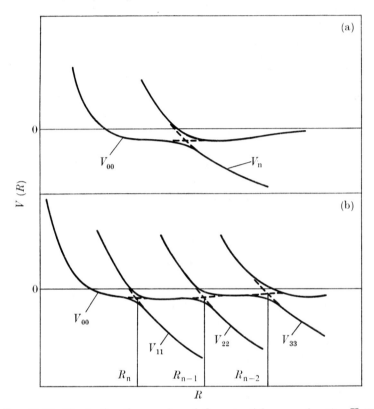

Fig. 27.27. Illustrating the crossing of the potential energy function $V_{00}(R)$ corresponding asymptotically to the system $(\mu^-p)+Z$ and functions $V_{ii}(R)$ corresponding asymptotically to the system $p+(\mu^-Z)_i$ with the mesic atom in the ith excited state (after Gershteĭn, loc. cit.).

asymptotic region, Gershteĭn obtained for the total transition cross-section, at collision energies $E \ll 8/Z^2$ (eV),

$$Q_{\mathrm{pZ}} = 3\pi \left(\frac{2}{M}\right)^{\frac{1}{2}} \frac{Z\mathscr{P}}{v}. \tag{112}$$

A comparison of the cross-sections given by eqn (112), with the experimental results, is shown in Fig. 27.19 and the agreement is satisfactory.

5.4. *Formation of muonic molecules*

The cross-sections for muonic molecule formation in the process

$$(\mu^-p)+H \to (p\mu^-p)+e,$$

and similar processes involving (μ^-d) and D, have been estimated by several investigators.† The most important contribution comes from an E1 transition of the (μ^-p)–p system from an incident s state to a bound p state of the (pμ^-p) molecular ion, with the radiationless transition of the electron into the continuum. The electron wave-function, in the initial state, allowed for the effect of the second proton in the target H_2 molecule, and the final state of the ejected electron was taken as a Coulomb function in the field of an effective nuclear charge. If terms of order M_μ/M are neglected the electric dipole moment (**d**) is given by

$$\mathbf{d} = -\frac{e}{2}\left\{\frac{M_2-M_1}{M_2+M_1}\mathbf{R}+(\mathbf{r}_1+\mathbf{r}_2)\right\}, \tag{113}$$

where **R** is the relative position vector of the two nuclei of masses M_1, M_2, and $\mathbf{r}_1$, $\mathbf{r}_2$ the position vectors of the two nuclei relative to the muon. The cross-section for muonic molecule formation in this process is then

$$Q_{\text{pp}}^{(M)} = \frac{16\pi e^2 a_\mu^3 \eta \zeta^2}{3 a_e^3 h v_0}\left(\sum_m |\langle \mathbf{d}\rangle|\right)^2, \tag{114a}$$

where
$$\eta = e^2/hv_e, \tag{114b}$$

$$\zeta = \left[\frac{2\pi\eta}{(1+\eta^2)(1-e^{-2\pi\eta})}\right]\exp(-2\eta \cot^{-1}\eta), \tag{114c}$$

and a_μ, a_e are the Bohr radii of the μ^- and electron, respectively, v_e the velocity of the conversion electron, v_0 the relative velocity of the (μ^-p)–p system, before the collision. The matrix element of the dipole moment is taken between wave-functions of the muonic molecule corresponding to a transition between an initial continuum s state, and a bound state with $L = 1$. The summation is taken over all possible projections of the orbital angular momentum in the final state, $m = 0, \pm 1$. The matrix elements of the first term in the dipole moment (eqn (113)) vanishes for $\Sigma_g \rightleftarrows \Sigma_u$ transitions, while the second term has non-vanishing elements only for such transitions. For the case $M_1 = M_2$, therefore, only $\Sigma_u \to \Sigma_g$ transitions are possible. In this case eqn (114a) has then to be multiplied by the statistical weight of the Σ_u state ($\frac{3}{4}$ for (pμ^-p), $\frac{1}{3}$ for (dμ^-d)).

Muonic molecular ion formation could also take place through a monopole (E0) transition. The cross-section, in this case, is given by

$$Q_{\text{pp}}^{(M)}(E=0) = \frac{8\pi}{9}\left(\frac{a_\mu}{a_e}\right)^5 \eta(1-e^{-2\pi\eta})^{-1}|\langle Q_0\rangle|^2, \tag{115}$$

† See, e.g., ZEL'DOVICH, YA. B. and GERSHTEĬN, S. S., loc. cit. COHEN, S., JUDD, D. L., and RIDDELL, R. J., loc. cit.

where the monopole moment,

$$Q_0 = \sum_i e_i r_i^2, \tag{116}$$

e_i being the charges, and r_i being the distances of the three particles from their centre of mass.

The value of $Q_{\text{pp}}^{(M)}$ calculated by Cohen et al., considering only E1 transitions, is given in Table 27.8, and is seen to agree with the experimental values to better than a factor of two. The cross-section for E0 transitions was smaller by a factor of more than 10^4, than that derived for the E1 transition. The probability of muonic molecule formation by radiative capture, without the ejection of the electron, was also calculated by Cohen et al. and came out comparable to the probability of the E0 transition, so that both are negligible in comparison with the E1 transition.

The calculated cross-section for (pμ^-d) formation is not in such good agreement, being less than half the value estimated experimentally, while that calculated for (dμ^-d) formation is too small by an order of magnitude. The observed cross-section in this case is, however, much smaller than that for (pμ^-p) formation, as is expected theoretically. The cross-section for the E0 transition process could be quite large, however, if the muonic molecule possessed a vibrational excited state ($L = 0, v = 1$) with small binding energy.

5.5. *Nuclear fusion rate in muonic molecules*

Although we are more concerned here to discuss the low energy collision properties of (μ^-p) and (μ^-d), and the process of formation of muonic molecules, it is of interest to compare, also, the estimates of nuclear fusion rates obtained in muon catalysed fusion studies with those expected from a knowledge of cross-sections that can be measured directly in nuclear physics experiments, together with theoretical considerations. Since the experimental nuclear fusion rates in muonic molecules have been obtained using methods that give at the same time (μ^-p) and (μ^-d) collision rates and the formation rate of (pμ^-d) and (dμ^-d) molecules, measurements of all these rates would be expected to have comparable reliability. Unfortunately, it is difficult to estimate reliably the expected nuclear fusion rates. They depend, critically, on a detailed knowledge of the nuclear barrier and the nuclear spin states in the muonic molecule. At best, an order of magnitude estimate can be expected. For the (pμ^-d) molecule in which the μ^- catalyses the reaction (see eqn (49))

$$p + d \rightarrow He^3 + \gamma,$$

Cohen et al.† concluded that the fusion reaction occurs predominantly when the two protons of the p+d system are in a relative singlet state, and that the dominant nuclear transition is a $\frac{1}{2} \rightarrow \frac{1}{2}$ magnetic dipole transition, with a transition

† loc. cit., p. 3334.

rate of $\sim 10^7$ s^{-1}, corresponding to an assumed value of 4×10^{-13} cm for the sum of the nuclear radii of the p and d. This rate will be reduced substantially when account is taken of the proportion of (pd) $J = \frac{1}{2}$ states in the (pμ^-d) molecule, and the proportion of these states corresponding to a singlet state of the two protons. Even so, the expected rate probably still will be somewhat larger than the observed rate of 3×10^5 s^{-1} (Table 27.7). On the other hand the estimate of Zel'dovich and Gershteĭn† for the rate of this reaction of $2 \cdot 6 \times 10^5$ s^{-1}, for a value of 5×10^{-13} cm for the sum of the proton and deuteron radii, is very close to the rate estimated experimentally.

The process of re-emission of the μ^-, without γ-radiation, appears to occur as a result of a nuclear monopole transition. Once again, this occurs in a $\frac{1}{2} \to \frac{1}{2}$ nuclear transition, with the two protons in a relative singlet state. Cohen et al. calculated a rate of 5×10^5 s^{-1} for this process, so that the ratio of μ^- re-emission probability to γ emission probability would be expected to be about 0·05. The experimental value given in Table 27.7 is about 0·18. In view of the uncertainty of the theoretical estimates, this must also be regarded as not unsatisfactory.

However, for the μ^--catalysed d–d fusion in the (dμ^-d) molecule, the situation is very unsatisfactory. Zel'dovich and Gershteĭn estimate a rate of $\sim 5 \times 10^{11}$ s^{-1} for this process, while both experimental estimates of the rate (Table 27.7) give a few times 10^5 s^{-1} for the rate. The theoretical estimate was based on the assumption that the nuclear reaction occurred from an S state, whereas the rotational state with angular momentum $L = 1$, for this muonic molecule, is metastable, so that the nuclear reaction must occur from a P state. The centrifugal barrier might be expected to decrease the reaction rate by a factor of 10^2 or 10^3, but not sufficiently to make it consistent with the experimental value. The experimental results on deuterium, however, are not so convincing as those for hydrogen, and there are large discrepancies between the results obtained in the bubble chamber and the diffusion chamber. Clearly, further work, both experimental and theoretical, is needed to resolve the discrepancy.

† loc. cit., p. 3335.

NOTES ON RECENT DEVELOPMENTS

In the period since the earlier volumes of this work were prepared, research has continued at the same or even increased tempo, so many new results have been obtained as well as new techniques, both experimental and theoretical, developed. Accordingly we include notes on this new material.

To avoid delay in publication we have omitted any diagrams. Although it has been necessary to condense the description as much as possible, advantage has been taken of the extensive background available in the published volumes. The subject matter in the notes has therefore been ordered in the same way as in these volumes. Rather than attempt a comprehensive reference summary, we have placed emphasis on providing an account of the progress made in the understanding and interpretation of the physical phenomena involved. Nevertheless it is hoped that few results of importance which have been published up to the middle of 1972 will have been overlooked.

1

THE PASSAGE OF ELECTRONS THROUGH GASES; TOTAL COLLISION CROSS-SECTION, ITS DEFINITION AND MEASUREMENT

RECENT measurements of the total cross-sections for scattering of low-energy electrons have been concerned with accurate and high resolution studies of details of the energy variation of the total cross-section. The low energy work (below the resonance region, see Chap. 1, §§ 4 and 6.1) has been critically reviewed by Bederson and Kieffer.† Absolute values of the e–H total cross-section between 0·7 and 10·5 eV have been measured by Eisner‡ using the crossed-beam recoil technique (Chap. 1, § 4.2.2). The angular resolution was less than 6° and the accuracy of the absolute determination is quoted as ± 25 per cent. Within this accuracy the cross-sections measured by Eisner agree with those of Brackmann et al. and of Neynaber et al. (Fig. 1.11).

For helium a time-of-flight method has been applied by Baldwin and Friedman.§ In this method a pulse of electrons was introduced into the scattering region at time $t = 0$, and they drifted over a distance of 46 cm to a grid where they were detected by a multiplier and a 256-channel pulse-height analyser, the drift time being measured by a delay time to pulse-height converter. The counting rate in the various channels was measured as a function of the helium pressure in the scattering region so that the total cross-section in helium for electrons in the energy range 0·1–2·0 eV could be obtained. The value derived for 1 eV electrons, $6·9 \pm 0·5 \times 10^{-16}$ cm², was higher than that obtained by other methods (see notes on Chap. 2).

Salop and Nakano‖ have used the apparatus shown schematically in Fig. 1.24 to measure total cross-sections for neon. Their results are discussed in relation to theory on p. 3403.

The atomic beam recoil technique has also been applied to measure

---

† BEDERSON, B. and KIEFFER, L. J., *Rev. mod. Phys.* **43** (1971) 601.
‡ EISNER, P. N., Thesis, New York University (1969), quoted by Bederson and Kieffer.
§ BALDWIN, G. C. and FRIEDMAN, S. I., *Rev. scient. Instrum.* **38** (1967) 519. See also BALDWIN, G. C., *7th int. conf. on the physics of electronic and atomic collisions, Amsterdam.* Abstracts, North Holland, 1971, p. 94.
‖ SALOP, A. and NAKANO, H. H., *Phys. Rev.* **A2** (1970) 127.

absolute total electron scattering cross-sections in K†,‡, Rb,† and Cs†, in the energy range 0·1–9 eV. The two independent determinations for K are in good agreement but about half the values obtained in the old work of Brode (Fig. 1.16). They also agree well with theoretical values in contrast to the earlier measurements (cf. Chap. 8, § 8, and p. 3404 of these notes).

It has been suggested§ that Brode's measurements in other vapours may be similarly incorrect since, as pointed out in Chapter 1, § 4.1, his method supposes the electron emission from the cathode to be unaffected by the metal vapour pressure in the apparatus. Ramsauer himself noted that the energy distribution of the emitted electrons might depend significantly on the vapour pressure.

Accurate studies of the detailed fine structure of the total scattering cross-sections for electrons in He, Ne, Ar, Kr, and Xe in the resonance region (cf. Chap. 1, § 6.2) have been carried out by Sanche and Schulz.|| They were concerned with obtaining precision determinations of the positions of the resonance structure and used an electron transmission technique. Their collision region was surrounded by an insulated cylinder to which was applied a small alternating potential (frequency 730 Hz) of amplitude 0·005–0·06 V with respect to the entrance and exit electrodes of the collision region. This produced a modulation in the electron energy. The resultant modulation of the transmitted current was detected synchronously. The output of this synchronous detection was then proportional to dQ/dE, the derivative of the total cross-section with respect to electron energy. Measurement of dQ/dE makes possible a more accurate measurement of the true position of the resonance. Also, by applying the modulation voltage only within the collision chamber itself, spurious surface and fringing effects near the entrance and exit slits should be eliminated. Sanche and Schulz suggested that some of the fine structure in the Q variation for He,†† which they were unable to reproduce, might have been due to instrumental effects near the slits. The new results of Sanche and Schulz for He were in very good agreement with the structure observed by Kuyatt et al. (see Chap. 1, § 6.2.3).

In addition to $2\,^2S$, $2\,^2P$, $2\,^2D$, $3\,^2S$, $3\,^2P$ resonances of the three-

† COLLINS, R. E., BEDERSON, B., and GOLDSTEIN, M., Phys. Rev. A3 (1971) 1976.
‡ VISCONTI, P. V., SLEVIN, J. A., and RUBIN, K., ibid. A3 (1971) 1310.
§ BEDERSON, B. and KIEFFER, L. J., loc. cit.
|| SANCHE, L. and SCHULZ, G. J., Phys. Rev. A5 (1972) 1672.
†† See for example GIBSON, J. R. and DOLDER, K. T., J. Phys. B (GB) 2 (1969) 741; GOLDEN, D. E. and ZECCA, A., Phys. Rev. A1 (1970) 241.

electron system, some of the observed structure was attributed to the occurrence of Wigner cusps (see p. 653) in the elastic cross-section associated with the opening up of $2\,^3S$ and $2\,^1S$ inelastic channels in He. Sanche and Schulz† obtained well-resolved structures at energies of $57\cdot16\pm0\cdot05$ eV and $58\cdot25\pm0\cdot05$ eV (see Fig. 1.22), identified as $(2s^22p)^2P$- and $(2s\,2p^2)^2D$-states of the three-electron system. These resonant states, which will be discussed further in relation to the topics of Chapters 3 and 5, can decay into various excited states, including ionizing states of He.‡ Resonances corresponding to similar states of Ne and Ar have also been observed by Sanche and Schulz.

† loc. cit., p. 3356.
‡ SIMPSON, J. A., MENENDEZ, M. G., and MIELCZAREK, S. R., *Phys. Rev.* **150** (1966) 76; BURROW, P. D. and SCHULZ, G. J., *Phys. Rev. Lett.* **22** (1969) 1271; GRISSOM, J. T., COMPTON, R. N., and GARRETT, W. R., *Phys. Lett.* **30A** (1969) 117. QUÉMÉNER, J. J., PAQUET, C., and MARMET, P., *Phys. Rev.* **A4** (1971) 494.

2

SWARM EXPERIMENTS WITH ELECTRONS IN ATOMIC GASES—MOMENTUM TRANSFER CROSS-SECTION

MEASUREMENTS of drift velocity, u, characteristic energy, ϵ_k, and magnetic drift velocity, $(F/H)u_y/u_x$, in helium have been extended by Crompton and his co-workers† in a low temperature experiment at 77 °K, thus enabling estimates of Q_d down to energies of 0·008 eV. The accuracy of swarm experiments has been discussed in review articles by Elford‡ and by Bederson and Kieffer.§ The measurements have been carried out with great care and Crompton and his collaborators claim that their method enables Q_d to be determined to an accuracy of ± 2 per cent over the range from 0·008 eV to 3 eV. The derivation of Q from Q_d requires an accurate knowledge of the angular distribution in the scattering, but after making allowance for uncertainty from this cause the accuracy in Q claimed for these swarm experiments is higher than the best accuracy claimed in a beam experiment, viz. that of Golden and Bandel,‖ also for He. The early measurements of Q in He by Ramsauer and Kollath and by Normand (see Fig. 1.23) suggested a complex structure in the variation of Q with electron energy below 2·0 eV. Crompton et al. pointed out that such a structure should have been readily discernible in the swarm experiments and their failure to detect it suggests that it is spurious. No such structure was observed in the measurements of Golden and Bandel. On the other hand the values of Q derived by Crompton et al. were substantially larger than those measured by Golden and Bandel. For example, for an electron energy of 1 eV where Q passes through a maximum, Crompton et al. deduce a value $6\cdot1(\pm 0\cdot1)\times 10^{-16}$ cm² compared with $5\cdot6\times 10^{-16}$ cm² of Golden and Bandel. The time-of-flight measurements of Baldwin (see notes on Chap. 1) gave a result even higher ($6\cdot9(\pm 0\cdot5)\times 10^{-16}$ cm²).

The results quoted here are based on the analysis discussed in the

† CROMPTON, R. W., ELFORD, M. T., and ROBERTSON, A. G., *Aust. J. Phys.* **23** (1970) 667.
‡ ELFORD, M. T., in *Case studies in atomic collision physics*, Vol. 2 (ed. E. W. McDaniel and M. R. C. McDowell, p. 94 (North Holland, Amsterdam).
§ loc. cit., p. 3355.   ‖ loc. cit., pp. 31–2.

review article of Bederson and Kieffer and differ somewhat from curve A of Fig. 2.16 which represented the best analysis obtained from the data of Crompton *et al.* in 1967.

At higher energies, values of Q derived from the swarm experiments of Crompton *et al.* decrease monotonically to $5 \cdot 7 \times 10^{-16}$ cm² at 4 eV and $4 \cdot 4 \times 10^{-16}$ cm² at 10 eV with the directly measured values of Golden and Bandel lying some 10–15 per cent lower. Comparison of the latter results with theoretical values is discussed on pp. 545–6. Results of comparable accuracy have been obtained by Robertson† for neon and are discussed in relation to theoretical values on p. 3403. By contrast the curve B of Fig. 2.16, derived from the swarm measurements of the Phelps group, lies much higher in the energy region around 1 eV.

A further method of measuring momentum transfer cross-sections for electrons has been developed in which the d.c. conductivity of a plasma containing argon at atmospheric pressure seeded with small amounts of a metal vapour is measured. The conductivity of such a plasma is given by eqn (53) of § 5. The method was first used by Rhoeling‡ using Cs seeding but has recently been developed into an apparently reliable method by Beynon and Brooker§ using K as the seeding metal. The plasma was confined to a length 33 cm of a tube 10 cm in diameter which could be maintained at a temperature of around 1300 °K. Argon gas of high purity was preheated in a copper tube 4·5 m long before passing through a stainless steel seed pot containing liquid potassium metal at an accurately controlled temperature. The residence time of the gas in the seed pot was 3 s and when it left it was saturated with potassium vapour. The seeded gas was then again heated before entering the region of the tube containing the plasma, the temperature of the seed pot, the seed line, and the main furnace being monitored at many places using thermocouples. The d.c. conductivity of the plasma was measured using two parallel circular electrodes, surrounded by a guard-ring device and set at an accurately known distance apart by means of a micrometer. These two electrodes were placed on either side of the flowing plasma and current–voltage characteristics determined for a number of different electrode separations. By using different separations the effect of the potential drop across the sheaths could be eliminated and the bulk conductivity of the plasma itself determined. The electron density, n, in eqn (53) was estimated from the temperature of the plasma, using

† ROBERTSON, A. G., *J. Phys. B (GB)* **5** (1972) 648.
‡ RHOELING, D., *Symposium on thermionic power conversion*, Colorado Springs (1962).
§ BEYNON, J. D. E. and BROOKER, P. G., *J. Phys. B(GB)* **5** (1972) 849.

Saha's equation. If n_A, n_K are respectively the concentrations and Q_A, Q_K the momentum transfer cross-sections for electrons of mean thermal velocity $\bar{v}$, the collision frequency, ν, can be written

$$\nu = \bar{v}(n_A Q_A + n_K Q_K).$$

Two experiments were carried out using very different relative n_K/n_A concentrations so that the mean cross-sections Q_A, Q_K could both be determined. For a mean electron energy of 0·17 eV they obtained $Q_A = 2·24(\pm 0·13) \times 10^{-17}$ cm², $Q_K = 2·66(\pm 0·11) \times 10^{-14}$ cm², in reasonable agreement with values obtained by Frost and Phelps for Ar (Fig. 2.22) and the recent values of Collins et al.† and Visconti et al.† for K (see notes on Chap. 1).

† loc. cit.

3

THE EXPERIMENTAL ANALYSIS OF THE CROSS-SECTIONS FOR IMPACT OF ELECTRONS WITH ATOMS AND IONS—IONIZATION CROSS-SECTIONS

1. Observed cross-sections for ionization of neutral atoms

Absolute values of total ionization cross-sections, Q_i, have been given for Ar,[†] Li,[‡] Cs,[§],[||] Mg,[††],[§§],[||||] Ca, Sr, and Ba.[‡‡],[§§],[||||],[†††]

There are large discrepancies between values of ionization cross-sections for the alkali and alkaline earth metals quoted by different authors. These discrepancies are most probably related to different methods used for determining the density of neutral atoms in the beam of metal atoms. McFarland used surface ionization. Okuno, and Rakhovsky and Stepanov used a condensation target forming a 'black body' atom trap. It has been suggested that the much larger cross-sections measured by McFarland could indicate an overestimate of surface ionization efficiency in his case.

This explanation cannot account for the discrepancies in the measured Q_i for Cs, however (see Table 3.2, p. 129), since both McFarland and Kinney[‡‡‡] and Heil and Scott,[§§§] in the work quoted in that table, used surface ionization to measure the beam intensity. The more recent measurements of Nygaard,[|||||] using a method similar to that of Tate and

---

[†] CROWE, A., PRESTON, J. A., and McCONKEY, J. W., *J. chem. Phys.* **57** (1972) 1620.
[‡] ZAPESOCHNYI, I. P. and ALEKSAKHIN, I. S., *Zh. eksp. teor. Fiz.* **55** (1968) 76 (*Soviet Physics, JETP* **28** (1969) 41).
[§] KORCHEVOI, YU. P. and PRZONSKI, A. M., *Zh. eksp. teor. Fiz.* **51** (1966) 1617 (*Soviet Physics, JETP* **24** (1967) 1089).
[||] NYGAARD, K. J., *J. chem. Phys.* **49** (1968) 1995.
[††] OKUNO, Y., OKUNO, K., KANEKO, Y., and KANOMATA, I., *J. phys. Soc. Japan* **29** (1970) 164.
[‡‡] OKUNO, Y., *J. phys. Soc. Japan* **31** (1971) 1189.
[§§] RAKHOVSKY, V. I. and STEPANOV, A. M., *6th int. conf. on the physics of electronic and atomic collisions, Cambridge, Mass.* Abstracts, M.I.T. Press, 1969, p. 621.
[||||] VAINSHTEIN, L. A., OCHKUR, V. I., RAKHOVSKY, V. I., and STEPANOV, A. M., *7th int. conf. on the physics of electronic and atomic collisions, Amsterdam.* Abstracts, North Holland, 1971, p. 880.
[†††] McFARLAND, R. H., *Phys. Rev.* **159** (1967) 20.
[‡‡‡] McFARLAND, R. H. and KINNEY, J. D., *Phys. Rev.* **137** (1965) A1058.
[§§§] HEIL, H. and SCOTT, B., *Phys. Rev.* **145** (1966) 279.   [|||||] loc. cit.

Smith and of Korchevoi and Przonski,[†] lie much closer to the values of McFarland and Kinney.

Analysis of the ion current to enable cross-sections for specific multiply-ionized states to be derived has been carried out by Crowe et al.[‡] for Ar, by Okudaira et al.[§] for Ar and Mg, by Okudaira[||] for Ca, Sr, and Ba, and by El-Sherbini et al. for Kr and Xe.[††]

The role of the Auger effect or the shake-off process in producing multiple ionization (§ 2.5.3.3) has been further studied by Abouaf for Mn,[‡‡] Cd,[‡‡] In,[§§] and Ag.[§§]

A careful study of the variation of Q_i near the ionization threshold for He and Ar has been made by Marchand et al.[||||] They used a crossed-beam technique with a cylindrical electrostatic velocity selection. The ionization current was measured at 10 mV intervals over an energy range of about 1 eV above the threshold. For He they found a threshold law $(E-E_i)^n$ with $n = 1.15$ over this energy range with a reproducibility of 3 per cent and thus consistent with the classical theoretical value ($n = 1.127$) predicted by Wannier (Chap. 3, § 2.5.3.1 and p. 3420). The best fitting power law extending over a range of 12 eV above threshold gave $n = 1.02$.

For Ar the fit is more difficult because of the onset of $Ar^+(^2P_{\frac{1}{2}})$ ionization at 0.2 eV above the ionization potential corresponding to $Ar^+(^2P_{\frac{3}{2}})$ ionization. For the energy region up to 1 eV above the $^2P_{\frac{1}{2}}$ ionization potential, $n = 1.34$, reproducible to 4 per cent. For the region of extent 0.2 eV between the two ionization potentials, $n = 1.3$ with an uncertainty of 15 per cent.

A great deal of recent work on the variation of Q_i with energy has been concerned with the accurate study of the fine structure in the variation and its interpretation in terms of the excitation of auto-ionizing states (Chap. 3, § 2.5.3.2). The main emphasis has been on the energies at which structure is observed rather than on the measurement of the absolute value of Q_i.

Much of the structure observed in the Q_i variation with energy arises from the excitation of auto-ionizing states of the target atom. Some of the most prominent features are associated, however, with the formation

[†] loc. cit. [‡] loc. cit.
[§] OKUDAIRA, S., KANEKO, Y., and KANOMATA, I., J. phys. Soc. Japan **28** (1970) 1536.
[||] OKUDAIRA, S., ibid. **29** (1970) 409.
[††] EL-SHERBINI, TH. M., VAN DER WIEL, M. J., and DE HEER, F. J., Physica, **48** (1970) 157.
[‡‡] ABOUAF, R., J. Phys. (Fr.) **31** (1970) 277.
[§§] ABOUAF, R., ibid. **32** (1970) 603.
[||||] MARCHAND, P., PAQUET, C., and MARMET, P., Phys. Rev. **180** (1969) 123.

of short-lived resonant states of the negative ion by the attachment of the incident electron to the target atom. Many of the low-lying resonances of this type discussed in Chapter 1 have insufficient energy to decay into ionizing states. However, the He resonances in the region 56–9 eV (see Chap. 1, Fig. 1.22) can decay in the process $He^- \rightarrow He^+ + 2e$ and so contribute to the fine structure of the Q_i variation in this energy region.

The most comprehensive work on the structure of the Q_i curve for He in the region 57–65 eV has been carried out by Quéméner et al.† using a crossed-beam method. Mono-energetic electrons were produced by a 127° electrostatic electron-energy selector and the ions produced were mass-analysed by a large quadrupole mass filter and then individually detected at the output of an electron multiplier. The fine structure observed by Quéméner was very complex, consisting of two dips corresponding to the formation of resonant He⁻ states and eight peaks corresponding to the excitation of auto-ionizing states of the neutral atom. The resolution of the apparatus was good enough to enable the natural line shapes of the resonances to be determined. They were fitted to the expression given by Fano and Cooper (Chap. 9, eqn (75)):

$$Q_i(\epsilon) = Q_i^a\{(q+\epsilon)^2/(1+\epsilon^2)\} + Q_i^b,$$

where $\epsilon = (E-E_r)/\tfrac{1}{2}\Gamma$ is the difference of the electron energy and resonance energy in units of the resonance half-width $\tfrac{1}{2}\Gamma$, Q_i^b the background (non-resonant) contribution to the cross-section, and q a shape factor.

For the He⁻ resonance of configuration $(2s^2 2p)^2P$ they derived the values $E_r = 57\cdot15 \pm 0\cdot04$ eV, $q = -0\cdot75 \pm 0\cdot12$ eV, $\Gamma = 0\cdot045 \pm 0\cdot007$ eV. The corresponding values for the $(2s2p^2)^2D$ were $E_r = 58\cdot23 \pm 0\cdot04$ eV, $q = -0\cdot95 \pm 0\cdot25$, $\Gamma = 0\cdot025 \pm 0\cdot010$ eV.

Peaks were found by Quéméner et al. at electron energies (in eV) of 58·40, 60·0, 62·73, 63·00, 63·60, 63·81, 64·44, and attributed to auto-ionizing states of neutral He. Studies of the fine structure of these ionizing states have been made for He using the trapped electron method by several other authors.‡ Similar measurements have been made for Ne§ (see also p. 3485) and Cs.∥

† QUÉMÉNER, J. J., PAQUET, C., and MARMET, P., Phys. Rev. A4 (1971) 494.
‡ BURROW, P. D. and SCHULZ, G. J., Phys. Rev. Lett. 22 (1969) 1271; GRISSOM, J. T., COMPTON, R. N., and GARRETT, W. R., Phys. Lett. 30A (1969) 117.
§ GRISSOM, J. T., GARRETT, W. R., and COMPTON, R. N., Phys. Rev. Lett. 23 (1969) 1011.
∥ HAHN, Y. B. and NYGAARD, K. J., Phys. Rev. A4 (1971) 125.

2. Further ionization of positive ions

Several recent measurements have been carried out of absolute cross-sections for further ionization of positive ions by electrons using the crossed-beam method for Na⁺,† K⁺,† Mg⁺,‡ Ba⁺,§,∥ Li⁺,§ C⁺,†† N⁺,†† O⁺‡‡ (Chap. 3, § 3.1.1). In the measurements of Peart et al.§§ on the further ionization of Ba⁺, the energy of the crossing electron beam was selected using a 127° electrostatic energy electron selector. The energy resolution was sufficiently good to enable four auto-ionization thresholds of Ba⁺ to be resolved in the energy range 16–22 eV.

The merging-beam technique which has been used by Neynaber∥∥ to study heavy particle collisions has been adapted by Mahadevan††† to study the form of the variation of Q_i with electron energy for the ionization of He⁺, N⁺, and O⁺ in the energy range up to 300 eV. The advantages of this method over the crossed-beam method include a longer interaction path and a negligible momentum change of the beam particles in ionizing collisions. Beam modulation and synchronous detection techniques were not then required in order to eliminate background processes.

Other measurements of the variation with energy of the cross-section for further ionization of Ne²⁺, Ne³⁺, Ar⁺, Ar²⁺, and Ar³⁺ using the ion trap method have also been reported by Hasted and Awad‡‡‡ (Chap. 3, § 3.1.2).

3. Ionization of metastable atoms

Recently measurements have been carried out on the cross-sections for ionization of metastable atoms by electron impact using modulated crossed-beam techniques. Dixon and Harrison§§§ have studied the ionization of metastable H(2s) atoms. A beam of protons of energy 2 keV crossed a Cs atomic beam and an atomic beam of H formed by charge exchange, 25 per cent of the atomic H being produced in the 2s metastable states. The proportion of H(2s) in the beam was measured by quenching in an electric field. The change in metastable flux when the neutral beam

---

† PEART, B. and DOLDER, K. T., *J. Phys. B (GB)* **1** (1968) 240.
‡ MARTIN, S. O., PEART, B., and DOLDER, K. T., ibid. **1** (1968) 537.
§ PEART, B. and DOLDER, K. T., ibid. **1** (1968) 872.
∥ FEENEY, R. K., HOOPER, J. W., and ELFORD, M. T., *Phys. Rev.* **A6** (1972) 1469.
†† AITKEN, K. L., HARRISON, M. F. A., and RUNDEL, R. D., *J. Phys. B (GB)* **4** (1971) 1189.
‡‡ AITKEN, K. L. and HARRISON, M. F. A., ibid. **4** (1971) 1176.
§§ PEART, B., STEVENSON, J. G., and DOLDER, K. T., ibid. **6** (1973) 146.
∥∥ NEYNABER, R. H., *Meth. expl Physics* **7A** (1968) 476.
††† MAHADEVAN, Y. P., *Proc. R. Soc.* **A327** (1972) 317.
‡‡‡ HASTED, J. B. and AWAD, G. L., *J. Phys. B (GB)* **5** (1972) 1719.
§§§ DIXON, A. J. and HARRISON, M. F. A., *7th int. conf. on the physics of electronic and atomic collisions, Amsterdam. Abstracts.* North Holland, 1971, p. 892.

was crossed by an electron beam enabled the ionization cross-section from the H(2s) state to be measured. In these experiments high-energy Rydberg states with a long life-time are also present in the beam. These must be removed. This was done by preliminary passage through an electrostatic field sufficient to ionize these states, but insufficient to give appreciable loss of 2s metastable atoms. The ions were then swept out of the beam electrostatically.

Relative ionization cross-sections of an unknown mixture of metastable states of the inert gases ($2\,^3S_0 + 2\,^1S_0$ for He, $3s\,^3P_2 + 3s\,^3P_0$ for Ne, and $4s\,^3P_0 + 4s\,^3P_2$ for Ar) have been measured by Shearer† using two electron beams coupled to a mass spectrometer. The first electron beam is used to excite the atoms of the atomic beam and moves parallel to the atomic beam (merging beam). Ions produced are removed from the beam which is then crossed at right angles by the second electron beam. The energy of the second beam is less than the ionization energy so that the measured ionization is attributed to ionization of metastable atoms in the beam.

4. Detachment of electrons from H⁻

Cross-sections for electron detachment from H⁻ ions by electron impact (Chap. 3, § 4) have been measured by Peart et al.‡ using a crossed-beam method in the energy range 12 to 1000 eV. The form of the variation with energy observed agrees reasonably well with that found by Dance et al. (Chap. 3, § 3.31) but the absolute values are about 20 per cent lower. Walton et al.§ have extended the cross-section measurements down to the threshold for the detachment process at 0·75 eV. It was not practicable to use the crossed-beam method at energies below 10 eV for accurate measurements because the electron energy resolution was not good enough. Instead an inclined-beam method was developed in which the electron beam intersects the ion beam at an angle of 20°. This method possesses many of the advantages of the merging-beam method but is technically less difficult. With this apparatus absolute cross-sections for detachment were determined between threshold and 30 eV with an energy resolution of 0·25 eV. The magnitudes obtained for energies above 12·5 eV agreed remarkably well with those measured with the crossed-beam apparatus. A striking feature of the results of Walton et al. was a marked structure for an electron energy around 14·2 eV which could indicate a resonance at this energy, or alternatively could be a

† SHEARER, W., private communication.
‡ PEART, B., WALTON, D. S., and DOLDER, K. T., J. Phys. B (GB) **3** (1970) 1346.
§ WALTON, D. S., PEART, B., and DOLDER, K. T., ibid. **4** (1971) 1343.

cusp associated with the opening up of the channel for double detachment at 14·35 eV.

Taylor and Thomas† interpreted the observed structure as a resonance with the three-electron configuration $(2s)^2(2p)[2P^0]$. Using an inclined-beam apparatus with an intersection angle of 10°, Peart and Dolder‡ observed further structures in the energy dependence of the detachment cross-section at 17·2 eV.

The double detachment process in which H⁻ ions are converted into H⁺ has been studied by Peart et al.§ using their crossed-beam apparatus for energies between threshold and 850 eV. The cross-section was found to reach a maximum value of $5·0 \times 10^{-17}$ cm at 70 eV, about 2 per cent of the value for single detachment. Between 100 and 850 eV the cross-section remains about 2·5 per cent of the single detachment cross-section.

Recent theoretical work on detachment of electrons from H⁻ is discussed on p. 3390.

5. Inner shell ionization

Turning next to cross-sections for inner shell ionization by electron impact (Chap. 3, § 5), while there has been great interest in recent years in inner shell ionization by heavy particles (summarized in Chap. 24, § 2.3.3), there have been few recent new studies of the cross-section for ionization by electrons. One can refer to the measurements of Hink and Paschke‖ on the K ionization of C while Hübner et al.†† have made similar measurements for thin targets of Ag and Cu for energies up to 150 keV. There have also been measurements in the high energy region for K ionization of a number of elements for energies up to 2 MeV by Scholz et al.‡‡ and up to 900 MeV by Quarles.§§

For the inner-shell ionization of light elements the fluorescence yield is very small and the primary ionization process is mostly followed by the emission of Auger electrons. Glupe and Mehlhorn‖‖ measured the variation with incident electron energy of the K-shell ionization cross-section for C, N, O, and Ne by measuring the K Auger electron intensity, using gaseous targets.

† TAYLOR, H. S. and THOMAS, L. D., Phys. Rev. Lett. **28** (1972) 1091.
‡ Private communication.
§ PEART, B., WALTON, D. S., and DOLDER, K. T., J. Phys. B (GB) **4** (1971) 88.
‖ HINK, W. and PASCHKE, H., Z. Phys. **244** (1971) 140.
†† HÜBNER, H., ILGEN, K., and HOFFMANN, K. W., ibid. **255** (1972) 269.
‡‡ SCHOLZ, W., LI-SCHOLZ, A., COLLÉ, R., and PREISS, I. L., Phys. Rev. Lett. **29** (1972) 761.
§§ QUARLES, C. A., Phys. Lett. **39A** (1972) 375.
‖‖ GLUPE, G. and MEHLHORN, W., ibid. **25A** (1967) 274.

Gerlach and du Charme† have made similar measurements using as target single layers of C, N, O, and Na atoms adsorbed on the (100) plane of a W crystal. Auger electrons emitted in the backward direction (between 128° and 148°) of the primary electron beam were observed. Since the angular distribution of the K Auger electrons should be isotropic it was possible to derive the total K ionization cross-section from these measurements. A description of the apparatus employed is given in the notes on Chapter V, p. 3380.

The measurements of Glupe and Mehlhorn which were relative and extended up to 15 times the K ionization energy showed better agreement with the classical theory of Gryzynski,‡ than with the Born approximation calculations of Burhop,§ although both theories were in good agreement with the experimental results above about 8 times the K ionization energy. The measurements of Gerlach and du Charme, on the other hand, which attempted to measure the absolute ionization cross-sections, were in good agreement with the Born approximation calculation up to 4 times the K ionization energy. Owing, however, to difficulty in estimating the number of adsorbed atoms the accuracy of their absolute cross-sections was not better than 50 per cent.

Studies of the polarization of characteristic X-radiation excited by electron impact have been reported by Mehlhorn.‖

† GERLACH, R. L. and DU CHARME, A. R., *Surf. Sci.* **32** (1972) 329.
‡ GRYZYNSKI, M., *Phys. Rev.* **138** (1965) A336.
§ BURHOP, E. H. S., loc. cit., p. 166.
‖ MEHLHORN, W., *Phys. Lett.* **26A** (1968) 166.

4

THE EXPERIMENTAL ANALYSIS OF THE CROSS-SECTIONS FOR IMPACT OF ELECTRONS WITH ATOMS AND IONS— CROSS-SECTIONS FOR PRODUCTION OF EXCITED ATOMS

As in the case of ionization most interest in recent studies of atomic excitation by electron impact has lain in the observation of fine structure in the excitation curves due to the formation of short-lived resonant states of the atomic negative ion. The same resonances often show up in several channels—total elastic scattering, ionization, excitation, and differential cross-section measurements, sometimes in all of them.

1. Excitation of atomic hydrogen

Several recent investigations have been concerned with the excitation of the $2p$ level of atomic hydrogen by studying the emission of Ly α radiation in a crossed-beam experiment (Chap. 4, § 1.4.1). Long et al.[†] have measured relative cross-sections to within an accuracy of 2 per cent normalized to the Born approximation value (see Table 7.1, p. 460) at 200 eV. With the same equipment cross-sections for H($2s$) relative to H($2p$) have also been measured.[‡] Ott et al.[§] have measured the polarization fraction P of the Ly α radiation.

Radiation emitted from the collision region in a direction mutually at right angles to both the electron and H beam was incident on a LiF crystal 1·5 mm thick, 2·5 cm diameter, set at the Brewster angle, and reflected radiation from the crystal was detected by an iodine vapour photon counter preceded by a dry oxygen filter. Both the analyser and photon counter were mounted on a platform capable of rotation about the direction of observation of the radiation. With the electron beam direction parallel to the plane of the crystal face the intensity of light entering the counter should be proportional to $I_{\|}$. On rotating the plat-

[†] LONG, R. L., COX, D. M., and SMITH, S. J., *J. Res. natn. Bur. Stand.* **72A** (1968) 521.
[‡] COX, D. M. and SMITH, S. J., *7th int. conf. on the physics of electronic and atomic collisions, Amsterdam.* Abstracts, North Holland, 1971, p. 707.
[§] OTT, W. R., KAUPPILA, W. E., and FITE, W. L., *Phys. Rev. Lett.* **19** (1967) 1361; *Phys. Rev.* **A1** (1970) 1089.

form through 90° so that the electron beam direction lay in the plane normal to the crystal face the counting rate should be proportional to $I_\perp$. Hence the polarization fraction, $P = \dfrac{(I_\parallel/I_\perp)-1}{(I_\parallel/I_\perp)+1}$ could be obtained. Phase-sensitive detection was used. The results obtained showed reasonable agreement with those expected theoretically, P rising from a value of about 0·12 at threshold, 10·2 eV, to a maximum of 0·25 at about 17 eV, then falling to 0 at about 200 eV and to $-0·07$ at 700 eV. The form of variation of total cross-section observed agreed quite well with that found by Chamberlain et al. (Chap. 4, Fig. 4.16).

In the same experiment Ott et al. also measured P for Ly α radiation emitted by metastable H(2s) atoms in electric fields of about 10 V cm^{-1} and obtained $P = -0·30 \pm 0·02$, independent of energy, in the range 12–200 eV, in close agreement with the theoretically expected value of $-0·323$.

McGowan et al.† have carried out a high resolution (0·07 eV) study of the excitation cross-section for Ly α in the energy range from threshold to 50 eV. Their measurements of the energy variation of the cross-section agree with those of Long et al. and the earlier measurements of Fite et al. (Chap. 4, § 1.4.1).

McGowan et al.† observed, in a range of 2 eV above threshold, five resonances at energies close to those predicted theoretically.

Excitation cross-sections for eleven Balmer lines have been measured by Walker.‡

Measurement of the total intensity of the radiation emitted is rendered difficult because the angular distribution depends on the polarization P, which is not usually well known. For electric dipole radiation the intensity, $I(\theta)$, per unit solid angle emitted in a direction making an angle θ with the electron direction can be written in terms of the total excitation cross-section Q_{ex}

$$I(\theta) = Q_{\text{ex}}(1-P\cos^2\theta)/4\pi(1-P/3). \tag{1}$$

Clout and Heddle§ pointed out that by measuring the intensity at an angle of $\theta = 54·5°$ for which $\cos^2\theta = 1/3$, $I(\theta_0) = Q_{\text{ex}}/4\pi$, so that Q_{ex} can be determined without knowledge of the polarization P.

Turning now to metastable H($2s_{\frac{1}{2}}$) production, further measurements have been reported by Kauppila et al.‖ and by Cox and Smith†† using

† McGowan, J. W., Williams, J. F., and Curley, E. K., Phys. Rev. 180 (1969) 132.
‡ Walker, J. D., Thesis, University of Oklahoma (1970).
§ Clout, P. N. and Heddle, D. W. O., J. opt. Soc. Am. 59 (1969) 715.
‖ Kauppila, W. E., Ott, W. R., and Fite, W. L., Phys. Rev. A1 (1970) 1099.
†† Cox, D. M. and Smith, S. J., loc. cit.

a similar H-beam technique to that described on p. 193. Kauppila *et al.* were concerned especially to use a more correct angular distribution of the Ly α radiation emitted following electrostatic quenching of the 2s metastable state than had been the case in previous work. They measured the ratio $R_{\pi/2} = I_{2s}(\pi/2)/I_{2p}(\pi/2)$ of the quenched Ly α radiation to the direct Ly α radiation emitted at right angles and found it to agree with that expected from Born approximation calculations for electron energies above 200 eV. The cross-section $Q_{\text{ex}}(2s)$ was then obtained from the expression $Q_{\text{ex}}(2s) = (1-P/3)R_{\pi/2}I_{2p}(\pi/2)$, derived from eqn (1) above where $P = -0.323$ is the polarization factor measured by Ott *et al.*[†] for quenched Ly α radiation. For $I_{2p}(\pi/2)$ they used the data of Long *et al.*[‡] above 15 eV and of McGowan *et al.*[§] below 15 eV. These data were normalized to the Born approximation values above 200 eV.

The results obtained were about 50 per cent larger than those of Hils *et al.*[||] in the electron energy region below 100 eV but agreed fairly well with the close-coupling theoretical calculations of Burke *et al.*[††] (see p. 528).

2. Excitation of He

Much recent work has been concerned with the measurement of excitation cross-sections in He. Several investigations have been concerned with the form of the excitation functions of ^3S, ^3P, and ^3D levels which are expected to occur primarily through electron exchange processes. Kay and Showalter[‡‡] found that in the energy range from 50 to 500 eV the excitation cross-section for the $4\,^3S$ level was proportional to E^{-n} with $n = 3.0 \pm 0.2$ as expected for a pure exchange mechanism (see p. 3389). The excitation curves for the $3\,^3P$ and $3\,^3D$ states, however, were found to fall off considerably less rapidly at high energies and Kay and Showalter were able to fit this to a combination of two components of the form $AE^{-n}+BE^{-1}$ with

$$n = 3.5 \pm 0.2, \quad A/B = 200 \quad \text{for } 3\,^3P,$$

and $\quad n = 3.7 \pm 0.5, \quad A/B = 10 \quad \text{for } 3\,^3D.$

They attribute the E^{-1} term to mixing between ^1F and ^3F levels with subsequent cascading $^3F \to {}^3D \to {}^3P$ (see p. 1759).

The excitation of $4F$ states of He has been studied by Jobe and

[†] OTT, W. R., KAUPPILA, W. R., and FITE, W. L., *Phys. Rev.* **A1** (1970) 1089.
[‡] LONG, R. L., COX, D. M., and SMITH, S. J., loc. cit.
[§] McGOWAN, J. W., WILLIAMS, J. F., and CURLEY, E. K., loc. cit.
[||] HILS, D., KLEINPOPPEN, H., and KOSCHMIEDER, H., *Proc. phys. Soc.* **89** (1966) 35.
[††] BURKE, P. G., ORMONDE, S., and WHITAKER, W., ibid. **92** (1967) 319.
[‡‡] KAY, R. B. and SHOWALTER, J. G., *Phys. Rev.* **A3** (1971) 1998.

St. John,† measuring the intensity of the $4F \to 3D$ infra-red radiation. A study of the pressure variation of the apparent cross-sections for $4\,^1P$ and $4F$ excitation provided direct evidence for the $4\,^1P$ to $4F$ excitation transfer process (Chap. 18, § 4.2.3.2), and enabled them to estimate both the cross-section for this process and for direct electronic excitation to the $4F$ state.

Recent work has extended excitation cross-section measurements for He up to electron energies of 5 keV.‡

Structure has been found in high resolution studies of the excitation of $3\,^3D$, $4\,^3D$, $4\,^1S$, $4\,^1D$ states of He for electron energies around 60 eV.§ These are correlated with the doubly excited states of He⁻ which decay by the process He⁻ $\to$ e+He($3\,^3D$), etc. Some evidence was also found for structure in the polarization of the emitted light near these resonances but the effects were smaller than those observed near the threshold (Chap. 4, § 1.4.2.3, Fig. 4.30).

Other recent excitation studies have been reported for Ne,‖,†† Ar,††,‡‡ Kr,‖ Na,§§ O,‖‖ N,‖‖ Cs††† and for simultaneous excitation and ionization of He,‡‡‡ Ne,§§§,‖‖‖,††††,‡‡‡‡ Ar,§§§,‡‡‡‡,§§§§,‖‖‖‖ Kr,§§§,‡‡‡‡ Xe,§§§,††††,‡‡‡‡ Cd,†††††,‡‡‡‡‡ Zn.‡‡‡‡‡

Lloyd et al.§§§§§ have measured excitation functions for the formation

† Jobe, J. D. and St. John, R. M., *J. opt. Soc. Am.* **57** (1967) 1449.

‡ Moustafa Moussa, H. R., de Heer, F. J., and Schutten, J., *Physica, 's Grav.* **40** (1969) 517; van Eck, J. and de Jongh, J. P., ibid. **47** (1970) 141.

§ Heideman, H. G. M. and Reman, R. J., *7th int. conf. on the physics of electronic and atomic collisions, Amsterdam.* Abstracts, North Holland, 1971, p. 318.

‖ Sharpton, F. A., St. John, R. M., Lin, C. C., and Fajen, F. E., *Phys. Rev.* **A2** (1970) 1305.

†† van Raan, A. F. J., de Jongh, J. P., and van Eck, J., *7th int. conf. on the physics of electronic and atomic collisions, Amsterdam.* Abstracts, North Holland, 1971, p. 704.

‡‡ Lloyd, C. R., Teubner, P. J. O., Weigold, E., and Hood, S. T., *J. Phys. B (GB)* **5** (1972) L44.

§§ Enemark, E. A. and Gallagher, A., *Phys. Rev.* **A6** (1972) 192; Gould, G., Thesis, University of New South Wales (1970).

‖‖ Stone, E. J. and Zipf, E. C., *Phys. Rev.* **A4** (1971) 610.

††† Zapesochny, I. P. and Aleksakhin, I. S., *Zh. eksp. teor. Fiz.* **55** (1968) 76 (*Soviet Physics, JETP* **28** (1969) 41).

‡‡‡ Anderson, R. J. and Hughes, R. H., *Phys. Rev.* **A5** (1972) 1194.

§§§ Luyken, B. F. J., de Heer, F. J., and Baas, R. Ch., *Physica, 's Grav.* **61** (1972) 200.

‖‖‖ Walker, K. J., Thesis, University of Oklahoma (1971).

†††† Hertz, H., *Z. Naturf.* **24A** (1969) 1937.

‡‡‡‡ van Raan, A. F. J., de Jongh, J. P., and van Eck, J., loc. cit.

§§§§ Clout, P. N. and Heddle, D. W. O., *J. Phys. B (GB)* **4** (1971) 483.

‖‖‖‖ Latimer, I. D. and St. John, R. M., *Phys. Rev.* **A1** (1970) 1612.

††††† Varshavskii, S. P., Mityureva, A. A., and Penkin, N. P., *Optika Spectrosk.* **29** (1970) 637 (*Optics Spectrosc.* N.Y. **29** (1970) 341).

‡‡‡‡‡ Aleinikov, V. S. and Ushakov, V. V., *Optika Spectrosk.* **29** (1970) 211 (*Optics Spectrosc.* N.Y. **29** (1970) 111).

§§§§§ Lloyd, C. R., Weigold, E., Teubner, P. J. O., and Hood, S. T., *J. Phys. B (GB)* **5** (1972) 1712.

of metastable He and Ar atoms in a collision chamber. The metastable atoms were detected by an electron multiplier and the time between collision and detection enabled them to be distinguished from photons. These measurements extended up to electron energies of 200 eV. The cross-section rises sharply to a maximum at a few volts above threshold (28 eV for He, 22 eV for Ar). The shape of the curve near threshold for He does not, however, agree very well with that obtained by Čermak (Chap. 4, § 2.2.2, Fig. 4.58 (b)).

Pichanick and Simpson† have used an 180° spherical monochromator to study metastable excitation of the inert gases close to threshold under conditions of very high resolution (0·035–0·050 eV). The metastable atoms were produced in a collision chamber and the metastable detector was a ring of gold foil of diameter 1·3 inch, 0·2 inch width. The high resolution enabled finer details of the structure to be observed than in earlier work (Chap. 4, § 2.2.3).

Borst‡ has carried out an RPD experiment to measure the excitation function of the $6\,^3P_2$ metastable state of Hg. In his apparatus metastable atoms striking the walls of the collision chamber eject electrons so that the positive current to the walls provides a measure of the metastable production. Borst's apparatus was operated as an electron trap device so that structure corresponding to the excitation of optically allowed and ionizing states as well as metastable states could be detected. Also, by monitoring trapped electrons near the threshold for the metastable state it was possible to obtain an absolute value of the cross-section. Borst was able to determine the form of metastable excitation function from 5 to 8·5 eV above which energy the apparatus detected photoelectrons from the excitation of radiating states and ions from ionizing states.

The experiments of Feldman and Nowick on the excitation of metastable auto-ionizing states of Li (Chap. 4, § 2.1.2.3, Fig. 4.47) have been extended to the other alkali metals. Structure in the excitation curve was observed corresponding to the excitation of states with two electrons outside closed sub-shells and a vacancy in one of the closed sub-shells. For example in the case of Rb for which the ground state configuration is $(4p^6 5s)$ the first metastable threshold at $15·8\pm0·3$ eV corresponds to a configuration $(4p^5 4d 5s)^4 F_{9/2}$ while structure giving a second threshold at 18·8 eV is attributed to a $(4p^5 5s\, 6s)^4 P_{\frac{3}{2}}$ state.

Celotta§ et al. have applied the atomic beam attenuation method

† PICHANICK, F. M. J. and SIMPSON, J. A., Phys. Rev. **168** (1968) 64.
‡ BORST, W. L., ibid. **181** (1969) 257.
§ CELOTTA, R., BROWN, H., MOLOF, R., and BEDERSON, B., ibid. A3 (1971) 1622.

(Chap. 1, § 4.2.2) to measure the total collision cross-section for the scattering of low-energy electrons by metastable argon. An atomic beam containing metastable Ar atoms was crossed by an electron beam. From the reduction of metastable flux as measured by an electron multiplier, due to the crossing electron beam, the total cross-section could be estimated. The measured cross-section fell from 8×10^{-14} cm^2 at 0·4 eV to 3×10^{-14} cm^2 at 1 eV and to 2×10^{-14} cm^2 at 7 eV, about twice as great as obtained from an adiabatic theory of Robinson.†

Peart and Dolder have also measured the cross-section for excitation of He$^+$ ions to the 2s state. As in the case of the experiment of Dance *et al.* (Chap. 4, § 3.2), a beam of He$^+$(1s) ions was crossed by a perpendicular electron beam. The He$^+$(2s) ions formed were, however, quenched by the motion of the ions in a magnetic field and the quenched resonance radiation detected by a specially designed U.V. photon detector. In this experiment the signal-to-background ratio was much better than in the experiment of Dance *et al.* and the effects of space–charge interaction and X-rays eliminated. It was possible to resolve structure due to autoionizing states of helium.

† ROBINSON, E. J., *Phys. Rev.* **182** (1969) 196.

5

THE EXPERIMENTAL ANALYSIS OF THE CROSS-SECTIONS FOR IMPACT OF ELECTRONS WITH ATOMS AND IONS—ANALYSIS OF ENERGY AND ANGULAR DISTRIBUTIONS

It has been seen in the notes on the earlier chapters that the problem of resonances in electron–atom scattering has dominated recent work in this field. In Chapters 3 and 4 we have been concerned with resonance phenomena as they show up in studies of ionization and excitation cross-sections. In this chapter we are concerned with the detection of resonances electronically by studies of the energy losses of electrons scattered inelastically, or by studies of elastic scattering through fixed angles.

Many of the resonances detected in this way will, of course, show up in the energy variation of ionization and excitation cross-sections. Selection rules may, however, forbid the decay of the resonances into ionized or excited states of the target atoms so that some resonances would be expected to appear only in measurements of elastic scattering.

1. Further applications of the electron trap method (Chap. 5, § 2.3)

A powerful technique for studying fine structure in inelastic scattering cross-sections is the electron trap method of Schulz (Chap. 5, § 2.3). Extensive recent investigations have used this method. In some cases, however, the SF_6 electron scavenger method developed by Curran† has been used. In this method, which has been adapted by Compton et al., the collision chamber contains SF_6 in addition to the substance under investigation. As in the electron trap method, the incident electrons pass along the axis of the collision chamber. If their energy is close to a critical energy then, after an inelastic scattering involving this critical energy loss, they will have small energy and can be readily 'scavengered' by SF_6 molecules contained in the collision chamber, to form SF_6^- (see Chap. 13, § 11.1). These slow SF_6^- ions are drawn out into a time-of-flight

† Curran, R. K., J. chem. Phys. **38** (1963) 780; Compton, R. N., Stockdale, J. A., and Reinhardt, P. W., Phys. Rev. **180** (1969) 111.

mass spectrometer. Each time the incident electron energy passes through a critical energy of the target gas the SF_6^- current rises. The effect is enhanced for an ionizing collision threshold since in this case two slow electrons are produced.

Burrow† has adapted the trapped electron method to study the excitation of atomic H near threshold and has observed sharp peaks just above the threshold for $n = 2, 3$ excitation as predicted by Taylor and Burke (see Chap. 9, p. 622). Most work with the trapped electron method has been concerned with He and other rare gases, however. Of particular interest is the application of the method to the region of the double excitation of the rare gases.‡ More structure in this region is observed using the trapped electron method than in studies of ionization or excitation cross-sections because the excitation of states which are inhibited from decaying by radiation or auto-ionization processes will be observed only in energy-loss studies. An example of such a state is the $(2p^2)^3P$ of He, the lowest doubly excited state of He which is metastable against auto-ionization. Burrow§ has measured the position of this state at $59 \cdot 64 \pm 0 \cdot 08$ eV, in good agreement with theoretical expectations and was even able to measure the slope of the total cross-section per unit energy for its excitation at threshold. A study of the excitation of 20 auto-ionizing levels of Cs in the region 12–20 eV has been carried out by Hahn and Nygaard.∥

2. Detection of superelastic collisions (see Chap. 5, § 2.4)

Further work on superelastic collisions between electrons and excited Hg atoms (Chap. 5, § 2.4) has been carried out by Burrow†† using a method essentially similar to that of Latyscheff and Leipunsky (Chap. 5, § 2.4) but with much better energy resolution of the incident electron beam (0·2 eV) than was available in the earlier work. With a retarding voltage of $-3 \cdot 2$ V relative to the cathode to permit only electrons that have gained more energy than this in superelastic collisions to reach the collector, Burrow studied the variation of the superelastic collision cross-section for incident electron energies between 0 and 2·5 eV in

---

† BURROW, P. D., *7th int. conf. on the physics of electronic and atomic collisions, Amsterdam*. Abstracts, North Holland, 1971, p. 316.

‡ BURROW, P. D. and SCHULZ, G. J., *Phys. Rev. Lett.* **22** (1969) 1271; GRISSOM, J. T., COMPTON, R. N., and GARRETT, W. R., *Phys. Lett.* **30A** (1969) 117; Oak Ridge National Laboratory Report TM–2618; GRISSOM, J. T., GARRETT, W. R., and COMPTON, R. N., *Phys. Rev. Lett.* **23** (1969) 1011.

§ BURROW, P. D., *Phys. Rev.* **A2** (1970) 1774.

∥ HAHN, Y. B. and NYGAARD, K. J., ibid. **A4** (1971) 125.

†† BURROW, P. D., ibid. **158** (1967) 65.

collision with Hg atoms excited to the $6\,^3P_1$ state. He observed a rapid fall in cross-section with increasing electron energy to reach a minimum at 0·55 eV followed by a rise to a maximum at 0·65 eV before the monotonic fall is resumed. This maximum is consistent with the sharp maximum found in the curve for the excitation of the Hg 2537 resonance line (Chap. 4, § 1.4.3, Fig. 4.33) which has been studied with high resolution by Zapesochnyi and Shpenik.† Indeed, applying detailed balancing to their excitation cross-section Burrow was able to reproduce quite well the form of his superelastic cross-section curve in this energy region. The behaviour of the cross-section at lower energies could be accounted for in terms of the effect of the sharp resonance observed by Zapesochnyi and Shpenik at 0·1 eV above threshold for the excitation process, assuming this resonance to be a $\text{Hg}^-(^4P_{9/2})$ negative ion state as proposed by Fano and Cooper (Chap. 9, p. 641). Burrow pointed out that the superelastic collisions must be attributed to the $6\,^3P_1$ excited state and not the $6\,^3P_0$ metastable state.

3. Measurements of differential cross-sections

Several recent measurements of differential cross-sections for elastic scattering (Chap. 5, § 5) have been carried out. Gibson and Dolder,‡ using a crossed-beam method have studied elastic scattering of electrons in He in the energy range 3–20 eV and angular range 25° to 145°. They observed a sharp P wave resonance at 19·45 eV. A crossed-beam method was also used by Andrick and Bitsch§ in the energy range 2–9 eV. They used their observed angular distributions to estimate the phases corresponding to the first four partial waves and used these to estimate values of total, momentum transfer, and differential cross-sections. Measurements at higher energies in helium have been made by Vriens, Kuyatt, and Mielczarek‖ who obtained relative cross-sections and by Bromberg†† who made absolute measurements for electrons of 500 eV energy.

Rubin et al.‡‡ obtained cross-sections from an experiment in which a Rb beam was crossed by electrons, by looking at the angular distribution of recoil particles as a slit of known width was swept across the beam.

---

† ZAPESOCHNYI, I. P. and SHPENIK, O. B., *Zh. eksp. teor. Fiz.* **50** (1966) 890 (*Soviet Phys. JETP* **23** (1966) 592).
‡ GIBSON, J. R. and DOLDER, K. T., *J. Phys. B (GB)* **2** (1969) 1180.
§ ANDRICK, D. and BITSCH, A., *7th int. conf. on the physics of electronic and atomic collisions, Amsterdam*. Abstracts, North Holland, 1971, p. 87.
‖ VRIENS, L., KUYATT, C. E., and MIELCZAREK, S. R., *Phys. Rev.* **170** (1968) 163.
†† BROMBERG, G. P., *J. chem. Phys.* **50** (1969) 3906; **52** (1970) 1243.
‡‡ RUBIN, K., VISCONTI, P., and SLEVIN, J., loc. cit. p. 91.

More conventional arrangements† have been used to measure elastic differential cross-sections in Na and K.

Chamberlain et al.‡ have used their apparatus which employs hemispherical electron spectrometers for energy selection of the incident electrons and energy analysis of the scattered electrons, to measure differential cross-sections for elastic scattering and for the excitation of the $2\,^1P$ and $2\,^1S$ states in He in the energy range 50–400 eV. The measured cross-sections for elastic scattering agreed with recent theoretical calculations. Cross-sections for excitation, however, were lower than the Born approximation values even at 400 eV (see p. 3387).

Recent measurements of differential cross-sections for inelastic scattering have been made by Williams§ for the excitation of the $n = 2$ level of H (electron energy range 50–200 eV, angular range 20–120°); in He for excitation of the $n = 2$ states in the energy range 19–23 eV, and angular range 0–110° by Ehrhardt et al.;∥ in He for the energy range near 60 eV and angular range 0–60° by Simpson et al.;†† in He for the excitation of the auto-ionizing states in the energy range 65–250 eV and angles 30° and 140° by Oda et al.;‡‡ for the He ($2\,^1S$) state in the energy range 26–56 eV and angular range 10–70° by Rice et al.;§§ for the He($2\,^3S$) state in the energy range 40–70 eV and angular range 25–150° by Crooks et al.;∥∥ for the He ($n = 2$ states) in the energy range 82–200 eV and angular range 30–150° by Opal and Beatty;††† and in Li,‡‡‡ Cd,§§§ Hg,∥∥∥ Na, K, Rb, and Cs.††††

† GEHENN, W. and REICHERT, E., Z. Phys. **254** (1972) 28; ANDRICK, D., EYB, M., and HOFMANN, M., J. Phys. B (GB) **5** (1972) 15; SLEVIN, J. A., Thesis, City University of New York (1970).
‡ CHAMBERLAIN, G. E., MIELCZAREK, S. R., and KUYATT, C. E., Phys. Rev. **A2** (1970) 1905.
§ WILLIAMS, K. G., 6th int. conf. on the physics of electronic and atomic collisions, Cambridge, Mass. Abstracts, M.I.T. Press, 1969, p. 731.
∥ EHRHARDT, H., LANGHANS, L., and LINDER, F., Z. Phys. **214** (1968) 179.
†† SIMPSON, J. A., MENENDEZ, M. G., and MIELCZAREK, S. R., Phys. Rev. **150** (1966) 76.
‡‡ ODA, N., NISHIMURA, F., and TAHIRA, S., Phys. Rev. Lett. **24** (1970) 42.
§§ RICE, J. K., TRUHLAR, D. G., CARTWRIGHT, D. C., and TRAJMAR, S., Phys. Rev. **A5** (1972) 762.
∥∥ CROOKS, G. B., DU BOIS, R. D., GOLDEN, D. E., and RUDD, M. E., Phys. Rev. Lett. **29** (1972) 327.
††† OPAL, C. B. and BEATY, E. C., J. Phys. B (GB) **5** (1972) 627.
‡‡‡ BURGESS, D. E., HENDER, M. A., SHUTTLEWORTH, T., and SMITH, A. C. H., 7th int. conf. on the physics of electronic and atomic collisions, Amsterdam. Abstracts, North Holland, 1971, p. 96.
§§§ NEWELL, W. R., ROSS, K. J., and WICKES, J. B. P., loc. cit. p. 99.
∥∥∥ EITEL, W., HANNE, F., and KESSLER, J., loc. cit. p. 102; SKERBELE, A. and LASSETTRE, E. N., J. chem. Phys. **52** (1970) 2708.
†††† HERTEL, I. V. and ROSS, K. J., J. Phys. B (GB) **1** (1968) 697; J. chem. Phys. **2** (1969) 285, 484.

Many of these authors express their results in terms of generalized oscillator strengths as a function of the magnitude of the momentum change and compare their results with Born approximation calculations (see Chap. 7, § 5.2.4 and p. 3391).

Of particular interest are the studies on He which show fine structure. Ehrhardt et al. reproduced the $2\,^3S$ resonance at 19·3 eV and the $2\,^1S$ resonance at 20·7 eV in agreement with theoretical calculations. Simpson et al. studied the effect of the He⁻ resonances at 57·1 and 58·2 eV on the ^3S, ^1S, and ^1P excitations. Oda et al. studied the energy dependence of the excitation cross-sections for the auto-ionizing states $(2s^2)^1S$ (57·9 eV); $(2s2p)^3P$ (58·3 eV); $(2p^2)^1D$ (59·9 eV); $(2s2p)^1P$ (60·31 eV), and three other states. The cross-sections for the optically forbidden transitions (to ^3P, ^1D states) decreased much more rapidly with increasing energy than those of the optically allowed transitions (to ^1P states).

Crooks et al. found evidence for a broad p-wave resonance of width about 15 eV at about 50 eV in the $2\,^3S$ channel.

4. Differential cross-sections in ionizing collisions

Ehrhardt et al.† have studied angular distributions in ionizing collisions of electrons in helium in great detail. Electrons of definite energy were selected using a 127° concentric cylindrical electron spectrometer. The two electrons that emerge after the ionizing collisions were detected in collectors, each of which is associated with a 127° concentric cylindrical energy analyser. In this way triple differential cross-sections,

$$\frac{\partial^3 I(E_0, E_a, \theta_a, \theta_b, \phi_b)}{\partial E_a \, \partial\Omega_a \, \partial\Omega_b}$$

for ionizing collisions in which an electron of energy E_0 is scattered with energy E_a into solid angle $d\Omega_a(\theta_a, 0)$, while the second electron (energy E_b) is ejected into solid angle $d\Omega_b(\theta_b, \phi_b)$, could be measured. Ehrhardt et al. studied the scattering of electrons of energy between 30 and 80 eV. Qualitatively they could distinguish two types of collision. One group of slow (ejected) electrons was found preferably in the angular range $\theta_b = -30°$ to $-110°$ with a maximum of the distribution close to $-60°$. In this group very little momentum was communicated to the ion so that it was termed the binary encounter peak. The other group of slow electrons was preferentially ejected into the backward direction and a backward maximum was found in the angular range between $\theta_b = +90°$ and $-120°$. The momentum transfer to the ion in these cases was large

† EHRHARDT, H., HESSELBACHER, K. H., JUNG, K., and WILLMANN, K., J. Phys. B (GB) 5 (1972) 1559.

and Ehrhardt et al. refer to this maximum as the 'recoil' peak. Calculations based on Born's approximation do not show the existence of this 'recoil' peak.

McConkey et al.[†] have carried out a complementary experiment in which the angular distribution of the recoiling ions in ionizing collisions in He and Ar for electron energies up to 300 eV have been studied. Their results are consistent with the results of Ehrhardt et al.

At higher energies the width of the maximum associated with the binary collisions becomes narrower and the shape of the maximum has been used to investigate the internal momentum distribution of the electrons in the scattering atom.[‡]

Extensive studies of small angle ionizing collisions of fast electrons in the rare gases have been made by the Amsterdam group.[§] A mass spectrograph, operated in coincidence with the scattered electrons, enabled collisions leading to multiple ionization to be studied. The incident electrons (of energy $\sim$ 10 keV) after passing through the collision region, entered an aperture of diameter 1 mm and were decelerated to an energy between 20 and 100 eV. They were then passed through two Wien filters in which they were analysed in energy with a resolution of 1–5 eV. Finally they were accelerated again to an energy of approximately 200 eV before detection by an electric multiplier. Electrostatic deflector plates between the collision region and the aperture enabled the electron beam to be swept across the aperture so that the angle of scattering of a detected electron, observed in coincidence with the ion produced, was known.

The electron beam crossed, at right angles, in the collision region, a neutral gas beam (1 mm $\times$ 8 mm) emerging from a multi-channel gas jet. The ions were extracted from the collision region by a homogeneous extraction field (50–600 V/cm) produced by electrodes with slits much larger than the ion beam dimensions, and so designed that all ions formed with thermal energies could be extracted. The mass spectrometer was designed to give 100 per cent transmission, in combination with the extraction system, for ions formed with zero excess energy.

The results were presented in the form of the generalized oscillator

[†] McConkey, J. W., Crowe, A., and Hender, M. A., *Phys. Rev. Lett.* **29** (1972) 1.

[‡] Amaldi, U. and Ciofi degli Atti, C., *Nuovo Cim.* **66A** (1970) 129; Camilloni, R., Giardini-Guidoni, A., and Tiribelli, R., *Phys. Rev. Lett.* **29** (1972) 618; Schmoranzer, H., Ulsh, R. C., and Bonham, R. A., *7th int. conf. on the physics of electronic and atomic collisions, Amsterdam.* Abstracts, North Holland, 1971, p. 866.

[§] van der Wiel, M. J., El-Sherbini, Th. M., and Vriens, L., *Physica, 's Grav.* **42** (1969) 411; van der Wiel, M. J., ibid. **49** (1970) 411; van der Wiel, M. J. and Wiebes, G. K., ibid. **53** (1971) 225; **54** (1971) 411.

strength $\mathrm{d}f(K)/\mathrm{d}E$ for energy loss E in terms of K the magnitude of momentum transfer,

$$\frac{\mathrm{d}f(K)}{\mathrm{d}E} = \frac{\mathrm{d}f}{\mathrm{d}E} + aK^2 + bK^4 + \ldots,$$

where $\mathrm{d}f/\mathrm{d}E$ is the optical oscillator strength. It was possible to compare the results of these experiments with those obtained using photo-ionization, using the latter results at a convenient energy loss, to normalize the results of the electron scattering experiments. The results of the two types of experiment agreed well for He. For high-energy transfers in Ne and Ar it was concluded that there must be an appreciable contribution from multiple ionization in the photo-ionization experiments. Structure was observed in the oscillator strength distribution with energy loss for these cases as well as for Kr and Xe ionization and this was attributed to threshold and resonance effects.

Differential cross-sections for electron back-scattering in the range 128–148° after producing K-shell ionization of C, N, O, Na, and Mg have been measured by Gerlach and du Charme.† Electrons were incident on absorbed layers of these atoms on a W (1, 0, 0) crystal surface. The back-scattered electrons in the specified angular range were energy-analysed using a concentric cylindrical electrostatic analyser. The back-scattered electrons were distinguished by modulating the potential of the electron gun while maintaining the analyser cylinder potentials constant. The back-scattered electrons showed a fixed energy loss while the Auger electrons had a fixed energy independent of the electron gun potential.

Corrections were made for double scattering of the electrons in the target but in this angular range the observed differential cross-sections in the energy range up to four times the K-ionization energy were between one and two orders of magnitude greater than those calculated using Born's approximation (p. 166).

5. Further experiments on spin polarization in scattering

The production of polarized electrons by elastic scattering has been known for a long time (Chap. 5, § 6.1). Recent work‡ has demonstrated that electrons scattered inelastically can also be polarized. Measurements of differential cross-sections and angular dependence of the

† GERLACH, R. L. and DU CHARME, A. R., *Phys. Rev. Lett.* **27** (1971) 290; *Phys. Rev.* A**6** (1972) 1892.

‡ EITEL, W. and KESSLER, J., *Z. Phys.* **241** (1971) 355; HANNE, G. F., JOST, K., and KESSLER, J., ibid. **252** (1972) 141.

polarization have been carried out for electron energies in the range 20–180 eV scattered in Hg with excitation of the $6\,^1P_1$ state (6·7 eV energy loss). At energies above 50 eV the angular distributions and spin polarizations were similar to those for elastic scattering. Spin polarization of electrons scattered by neon has also been studied.†

Collins et al.‡ have extended and greatly improved the measurements of spin-exchange elastic scattering cross sections in K using the recoil analysis technique (Chap. 5, § 7.2). They have obtained differential cross-sections for the spin-exchange cross-section in the energy range 0·4–1·2 eV which are in reasonable agreement with the close-coupling calculations of Karule and Peterkop§ (see p. 3404). The ratio $Q_{ex}(\theta)/Q(\theta)$, where θ is the electron scattering angle, was found to attain its theoretical maximum value of 4/3 for $\theta > 145°$ for electron energies of 1·0 and 1·2 eV.

Kriščiokaitis and Tsai‖ have suggested a method of obtaining high-intensity sources of polarized electrons by elastic spin-exchange collisions of electrons with selected neutral hydrogen atoms in the $F = 1, M_F = 1$ state in a hydrogen maser.

Hils et al.†† have studied the direct (non spin-flip) cross-section for electrons of 3–4 eV energy by measuring the polarization of the electrons after elastic scattering from a polarized K atomic beam. The scattered electrons were accelerated and their polarization measured by Mott scattering. If f, g are respectively the direct and spin exchange scattering amplitudes so that the differential cross-section is given by

$$I = \tfrac{1}{2}\{|f|^2 + |g|^2 + |f-g|^2\},$$

these experiments measure $\dfrac{f^2}{I} = 1 - \dfrac{P_e}{P_A}$ where P_e, P_A are respectively the polarization of the scattered electrons and the atomic beam. The experiments of Collins et al. on the other hand measured the quantity g^2/I so that the two experiments are complementary.

Hils et al.†† found that for 4 eV electrons, P_e went from $-0·013 \pm 0·023$ to $+0·193 \pm 0·073$ between $\theta = 20°$ and $40°$. Clearly the accuracy obtained using the atomic recoil technique of Bederson and his co-workers is considerably better.

† GUPTA, R., Thesis, Boston University, Mass. (1970).
‡ COLLINS, R. E., BEDERSON, B., and GOLDSTEIN, M., Phys. Rev. **A3** (1971) 1976.
§ KARULE, E. M. and PETERKOP, R. K., in *Atomic collision III* (editor Veldre, Y. I.) (Latvian Acad. Sci., Riga, U.S.S.R., 1965); *JILA Information Centre Report No. 3*, University of Colorado, Boulder, Colorado, 1–27; KAULE, E. M., ibid. 29–48.
‖ KRIŠČIOKAITIS, R. J. and TSAI, WU-YANG, *Nucl. Instrum. Meth.* **83** (1970) 45.
†† HILS, D., MCCUSKER, M. V., SMITH, S. J., and KLEINPOPPEN, H. A., *7th int. conf. on the physics of electronic and atomic collisions, Amsterdam*. Abstracts, North Holland, 1971, p. 354.

Bederson et al.† have applied the atomic beam recoil technique to measure spin-exchange scattering in inelastic electron collisions using a polarized K atomic beam with the excitation $4\,^2S_{\frac{1}{2}}$–$4\,^2P_{\frac{1}{2},\frac{3}{2}}$. The differentially scattered atomic beam was spin-analysed, and discrimination against elastic scattering and scattering from other inelastic channels was possible from kinematics. An electron energy range from 3·0–10·0 eV was covered and an electron scattering angular range from 0–20°. The quantity measured was $R(\theta) = Q_{SP}/Q$ where Q is the full differential cross-section for $4S$–$4P$ scattering and Q_{SP} is the differential cross-section for scattering with change of the quantum number M_S of the recoil atom. The observed spin-flip is a combination of a contribution occurring during excitation and another as a result of spin–orbit effects in the excited state. One has‡

$$R(\theta) = \frac{4}{9} + \frac{1}{18I}\{I_0(\text{ex}) + 10I_1(\text{ex}) - 8I_1\},$$

where $I = I_0 + 2I_1$ is the full differential cross-section per unit solid angle, the suffixes 0, 1 specify the change $\Delta M_e = 0, \pm 1$ respectively of the orbital magnetic quantum number, M_e, in the excitation, so that

$$I_0(\text{ex}) = |g_0|^2, \qquad I_1(\text{ex}) = |g_1|^2$$

and $\qquad I_0 = |f_0|^2 + |g_0|^2, \qquad I_1 = |f_1|^2 + |g_1|^2,$

where $f_{0,1}$, $g_{0,1}$ are the direct and exchange amplitudes respectively. The results obtained were in reasonable agreement with the close-coupling calculations of Karule and Peterkop§ (see p. 3404).

† BEDERSON, B., KASDAN, A., and GOLDSTEIN, M., *7th int. conf. on the physics of electronic and atomic collisions, Amsterdam*. Abstracts, North Holland, 1971, p. 352.
‡ RUBIN, K., BEDERSON, B., GOLDSTEIN, M., and COLLINS, R. E., *Phys. Rev.* **182** (1969) 201; BEDERSON, B., *Comments on Atomic and Molecular Physics*, B**1** (1971) 160.
§ loc. cit.

6

ELECTRON COLLISIONS WITH ATOMS— THEORETICAL DESCRIPTION—GENERAL AND SEMI-EMPIRICAL THEORY OF ELASTIC SCATTERING

1. Quantum defect theory

ON p. 400 the extrapolation of quantum defects for a Rydberg sequence of bound states into the continuum to obtain phase shifts for scattering of electrons by the positive ion core was discussed. Application to He$^+$ is described on p. 536. Doughty, Seaton, and Sheorey† considered double extrapolation across both Rydberg series and isoelectronic sequences as a possible way of obtaining phase shifts for scattering by neutral atoms. They found that the experimental data are not sufficiently complete or accurate and that more explicit account should be taken of polarization. However, they were able to derive a model scattering potential for helium involving the known dipole and quadrupole polarizabilities of the atom and isoelectronic ions, and a cut-off parameter r_0. The latter was obtained for each ion in the sequence so as to give agreement with the phases derived from the appropriate quantum-defect extrapolation. Extrapolation along the sequence to the neutral atom case gave r_0 for helium so that the scattering phases could be calculated. In this way the zero-energy scattering length for helium was found to be $-1\cdot 195a_0$,† which is close to values derived in other ways (see p. 544).

Sheorey‡ also applied this procedure to neon and argon. For the former he found a scattering length of $0\cdot 355a_0$, somewhat larger than that calculated directly by Thompson (see p. 3403). For argon less satisfactory results were obtained probably due to poor quantum defect data for ScIII.

2. Relativistic effects including spin polarization (see Chap. 6, § 6)

A number of further calculations of relativistic effects on scattering of electrons by atoms and on the production of electron polarization (see Chap. 6, § 6) have been carried out. The subject has been reviewed

† DOUGHTY, N. A., SEATON, M. J., and SHEOREY, V. B., *J. Phys. B* (*GB*) **1** (1968) 802.
‡ SHEOREY, V. B., ibid. **2** (1969) 442.

by Walker† and the relation between non-relativistic and relativistic theoretical predictions is now more thoroughly understood. Thus in Table 6.4 of Chapter 6 comparison is made between total cross-sections calculated for impact of electrons of 2, 20, and 200 eV energy with He, Kr, Cs, and Hg (a) using the non-relativistic Hartree–Fock field and non-relativistic scattering theory and (b) using the same field and relativistic scattering theory. Since the Hartree–Fock field includes exchange effects, a more realistic comparison should be between calculations both including exchange effects in the scattering, one non-relativistic using the non-relativistic Hartree–Fock field and one relativistic, using the relativistic Hartree–Fock field. When this is done little difference between the two calculated cross-sections is found at any of the impact energies for any of the atoms.

The earlier calculations of differential cross-sections and polarizations were carried out using a relativistic Hartree field. In general the relativistic Hartree–Fock field‡ is somewhat weaker while inclusion of exchange between the incident and atomic electrons increases the lower-order phase shifts. The net result is that, at energies of 300 eV in mercury, the first few phases, including all exchange effects, are not very different from those calculated using a relativistic Hartree field but as the relative angular momentum increases the phases are reduced. At lower energies the first few phases are larger than those given by the relativistic Hartree field. Exchange effects also tend to reduce the spin polarization.

Calculations for mercury, including exchange, have been carried out by Walker§ for electron energies from 0·5 to 1500 eV above which energy exchange effects are unimportant. At least at energies above a few tens of eV, inclusion of exchange merely introduces an angular displacement of a few degrees in the differential scattering or spin polarization distribution.

Surprisingly good agreement is obtained with observed data‖ down to energies as low as 3·5 eV, for the differential and total cross-sections and the polarization.

The possible effects of atomic distortion at low energies are discussed on p. 3409.

† WALKER, D. W., *Adv. Phys.* **20** (1971) 257.
‡ COULTHARD, M. A., *Proc. phys. Soc.* **91** (1967) 44.
§ WALKER, D. W., *J. Phys. B (GB)* **2** (1969) 356.
‖ DEICHSEL, H., REICHERT, E., and STEIDL, H., *Z. Phys.* **189** (1966) 212 (see Chap. 5, p. 353); BROMBERG, J. P., *J. chem. Phys.* **51** (1969) 4117; EITEL, W., JOST, K., and KESSLER, J., *Phys. Rev.* **159** (1967) 47; *Z. Naturf.* **A23** (1968) 2122.

Walker† has also carried out relativistic exchange calculations for He, Ne, Ar, Kr, and Xe. Tables giving the energies at which the spin polarization of the scattered electron is complete have been drawn up for Hg, Ar, Kr, and Xe, as well as the energy and angular widths within which the magnitude of the polarization remains greater than 0·9.

† loc. cit., p. 3384.

7

ELECTRON COLLISIONS WITH ATOMS— THEORETICAL DESCRIPTION—BORN'S APPROXIMATION

Since this chapter was written further evidence has been obtained about the region of applicability of the Born and Born–Oppenheimer approximations. The Born approximation has also been used to calculate cross-sections for ionization of a number of atoms and positive ions. We discuss the recent work in relation to scattering by different atoms and ions.

1. Atomic hydrogen

1.1. *Excitation*

Total cross-sections for excitation of H(2s) (see p. 3368) measured by Cox and Smith, which are normalized from the absolute value of the cross-section for H(2p) excitation at 500 eV calculated by Geltman and Hidalgo (see p. 3394) show good agreement with Born's approximation at energies > 200 eV. At lower energies the theoretical values become rapidly too large, being about twice the observed at 30 eV.

The angular distribution of electrons scattered after exciting the $n = 2$ levels of atomic hydrogen, measured by Williams (see p. 3377) for impact energies of 50, 100, and 200 eV and angles $> 30°$, falls very much less rapidly with increasing angle than that calculated by the Born (or Born–Oppenheimer) approximation. This does not mean that these approximations predict too small a total excitation cross-section because scattering at angles $> 30°$ contributes very little to the total. Further discussion of these experimental results is given on p. 3401 in connection with more elaborate approximations.

1.2. *Ionization of* H(2s)

The cross-section measured by Dixon and Harrison for ionization of H(2s) (see p. 3364) agrees with the predictions of Born's approximation for electron energies > 200 eV. At smaller energies, as usual, the approximation gives results which are too large.

2. Helium

2.1. Excitation of singlet states

Bell, Kennedy, and Kingston† have calculated cross-sections for excitation of the $n\,^1S$ ($n = 2$ to 7), $n\,^1P$ ($n = 2$ to 4), and $3\,^1D$ states by Born's approximation, using very accurate atomic wave-functions in the form

$$\psi_{nL}(\mathbf{r}_1, \mathbf{r}_2) = (1+P_{12}) \sum_t \exp(-\alpha_t r_1 - \beta r_2) \sum_{ijk} c_{tijk}\, r_1^i\, r_2^j\, r_{12}^k,$$

where P_{12} is the permutation operator. The coefficients α_t, β, and c_{tijk} were taken to be those obtained by Perkins for the $n\,^1S$ ($n = 3$ to 7) states and by Weiss for the $1\,^1S$, $2\,^1S$, $n\,^1P$ ($n = 2$ to 4), and $3\,^1D$ states. Except for $4\,^1P$, in all cases at least forty terms were included in the expansion. Kim and Inokuti‡ have calculated cross-sections for excitation of $2\,^1P$, $3\,^1P$, $2\,^1S$, and $3\,^1S$ using Weiss functions throughout and there is little disagreement of their results with those of Bell et al. The results obtained should be of very high accuracy, within the range of validity of Born's approximation, and should be taken as replacing those given in Chapter 7, Table 7.3, which were obtained with much simpler wave-functions. In Table 7A.1 the new values are given and, for comparison, the less accurate values of Table 7.3.

TABLE 7A.1

Total cross-sections in 10^{-20} cm² for excitation of various singlet levels in helium

Electron energy		$2\,^1S$	$3\,^1S$	$4\,^1S$	$2\,^1P$	$3\,^1P$	$4\,^1P$	$3\,^1D$
100 eV	(a)	203	—	—	1370	339	134	8·2
	(b)	195·7	43·95	16·82	1306	322·4	129·5	9·070
200 eV	(a)	110	—	—	980	243	96	4·6
	(b)	103·4	23·27	8·911	940·2	232·6	93·41	5·146
400 eV	(a)	—	—	—	635	159	62·5	2·5
	(b)	52·04	11·94	4·574	615·3	152·1	61·04	2·727

(a) As in Table 7.3.   (b) Calculated by Bell et al.

Bell et al. also calculated cross-sections for $5\,^1P$, $6\,^1P$, and $n\,^1D$ ($n = 4, 5, 6$) using less accurate but still quite elaborate wave-functions.

The more precise values of the theoretical cross-sections do not differ from the earlier ones by as much as the differences between the absolute values of measured cross-sections so that the conclusions of Chapter 7,

† BELL, K. L., KENNEDY, D. J., and KINGSTON, A. E., J. Phys. B (GB) 2 (1969) 26.
‡ KIM, Y. K. and INOKUTI, M., Phys. Rev. 175 (1968) 176.

§§ 5.6.7 and 5.6.9 are unchanged. More recent experimental data (see p. 3370), although still far from precise, nevertheless confirm and to some extent amplify the earlier observations. To summarize, it appears that for ^1P excitation Born's approximation gives good results for electron energies > 250 eV and for ^1S at somewhat greater energies. At lower energies it gives cross-sections which are too large, particularly for ^1S (cf. the corresponding results for H($2s$), p. 3386).

For the ^1D states the situation is quite different and, at least for $n < 6$, it seems clear that Born's approximation underestimates the cross-section even up to energies of 5000 eV. Thus for $4\,^1D$ at 108 eV Bell et al. give a theoretical cross-section of $4{\cdot}76 \times 10^{-20}$ cm^2. In the same units observed values are 12 (Gabriel and Heddle; St. John, Miller, and Lin); 19 (Zapesochnyi) all given in Table 4.1, p. 202; 10·5 (de Heer, Vroom, de Jong, van Eck, and Heideman);† and 11·5 (McConkey and Woolsey).‡ At 2000 eV, according to McConkey and Woolsey the observed value is 0·65, still more than twice the calculated (0·30). For $3\,^1D$ the discrepancy, in the same sense, is even more marked, the comparison being made with the observations of Moussa et al.§ There is some evidence that as n increases Born's approximation gives results which are closer to experiment but this is not fully established.

Further evidence about the validity of Born's approximation is obtained from differential cross-section measurements. The absolute measurement of the differential cross-section for excitation of the $2\,^1P$ states by electrons scattered through 5°, made by Chamberlain et al.∥ (see p. 3377) gives values which at electron energies of 400, 200, 100, and 50 eV respectively are 0·905, 0·825, 0·685, and 0·375 times that given by Born's approximation as calculated by Kim and Inokuti. Relative measurements over the angular range 5–20°†† indicate that the approximation is not adequate for electron energies below 500 eV and momentum transfers less than 2·0 a.u.

This is consistent with the evidence from the total cross-sections. On the other hand for $2\,^1S$ Chamberlain et al.∥ (see p. 3377) find that, at 5°, the ratio of measured scattering to that given by Born's approximation

† DE HEER, F. J., VROOM, F. O. M., DE JONG, J. P., VAN ECK, J., and HEIDEMAN, H. G. M., 6th int. conf. on the physics of electronic and atomic collisions, Cambridge, Mass. Abstracts, M.I.T. Press, 1969, p. 350.

‡ MCCONKEY, J. W. and WOOLSEY, J. M., ibid. p. 355.

§ MOUSTAFA MOUSSA, H. R., DE HEER, F. J., and SCHUTTEN, J., Physica, 's Grav. **40** (1969) 517.

∥ CHAMBERLAIN, G. E., MIELCZAREK, S. R., and KUYATT, C. E., Phys. Rev. A**2** (1970) 1905.

†† SIMPSON, J. A. and MIELCZAREK, S. R., 6th int. conf. on the physics of electronic and atomic collisions, Cambridge Mass. Abstracts, M.I.T. Press, 1969, p. 344.

remains close to 0·8 over the entire energy range from 50 to 400 eV. This does not seem to be consistent with the data from total cross-sections.

At large angles of scattering ($> 40°$) the experiments of Opal and Beaty (see p. 3377)† show that, at 200 eV, the differential cross-sections for excitation of the $2\,^1P$ and $2\,^1S$ states are much greater than given by Born's approximation (see also the situation for H, p. 3386). This is not inconsistent with the conclusions derived from the total cross-section measurements because very little contribution to the total cross-section comes from scattering at these angles.

It seems likely that the reason why Born's approximation gives results which are too low for the weakly excited ^1D levels is because no allowance is made for population of these levels by double transitions through the strongly excited ^1P states. Confirmation of this has been provided by close-coupling calculations carried out by Chung and Lin‡ (see p. 3402).

Discussion of the applicability of Born's approximation to calculate cross-sections for excitation by proton impact, based on the use of the same helium wave-functions as used by Bell, Kennedy, and Kingston for electron impact, is given in Chapter 23, p. 2423 and Chapter 25, p. 3118.

2.2. *Excitation of triplet states*

It was pointed out in Chapter 7, p. 493 that experimental confirmation of the very rapid fall with electron energy above 100 eV of the cross-sections for excitation of triplet states predicted by the Born–Oppenheimer approximation, is very difficult because of the need to eliminate secondary effects. Some supporting evidence for excitation of $2\,^3P$ (see Fig. 7.29) had been obtained. As pointed out on p. 3370 further evidence is now forthcoming from the experiments of Kay and Shwalter.§ By working at pressures low enough (down to 2×10^{-6} torr) to ensure a linear relationship with the cross-section they obtained relative values for $4\,^3S$ between 50 and 200 eV impact energy which fell off at the fast rate predicted from theory (as E^{-3} where E is the impact energy). For the $3\,^3P$ and $3\,^3D$ states there appear to be complications arising from cascade effects from F levels (see p. 3572 and Chap. 18, § 4.2.3.2) but these may be approximately disentangled (see p. 3370) leaving residual cross-

† OPAL, C. B. and BEATY, E. C., *J. Phys. B (GB)* **5** (1972) 627.
‡ CHUNG, S. and LIN, C. C., *6th int. conf. on the physics of electronic and atomic collisions, Cambridge, Mass.* Abstracts, M.I.T. Press, 1969, p. 363.
§ loc. cit. p. 3370.

sections which fall off rapidly with E though still not so rapidly as predicted. On the whole it seems unlikely that any real discrepancy exists.

2.3. *Ionization* (see pp. 487–91). Bell and Kingston† have recalculated the ionization cross-section of helium using the same wave-functions for the continuum as did Sloan (see p. 487) but a more elaborate ground-state function (the six-parameter function of Stewart and Webb). The final results differ from those of Sloan by not more than 1 per cent, so the comparison between theory and experiment shown in Chapter 7, Fig. 7.25 remains unchanged.

Energy distributions of electrons ejected in ionizing collisions with helium atoms measured by Beaty, Opal, and Peterson‡ agree quite well with the predictions of the Born approximation for incident electrons of 1000 eV energy. At 100 eV, for which the Born-exchange approximation gives a total cross-section about 30 per cent too large, it nevertheless gives quite good results for the shape of the energy distribution of secondary electrons, predicting much smaller production of electrons with energies > 20 eV than does the Born approximation.

Gillespie§ has calculated cross-sections for simultaneous excitation and ionization of He in which the He$^+$ ion is left in the state nl with $l = 0, 1$ and $n = 2, 3, 4$.

3. Electron detachment from H by electron impact

Faisal and Bhatia‖ have calculated cross-sections for the detachment processes
$$\text{H}^- + \text{e} \rightarrow \text{H}(1s) + \text{e} + \text{e},$$
$$\rightarrow \text{H}(2s) + \text{e} + \text{e},$$
by an impact parameter method (see Chap. 7, § 5.3.2) in which the colliding electron is taken to pursue a hyperbolic trajectory in the Coulomb field of the negative ions. The interaction taken is confined to the asymptotic dipole term in the multipolar expansion of the full interaction, cut off at close distances by ignoring all collisions in which the impact parameter p is less than a certain value p_0. This was taken to be $3 \cdot 42 a_0$, the approximate radius of H$^-$. Quite good agreement is obtained with the experimental results, including particularly the recent observations of Walton, Peart, and Dolder (see p. 3365) (apart from the resonance effect near 14·2 eV, see p. 3365).

† BELL, R. L. and KINGSTON, A. E., *J. Phys. B (GB)* **2** (1969) 1125.
‡ BEATY, E. C., OPAL, C. B., and PETERSON, W. E., *7th int. conf. on the physics of electronic and atomic collisions, Amsterdam.* Abstracts, North Holland, 1971, p. 872.
§ GILLESPIE, E. S., *J. Phys. B (GB)* **5** (1972) 1916.
‖ FAISAL, F. H. M. and BHATIA, A. K., *Phys. Rev.* **A5** (1972) 2144.

4. Alkali metal atoms—excitation

The measurements of Hertel and Ross† enable generalized oscillator strengths for the excitation of various alkali metal atoms to be determined relative to those for the resonance transitions. Assuming that at an energy ten times the threshold value Born's approximation gives good results for the resonance $3\,^2S \rightarrow 3\,^2P$ transition at small angles, normalized absolute values can be obtained for other transitions. It is then found that for excitation of $3\,^2D$ Born's approximation predicts much smaller values than are observed, even for 100 eV electrons. This is consistent with the corresponding comparison for the excitation of $3\,^1D$ in helium (see p. 3388) and presumably can be understood in the same way. For excitation of $4\,^2S$ Born's approximation gives much better results, while for $4\,^2P$ there is evidence that as the momentum transfer (or angle of scattering) increases, the good agreement found for small transfers begins to fail, the theoretical values being too small. However, too much weight should not be given to the detailed comparison as only simple one-electron atomic wave-functions were used.

5. Ionization of other atoms

Cross-sections for ionization of a great number of atoms and positive ions have been calculated using the Born, or when appropriate, the Coulomb–Born approximation. For neutral atoms with outer $2p$ and $3p$ electrons Peach‡ has calculated cross-sections for an electron energy range from threshold to 5000 eV using the same wave-functions as for ionization by proton impact, described in Chapter 23, p. 2429.

Comparison with observed data is complicated by the considerable differences in the absolute values of the cross-sections obtained by different experimenters (see Chap. 3, Tables 3.1 and 3.2). For neon the observed results are not too inconsistent with each other and with the calculated values, whereas for argon the calculated values agree well with the experimental data of Smith (P. T.), Tozer and Craggs, and Asundi and Kurepa, which are, at energies near the maximum, about 25 per cent higher than the observed values of Rapp and Englander-Golden. Again, for atomic nitrogen the calculated values agree better with the observations of Peterson than those of Smith (A. C. H) *et al.* while for atomic oxygen the calculated values fall below all the observed results.

† HERTEL, I. V. and Ross, K. J., *6th int. conf. on the physics of electronic and atomic collisions, Cambridge, Mass.* Abstracts, M.I.T. Press, 1969, p. 280.

‡ PEACH, G., *J. Phys. B (GB)* **4** (1971) 1670—this paper corrects a numerical error in the earlier work by the same author in the same journal, **1** (1968) 1088 and **3** (1970) 328.

Peach† has also calculated cross-sections for further ionization of Li^+, Mg^+, K^+, and the positive ions isoelectronic with Ne, using Born's approximation throughout. Cross-sections for Li^+ have also been calculated by Economides and McDowell‡ again using this approximation but representing the wave-function of the ejected electron by the polarized orbital approximation. Moores and Nussbaumer§ have used the Coulomb–Born approximation and represented the wave-functions of the active atomic electron by regarding it as moving in a core potential of scaled Fermi–Thomas form (see p. 3482). They have applied this technique to both Li^+ and Mg^+.

For Li^+ the calculated values do not differ greatly at energies greater than 4 times the threshold value, at which they exceed the observed (see p. 3364) by about 30 per cent. On the other hand the observed and calculated values are in agreement at energies greater than 20 times the threshold value. The situation is rather similar for Na^+ and Mg^+ while for Mg^{++} the calculated values are rather closer to the observed—even at the peak the observed value is only about 15 per cent smaller than the calculated. For K^+ the calculated values are 50 per cent greater than the observed even at energies as high as 100 times the threshold.

Calculations‖ have also been carried out for B and the isoelectronic ions C II, N III, and O IV by the truncated Born (or Coulomb–Born) (see Chap. 7, p. 465) and the Born (or Coulomb–Born) exchange approximations (Chap. 7, p. 465 (188)).

McFarlane†† has calculated the polarization of $L_{\alpha 1}$, $L_{\alpha 2}$, and L_l X-radiation arising from electron impact, using a hydrogenic model. For electron energies greater than 4 times the threshold value the magnitude of the calculated polarization is less than 3 per cent.

† loc. cit., p. 3391.
‡ Economides, D. G. and McDowell, M. R. C., *J. Phys. B (GB)* **2** (1969) 1323.
§ Moores, D. C. and Nussbaumer, H., ibid. **3** (1970) 161.
‖ Stingl, E., ibid. **5** (1972) 1160.
†† McFarlane, S. C., ibid. **5** (1972) 1906.

8

ELECTRON COLLISIONS WITH ATOMS—THEORETICAL DESCRIPTION—ANALYTICAL THEORY FOR SLOW COLLISIONS

1. General theoretical developments

THE use of variational trial functions for the calculation of elastic and inelastic cross-sections for collisions of electrons with light atoms, and particularly hydrogen, has developed in some new directions since Vol. 1 was written. This concerns the choice of trial functions to lead to satisfactory results with more economical use of computer time so that more complicated problems may be tackled.

The success which has attended the use of trial functions for atomic and molecular structure calculations in the form of suitable combinations of Slater-type orbitals has opened the possibility for similar applications to collision problems. Essentially this involves trial functions of the form

$$\Psi_t = \Psi_c + \Phi, \tag{1}$$

where Ψ_c is chosen simply to give the correct asymptotic form while Φ is a suitable combination of configurations in which the individual orbitals are of Slater type. Already good results have been obtained in this way (see p. 3399).

The second type of trial function which attempts to represent as directly as possible the main physical features of a particular problem is that in which so-called pseudo-state terms are added to a truncated close-coupling expansion (see Chap. 8, § 2.2.3). Thus we take

$$\Psi^t = \sum_{n=1}^{s} \psi_n(\mathbf{r}_a) F_n(\mathbf{r}) + \sum_{p=1}^{q} \bar{\psi}_p(\mathbf{r}_a) F_p(\mathbf{r}), \tag{2}$$

where the ψ_n are accurate atomic functions and the $\bar{\psi}_p$ are pseudo-functions, which are orthonormal with respect to each other and to the ψ_n. The $\bar{\psi}_p$ are chosen so as to include terms in Ψ^t which represent as concisely as possible the main features of the situation which are not included in the truncated expansion itself. These features may, for example, be the polarizability of the atom or correlation effects. Considerable success has been attained with such expansions both for calculating cross-sections for impact energies near thresholds, in which polarization is important, and for energies intermediate between these and the considerably higher energies for which Born's approximation is valid. For these cases account must be taken of correlation (see pp. 3396, 3400).

A quite different semiclassical approximation due to Glauber, has also been found to give good results at intermediate energies for e–H collisions though it is not clear how readily it may be applied in practice to more complicated systems (see p. 3398). Other approximations† to apply to this energy range have also been suggested.

As far as detailed applications of theory to particular atoms is concerned the most striking development has been the clarification of the confused situation for the scattering of electrons by alkali metal atoms which existed when Vol. 1 was written. It appears that the experimental data available at the time were not accurate and more recent measurements give results in remarkably good agreement with calculated values. This refers not only to total and differential elastic cross-sections but also to spin-exchange in elastic and inelastic scattering (see p. 3405).

The polarized orbital method has been modified and extended in various ways. Of particular interest are the applications of the method to collisions of electrons with atoms of open-shell structure in which close-coupling exists between channels involving the terms of the ground configuration, and the extension to higher energies by incorporating non-adiabatic terms.

We first describe in a little more detail the general nature of the theoretical developments and then discuss their application, together with any further application of methods previously described, to individual atoms.

2. Pseudo-state expansions

The truncated eigenfunction expansion

$$\Psi^t(\mathbf{r}_1, \mathbf{r}_2) = \sum_{n=0}^{s} F_n(\mathbf{r}_1)\psi_n(\mathbf{r}_2) \tag{3}$$

discussed in Chapter 8, § 2.2.3 (see (25) of that section) is limited in practice by the computational time required if s is not small. In particular, to take account of dipole polarization of the atoms by the incoming electron, terms in (3) associated with continuum states ψ_K of the atom should be included. The procedure adopted by Burke and Taylor of adding terms of the form $\chi(r_1, r_2, r_{12})$ attempts to make up for the terms omitted in (3)

† GELTMAN, S. and HIDALGO, M. B., *J. Phys. B (GB)* **4** (1971) 1299; **5** (1972) 617.

by adding what are effectively pseudo-states. In this particular example the pseudo-states allow, through the presence of explicit terms in r_{12}, for correlation effects between the electrons. Other pseudo-states may be introduced designed to allow for other features which may be of particular importance in a particular energy range.

As an illustration, suppose it is desired to make adequate allowance for dipole distortion of the atom by appropriate addition of a pseudo-state function to a two-term expansion

$$F_{1s}(\mathbf{r}_1)\psi_{1s}(r_2) + F_{2s}(\mathbf{r}_1)\psi_{2s}(r_2) \tag{4}$$

applied to calculating the elastic scattering below the excitation threshold.

The polarizability of the atom in its ground state is given in atomic units by

$$\alpha(1s) = 2 \sum_{n=2}^{\infty} \langle \psi_{1s} r \cos\theta \, \psi_{np}\rangle^2/(E_n - E_1) \tag{5}$$

where $E_n = -1/2n^2$. The sum may be replaced by a single term

$$\frac{3}{2\pi} \langle \psi_{1s} r \cos\theta \, \bar{\psi}_{2p}\rangle^2/(\bar{E}_{21} - E_1), \tag{6}$$

where

$$\bar{E}_{21} = \langle \bar{\psi}_{2p} H \bar{\psi}_{2p}\rangle, \tag{7}$$

H being the atomic Hamiltonian, by appropriate choice of $\bar{\psi}_{2p}$.

Damburg and Karule† showed that the pseudo $2p$ function

$$\bar{\psi}_{2p} = \left(\frac{32}{43\pi}\right)^{\frac{1}{2}} e^{-r}(r^2 + \tfrac{1}{2}r^3)\cos\theta \tag{8}$$

in atomic units, gives the correct value of α when substituted in (6). Good results, as far as allowance for dipole distortion is concerned, should therefore be expected if to (4) is added a pseudo-state term

$$\bar{F}_{2p}(\mathbf{r}_1)\bar{\psi}_{2p}(\mathbf{r}_2). \tag{9}$$

It would not be expected, however, that this method would give good results at energies just below the $n = 2$ threshold—the pseudo-threshold corresponding to the pseudo $2p$ function will not be equal to that for the real $2p$ function, a defect which is likely to be important as the real threshold is approached.

This is to be contrasted with the polarized-orbital procedure which introduces into the expansion a term of form

$$F_{1s}(\mathbf{r}_1)\phi(\mathbf{r}_1, \mathbf{r}_2), \tag{10}$$

where $\phi(\mathbf{r}_1, \mathbf{r}_2)$ is obtained as described in Chapter 8, p. 512. As F_{1s} is

† DAMBERG, R. J. and KARULE, E., *Proc. phys. Soc.* **90** (1967) 637.

oscillatory while $\overline{F}_{2p}$ is a damped, closed-channel, function the two procedures represent the effect of dipole distortion in rather different ways.

This procedure can be extended to include a pseudo-state which gives the correct quadrupole polarizability and so on. It may also be generalized to electron energies between the $n = 2$ and $n = 3$ thresholds. In this case there will be three terms in the eigenfunction expansion which are associated with oscillatory functions F_n which represent wave motion. It is then appropriate to introduce three pseudo-states, $\bar{\psi}_{3s}$, $\bar{\psi}_{3p}$, and $\bar{\psi}_{3d}$ which satisfy conditions of the form (6) corresponding to different initial and final 2-quantum states and which are orthogonal to all the real states present in the expansion. Again this would not be expected to give good results close to the $n = 3$ threshold.

A further example of the use of pseudo-states is to represent the loss of flux from the elastic channel due to inelastic collisions as in the calculations of Burke and Webb discussed on p. 3400. They added so-called $3\bar{s}$ and $3\bar{p}$ pseudo-states to a $1s$, $2s$, and $2p$ expansion. These pseudo-states were chosen to be orthogonal to the $1s$, $2s$, and $2p$ states and both to have a threshold equal to the ionization energy of hydrogen.

Results obtained by use of pseudo-state expansions are discussed on pp. 3398–400. The use of pseudo-state expansions in the calculation of cross-sections for inelastic collisions of protons with hydrogen atoms is discussed in Chapter 23, p. 2613 and Chapter 25, p. 3115.

3. Extended polarized-orbital method

The original polarized-orbital method has been extended in two main directions. In the formulation discussed in Chapter 8, § 2.4, account was taken only of dipole distortion terms in deriving the polarization potential V_p. La Bahn and Callaway extended this and included monopole, quadrupole, and other terms so that V_p represented the full perturbation due to distortion of the atomic target under adiabatic conditions (cf. Drachman's analysis of positron H atom scattering described in Chapter 26, p. 3173).

The second further extension is the inclusion of terms arising from first-order corrections to the adiabatic approximation due to the finite kinetic energy of the incident electron. If the atomic wave-function is modified from the undisturbed form $\psi_0(r_2)$ say, by the presence of a stationary electron at $\mathbf{r}_1$, to the form

$$\psi = \psi_0(r_2) + \chi(\mathbf{r}_1, \mathbf{r}_2), \tag{11}$$

normalized to unity, then the correction involves addition of a further effective scattering potential†

$$V_d(r_1) = \frac{\hbar^2}{2m} \int |\nabla_1 \chi(\mathbf{r}_1, \mathbf{r}_2)|^2 \, d\mathbf{r}_2, \qquad (12)$$

where m is the electron mass.

It may be shown that, for scattering by a spherically symmetrical atom, V_d falls off asymptotically as βr_1^{-6}. In particular, for atomic hydrogen $\beta = 129/4$ a.u. Values of β for other atoms have been given by Dalgarno, Drake, and Victor.‡

Callaway et al. tested the use of this additional term by comparing phase shifts for scattering of electrons with energies less than 30 eV by atomic hydrogen, calculated by the polarized orbital method with V_d included, with accurate values obtained using variational methods. Moderately satisfactory results were obtained. Better results were obtained for scattering of electrons by helium over a wide energy range (see p. 3401).

4. Collisions at intermediate energies—the Glauber method§

An approximation, due to Glauber, which has received application in nuclear physics, has been applied with considerable success to the calculation of cross-sections at intermediate impact energies, too low for Born's approximation to be valid and yet too high for close-coupling techniques to be practicable. The approximation is essentially a semi-classical one and gives for the amplitude for elastic scattering of a particle of mass m by a centre of force of potential $V(r)$, in which the momentum is changed from $\mathbf{k}_i \hbar$ to $\mathbf{k}_f \hbar$,

$$f_G(\mathbf{k}_i, \mathbf{k}_f) = \frac{ik_i}{2\pi} \int e^{i\mathbf{K}\cdot\boldsymbol{\rho}} \{1 - e^{i\chi(\boldsymbol{\rho})}\} \, d\boldsymbol{\rho}. \qquad (13)$$

Here $\mathbf{K} = \mathbf{k}_i - \mathbf{k}_f$, $\boldsymbol{\rho}$ is a vector in the plane perpendicular to $\mathbf{k}_i + \mathbf{k}_f$ and

$$\chi(\boldsymbol{\rho}) = -\frac{1}{\hbar v_i} \int_{-\infty}^{\infty} V(\boldsymbol{\rho} + z\hat{\mathbf{x}}) \, dz, \qquad (14)$$

where $\hat{\mathbf{x}} = (\mathbf{k}_i + \mathbf{k}_f)/|\mathbf{k}_i + \mathbf{k}_f|$ and $v_i = \hbar k_i/m$.

† DALGARNO, A. and McCARROLL, R. W., *Proc. R. Soc.* **A237** (1956) 383; KLEINMAN, C. J., HAHN, Y., and SPRUCH, L., *Phys. Rev.* **165** (1968) 53; OPIK, U., *Proc. phys. Soc.* **92** (1967) 573; CALLAWAY, J., LA BAHN, R. W., PU, R. T., and DUXLER, W. M., *Phys. Rev.* **168** (1968) 12.

‡ DALGARNO, A., DRAKE, G. W. F., and VICTOR, G. A., *Phys. Rev.* **176** (1968) 194.

§ For a recent review see GERJUOY, E., *The physics of electronic and atomic collisions* (ed. T. R. Govers and F. J. de Heer), North Holland, 1972, p. 247.

In this form the approximation is based on the supposition that the conditions are nearly classical and the collision develops by the almost undisturbed passage along a ray path in the form of a straight line making equal angles with the initial and final directions of motion. The choice of direction of this ray is not unambiguous but, as selected, the amplitude satisfies detailed balance and preserves the conservation of particle flux. At high energies it tends to the Born approximation— when χ is small, so $\exp\{i\chi(\mathbf{\rho})\} \simeq 1+i\chi(\mathbf{\rho})$, (13) is equal to the Born amplitude.

To extend this formula to deal with scattering by targets with structure, the sudden approximation (see for example Chap. 17, p. 1590) is used. Thus for collisions with H atoms $f_G(\mathbf{k}_i, \mathbf{k}_f, \mathbf{r}_2)$ is first evaluated from (13) with V taken as the interaction when the atomic electron is fixed at $\mathbf{r}_2$ relative to the nucleus. The amplitude for a collision in which the state n is excited is then assumed to be given by

$$f_{G,n}(\mathbf{k}_i, \mathbf{k}_f) = \int \psi_n^*(\mathbf{r}_2) f_G(\mathbf{k}_i, \mathbf{k}_f, \mathbf{r}_2) \psi_0(r_2) \, d\mathbf{r}_2. \tag{15}$$

Application of this formula to the excitation of the 2s and 2p states of atomic hydrogen is discussed on p. 3401.

5. Collisions with hydrogen atoms

5.1. *Elastic scattering at energies below the excitation threshold* (see Chap. 8, § 2).

Burke, Gallaher, and Geltman† have used two pseudo-state expansions to calculate phase shifts for elastic scattering of electrons by hydrogen atoms at energies below the excitation threshold. One of these expansions, referred to as I, includes 1s and 2s exact atomic functions and a pseudo-state function $2\bar{p}$ as described on p. 3395. The second, referred to as II, includes 1s, 2s, and 2p exact atomic functions and a pseudo-state function $2\bar{\bar{p}}$ which is a linear combination of 2p and $2\bar{p}$ states orthogonal to 2p and normalized. Comparison is made with the results of accurate variational calculations for the phase shifts η_0, η_1, and η_2. Both pseudo-state expansions give better results than the 1s, 2s, 2p expansion as may be seen from Table 8A.1 for the zero-order phase shifts.

Table 8A.1 also includes phase shifts calculated by Callaway, Oberoi, and Seiler‡ using a trial function of the form (1) with Φ given by

$$\Phi = \psi_0(r_a)\phi(r), \tag{16}$$

† BURKE, P. G., GALLAHER, D. F., and GELTMAN, S., *J. Phys. B* (*GB*) **2** (1969) 1142.
‡ CALLAWAY, J., OBEROI, R. S., and SEILER, G. J., *Phys. Lett.* **31A** (1970) 547.

with $\phi(r)$ expanded in a set of fifteen basic Slater-type functions. The results are very close to those given by the $1s$, $2s$, $2p$ close-coupling approximation. Comparison with the accurate variational results of Schwartz (see Chap. 8, p. 516) shows that the pseudo-state methods give somewhat better results than the close-coupling approximation, especially for the singlet phases.

TABLE 8A.1

Zero-order phase shifts for scattering of electrons by hydrogen atoms calculated with different approximations

Electron energy (ryd)	0·01	0·09	0·25	0·49	0·64
	η_0 Singlet (radians)				
Schwartz	2·553	1·696	1·202	0·930	0·886
$1s\ 2s\ 2p$	2·492	1·596	1·093	0·817	0·773
$1s\ 2s\ 2\bar{p}$	2·529	1·657	1·155	0·875	0·823
$1s\ 2s\ 2p\ 2\bar{\bar{p}}$	2·532	1·663	1·162	0·881	0·832
Expansion in Slater orbitals	2·492	1·595	1·092	0·816	0·772
	η_0 Triplet (radians)				
Schwartz	2·939	2·500	2·105	1·780	1·643
$1s\ 2s\ 2p$	2·936	2·497	2·097	1·767	1·633
$1s\ 2s\ 2\bar{p}$	2·937	2·498	2·102	1·777	1·641
$1s\ 2s\ 2p\ 2\bar{\bar{p}}$	2·937	2·498	2·102	1·777	1·642
Expansion in Slater orbitals	2·936	2·496	2·096	1·767	1·630

Comparison of the pseudo-state methods II with other methods for calculating the positions and widths of ^1S and ^3P resonances of H^- below the $n = 2$ threshold is discussed on p. 3412.

5.2. *Elastic and inelastic scattering at energies above the 2-quantum excitation threshold*

Geltman and Burke[†] have used a $1\bar{s}$, $2\bar{s}$, $2p$, $3\bar{s}$, $3\bar{p}$, $3\bar{d}$ expansion (see pp. 3395–6) to calculate cross-sections for elastic collisions and for collisions in which the $2s$ and $2p$ states are excited by impact of electrons with energies between the $n = 2$ and $n = 3$ excitation thresholds. Comparison with results obtained using $1s$, $2s$, $2p$, $3s$, $3p$, $3d$ and $1s$, $2s$, $2p$ expansions with correlation (see pp. 528 and 514) shows that, while the latter gives slightly better results as judged by a multi-channel minimum principle, the pseudo-state expansion gives better results than the 6-state atomic expansion except at energies close to the $n = 3$ threshold—

[†] GELTMAN, S. and BURKE, P. G., *J. Phys. B (GB)* 3 (1970) 1062.

the pseudo-state expansion cannot yield the resonance features which occur in this energy range.

Comparison with experimental data (for 2s, Kauppila *et al.*, see p. 3370, and for 2p, McGowan *et al.*, see p. 3369) shows that, while the relatively small variation of the inelastic cross-section with energy in this range is consistent with these data the observed values are about 20 per cent smaller than those calculated. On the other hand the ratio Q_{2s}/Q_{2p} of the 2s and 2p cross-sections calculated directly by Ott, Kauppila, and Fite (see p. 3370) agrees very well with that calculated over the whole energy range concerned.

Comparison of the pseudo-state expansion involving $3\bar{s}$, $3\bar{p}$, and $3\bar{d}$ with other methods for calculating positions and widths of ^1S and ^3P resonances of H⁻ below the $n = 2$ threshold is discussed on p. 3413.

5.3. *Elastic and inelastic scattering at intermediate energies*

We now consider calculations directed towards evaluating cross-sections for collisions with electrons with energy intermediate between the high energies for which Born's approximation is valid (see Chap. 7, § 5.6.4) and the energies comparable with the ionization threshold and below for which close-coupling methods are readily practicable. Two new types of calculation have been carried out, one using a suitable pseudo-state expansion (see p. 3396) and the other the Glauber method (see p. 3397).

Burke and Webb† used a 1s, 2s, 2p, $3\bar{s}$, $3\bar{p}$ expansion in which the pseudo-states $3\bar{s}$, $3\bar{p}$ were chosen to represent the loss of elastically scattered flux due to inelastic collisions, as described on p. 3396. They calculated cross-sections for impact energies from 16 to 54 eV and obtained encouraging results for both the 2s and 2p excitations. Thus for 2p the 1s, 2s, 2p expansion gives cross-sections which are much larger than those measured by McGowan *et al.* and by Long *et al.* (see pp. 3368–9) but the pseudo-state expansion gives quite good agreement over the whole energy range covered. For 2s the experimental situation is less clear (see p. 3370). For this state the 3-state expansion again gives results which exceed the larger of the observed cross-sections (Kauppila *et al.*) by a factor of two or so while the pseudo-state agrees quite well with these measured values except at the lowest energy where it is too large.

The Glauber approximation gives quite good results‡ for both 2p and

† BURKE, P. G. and WEBB, T. G., *J. Phys. B* (*GB*) **3** (1970) L131.
‡ TAI, H., BASSEL, R. H., GERJUOY, E., and FRANCO, V., *Phys. Rev.* **A1** (1970) 1819.

2s (assuming the observed values of Kauppila *et al.*) down to energies of about 25 eV. It was pointed out on p. 3386 that Born's approximation gives much too little scattering at 30° angles for electrons scattered after exciting the $n = 2$ states as observed by Williams *et al.* The Glauber approximation gives much better results. Thus for 200 eV electrons it agrees with the observed results out to 80°, and while the agreement is less close at low energies it is much superior to that achieved with the Born approximation.

6. Collisions with helium atoms

6.1. *Elastic scattering*

Bransden and McDowell[†] have attempted a phase-shift analysis of observed data on the total elastic and momentum transfer cross-sections and the differential cross-section, combined with that on the forward scattering derived from dispersion relations. The latter derivation followed the same lines as that of Lawson *et al.* described in Chapter 8, pp. 541–3 but with some improvement in the helium wave-functions used.

La Bahn and Callaway[‡] have applied their extended polarized orbital method to calculate differential cross-sections for electron energies between 1 and 500 eV. Special care has been taken to include all the phase shifts necessary to give an accurate value for the partial wave sum. Thus for non-zero scattering angles in the 100–500 eV energy range they found fifty phase shifts were necessary to give an accuracy of 1·5 per cent but for zero angle convergence is so slow that they included all phases of order up to 10 051.

Very good agreement is obtained with observed data in the 1–100 eV energy range including the recent absolute measurements of Gibson and Dolder (see Chap. 8, pp. 546–8 and also p. 3376) with the exception of those due to Ramsauer and Kollath which give inexplicably large small-angle scattering at low energies (see Chap. 8, Fig. 8.14). In this energy range the agreement of the calculated s and p phases with those derived by Bransden and McDowell[†] is good. It is less so for the d wave but this may well be due to the fact that, in the phase shift analysis, full weight was given to the doubtful small angle data of Ramsauer and Kollath and only three phase shifts were included.

Comparison of the differential cross-sections calculated by La Bahn

[†] BRANSDEN, B. H. and McDOWELL, M. R. C., *J. Phys. B (GB)* **2** (1969) 1187.
[‡] LA BAHN, R. W. and CALLAWAY, J., *Phys. Rev.* **180** (1969) 91.

and Callaway† in the energy range between 100 and 500 eV also gives quite good agreement with the observed data including the relative measurements of Vriens, Kuyatt, and Mielczarek (see p. 3376) and absolute measurements by Bromberg (see p. 3376). In particular the large excess of scattering at small angles above that predicted by Born's approximation (see p. 548) is reproduced. However, at zero angle of scattering there remains a substantial disagreement. Thus the extended polarization method necessarily gives an incorrect value for the imaginary part of the forward scattered amplitude as a major contribution comes, at these energies, from inelastic collisions (see Chap. 8, § 2.2.2 and (23 b)) which are not taken account of in the theory. The real part determined from the dispersion relation‡ also disagrees with that given by the method. The net effect is that the forward-scattered intensity obtained from the dispersion relations decreases from $4 \cdot 2 a_0^2$ at 100 eV to $2 \cdot 2 a_0^2$ at 500 eV while La Bahn and Callaway give nearly constant values falling only from $2 \cdot 8 a_0^2$ to $2 \cdot 6 a_0^2$. Extrapolation of the observed differential cross-sections favours the variation derived from the dispersion relation but the absolute values are still uncertain.

6.2. *Inelastic scattering*

As discussed on p. 3388 there is now clear evidence that Born's approximation gives cross-sections which are too small for excitation of ^1D levels of helium, even at energies of a few hundred eV. The possibility was raised that this arises because the approximation makes no allowance for the population of ^1D levels through intermediate ^1P states which are strongly excited. Chung and Lin§ have carried out close-coupling calculations which strongly support this suggestion.

When $1\,^1S$, $3\,^1S$, $3\,^1P$, and $3\,^1D$ states were included in the close-coupling expansion with Hartree–Fock wave-functions the cross-sections for excitation of $3\,^1S$, $3\,^1P$, and $3\,^1D$ states by electrons of energy 100 eV were found to be, in units 10^{-18} cm², 0·51, 3·65, and 0·32 as compared with 0·15, 2·70, and 0·044 according to Born's approximation. Again, an expansion including $1\,^1S$, $4\,^1S$, $4\,^1P$, $4\,^1D$, and $4\,^1F$ states, with hydrogenic wave-functions gave 0·0487, 1·15, 0·106, and 0·0054 for the cross-section for excitation of the respective 4-quantum states as compared with 0·0521, 1·09, 0·025, and 0·00014. While these results show quite clearly that coupling through the P state of the same total quantum number is very important in determining the excitation cross-sections

---

† LA BAHN, R. W. and CALLAWAY, J., *Phys. Rev.* **188** (1969) 520.
‡ BRANSDEN, B. H. and McDOWELL, M. R. C., *J. Phys. B (GB)* **3** (1970) 29.
§ loc. cit., p. 3389.

for the D and F states, Chung and Lin found that coupling with lower P states is far from negligible so that the results quoted above cannot be taken as accurate.

7. Scattering by rare gas atoms

Thompson[†] has repeated his earlier calculations on the elastic scattering of electrons by neon and argon atoms (Chap. 8, § 7.1.2, p. 555) using the Pople–Schofield method[‡] for calculating the adiabatic polarization interaction. For neon this has the advantage of giving a polarizability in close agreement with that observed. Remarkably good agreement is obtained with observed total cross-sections, including the recent measurements of Salop and Nakano (see p. 3355) and also with the recent momentum transfer cross-sections measured by Robertson (see p. 3359), at least down to energies > 0.2 eV. The theoretical value of the zero-energy scattering length, $0.174a_0$, seems, however, to be somewhat too small. Although it cannot be determined with precision from available experimental data, Robertson estimates it to be $0.24a_0$, close to that derived by O'Malley by effective range extrapolation (see Chap. 8, Table 8.9). Comparison of calculated differential cross-sections with the rather meagre measurements available (see p. 335) also reveals quite reasonable agreement.

For argon the calculated polarizability is somewhat higher than that observed so the derived polarization interaction was renormalized to give the correct asymptotic behaviour. Quite good agreement is obtained with the total cross-sections observed by Aberth, Sunshine, and Bederson[§] which are somewhat larger than those measured by Brüche (Chap. 1, p. 24) and by Golden and Bandel (Chap. 1, p. 19). The calculated differential cross-sections at 8·7, 19·6, and 39·3 eV agree remarkably well with those observed, particularly when allowance is made for finite energy and angular resolution.

Sawada, Purcell, and Green[‖] have carried out calculations by the distorted-wave method of differential and total cross-sections for excitation of rare-gas atoms from (np^6) to $(np^5)n'l$ configurations. They use as distortion potential an empirical form which for an atom of

[†] THOMPSON, D. G., *J. Phys. B (GB)* **4** (1971) 468; see also GARBATY, E. A. and LA BAHN, R. W., *Phys. Rev.* A**4** (1971) 1425.
[‡] POPLE, J. A. and SCHOFIELD, R., *Phil. Mag.* **2** (1957) 591.
[§] ABERTH, W., SUNSHINE, G., and BEDERSON, B., *Proc. 3rd int. conf. on the physics of electronic and atomic collisions, London* (ed. M. R. C. McDowell), North Holland, 1964, p. 53.
[‖] SAWADA, T., PURCELL, J. E., and GREEN, A. E. S., *Phys. Rev.* A**4** (1971) 193.

N electrons and atomic number Z is given in atomic units by[†]

$$V(r) = 2r^{-1}\{N\gamma - Z\}, \qquad (16)$$

with $\gamma = 1 - \{(e^{r/d} - 1)H + 1\}^{-1}$, the parameters d and H being chosen to give good agreement with observed differential cross-sections for elastic scattering by the atom concerned.[‡] A similar potential was used to calculate the wave-functions for the active atomic electron, the parameters being chosen to give a good fit to the observed energies of excited states. Detailed applications were made to the Ne $2p$–$3s$ and Ar $3p$–$4s$ transitions and results are compared with angular distributions observed by Nicoll and Mohr (see Chap. 5, p. 343 and Chap. 8, § 7.2). Reasonable agreement is obtained.

8. Scattering by alkali metal atoms (Chap. 8, § 8)

The situation about the scattering of electrons by alkali metal atoms, which was very confused indeed at the time of writing of Chapter 8, has greatly clarified in the last few years. Although there seemed every reason to believe that a close-coupling expansion based on a closed core plus outer ns and np wave-functions (with n ranging from 2 to 6 in going from Li to Cs) should give good results for the calculation of differential and total cross-sections for electron energies up to and even beyond the np excitation threshold, these results disagreed strongly with the total cross-section observations of Brode which were the only absolute measurements available. However, in 1971, as described on p. 3356, Collins, Bederson, and Goldstein[§] made new absolute measurements for K which showed good agreement with the close-coupling calculations of Karule and of Karule and Peterkop (see p. 562). A little later Visconti, Sliven, and Rubin[||] measured absolute cross-sections for K, Rb, and Cs which also agreed well with the results of these calculations. There seems little doubt that the earlier measurements overestimated the cross-section by a factor close to two and also showed more structure than is actually present. This also affects the observations for Li which were normalized with respect to Brode's absolute values for Na. The renormalized results are no longer in disagreement with the calculations of Karule (p. 562) and of Burke and Taylor,[††] using the close-coupling method including the states with ns and np outer orbitals.

At very low impact energies, for which cross-sections can only be measured by swarm methods, the experimental results discussed in

[†] GREEN, A. E. S., SELLIN, D. L., and ZACHOR, A. S., *Phys. Rev.* **184** (1969) 1.
[‡] BERG, R. A., PURCELL, J., and GREEN, A. E. S., ibid. **A3** (1971) 508.
[§] loc. cit., p. 3356.   [||] loc. cit., p. 3356.
[††] BURKE, P. G. and TAYLOR, A. J., *J. Phys. B (GB)* **2** (1969) 869.

Chapter 2, § 7.4 for caesium were quite chaotic (see, for example, Fig. 2.28). While it cannot be said that this situation has been clarified the measurements made by Beynon and Brooker for potassium at a mean electron energy of 0·17 eV (see p. 3359) are not inconsistent with the predictions of the close-coupling approximation. It is too early to decide whether this is a coincidence until data are available for other alkali metal atoms.

Further evidence is forthcoming from measurements of absolute and of relative differential elastic scattering cross-sections. Recent measurements for sodium (Andrich et al., Gehenn and Reichert, see p. 3377) and for potassium (Slevin et al., Gehenn and Wilmann, Hils et al., see p. 3381) give results which agree reasonably well with the close-coupling calculations. The sodium measurements show evidence of the existence of a cusp at the excitation threshold, in the observed variation of the differential cross-section with energy at a fixed scattering angle, as predicted by the theory (see Chap. 9, p. 653).

It is also possible to compare calculated and observed cross-sections for scattering of polarized electrons by unpolarized atoms or vice versa. Thus if $f(\theta)$ and $g(\theta)$ are the amplitudes for direct and exchange scattering through an angle θ the differential elastic cross-section for scattering of unpolarized electrons by unpolarized atoms is given by

$$I(\theta)\,d\omega = \{\tfrac{3}{4}|f(\theta)-g(\theta)|^2+\tfrac{1}{4}|f(\theta)+g(\theta)|^2\}\,d\omega \qquad (17\,\text{a})$$

$$= \tfrac{1}{2}\{|f|^2+|g|^2+|f-g|^2\}\,d\omega \qquad (17\,\text{b})$$

and the differential cross-section for spin exchange by

$$I_{\text{se}}\,d\omega = \tfrac{1}{2}|g|^2\,d\omega. \qquad (18)$$

Consider now the scattering of a partially polarized electron beam (polarization P_{e}^0) by an atom beam partially polarized (polarization P_{a}^0) about the same direction. It may then be shown† that, if $P_{\text{e}}(\theta)$, $P_{\text{a}}(\theta)$ are the polarizations of the electrons scattered through an angle θ and of the corresponding recoil atoms respectively, then

$$I(\theta)P_{\text{e}}(\theta) = P_{\text{e}}^0\{I(\theta)-|g|^2\}+P_{\text{a}}^0\{I(\theta)-|f|^2\}, \qquad (19\,\text{a})$$

$$I(\theta)P_{\text{a}}(\theta) = P_{\text{a}}^0\{I(\theta)-|g|^2\}+P_{\text{e}}^0\{I(\theta)-|f|^2\}. \qquad (19\,\text{b})$$

Thus by observing the polarizations $P_{\text{e}}(\theta)$, $P_{\text{a}}(\theta)$, as well as $I(\theta)$, when $P_{\text{e}}^0 = 0$, for example, it is possible to determine $|f|^2$ and $|g|^2$ and $|f-g|^2$ separately.

† BURKE, P. G. and SCHEY, H. M., Phys. Rev. 126 (1962) 163. See MOTT, N. F. and MASSEY, H. S. W., The theory of atomic collisions (3rd edn) (Clarendon Press, Oxford, 1965), Chapter XVII, § 2.8 and Chapter X, § 6.

Such measurements have been carried out for potassium (see p. 3381) by Hils et al., measuring $P_\mathrm{a}(\theta)$ for 3·3 eV electrons, and by Collins et al., measuring $P_\mathrm{a}(\theta)$ for electrons in the energy range 0·5 to 1·2 eV. Comparison may be made with the predictions of the close-coupling approximations and, when allowance is made for the velocity distributions of the electrons and atoms, there is quite good agreement.

Moores and Norcross[†] have investigated the convergence of the close-coupling expansion for sodium by including also the states with outer $3d$ and $4s$ orbitals. They find little effect from these additional states.

The close-coupling calculations extend beyond the excitation threshold so that they also yield cross-sections for the ns–np excitation. The measurable cross-sections in this case have been discussed by Kleinpoppen[‡] and by Rubin et al.[§] in terms of the direct and exchange amplitudes f_M, g_M for excitation of states with the z-component of angular momentum $M\hbar$. Since $f_M = f_{-M}$, $g_M = g_{-M}$ we have four independent amplitudes involved. Additional information is available from observations of the polarization of the emitted radiation (see Chap. 8, § 10).

For sodium the total excitation cross-sections and the polarization of the impact radiation observed by Enemark and Gallagher and by Gould (see p. 3371) agree quite well with each other and with the calculated. Earlier observations of the total cross-sections by Zapesochnyi (see Chap. 4, § 1.4.4.2) gave values about half as large while the earlier measurements of the polarization by Hafner and Kleinpoppen[||] gave polarizations about 60 per cent larger than observed. For lithium also the polarization of the resonance radiation observed by the same authors is about 15 per cent higher than that calculated by Karule[††] and by Burke and Taylor.[‡‡]

Bederson, Kasdan, and Goldstein (see p. 3382) have measured the depolarization of polarized potassium atoms recoiling from collisions with unpolarized electrons which produce excitation of the $4p$ state. The depolarization is measured after the excited atoms have undergone radiative transitions to the ground state and is given by

$$\frac{4}{9} + \frac{10|g_1|^2 + |g_0|^2 - 8I_1(\theta)}{8\{2I_1(\theta) + I_0(\theta)\}}. \tag{20}$$

[†] MOORES, D. L. and NORCROSS, D. W., J. Phys. B (GB) 5 (1972) 1482.
[‡] KLEINPOPPEN, H., Phys. Rev. A3 (1971) 2015.
[§] loc. cit., p. 3382.
[||] HAFNER, H. and KLEINPOPPEN, H., Z. Phys. 198 (1967) 315.
[††] KARULE, E. M., J. Phys. B (GB) 3 (1970) 860.
[‡‡] loc. cit., p. 3404.

Measurements were made for an energy range from 3 to 10 eV and electron scattering angles from an angle $\simeq$ 0 to 20°. Good agreement is obtained with close-coupling calculations.

9. Scattering of electrons by neutral and ionized carbon, oxygen, and nitrogen (see Chap. 8, § 9)

Saraph, Seaton, and Shemming[†] have recalculated cross-sections for excitation of forbidden transitions between the terms of the ground configurations of ions with the structure $1s^2 2s^2 2p^q$. They used an improved version of the method described in Chapter 8, § 9. In particular, variational techniques were used to improve the approximations while certain inconsistencies, including 'post-prior' discrepancies, were removed by requiring the orbitals for the colliding electrons to be orthogonal to those for the ionic electrons. The calculations were extended to collisions with ions in $3p^3$ configuration by Czyzak et al.[‡]

The cross-sections, for elastic scattering of electrons by O, N, and C atoms and for collisions involving excitation of transitions within the ground configurations, calculated, using close-coupling expansions as discussed in Chapter 8, § 9.2 (see Figs. 8.29 and 8.30), were found to be somewhat incorrect due to an algebraic error in the formulation. This was corrected by Henry, Burke, and Sin Fai Lam.[§] The revised cross-sections for the inelastic collisions in O now agree rather better, to about 10 per cent, with those calculated by Saraph, Seaton, and Shemming using the method outlined in Chapter 8, § 9.1.

The corrections introduce little change in the calculated total cross-sections for O over the whole energy range of the calculations and in N and C at energies above 3 eV. Below this energy sharp maxima are introduced rising to 6.5×10^{-15} cm^2. There is thus no improvement in the comparison with observed data given in Chapter 8, Fig. 8.31, for O and N. However, Vo Ky Lan, Feautrier, Le Dourneuf, and van Regemorter[||] have carried out a polarized orbital calculation for O in which the wavefunction for each term of the ground configuration includes a dipole

[†] SARAPH, H. E., SEATON, M. J., and SHEMMING, J., *Phil. Trans. R. Soc.* A**264** (1969) 77.
[‡] CZYZAK, S. J., KRUEGER, T. K., MARTINS, P. DE A. P., SARAPH, H. E., and SEATON, M. J., *Mon. Not. R. Astron. Soc.* **148** (1970) 361.
[§] HENRY, R. J. W., BURKE, P. G., and SIN FAI LAM, A-L, *Phys. Rev.* **178** (1969) 218. Corrected calculations were also made by SMITH, K., CONNEELY, M. J., and MORGAN, L. A., ibid. **177** (1969) 196, which agree with those of Henry et al. except for scattering by O for which their results are in error due to use of an incorrect fractional percentage coefficient. (LE DOURNEUF, M., SARAPH, H. E., SEATON, M. J., and VO KY LAN, *J. Phys. B (GB)* **5** (1972) L87.)
[||] VO KY LAN, FEAUTRIER, N., LE DOURNEUF, M., and VAN REGEMORTER, H., *J. Phys. B (GB)* **5** (1972) 1506.

distortion term (see p. 512). They find that this has little effect on the inelastic cross-sections but reduces the elastic cross-section at low energies, giving values which are consistent with the rather widely spread observed data (see Fig. 8.31, Chap. 8).

Henry, Burke, and Sin Fai Lam have also calculated collision strengths for inter-term transitions within the ground configurations of O^+, O^{++}, and N^+ and obtained results in good agreement with those given in Tables 8.15 and 8.16 of Chapter 8 which were obtained by the method described in Chapter 8, § 9.1.

Ormonde, Smith, Torres, and Davies† have extended their calculations by including in the close-coupling expansion terms arising from excited configurations which are likely to interact strongly with the ground terms. Thus for NI they include the terms arising from the ground and first excited configuration $(1s^2\,2s^2\,2p^2)(^3P)3s(^{4,2}P)$ and in some cases also the first ^4P term of the $1s^2\,2s\,2p^4$ configuration. The most important new features which appear are a number of auto-ionizing states which are discussed on p. 3490.

Considerable interest is attached to the calculation of collision strengths of certain intercombination transitions in ionized C, N, and O atoms. These lead to excitation of the upper states of certain semi-forbidden lines which are strong features of the emission spectra of many quasars. For example, the $\lambda 1909$ line of C^{2+} arises from the transition between the $2s\,2p\,^3P_1$ state and the ground $2s^2\,^1S_0$ state. Osterbrock‡ has calculated collision strengths for excitation of the $^3P_1$ and $^1P_1$ states of C^+ and the isoelectronic ions B^+, N^{3+}, O^{4+}, and Ne^{6+} using a truncated close-coupling expansion which includes the terms of the ground and excited configurations.

Jackson§ has calculated the collision strength for excitation of the $2s\,2p^2\,^4P$ level from the ground $2s^2\,2p\,^2P^0$ level of C^+. He uses a close-coupling expansion including the terms of ground and excited configuration to which is added a linear combination of terms arising from the $2s^2\,2p^2$, $2s\,2p^3$, and $2p^4$ configuration of the C^+e system.

10. Scattering of electrons by other atoms

Burke and Moores‖ have used close-coupling expansions to calculate cross-sections for elastic scattering and for excitation of np and nd states of Mg^+ ($n = 3$) and Ca^+ ($n = 4$) by electron impact. For Mg^+, $3s$, $3p$,

† ORMONDE, S., SMITH, K., TORRES, B. W., and DAVIES, A. R., Preprint 1972.
‡ OSTERBROCK, D. E., *J. Phys. B (GB)* **3** (1970) 149.
§ JACKSON, A. R. C., ibid. **5** (1972) L83.
‖ BURKE, P. G. and MOORES, D. L., ibid. **1** (1968) 575.

and 3d states for the outer electron and for Ca$^+$, the 4s, 3d, and 4p states were included in the expansion.

For the elastic scattering the main interest is in the location of resonances and determination of the resonance parameters and this is referred to again on p. 3418. The excitation cross-sections in general do not differ greatly from those calculated by the Coulomb–Born approximation† (see Chap. 7, p. 428).

van Blerkom‡ has calculated elastic and inelastic cross-sections for collisions with Mg atoms involving the $3\,^1S$, $3\,^1P$, and $3\,^3P$ states. A close-coupling expansion was used including all wave-functions associated with these states. The atomic wave-functions were obtained by assuming a fixed core potential and neglecting exchange effects between the core and the two outer electrons.

Attempts have been made§ to allow for atomic distortion effects in the scattering of slow electrons by mercury atoms but with little success. On the other hand Carse‖ has obtained interesting results by carrying out a close-coupling calculation of differential and total cross-sections for excitation from a 6s to a 6p orbital. The eigenfunction expansion used included only the atomic states involved in the transition. Comparison of the differential cross-sections with those observed by Gronomeier†† for energies from 20 to 30 eV, over a wide angular range, while not revealing detailed agreement nevertheless indicates that the calculated structure arises from a similar number of harmonics and varies in much the same way with impact energy. At higher energies (40–200 eV) the calculated elastic and inelastic differential cross-sections are not so similar as for sodium (see p. 3405). It is rather remarkable that for an atom as heavy as mercury such promising results have been obtained.

11. Excitation of fine structure transitions (see Chap. 8, § 9.4)

Martins, Saraph, and Seaton‡‡ have calculated cross-sections for excitation of transitions between fine structure levels of ions with $1s^2\,2s^2\,2p^3$ configurations, neglecting the effect of coupling to other configurations.

---

† BELY, O. and PETRINI, D., *Phys. Lett.* **23** (1966) 442.
‡ VAN BLERKOM, J. K., *J. Phys. B (GB)* **3** (1970) 932.
§ WALKER, D. W., ibid. **2** (1969) 356; MOHR, C. B. O., ibid. **5** (1972) 825.
‖ CARSE, G. D., ibid. **5** (1972) 1928.
†† GRONOMEIER, K-H, *Z. Phys.* **232** (1970) 483.
‡‡ MARTENS, P. DE A. P., SARAPH, H. E., and SEATON, M. J., *J. Phys. B (GB)* **2** (1969) 427; MARTENS, P. DE A. P. and SEATON, M. J., ibid. 333.

9
RESONANCE PHENOMENA—THRESHOLD BEHAVIOUR

1. The calculation of resonance energies and level widths (see Chap. 9, § 3)

THERE have been further developments of the technique for calculating not only the energies of resonance levels but also the level widths and shape parameters.

At the time of writing of Vol. 1 the only method used for the calculation of level widths as well as resonance energies was that of the numerical solution of suitable close-coupling expansions. The projection method of O'Malley and Geltman (see p. 621) gave values for the energies which did not include the level shift (see eqn (22), p. 600). This method is essentially one in which the open-channel states are projected out of the trial function which is then used in a normal Rayleigh–Ritz procedure to determine the undisplaced eigenvalues which give the resonance energies. Bhatia and Temkin† not only applied this method to the ^1P doubly excited states of He using elaborate trial functions but then proceeded to use the eigenfunctions which they obtained to calculate the level shift ΔE from eqn (22), p. 600 and level width Γ from eqn (31) of p. 601, using the formula (14) of p. 599 for the matrix element $V_E \phi$. In this formula ϕ is taken to be the variationally determined function while ψ_E is the wave-function representing the collision of a free electron with the target, calculated by the polarized orbital approximation (see Chap. 8, p. 510). Detailed results of applications to He and H⁻ are described on pp. 3412–15.

An alternative procedure for determining resonance energies has been introduced by Taylor and his collaborators‡ which is usually referred to as the stabilization method. This depends on the choice of a trial function for the system of target plus colliding electron as a combination of a set of basic functions which are all quadratically integrable and based on a

---

† BHATIA, A. K. and TEMKIN, A., *6th int. conf. on physics of the electronic and atomic collisions, Cambridge, Mass.* Abstracts, M.I.T. Press, 1969, p. 926.
‡ TAYLOR, H. S., *Adv. Chem. Phys.* **18** (1970) 91; TAYLOR, H. S. and WILLIAMS, J. K., *J. chem. Phys.* **42** (1965) 4063; ELIEZER, I., TAYLOR, H. S., and WILLIAMS, J. K., ibid. **47** (1967) 2165.

core which provides a good representation of the target state. The eigen-energies corresponding to the trial function are determined in the usual way. Those which lie above the energy of the ground state of the target and are found to be insensitive to improvements in the basic set are assumed to correspond to auto-ionizing states, an assumption found to be valid in model calculations. A modified procedure has been introduced by Nesbet and Lyons† which makes possible the determination of the parameters of the resonance states as well as the resonance energy.

This method is based on the type of trial function already referred to on p. 3393. Thus in the general case it is given by

$$\Psi = A \sum_p \psi_p F_p + \sum_\mu c_\mu \Phi_\mu. \tag{1}$$

ψ_p is a normalized wave-function for the stationary state of the target corresponding to the open channel p, F_p is the single electron continuum wave-function for this channel, and Φ_μ is one of an orthonormal set of Slater determinants for the total system of target plus colliding electron. A is the anti-symmetrizing operator. Φ_μ is chosen as in atomic structure calculations. The eigenvalues of the matrix $M_{\mu\nu}$, where

$$M_{\mu\nu} = \Phi_\mu H \Phi_\nu - E \delta_{\mu\nu}, \tag{2}$$

H being the total Hamiltonian and E the total energy, are then determined in the usual way. These may be used to select the ranges of energy in which to apply variational procedures using the full wave-function (1) to obtain the scattering phase shifts (or eigenphases in a multi-channel problem) spanning the width of a resonance. Results for certain levels of H⁻ are given on p. 3414.

Bottcher‡ has called attention to the variational method introduced much earlier by Bransden and Dalgarno§ for dealing with auto-ionizing states and has applied it to calculate resonance energies and level widths for comparatively complex systems. This depends on the use of a trial function of the form
$$\Psi_t = \phi + \lambda \psi^{(+)}, \tag{3}$$

where ϕ is a quadratically integrable function representing the quasi-bound auto-ionizing state and $\psi^{(+)}$ has the form

$$\psi^{(+)} = A F^{(+)}(\mathbf{r}_1) \phi_c(\mathbf{r}_2,..., \mathbf{r}_N), \tag{4}$$

where A is the antisymmetrizing operator, $\phi_c(\mathbf{r}_2,..., \mathbf{r}_N)$ is the normalized wave-function for the atomic core, and $F^{(+)}(\mathbf{r}_1)$ represents an outgoing

† NESBET, R. K. and LYONS, J. D., *Phys. Rev.* **A4** (1971) 1812.
‡ BOTTCHER, C., *J. Phys. B (GB)* **2** (1969) 766.
§ BRANSDEN, B. H. and DALGARNO, A., *Phys. Rev.* **88** (1952) 148; *Proc. phys. Soc.* **A66** (1953) 911.

wave corresponding to flow of a single particle per second across the infinite sphere. Thus, if the core is uncharged,

$$F^{(+)}(\mathbf{r}_1) \sim v^{-\frac{1}{2}} r_1^{-1} \exp\{i(kr_1 - \tfrac{1}{2}l\pi)\} Y_{lm}(\theta_1, \phi_1), \quad (5)$$

where k is the wavenumber, v the velocity, and l, m the angular momentum quantum numbers of the motion of the outgoing electron whose coordinates are (r_1, θ_1, ϕ_1). If the core is charged Coulomb distortion terms must be included (see Chap. 6, p. 396). It may then be shown that, if

$$K = \int \Psi_t (H - E) \Psi_t \, d\tau, \quad (6)$$

where H is the Hamiltonian and E the energy of the total system, then for variation of $\Psi^{t(t)}$

$$\delta K = i\lambda \, \delta\lambda^*, \quad (7)$$

and

$$\delta |K|^2 = 0. \quad (8)$$

Since with the normalization (5) the number of particles ejected per second is $|\lambda|^2$, we have

$$\Gamma = \hbar |\lambda|^2 \quad (9)$$

where Γ is the level width.

In practice ϕ and ϕ_c are represented by Hartree–Fock approximations and a further quadratically integrable trial function $\sum c_i \chi_i$ may be added, the χ_i being Slater-type functions and the c_i linear parameters to be determined. The χ_i may also include non-linear parameters α_i. It is then convenient to work in terms of real quantities $K_r = \mathrm{re}\, K$ and μ, ν where $\lambda = \mu + i\nu$. The variational equations may then be taken as

$$\frac{\partial K_r}{\partial \mu} = -\nu, \quad \frac{\partial K_r}{\partial \nu} = \mu, \quad \frac{\partial K_r}{\partial c_i} = 0, \quad \frac{\partial}{\partial c_i}|K|^2 = 0, \quad (10)$$

with

$$K_r(E) = E_\phi - E + \Delta E_\phi, \quad (11\,\mathrm{a})$$

$$\Gamma = \hbar(\mu^2 + \nu^2). \quad (11\,\mathrm{b})$$

$E_\phi + \Delta E_\phi$ is the resonance energy including the level shift ΔE_ϕ.

Examples of the application of this method to electron collisions with H atoms are given on p. 3414 and to heavier atoms on p. 3418.

1.1. Detailed applications

1.1.1. H⁻ and He

In Table 9A.1 comparison is made between results obtained, by the different methods referred to earlier, for the energies and level widths of resonance states of H⁻. These include the pseudo-expansion method of Burke, Gallaher, and Geltman (p. 3398), the multi-configuration method in which the configurations are represented by Slater-type orbitals (Seiler et al., p. 3398), the search procedures associated with this method (Nesbet and Lyons, p. 3411), the projection method as used by Bhatia and Temkin (see p. 3410), and the variational method of Bottcher. Results obtained by the presumably accurate procedure of Taylor and Burke, in which explicit account is taken of correlation effects, are included for comparison. It will be seen that most methods give good results even for the widths of narrow levels. The comparatively simple variational method of Bransden and Dalgarno, as developed by Bottcher, while not yielding satisfactory results for resonance energies is not too unsatisfactory for level widths.

The energies calculated by Bhatia and Temkin (see p. 3410) for the three lowest ^1P and for the lowest ^3P resonance levels of He agree quite well with those calculated by close-coupling and quantum defect methods (see Tables 9.3 and 9.4 of Vol. 1). The level widths, calculated only for the ^1P-levels, agree less well but are still quite close. Bottcher† has checked the effectiveness of the variational method described on p. 3411 for the calculation of level widths by comparison with results obtained by close-coupling methods for the lowest states of the ^1S, ^1P, ^3P, and ^1D series. While the agreement is not as close as obtained by more elaborate methods, Bottcher's results are not in error by as much as a factor of two. This suggests the method could be used with advantage to determine to a useful accuracy the level widths of heavier atoms (see p. 3418).

Energies of auto-ionizing states of He which converge to the $n = 3$ and $n = 4$ thresholds have been calculated by Holøien and Midtdal‡ and by Oberoi§ using the projection method. The former authors used trial functions of Sturmian form while Oberoi used an expansion in products of hydrogenic functions by exponentially decaying functions expressed in terms of Slater-type orbitals. Comparison of these results for ^1S states, which do not allow for the level shift ΔE, with results obtained by close-coupling calculations which do, is given in Table 9A.2. For the ^1P levels experimental data are available from the absorption measurements of Madden and Codling (see p. 1117). Comparison with the calculated values of Oberoi, given in Table 9A.3, shows very good agreement.

Oberoi§ has also calculated the energies for a number of corresponding states of H$^-$ and, where comparison is possible (for certain states associated with the $n = 4$ threshold) they agree well with those calculated

† BOTTCHER, C., J. Phys. B (GB) **3** (1970) 1201.
‡ HOLØIEN, E. and MIDTDAL, J., ibid. **5** (1972) 1111.
§ OBEROI, R. S., ibid. p. 1120.

TABLE 9 A.1

Resonance levels of H⁻ associated with the $n = 2$ threshold

Resonance state	$^1S(1)$	$^1S(2)$	$^1S(3)$	$^3S(1)$	$^3S(2)$	$^3S(3)$	$^1P(1)$	$^1P(2)$	$^3P(1)$	$^3P(2)$	^1D
Resonance energies (eV)											
(a)	9·560	10·178	—	10·150	—	—	10·177	—	9·740	—	10·125
(b)	9·587	—	—	—	—	—	—	—	9·759	—	—
(c)	9·574	10·178	10·203	10·151	10·201	10·204	10·185	10·204	9·768	10·202	10·160
(d)	—	10·183	10·150	—	—	—	—	—	9·720	—	—
(e)	9·571	—	—	—	—	—	—	—	—	—	—
Level widths (eV)											
(a)	$4·75 \times 10^{-2}$	$2·19 \times 10^{-3}$	—	$2·06 \times 10^{-5}$	—	—	$4·0 \times 10^{-5}$	—	$5·94 \times 10^{-3}$	—	$8·8 \times 10^{-3}$
(b)	$5·01 \times 10^{-2}$	—	—	—	—	—	—	—	$5·71 \times 10^{-3}$	—	—
(c)	$5·44 \times 10^{-2}$	$2·31 \times 10^{-3}$	$1·3 \times 10^{-4}$	$1·90 \times 10^{-5}$	$1·36 \times 10^{-6}$	$8·16 \times 10^{-8}$	$2·42 \times 10^{-5}$	$2·06 \times 10^{-7}$	$7·98 \times 10^{-3}$	$4·28 \times 10^{-5}$	$7·74 \times 10^{-3}$
(d)	$4·918 \times 10^{-2}$	$3·615 \times 10^{-3}$	$1·949 \times 10^{-5}$	—	—	—	—	—	$6·3 \times 10^{-3}$	—	—
(e)	$3·47 \times 10^{-2}$	—	—	—	—	—	—	—	—	—	$8·8 \times 10^{-3}$
(f)	—	—	—	—	—	—	—	—	—	—	—

(a) TAYLOR, A. J. and BURKE, P. G., *Proc. phys. Soc.* **92** (1967) 336.
(b) BURKE, P. G., GALLAHER, D. F., and GELTMAN, S., loc. cit., p. 3398.
(c) SEILER, G. J., OBEROI, R. S., and CALLAWAY, J., loc. cit., p. 3398.
(d) BHATIA, A. K. and TEMKIN, A., loc. cit., p. 3410.
(e) NESBET, R. K. and LYONS, J. D., loc. cit., p. 3411.
(f) BOTTCHER, C., loc. cit., p. 3411.

RESONANCE PHENOMENA—THRESHOLD BEHAVIOUR

TABLE 9 A.2

^1S resonance levels of He associated with the $n = 3$ threshold

State	^1S (1)	(2)	(3)	(4)	(5)	(6)	(7)	(8)	(9)	(10)
				Energy below $n = 3$ threshold (Z^2 ryd)						
Close-coupling(a)	0·1767	0·1583	0·1400	0·1308	—	0·1277	0·1231	—	0·1217	—
Oberoi(b)	0·1772	0·1581	0·1407	0·1315	0·1287	0·1281	0·1232	0·1221	0·1219	0·1192
Holøien and Midtdal(c)	0·1756	0·1556	0·1389₅	0·1287	0·1277	—	0·1218₅	0·1162₅	—	—

(a) ORMONDE, S., WHITAKER, W., and LIPSKY, L., *Phys. Rev. Lett.* **19** (1967) 1161.
(b) loc. cit., p. 3413.
(c) loc. cit., p. 3413.

TABLE 9 A.3

Comparison of observed and calculated energies of $^1P^0$ resonance levels of He associated with the $n = 3$ and $n = 4$ threshold

	Below $n = 3$ threshold (in Z^2 ryd)					
State	$^1P^0$ (1)	(4)	(8)	(12)	(16)	(20)
Observed (Madden and Codling(a))	0·16665	0·13507	0·12510	0·12030	0·11765	0·11592
Calculated (Oberoi(b))	0·16783	0·13555	0·12529	0·12037	0·11787	0·11630
Calculated, close-coupling (Burke and Taylor(c))	0·1670	—	—	—	—	—
	Below $n = 4$ threshold (in Z^2 ryd)					
State	$^1P^0$ (1)	(4)	(9)			
Observed (Madden and Codling(a))	0·096532	0·080294	0·073739			
Calculated (Oberoi(b))	0·097442	0·080324	0·073234			

(a) loc. cit., p. 1111.
(b) loc. cit., p. 3413.
(c) BURKE, P. G. and TAYLOR, A. J., *J. Phys. B (GB)* **2** (1969) 44.

by Burke, Ormond, Taylor and Whitaker using a close-coupling expansion (see p. 638).

1.1.2. He⁻

Burke, Cooper, and Ormonde† have extended the preliminary calculations reported in Chapter 8, pp. 550–3, so as to make a thorough study of resonance effects occurring in the neighbourhood of the thresholds for excitation of the two-quantum states.

The preliminary calculations were carried out using a four-state close-coupling expansion involving the four two-quantum excited states $2\,^3S$,

† BURKE, P. G., COOPER, J. W., and ORMONDE, S., *Phys. Rev.* **183** (1969) 245.

$2\,^1S$, $2\,^3P$, $2\,^1P$ only. In the later work the $1\,^1S$ state was included so that the resonance below the $2\,^3S$ threshold could be investigated and cross-sections calculated for excitation of the two-quantum states from the ground state. For this work an orthogonal set of bound-state wave-functions was used which were of the same form as for the earlier calculations, consisting of a properly symmetrical combination of terms of the form

$$\psi_t(r_1)\phi_{nl}(\mathbf{r}_2)Y_{l,m_l}(\theta_2,\phi_2)\chi_{s,m_s}(1,2), \qquad (12)$$

where $\psi_t(r_1)$ is the ground-state wave-function of a helium ion, Y_{l,m_l} a spherical harmonic, and χ_{s,m_s} the spin function. However, instead of representing ϕ_{nl} as a combination of Slater-like orbitals it was obtained from the solution of the differential equation which results when the symmetrized form of (12) is substituted in the variational principle. The same procedure was used for the ground state so that all the functions were orthogonal. In these calculations the thresholds were taken at the calculated instead of experimental values. They covered the energy range up to about 21·6 eV above the ground state and partial cross-sections were evaluated for the different elastic and inelastic processes with total orbital angular momentum quantum number $L = 0, 1$, and 2.

^2S, ^2P, ^2D, and ^2F resonance states of He⁻ were found. Of these the ^2S falls below the lowest excitation threshold by an amount which agrees closely with the observed resonance at 19·3 eV in the elastic scattering. The calculated width, 0·039 eV, is certainly too large which means that the coupling between ground and excited states for $L = 0$ is actually less than the theory predicts. Predicted cross-sections for $L = 0$ are therefore likely to be too large.

The other three resonance states ^2P, ^2D, and ^2F lie above the $2\,^3S$ threshold at energies of 20·2, 21·0, and 22 eV respectively above the ground state of He. The level widths, approximately 0·4, 0·5, and 1 eV respectively, are all much larger than for the ^2S level.

This system of resonance levels arises from the strong interactions between the 2-quantum levels. But for the level separations of order 1 eV the situation would be similar to that for H⁻ in which case an interaction falling off as slowly as r^{-2} arises through the orbital degeneracy of the $2s$ and $2p$ levels. Long-range forces arise in a more complicated fashion for He⁻ but nevertheless play an important role.

Calculated cross-sections were obtained for fifteen elastic and inelastic processes. Apart from the ^2S resonance, comparison with other experimental data is possible. Thus, for the excitation of the $2\,^3S$ state, measured total excitation function and differential cross-sections are available. The former (see Chap. 4, p. 254 and Fig. 4.57) includes two maxima near threshold which may be ascribed to the ^2P and ^2D resonances. This is confirmed by comparison with the differential cross-section measurements of Ehrhardt and Willmann (see Chap. 8, p. 348) and of Ehrhardt, Langhans, and Linder† when allowance is made for theoretical overestimation of $L = 0$ cross-sections and neglect of contributions from $L > 2$. Calculated differential cross-sections for $2\,^1S$ excitation, when the same allowance is made, agree quite well with observed results.†

A further check with experiment concerns the polarization of light emitted through impact excitation of the $2\,^3P$ levels. According to the theory, corrected for the overestimated contribution from $L = 0$, the polarization of $2\,^3P$–$2\,^3S$ radiation falls from a threshold value of 38 per cent to a minimum of about 16 per cent before rising to a second maximum at an energy about 1·2 eV above the threshold. This agrees quite well with the observations of Whitteker and Dalby.‡

Smith (K.), Golden, Ormonde, Davies, and Torres§ have carried out close-coupling calculations to search for the triply-excited resonance states of He⁻, first observed by Kuyatt et al. near 58 eV in the transmission of electrons through helium (see Chap. 9, p. 639) and since observed in a variety of other experiments (see p. 3357).

They found a resonance at 56·48 eV with a half-width of 1·2 meV for the scattering state with total orbital angular momentum quantum number $L = 1$ when they included $1s^2\,^1S$, $1s\,2s\,^3S$, and $2s^2\,^1S$ atomic states in the close-coupling expansion. This identifies the resonance state as $(2s^2\,2p)^2P_0$. By adding the additional state $1s\,2s\,^1S$ and replacing $2s^2\,^1S$ by $2s\,2p\,^3P$ they found the second resonance for the scattering state with $L = 2$ at 58·34 eV with a half-width of 13·3 meV. This establishes the $(2s\,2p)^2\,^2D$ character.

1.1.3. *Other atoms and ions*

Cooper, Conneely, Smith (K.), and Ormonde∥ have carried out close-coupling calculations to investigate resonant structure in the absorption spectrum of lithium at quantum energies between the $2\,^3S$ and $2\,^1P$ thresholds of Li⁺, between 60 eV and 70 eV above the ground state of Li. Although the experimental data concerned are discussed in the notes on

† EHRHARDT, H., LANGHANS, L., and LINDER, F., *Z. Phys.* **214** (1968) 179.
‡ WHITTEKER, J. H. and DALBY, F. W., *Can. J. Phys.* **46** (1968) 193.
§ SMITH, K., GOLDEN, D. E., ORMONDE, S., DAVIES, A. R., and TORRES, B. W., preprint, 1972.
∥ COOPER, J. W., CONNEELY, M. J., SMITH, K., and ORMONDE, S., *Phys. Rev. Lett.* **25** (1970) 1540.

Chapter 14 we briefly describe the calculations here as they follow on similar lines to the corresponding calculations for elastic scattering of electrons by helium atoms (see Chap. 9, § 5, p. 638 and Chap. 8, § 6.2, p. 551).

A close-coupling expansion was used including the four $n = 2$ states of Li$^+$, $1s\,2s\,^{1,3}S$ and $1s\,2p\,^{1,3}P$. The bound-state wave-functions used were those given by Morse, Young, and Haurwitz.†

The results obtained are compared on p. 3488, with experimental information available from optical absorption experiments.

Auto-ionizing levels of Be arising from doubly excited $2p\,2p$ and $2p\,3s$ configurations have been investigated by close-coupling expansions.‡ The $2p\,3s\,^3P$ resonance has a calculated width of about 2×10^{-5} eV consistent with optical observations§ of the line-width associated with the $2p\,3s\,(^3P)$–$2s\,3s\,(^1S)$ transition.

Close-coupling methods have also been applied to investigate auto-ionizing states of Ne∥ and of Ar.†† For the former atom the expansion included the following terms of Ne$^+$—$1s^2\,2s^2\,2p^5(^2P^0)$, $1s^2\,2s\,2p^6(^2S)$, $1s^2\,2s^2\,2p^4(^3P)\,3s\,(^4P,\,^2P)$, $1s^2\,2s^2\,2p^4\,(^3P)\,3p\,(^4P_0)$. Similar terms were included for Ar. Comparison of the results obtained for Ar with observed data on optical absorption is discussed on p. 3486.

Auto-ionizing states arising between the terms of the ground configurations of O, N, and C have been found in the course of the close-coupling calculations involving terms of the ground configurations discussed on p. 3407. Multiconfiguration close-coupling expansions, referred to on p. 3408, have yielded further information on higher auto-ionizing levels.

Further calculations which have provided information on auto-ionizing levels are those of Burke and Moores for Mg and Ca (see p. 3408) and of van Blerkom for Mg$^-$ (see p. 3409).

Bottcher‡‡ has applied the variational method outlined on p. 3411 to calculate the position and widths of resonances in Be, O, C, and F. Comparison is made, as far as level widths are concerned, with close-coupling calculations for Be and O and reasonably good agreement is obtained. For this purpose the method seems to be quite econo-

† MORSE, P. M., YOUNG, L. A., and HAURWITZ, E. S., *Phys. Rev.* **48** (1935) 948.
‡ NORCROSS, D. W., *J. Phys. B (GB)* **2** (1969) 1300.
§ PASCHEN, VON F. and KRUGER, P. G., *Ann. Phys. Lpz.* **8** (1931) 1005.
∥ SMITH, K. and ORMONDE, S., *6th int. conf. on the physics of electronic and atomic collisions, Cambridge, Mass.* Abstracts, M.I.T. Press, 1969, p. 920.
†† ORMONDE, S., SMITH, K., and TORRES, B. W., *7th int. conf. on the physics of electronic and atomic collisions, Amsterdam.* Abstracts, North Holland, 1971, p. 1025.
‡‡ BOTTCHER, C., *J. Phys. B (GB)* **3** (1970) 1201.

mical and effective but it does not give good results for the resonance energies.

2. Behaviour of cross-sections at thresholds (see Chap. 9, § 10)

According to the analysis presented on pp. 652–3 the elastic cross-section may exhibit a cusp-like singularity at an excitation threshold leading to production of an outgoing s electron. If it can be assumed that the existence of long-range interactions does not modify the argument significantly, no such effects would be expected with outgoing electrons of finite orbital angular momentum. However, the first experimental evidence of the presence of a cusp, observed by Andrick, Eyb, and Hofmann† in the energy variation of the differential elastic cross-section for scattering of electrons by sodium atoms, at different angles of scattering, seemed to suggest that it is associated with d-wave elastic scattering. This could not be associated with an outgoing s-wave in the inelastic channel involving $3\,^2P$ excitation. Moores and Norcross‡ have cleared up this difficulty by showing from their close-coupling calculations (see p. 3406) that, very close to the $3\,^2P$ threshold at $2\cdot1$ eV, there is a d-wave resonance, the effect of which is superimposed on the cusp. They expect that similar effects will occur in scattering by Li and by K atoms.

2.1. *The S, T, R, and M matrices*

In Chapter 9, p. 600 it was shown that for single open-channel scattering the total phase shift for elastic collisions is given, at impact energy E, close to a resonance energy E_r, by

$$\zeta(E) = \eta(E) - \operatorname{arc cot}\{(E-E_r)/\tfrac{1}{2}\Gamma\}, \tag{13}$$

where $\eta(E)$ is the slowly varying background phase shift and Γ is the level width.

This may be generalized to collisions in which there are n open channels by using the analysis of Chapter 9, §§ 10.1–10.3. In terms of the notation of these sections, let S be the open-channel S-matrix (see § 10.3) and S_0 the corresponding S-matrix for the background scattering. Following (181) on p. 649 we define the eigenphase matrices in terms of the transformations

$$\mathbf{S} = \mathbf{U}^\dagger e^{2i\zeta}\mathbf{U}, \qquad \mathbf{S}_0 = \mathbf{U}_0^\dagger e^{2i\eta}\mathbf{U}_0 \tag{14}$$

which diagonalize $\mathbf{S}$ and $\mathbf{S}_0$, $\mathbf{U}$ and $\mathbf{U}_0$§ being unitary matrices.

---
† ANDRICK, D., EYB, M., and HOFMANN, M., *J. Phys. B (GB)* **5** (1972) L15.
‡ MOORES, D. and NORCROSS, D. W., private communication.
§ MACEK, J., *Phys. Rev.* A2 (1970) 1101.

We also introduce the column matrix γ whose elements give the probability amplitude for a resonance state to break up via one or other open channel. Then if $\mathbf{y}$ is the row matrix defined by

$$\mathbf{y} = \mathbf{U}_0^\dagger \gamma, \tag{15}$$

we have†
$$E - E_r = \tfrac{1}{2}\Gamma \sum_{i=1}^{n} y_i^2 \cot\{\eta_i - \zeta(E)\}, \tag{16}$$

the n roots ζ_j ($j = 1,...,n$) of which give the eigenphases as functions of E.

2.2. *Threshold law for ionization*

The theoretical position regarding the threshold law for ionization is still not definite. It has been found that the analysis of Rudge and Seaton (see p. 663) is not valid right to the threshold so their deduction of a linear threshold law (see (249), p. 663) is not valid. Peterkop‡ has obtained Wannier's result

$$Q_i = B(E - E_i)^{1.127},$$

for the variation of the ionization cross-section Q_i with impact energy E close to the threshold E_i, by a semi-classical calculation. Recent experimental evidence (see p. 3362) supports this result.

The last paragraph of p. 664 is not correct as stated. It refers to the cross-section for $(n+1)$-fold ionization in a *single* collision, not to the cross-section for further ionization of an already n-fold ionized atom as stated. In any case the result given in (250) is not firmly established as similar uncertainty prevails as for single ionization.

† MACEK, J., loc. cit., p. 3419.
‡ PETERKOP, R. V., *J. Phys. B* (*GB*) **4** (1971) 513.

10

ELECTRON COLLISIONS WITH MOLECULES —TOTAL AND ELASTIC SCATTERING

1. Diffraction of electrons by molecules (see Chap. 10, § 1)

THE precision of measurements of differential cross-sections for scattering of electrons of some tens of keV energy by molecules has continued to increase. For example, Fink and Bonham† have developed equipment in which the intensity of 40 keV electrons scattered by a beam of gas molecules is measured with a plastic scintillator and photomultiplier. The relative accuracy of the intensity measurements is as high as 0·1 per cent while the scattering angle is measured to better than 0·25 per cent. Measurements have been made on N_2‡ over the range $2 < s < 60$, $\times 10^8$ cm^{-1}, of the quantity $s = 4\pi \sin \frac{1}{2}\theta/\lambda$, where λ is the electron wavelength and θ the angle of scattering. When these are analysed, taking into account the failure of Born's approximation for the elastic scattering (see Chap. 10, § 1.7.1), the results show clearly that bonding effects are present (see Chap. 10, § 1.7.3).

Similar measurements in CH_4§ have been used to test the wavefunctions obtained by Moccia‖ based on a single-centre multiconfiguration model. In the range $0·2 < s < 1$, $\times 10^9$ cm^{-1} the theoretical results were found to be unsatisfactory just as were those obtained by treating the atoms as independent scatterers.

On p. 685 of Chapter 10 reference was made to the possible importance of triple scattering effects within the molecule in the interpretation of precise data on molecular interference scattering. Liu and Bonham†† have worked out in detail the contribution of such effects for diffraction by the ReF_6 molecule studied experimentally by Jacob and Bartell.‡‡ It was found that this contribution was such as to explain the small difference between the observed data and those derived theoretically

---

† FINK, M. and BONHAM, R. A., *Rev. scient. Instrum.* **41** (1970) 389.
‡ FINK, M., LEE, J., and BONHAM, R. A., *6th int. conf. on the physics of electronic and atomic collisions, Cambridge, Mass.* Abstracts, M.I.T. Press, 1969, p. 850.
§ BONHAM, R. A. and FINK, M., ibid., p. 848.
‖ MOCCIA, R., *J. chem. Phys.* **40** (1964) 2164.
†† LIU, J. W. and BONHAM, R. A., *J. mol. Struct.* **11** (1972) 297.
‡‡ JACOB, E. J. and BARTELL, L. S., *J. chem. Phys.* **53** (1970) 2231.

2. Scattering of electrons of medium speed by molecules (see Chap. 10, § 2)

Williams† (see p. 3377) has measured angular distributions of electrons with energies of 100 and 200 eV elastically scattered between angles of 20° and 130°. The equipment is the same as that used to obtain corresponding measurements for helium (p. 3377). Khare and Shobha‡ have calculated the differential cross-sections using the Born–Ochkur approximation (see Chap. 7, p. 455) modified by the addition of a second-order amplitude arising from adiabatic and dynamic polarization rather similar to that introduced by La Bahn and Callaway (see p. 3396) in their work on the scattering of medium energy electrons in helium. The Born–Ochkur amplitudes were calculated using an approximate H_2 wave-function due to Huzinaga.§ The second-order amplitude was calculated by Born's approximation assuming cut-off potentials corresponding to the dipole and quadrupole polarizabilities and the non-adiabatic interaction falling off as r^{-6}‖ (see p. 3397). Although no absolute measurements are available, quite good agreement is obtained between the forms of the calculated and observed distributions. The second-order amplitude is only important at angles less than 15° for 200 eV electrons and less than 25° for 100 eV electrons.

Bromberg†† has measured absolute differential cross-sections for elastic scattering of electrons of 500 eV energy by N_2 and CO over the angular range 2–65° using the same equipment as for the corresponding measurements in He (p. 3376).

Hilgner, Kessler, and Steeb‡‡ made the first observations of spin polarization by electron scattering from molecules using the same technique as for corresponding experiments for scattering by atoms (see Chap. 5, § 6 and pp. 3380–2). They used 300 eV electrons with C_2H_5I and I_2 as target molecules. Further experiments§§ have since been carried out using electrons of energy ranging from 60 to 1600 eV and $Bi(C_6H_5)_3$, C_6H_6, H_2O, and CCl_4 as target molecules.

† WILLIAMS, K. G., *6th int. conf. on the physics of electronic and atomic collisions, Cambridge, Mass.* Abstracts, M.I.T. Press, 1969, p. 735.
‡ KHARE, S. P. and SHOBHA, P., *J. Phys. B (GB)* **5** (1972) 1938.
§ HUZINAGA, S., *Prog. theoret. Phys. Kyoto* **17** (1957) 162.
‖ DALGARNO, A., DRAKE, G. W. F., and VICTOR, G. A., *Phys. Rev.* **176** (1968) 194.
†† BROMBERG, J. P., loc. cit., p. 3376.
‡‡ HILGNER, W., KESSLER, J., and STEEB, E., *Phys. Rev. Lett.* **18** (1967) 983; *Z. Phys.* **221** (1969) 324.
§§ HILGNER, W., KESSLER, J., and STEEB, E., *Z. Phys.* **221** (1969) 324.

3. Collisions of slow electrons with molecules
3.1. *Broad features* (see Chap. 10, § 3)
3.1.1. *Experimental data*

For H_2 new measurements have been made of differential cross-sections. These have included the remarkable experiments of Linder and Schmidt[†] who have investigated the scattering of electrons in the energy range 0·3 to about 15 eV using such high energy resolution ($\simeq$ 30 meV) that at each scattering angle they have been able to resolve energy losses due to rotational excitation (further details of their equipment are given on p. 3437). In this way they have been able to measure the angular distributions of electrons which have strictly suffered no energy loss apart from elastic momentum transfer. Measurements were made over an angular range from 20 to 120° and normalized so that the total cross-section for all processes, elastic and inelastic, integrated over the angular range (including extrapolated values between 0 and 20° and between 120 and 180°) agreed with the absolute measurement of the total cross-section made by Golden, Bandel, and Salerno (see p. 698).

Trajmar, Truhlar, and Rice[‡] have also made measurements of angular distributions of electrons of energies ranging from 7 to 81·6 eV over an angular range 10 to 80°. They used equipment similar to that of Simpson—described on p. 20—but were unable to resolve energy loss due to rotation. Their measurements therefore include not only elastic scattering but also inelastic collisions leading to rotational excitation as is also true of earlier measurements and those of Williams (see above), which extend down to 30 eV. Good agreement is found between most of these measurements. When allowance is made for rotational excitation at the lower energies, the measurements of Trajmar *et al.* agree well with those of Linder and Schmidt.

Crompton, Gibson, and McIntosh[§] have derived momentum-transfer cross-sections for electrons of energies less than 2 eV in H_2 from analysis of new, precise measurements of electron drift and diffusion in para H_2. These are quite close to those derived by Frost and Phelps from earlier swarm data (see Fig. 11.24, p. 772).

Linder and Schmidt[||] have carried out corresponding measurements for O_2, the electron energy range being from 0 to 4 eV and scattering angles from 0 to 120°. The results are of special interest in connection

---
[†] LINDER, F. and SCHMIDT, H., *Z. Naturf.* **26A** (1971 1603).
[‡] TRAJMAR, S., TRUHLAR, D. G., and RICE, J. K., *J. chem. Phys.* **52** (1970) 4502.
[§] CROMPTON, R. W., GIBSON, D. K., and McINTOSH, A. I., *Aust. J. Phys.* **22** (1969) 715.
[||] LINDER, F. and SCHMIDT, H., *Z. Naturf.* **26A** (1971) 1617.

with the excitation of vibration through formation of long-lived complexes, details of which are described and discussed on p. 3443. Resonance effects in elastic scattering arising from these complexes are clearly observed at impact energies below 0·75 eV (see Fig. 10.37 of Chap. 10). Elastic differential cross-sections measured at higher energies were calibrated from the recent measurements by Salop and Nakano† of the total cross-section extrapolated to energies below 2 eV.

Trajmar, Cartwright, and Williams‡ have measured differential cross-sections for elastic scattering of electrons by O_2 for incident energies ranging from 4 to 45 eV and scattering angles from 10 to 90°. Comparison is only possible with the data of Linder and Schmidt at 4 eV. Between 60 and 90° the agreement is quite good but the maximum observed by the latter authors near 50° does not appear in the measurements of Trajmar et al.

Shyn, Stolarski, and Carignan§ have measured angular distributions of electrons scattered from N_2 using a crossed-beam technique. The incident electrons were rendered homogeneous in energy to 0·06 eV by passage through a 127° electrostatic analyser of the type developed by Marmet and Kerwin (see Chap. 3, p. 113) and the energy distribution of the scattered electrons was analysed in a similar way. Measurements were made over an incident energy range from 5 to 90 eV and an angular range from -114 to $+160°$. With the available energy resolution it was possible to separate inelastic collisions involving vibrational but not rotational excitation. The measurements agreed well with earlier measurements (see Chap. 10, p. 703), particularly at energies of 50 eV and below. By integration of the angular distribution extrapolated to 0° and 180° the total 'elastic' cross-section, including inelastic collisions involving rotational excitation, was obtained as a function of electron energy. When normalized at 2·3 eV to agree with the absolute value of the total cross-section observed by Bruche, good agreement is obtained with other absolute measurements (see Chap. 10, p. 699) particularly when allowance is made for the fact that these measurements include inelastic collisions of all kinds.

Brooker and Beynon‖ have measured the mean momentum-transfer cross-section Q_d for electrons in N_2 at 1300 °K (0·18 eV). They worked with a plasma produced by heating N_2 gas at around atmospheric pressure, seeded with about 0·1 per cent of potassium vapour, to 1300°.

† Salop, A. and Nakano, H. H., *Phys. Rev.* A2 (1970) 127.
‡ Trajmar, S., Cartwright, D. C., and Williams, W., ibid. A4 (1971) 1482.
§ Shyn, T. W., Stolarski, R. S., and Carignan, G. R., ibid. A6 (1972) 1002.
‖ Brooker, P. G. and Beynon, J. D. E., *J. Phys. B (GB)* 5 (1972) 1186.

Two types of measurements were made. In one the effect of the plasma on the propagation characteristic of microwaves in the waveguide formed by the furnace tank was measured (see Chap. 2, § 5) and in the other the d.c. conductivity of the plasma, as in the experiments described on p. 3359 to measure the momentum-transfer cross-section of potassium. The former experiment gave $Q_d = 7 \cdot 5 \pm 0 \cdot 6 \times 10^{-16}$ cm² and the latter $6 \cdot 8 \pm 0 \cdot 3 \; 10^{-16}$ cm². These agree quite well with the value derived by analysis of swarm data in N_2 (see Chap. 11, p. 777).

Truhlar, Williams, and Trajmar[†] have also measured differential elastic cross-sections for scattering of electrons by CO, in the impact energy range 10–80 eV and angular range 15–85°. Comparison is made with much earlier measurements of Arnot (see Chap. 10, p. 703).

3.1.2. Theoretical results

Since Vol. 2 was written there has been considerable progress in applying the techniques found to be suitable for electron–atom collisions, at least to the simpler molecules. Much of this is based on the adiabatic approach in which the scattering is calculated ignoring rotation and vibration, but, for H_2, rotation has been taken into account in a close-coupling treatment.

At the time of writing Vol. 2 all the calculations of the scattering of electrons by two fixed centres of force were carried out in spheroidal coordinates centred on the two scatterers. Hara[‡] has extended the calculations described in Chapter 10, p. 711 so that they are essentially the same as for the adiabatic-exchange calculations carried out for atoms (see Chap. 8, p. 510). He finds good agreement with the experimental total cross-sections over the energy range from 1 to 10 eV.

Although the problem as solved by Hara is not separable in spheroidal coordinates the analysis of Chapter 10, pp. 703–6 still applies to a good approximation so we may refer to the η_{lm} phase shifts. For the H_2 case the important phase shifts in the energy range concerned are η_{00} and η_{10}.

Temkin and Vasavada[§] called attention to the possibility of carrying out detailed calculations of the scattering of electrons by axially symmetrical fields of force in terms of spherical polar rather than spheroidal coordinates. The amplitude for a collision in which the electron is scattered into the solid angle between Ω and $\Omega + d\Omega$ from a diatomic molecule whose orientation relative to the laboratory frame of reference

---

[†] TRUHLAR, D. G., WILLIAMS, W., and TRAJMAR, S., *J. chem. Phys.* **57** (1972) 4307.
[‡] HARA, S., *J. phys. Soc. Japan* **27** (1969) 1009, 1262.
[§] TEMKIN, A. and VASAVADA, K. V., *Phys. Rev.* **160** (1967) 109.

is specified by the Euler angles α, β, γ can be expanded in the form

$$f(\alpha,\beta,\gamma;\Omega) = \sum_{mm'} \sum_{l_i l_j} a_{l_i l_j m} \mathscr{D}^{(l_i)}_{m'm} \mathscr{D}^{(l_j)*}_{0m} Y_{l_j m'}(\Omega), \qquad (1)$$

where the Y_{lm} are the usual spherical harmonics and the $\mathscr{D}$ functions are known functions of α, β, γ. The $a_{l_i l_j m}$ are independent of α, β, γ, and Ω so that the average of $|f|^2$ over all molecular orientations may be carried out without difficulty.

It is often a good approximation to take

$$a_{l_i l_j m} = \delta_{l_i l_j}\{4\pi(2l_i+1)\}^{\frac{1}{2}} k^{-1} \exp(i\eta_{l_i m}) \sin \eta_{l_i m}, \qquad (2)$$

where $\eta_{l_i m}$ is nearly equal to the corresponding phase shift obtained from a calculation using spheroidal coordinates with the same assumption about the molecular wave-functions and their distortion by the colliding electron. k is as usual the electron wave-number.

The relative advantages and disadvantages of the use of spherical polar and of spheroidal coordinates are not yet fully clear. For real problems the Schrödinger equation is not separable in either set of coordinates. The spheroidal system may nevertheless be more natural. On the other hand the spherical harmonics are such relatively simple and thoroughly familiar functions that their manipulation even in a large expansion offers little difficulty. Moreover, in contrast to spheroidal harmonics, they are independent of the electron energy.

The first direct application of so-called single-centre expansions was made[†] to the scattering of electrons by H_2^+ molecules. Tully and Berry[‡] extended this to the e–H_2 problem, taking into account electron exchange but not distortion of the molecule. Their results for the phase shifts η_{lm} are qualitatively similar to those obtained by Hara who, however, included dipole distortion effects.

The most elaborate calculations are those in which the wave-function describing the collision is expanded in terms of an expansion in molecular wave-functions calculated according to the Born–Oppenheimer approximation (see Chap. 12, pp. 827–9) which include the dependence on the rotational angular coordinates.[§] In the latest calculations only terms[∥] involving the ground electronic state are included but full allowance is made for exchange and for rotational excitation. Dipole distortion of the molecule is included by a procedure similar to that used in the adiabatic-exchange approximation for atoms. The results obtained for

[†] TEMKIN, A. and VASAVADA, K. V., loc. cit., p. 3425.
[‡] TULLY, J. C. and BERRY, R. S., *J. chem. Phys.* **51** (1969) 2056.
[§] LANE, N. F. and GELTMAN, S., *Phys. Rev.* **160** (1967) 53.
[∥] HENRY, R. J. W. and LANE, N. F., ibid. **183** (1969) 221; HENRY, R. J. W., ibid. **A2** (1970) 1349.

the total cross-section are not quite so good as those obtained by Hara, being a little too large, possibly because the effective polarization potential is too big. However, they are by no means unsatisfactory and yield good results for rotational excitation as discussed on p. 3438.

A very interesting application of single-centre expansion methods is the calculation by Burke and Sin-Fai Lam† of cross-sections for elastic scattering of electrons with N_2 molecules, covering the energy range from 0 to 27 eV. They allowed for exchange but not for molecular distortion. The static interaction was based on that obtained by using the Hartree–Fock–Roothan wave-functions for N_2 given by Nesbet.‡ The first two terms in the expansion§ of the latter in even spherical harmonics round the centre of the molecule were retained. It was found that exchange is of considerable importance. When it is included the $^2\Pi_g$ scattering function leads effectively to an η_{20} phase shift, in the notation of (2), which shows a resonance behaviour, rising from 0·2 to 2·7 radians as the electron energy increases from 1·1 to 3·8 eV. This corresponds to a time delay $\hbar\, \partial\eta/\partial E$ (see p. 644) of nearly 10^{-15} s or an effective level-width of 0·76 eV. The cause of this hold-up time must be ascribed to temporary capture of the electron into a π_g valence orbital behind the centrifugal barrier associated with an electronic angular-momentum quantum number of 2. The plausibility of this was established earlier by Krauss and Mies.∥

This long hold-up time associated with the $^2\Pi_g$ scattering state must be regarded as the origin of the resonance-like effects which occur in the elastic and vibrational excitation of N_2 (see Chap. 10, pp. 717–20; Chap. 11, pp. 738–42; and p. 3446). Whether or not the η_{20} phase-shift attains a value of π or close to it is still uncertain, for the total cross-section calculated by Burke and Sin-Fai Lam attains a maximum value about twice as large as that observed. This is because of the large contribution made by the $^2\Pi_g$ state. Nevertheless there seems little doubt that the hold-up time associated with $^2\Pi_g$ scattering is relatively long—all that is really necessary is that the maximum value of $\partial\eta/\partial E$ is large enough.

Too much weight should not be given to the quantitative details of this calculation as it is likely that molecular polarization, which was not included, can produce considerable modifications. In a later calculation

† BURKE, P. G. and SIN-FAI LAM, A.-L., *J. Phys. B* (*GB*) **3** (1970) 641.
‡ NESBET, R. K., *J. chem. Phys.* **40** (1964) 3619.
§ FAISAL, F. H. M., *J. Phys. B* (*GB*) **3** (1970) 636.
∥ KRAUSS, M. and MIES, F. H., *Phys. Rev.* **A1** (1970) 1592.

Burke and Chandra† calculated the scattering due to an interaction of the form

$$V(r) = V_s(r) + V_{\text{pol}}(r), \qquad (3)$$

where V_s is the static interaction and V_{pol} a semi-empirical polarization potential given by

$$V_{\text{pol}}(r) = -\tfrac{1}{2}e^2 r^{-4}\{\alpha + \alpha' P_2(\cos\theta)\}[1-\exp\{-(r/r_0)^6\}]. \qquad (4)$$

α and α' are the polarizabilities of N_2 (see Chap. 11, p. 748) and r_0 is an adjustable cut-off parameter. The axis of polar coordinates is taken along the molecular axis. With suitable choice of r_0 the peak of the total cross-section could be adjusted to agree with observation but it remained about twice too large.

4. Fine structure in total cross-sections for electron–molecule collisions

Sanche and Schulz‡ have developed an improved electron transmission spectrometer of high energy resolution (energy spread at half minimum between 20 and 30 meV) and high sensitivity, which they have applied to observe the fine structure of the total cross-sections for electrons in H_2, D_2, CO, N_2, NO, and O_2. They have compared their results for the location of the resonances and for the vibrational structure of the resonance states with those obtained in earlier experiments, some of which were discussed in Chapter 10. We discuss briefly these various observations in relation to the particular molecules concerned.

4.1. H_2

Since the final observations of fine structure in the total cross-sections made by Kuyatt, Mielczarek, and Simpson (see Chap. 10, p. 714 and Chap. 13, p. 937) a number of experiments have been carried out. Some have observed electron transmission while others have observed the fine structure in cross-sections for vibrational or electronic excitation. From these experiments, which give remarkably consistent results, at least four band systems corresponding to different autodetaching resonance states of H_2^- have been identified. Many corresponding bands have been observed for D_2^- and in some cases also HD^-. Table 10 A.1 summarizes the observed data.

It is noteworthy that the four apparent levels of the f band lie above the energy of H_2^+.

† BURKE, P. G. and CHANDRA, N., J. Phys. B (GB) 5 (1972) 1696.
‡ SANCHE, L. and SCHULZ, G. J., Phys. Rev. Lett. 26 (1971) 943; Phys. Rev. A6 (1972) 69.

TABLE 10 A.1

Band systems of autodetaching states of H_2^-

Band designation	a and d	b	c and e	f	g
Energy (eV) above the ground state of H_2 of level with					
$v = 0$	11·32,[a] 11·28,[b] 11·30,[c] 11·30[d]	11·27[c]	11·43,[a] 11·46,[b] 11·37,[c] 11·50[d]		13·66,[a] 13·62,[e] 13·63[d] 15·09[a]
$v = 1$	11·62,[a] 11·56,[b] 11·62,[c] 11·62[d]	11·47[c]	11·74,[a] 11·72,[b] 11·71,[c] 11·79[d]		13·94,[a] 13·91,[e] 13·93[d] 15·32[a]
Highest v observed	8[c]	6[c]	6[a,b]	11[e]	

(a) SANCHE, L. and SCHULZ, G. J., loc. cit. (transmission).
(b) KUYATT, G. E., SIMPSON, J. A., and MIELCZAREK, S. R., J. chem. Phys. **44** (1966) 437 (transmission).
(c) COMER, J. and READ, F. H., J. Phys. B (GB) **4** (1971) 368 (excitation of vibration, see p. 3440).
(d) WEINGARTSHOFER, A., EHRHARDT, H., HERMAN, V., and LINDER, F., Phys. Rev. A**2** (1970) 294 (electronic excitation).
(e) GOLDEN, D. E., Phys. Rev. Lett. **27** (1971) 227.

Eliezer, Taylor, and Williams† have applied the stabilization method (see p. 3410) to calculate the potential energy curves of the H_2^- autodetaching states to be expected. They interpret the band systems labelled 'a and d' and 'c and e' respectively as arising essentially from attachment of an electron to the $c^3\Pi_u$ and $C^1\Pi_u$ states of H_2 (see Fig. 13.1). The calculated energies of the $v = 0$ levels in the respective states are 11·07 and 11·46 eV and the separations between the $v = 0$ and $v = 1$ levels are 0·30 and 0·29 eV respectively. In view of the necessarily approximate nature of the calculations the agreement with the observed data is surprisingly good.

In D_2 band (a) has been observed by Sanche and Schulz, Kuyatt et al., and Comer and Read (referred to as (a), (b), (c) of Table 10 A.1), band (c) by Sanche and Schulz and by Comer and Read, and bands (f) and (g) by Sanche and Schulz.

4.2. O_2

New results at low energies are best discussed in relation to vibrational excitation and so are considered on p. 3443.

At impact energies between 8 and 12 eV a large number of structural features were observed by Sanche and Schulz‡ which could be interpreted as seventeen resonances. At 11·81±0·05 eV a regular progression appears which is referred to as band 'a'. Not only are the vibrational spacings close to those of the $a\,^4\Pi_u$ states of O_2^+ but the Franck–Condon probabilities for transitions from the ground vibrational state of the ground ($X^3\Sigma_g^-$) electronic state of O_2 are also very close. Sanche and Schulz‡ produce evidence which suggests that the O_2^- state is a $^4\Pi_u$ state formed by addition of two $3s\sigma_g$ electrons to the O_2^+ core. A second band labelled (b) is observed beginning at 14·26 eV above the energy O_2^+ with vibrational spacings close to those of $O_2^+(b\,^4\Sigma_g^-)$.

4.3. N_2

Results at energies below 7 eV are discussed on p. 3446 in connection with vibrational excitation.

Between 7 and 11 eV a number of features appear. The first progression may be interpreted as arising from excitation of different vibrational levels of the $B\,^3\Pi_g$ state (see Fig. 13.46, p. 958) but the next, quite regular, progression, referred to as band system (a), beginning at 9·25 eV, has been traced out to 11·02 eV, including 18 vibrational levels

† ELIEZER, I., TAYLOR, H. S., and WILLIAMS, J. K., J. chem. Phys. **47** (1967) 2165.
‡ loc. cit., p. 3428.

with spacing varying smoothly from 0·13 eV between $v = 1$ and 0 to 0·085 eV between $v = 16$ and 17. The source of this progression is not clear. It cannot arise from attachment of an electron to a Rydberg state.

A possible band system, labelled (b), is observed beginning at 11·48 eV, by Heideman et al. (see p. 962) and by Comer and Read,† and between 11·47 and 11·51 eV by Sanche and Schulz.‡ From the observed angular distribution of scattered electrons it appears that s electrons are involved, suggesting that the N_2^- state is a $^2\Sigma_g$ state arising from attachment to the $E\,^3\Sigma_g^+$ and the $a''\,^1\Sigma_g^+$ state of N_2. The vibrational spacing suggests association with the $X\,^2\Sigma_g^+$ ground state of N_2^+, presumably by binding two $3s\sigma_g$ electrons.

A further strong resonance feature is observed at 11·87 eV†, ‡, § (see Chap. 13, p. 963) and a number of others which are difficult to disentangle in terms of vibrational progressions because of overlap.

4.4. CO

The features present in the transmitted current in CO are very similar to those observed in N_2 except that there is no level corresponding to the mysterious band (a) which begins at 9·23 eV. Corresponding to the very strong resonance at 11·48 eV in N_2 there is a resonance of similar strength located at 10·02 eV and between 9·98 and 10·04 eV by Sanche and Schulz‡ and by Comer and Read respectively. A regular progression beginning at 13·95±0·05 eV with initial spacing of 0·205 eV was observed out to the sixth vibrational level by Sanche and Schulz.

4.5. NO

In the energy range from 5 eV to 7·5 eV four band systems are observed. The first three systems starting at 5·04, 5·41, and 5·46 eV each contain four vibrational levels while the fourth, starting at 6·45 eV, includes six levels. In all systems the vibrational spacings agree well with the corresponding separations for the ground $X\,^1\Sigma_u^+$ state of NO^+. Similar agreement is found for the Franck–Condon transition probabilities from the ground state of NO. It seems likely then that all four systems arise from states of NO^- in which two electrons are bound in Rydberg orbits to the ground state of NO^+ as a core.

Sanche and Schulz‡ searched for resonances arising from NO^- states in which two electrons are attached to excited states of NO^+ with nuclear

---

† COMER, J. and READ, F. H., *J. Phys. B (GB)* **4** (1971) 1055.
‡ loc. cit., p. 3428.
§ HALL, R. I., MAZEAU, J., REINHARDT, J., and SCHERMANN, G., *J. Phys. B (GB)* **3** (1970) 991.

separation close to that of the ground state of NO, namely the $b\,^3\Pi$, $A\,^1\Pi$, $C\,^3\Pi$, and $D\,^1\Pi$ states. They found structure, at 12·36, 14·19, and 17·51±0·05 eV, close to the expected energies for all but $C\,^3\Pi$.

4.6. *Summarizing remarks*

The evidence from these experiments provides strong support to the suggestion put forward by Krauss and Mies,† on theoretical grounds, that the Type I resonance states (see Chap. 12, p. 815) of diatomic molecules arise from attachment of electrons to excited Rydberg states of molecules rather than to valence states. The states therefore essentially involve attachment of two electrons in Rydberg orbitals to a positive ion core. In all the cases referred to above, for O_2, N_2, CO, and NO for which sufficient data are available to attempt identification of the negative ion state concerned it turns out to be of the suggested type except for the state in N_2 at 9·25 eV.

† loc. cit., p. 3427.

11

COLLISIONS OF ELECTRONS WITH MOLECULES—EXCITATION OF VIBRATION AND ROTATION—ANALYSIS OF SWARM EXPERIMENTS AT LOW ENERGIES

SINCE Chapter 11 was written one of the chief theoretical developments has been the application of the sudden approximation (see Chap. 18, p. 1590) to calculate rotational excitation cross-sections for H_2 and N_2. For the former molecule the elaborate close-coupling technique referred to on p. 3426 has also been applied. The sudden approximation has even been used to estimate vibrational excitation cross-sections for H_2.

There has been progress in interpreting the role of resonance or near-resonance effects in determining inelastic cross-sections for H_2, N_2, and O_2.

A considerable amount of attention has also been devoted to the theoretical study of the collisions of electrons with polar molecules, in which rotational excitation plays a primary role.

On the experimental side a number of new results have been obtained, thanks to improvements in energy and angular resolution and increased precision generally. For H_2 the energy resolution is high enough to make possible the measurement of differential cross-sections for inelastic collisions involving single rotational transitions while for heavier molecules there is no difficulty in studying the excitation of specific vibrational transitions with even higher precision than described, for example, in Chapter 11, § 3.

Improved accuracy in measurement of the transport properties of electrons in H_2 and D_2 at low mean energies has made it possible to derive, with high accuracy, cross-sections for rotational excitation. One of the most puzzling observations is that of a dependence of the mobility of electrons in H_2 and O_2 on the pressure p for a fixed value of F/p where F is the electric field strength. The dependence is in the same sense as would arise if, under certain circumstances, an electron is held up for an appreciable time on collision with a molecule.

We now consider these different developments in more detail.

1. Application of the sudden approximation

According to this approximation, if $f(\alpha,\beta,\gamma;\Omega;R)$ is the amplitude for elastic scattering of an electron into the solid angle between Ω and $\Omega+\mathrm{d}\Omega$ by a molecule when the nuclei are fixed in space at a distance R apart and such that the Euler angles defining the orientation of the reference frame fixed in the molecule relative to that fixed in the space in which the axis of Oz lies along the direction of incidence of the colliding electron are α, β, γ, the differential cross-section for a rotational transition from a state with rotational quantum numbers J, M to one with numbers J', M' is given by

$$\left|\int \chi_{JM}(\alpha,\beta,\gamma)f(\alpha,\beta,\gamma;\Omega;R_0)\chi^*_{J'M'}(\alpha,\beta,\gamma)\,\mathrm{d}\omega_{\alpha\beta\gamma}\right|^2 \mathrm{d}\Omega. \tag{1}$$

χ_{JM}, $\chi_{J'M'}$ are the initial and final rotational wave-functions, R_0 is the equilibrium separation, and the integral is taken over all orientations of the molecule.

The derivation of this formula follows exactly the same lines as the corresponding formula (232) of Chapter 17 derived for collisions between heavy particle systems. It is a good approximation under the same conditions—the collision is a sudden one in that the electron is moving much faster than the rotating atoms and the energy lost in the collision is only a small fraction of the total energy.

Using the form (1) on p. 3426 for $f(\alpha,\beta,\gamma;\Omega;R_0)$ Chang and Temkin[†] have calculated cross-sections for excitation of transitions between rotational states of H_2 while Burke and Sin-Fai Lam[‡] have carried out corresponding calculations for N_2, using the amplitudes obtained in their elastic scattering calculations described on p. 3427.

Independently Hara[§] has used the elastic amplitudes he obtained for H_2 using a spheroidal harmonic expansion (see p. 3425) to calculate the rotational cross-sections.

Comparison of these results with the close-coupling calculations of Henry and Lane (see p. 3426) and with observed values is discussed on p. 3438.

While a corresponding formula to (1) may be obtained for excitation of vibrational transitions the practical utility is much less because the integral involved, which is now over the nuclear separation R, can only be evaluated if the amplitude f is known as a function of R and not

---

[†] CHANG, E. S. and TEMKIN, A., *Phys. Rev. Lett.* **23** (1969) 399; *J. phys. Soc. Japan* **29** (1970) 172.
[‡] loc. cit., p. 3427.
[§] HARA, S., *J. phys. Soc. Japan* **27** (1969) 1592.

merely at the equilibrium separation R_0. Nevertheless, Faisal and Temkin† have made a semi-quantative study of the possibilities for H_2.

Working from the amplitude (1) together with (2) of p. 3426, the problem is to evaluate the phase shifts η_{lm} as functions of the nuclear separation R. Faisal and Temkin assumed that

$$\eta_{lm}(R) \simeq \eta_{lm}(R_0) + \left(\frac{\partial \eta_{lm}}{\partial R}\right)_{R_0} (R-R_0), \tag{2}$$

where

$$\left(\frac{\partial \eta_{lm}}{\partial R}\right)_{R_0} \simeq R_0^{-1}\{\eta_{lm}(R_0) - \eta_{lm}(0)\}, \tag{3}$$

with $\eta_{lm}(0)$ the appropriate phase-shift for scattering by He calculated by Duxler, Poe, and La Bahn (exchange adiabatic approximation, see p. 3397). For $\eta_{lm}(R_0)$ they used the values calculated by Tully and Berry (see p. 3426) with estimated additional contributions from long-range quadrupole and polarization forces which were not otherwise included. Further, they took account of the importance of the $\eta_{10}(\Sigma_u)$ phase shift by introducing a scaling factor $g(E)$, depending on the impact energy E, to multiply the estimated value of $(\partial \eta_{10}/\partial R)_{R_0}$ so as to obtain the best agreement with observation. The results are discussed on p. 3439.

2. Collisions with polar molecules

In Chapter 11, p. 748 it was pointed out that an electron colliding with a weak permanent rotating dipole of moment μ can cause a transition in which the rotational quantum number changes by unity. Inelastic transitions of this kind will be much more probable than elastic collisions. These conclusions are certainly valid when

$$2m\mu e^2/\hbar^2 \ll 1 \quad \text{or} \quad \mu \ll \tfrac{1}{2}ea_0, \tag{4}$$

m being the electron mass assumed small compared with that of the rotator. Under these conditions the rotational excitation cross-section may be calculated by Born's first approximation, in which the elastic scattering vanishes and only appears in the second approximation.

A similar situation arises in the collisions between two polar molecules discussed in Chapter 17, p. 1606.

Crawford and Dalgarno‡ have carried out calculations for the scattering of thermal electrons by CO molecules which have a permanent dipole moment $\simeq 0.04ea_0$, satisfying (4). In addition to the dipole term, empirical terms representing the static polarization were also included. Cut-off distances r_d, r_{p0}, and r_{p2} were introduced in these terms, r_{p0} and

† FAISAL, F. H. M. and TEMKIN, A., *Phys. Rev. Lett.* **28** (1972) 203.
‡ CRAWFORD, O. H. and DALGARNO, A., *J. Phys. B (GB)* **4** (1971) 494.

r_{p2} applying to the isotropic and second harmonic terms respectively in the polarization potential. r_{p0} was determined to fit the observed momentum-transfer cross-section (see Chap. 11, p. 782 and Fig. 11.36) and r_d, r_{p2} taken as $\frac{1}{2}r_e$ and $\frac{3}{4}r_e$ respectively where r_e is the equilibrium separation. A close-coupling calculation was then carried out. At an energy of 0·03 eV the cross-section for excitation of the $0 \to 1$ rotational transition is $1·12 \times 10^{-15}$ cm² as compared with that for elastic scattering of $1·78 \times 10^{-16}$ cm². The latter is relatively smaller at lower energies, being about 0·027 of the excitation cross-section at 0·01 eV and 0·009 at 0·005 eV. On the other hand, at higher energies the departure from the pure dipole field becomes progressively more important so that at 0·1 eV the elastic cross-section slightly exceeds that for the rotational transition.

When (4) is not satisfied a close-coupling calculation is required even to calculate the rotational excitation. In this connection it is of interest to note that there are no bound states for an electron in the field of two fixed equal and opposite charges q at a distance d apart unless†

$$qd \geqslant 0·6393ea_0, \qquad (5)$$

corresponding to a dipole moment $\mu_0 = 0·6393ea_0$. This critical moment is close to that involved in (4) and, in fact, no solution exists for the scattering of an electron in the field of a static dipole of moment μ_0. However, this conclusion does not apply if the dipole is allowed to rotate. Calculations carried out by close-coupling methods for molecules with dipole moments $\mu > \mu_0$ show that the conclusions derived for cases where $\mu < \mu_0$ largely remain valid, the inelastic scattering greatly exceeding the elastic.

Itikawa and Takayanagi‡ have carried out close-coupling calculations for collisions of electrons with CN radicals to which a dipole moment ea_0 ($> \mu_0$) was assigned. The interaction was taken to be of the form

$$V_C(r_C) + V_N(r_N) - \{\mu e r^{-2}\cos\theta + \tfrac{1}{2}\alpha e^2 r^{-4}\}[1 - \exp\{-(r/r_0)^6\}], \qquad (6)$$

where α is a mean polarizability taken as $15·0ea_0^3$, r_0 a cut-off parameter, and $V_C(r_C)$, $V_N(r_N)$ are the interactions with the undistorted carbon and nitrogen atoms, r_C, r_N being the distances from the respective nuclei. The calculated cross-sections for elastic scattering and for the $0 \to 1$ excitation are given in Table 11 A.1, r_0 being taken as $1·5a_0$. Even at 1 eV the latter is nearly 5 times greater than the former.

It was found in these calculations that, even though (4) is not satisfied,

† Wightman, A. S., *Phys. Rev.* **77** (1950) 521.
‡ Itikawa, Y. and Takayanagi, K., *J. phys. Soc. Japan* **26** (1969) 1254.

Born's first approximation gives quite good results for $Q(0 \to 1)$. Accordingly, it was used in later calculations by Itikawa to obtain cross-sections for rotational excitation of NH_3† ($\mu = 0.578 ea_0$), H_2O† ($\mu = 0.728 ea_0$), and H_2CO‡ ($\mu = 0.921 ea_0$).

TABLE 11 A.1

Calculated cross-sections $Q(0 \to 0)$ and $Q(0 \to 1)$, in πa_0^2, for elastic scattering and for rotational excitation of 'CN' by electron impact

Incident energy (eV)	0.01	0.03	0.1	0.3	1.0
$Q(0 \to 0)$	1.390×10^3	2.29×10^2	79.8	90	60.1
$Q(0 \to 1)$	1.326×10^4	5.82×10^3	2.20×10^3	8.52×10^2	2.91×10^2

No satisfactory explanation is yet forthcoming for the large momentum transfer cross-sections derived from drift velocity data for H_2O and NH_3 (see Chap. 10, Fig. 10.23).

3. Experimental results

3.1. *Rotational excitation of* H_2

The first measurements in which energy losses due to rotational excitation were separately resolved in a crossed-beam experiment were carried out by Ehrhardt and Linder.§ A thorough study using the same equipment has since been carried out by Linder and Schmidt.‖ The data obtained for elastic scattering have already been referred to on p. 3423.

Briefly, the apparatus consisted of an electron beam, velocity-selected by passage through a 127° electrostatic analyser, which intersected a molecular beam at 90°. Electrons scattered into a certain solid angle were velocity-analysed by a second 127° analyser, and detected with a multiplier and pulse-counting electronics. The collector system could be rotated from 0° to 120° with respect to the incident electrons. With an energy resolution of 30–40 meV, rotational transitions in H_2 could be clearly distinguished in the energy loss spectrum. Checks on the observed ratio of inelastic to elastic scattering were made. Thus the transmission of the apparatus as a function of energy was calibrated by comparison with data available from other experiments on the differential elastic scattering of electrons by He.

† ITIKAWA, Y., *J. phys. Soc. Japan* **30** (1971) 835.
‡ IDEM, **32** (1972) 217.
§ EHRHARDT, H. and LINDER, F., *Phys. Rev. Lett.* **21** (1968) 419.
‖ loc. cit., p. 3423.

Differential cross-sections for collisions involving the rotational transition $J = 1 \to 3$ were measured at energies of 2·5, 3·5, 4·5, 8·0, and 10·8 eV. The angular variation of the cross-section is not very marked, being only about 30 per cent over the observed angular range. The results agree quite well in general with the theoretical values given by Chang and Temkin (sudden approximation, spherical polar coordinates, p. 3434), Hara (sudden approximation, spheroidal coordinates, p. 3434), and Henry and Lane (close-coupling, p. 3434).

The total cross-section for the $J = 1 \to 3$ transition, measured over the energy range from 0·3 to 10 eV, rises to a maximum of $1 \cdot 1 \times 10^{-16}$ cm² near 4 eV after which it falls steadily to a little less than $0 \cdot 8 \times 10^{-16}$ cm² at 10 eV. The agreement with the close-coupling calculations of Henry and Lane is very good up to the maximum but at higher energies the theoretical results fall below the observed. Hara's theoretical values fall about 10 per cent below the observed near the maximum but are much closer near 10 eV. The evidence strongly suggests that either theoretical approach gives good results.

New data on the cross-sections for excitation of the $J = 0 \to 2$ transitions at energies below 0·5 eV is forthcoming from analysis[†] (see p. 3423) of the precise measurements of swarm data for electrons in para H_2, made by Crompton and McIntosh.[‡] The derived cross-sections in 10^{-16} cm² are 0·027, 0·074, 0·099, 0·120, 0·160, 0·210, and 0·263 at energies of 0·050, 0·10, 0·15, 0·20, 0·30, 0·40, and 0·50 eV respectively, values which are in excellent agreement with the close-coupling calculations of Henry and Lane. For comparison, the earlier calculations of the cross-section, by Gerjuoy and Stein, using Born's approximation (Chap. 11, p. 749), gave a nearly constant value of $0 \cdot 06 \times 10^{-17}$ cm² over the whole range from 0·01 to 0·50 eV.

Gibson[§] has used the new data on the $J = 0 \to 2$ transitions to re-analyse earlier swarm data in normal hydrogen, particularly with the aim of determining more accurately the cross-sections for $J = 1 \to 3$ transitions in the energy range up to 0·5 eV. The resulting values for this cross-section again agree very well with the close-coupling calculations of Henry and Lane. Comparison with the results of Linder and Schmidt is not very useful as their data only extended down to 0·3 eV, but their values at 0·3 and 0·5 eV are a little larger than those derived by Gibson.

[†] CROMPTON, R. W., GIBSON, D. K., and McINTOSH, A. J., *Aust. J. Phys.* **22** (1969) 715.
[‡] CROMPTON, R. W. and McINTOSH, A. J., ibid. **21** (1968) 637.
[§] GIBSON, D. K., ibid. **23** (1970) 683.

The cross-sections for the 0 → 2 and 1 → 3 transitions, recently derived, differ appreciably from those used by Frost and Phelps and by Engelhardt and Phelps in their analyses of earlier data described in Chapter 11, Fig. 11.24.† Below 0·15 eV the new values are somewhat larger (by about 15 per cent at 0·1 eV).

3.2. *Vibrational excitation (with and without rotational excitation) for* H_2

The experiments of Linder and Schmidt also provide much new information about cross-sections for excitation of vibration (with and without rotation). Differential cross-sections have been measured over the energy range 2·5 to 10 eV for the pure vibrational transitions $v = 0 \to 1$ and for those which also involve the rotational transition $J = 1 \to 3$. The latter are very nearly the same as for the $J = 1 \to 3$ transitions with no accompanying vibrational excitation (see above) but for the pure vibrational transition the distribution is strongly peaked in the forward direction with a minimum near 90°. Preliminary close-coupling calculations at 4·42 eV by Henry, in which vibrational as well as rotational motion is included, give generally similar results but are less strongly peaked in the forward direction. In general terms the shape of the distribution is what would be expected if the vibrational excitation occurred mainly through p_u scattering.

The total cross-sections for excitation of the $v = 0 \to 1$ ($\Delta J = 0$) transition rises from threshold to a maximum a little greater than $0·5 \times 10^{-16}$ cm², near 3 eV, and then falls quite rapidly to a little over $0·1 \times 10^{-16}$ cm² at 10 eV. For the $v = 0 \to 1$, $J = 1 \to 3$ transition, the cross-section is similar in shape but about one-half as large.

Remarkably good agreement with these results has been obtained by Faisal and Temkin with their semi-empirical adiabatic theory (see p. 3435) taking the adjustable scaling factor $g(E)$ to vary smoothly from 3·7 at 1·22 eV to 1·2 at 8·70 eV. The agreement with the observed angular distribution is much better than with the close-coupling results and is also very satisfactory for the total cross-sections. More elaborate calculations, of a less empirical character, by this method are clearly called for.

New measurements of the cross-section for $v = 0 \to 1$ transitions in H_2 at energies close to the threshold have been carried out by Burrow and Schulz‡ using the trapped electron method (see Chap. 5, p. 284)

---

† The cross-sections Q_0^2 and Q_1^3 shown in Fig. 11.24 are the assumed cross-sections multiplied by the fractional population of the initial state at 300 °K in each case.
‡ BURROW, P. D. and SCHULZ, G. J., *Phys. Rev.* **187** (1969) 97.

adapted for use at low impact energies. This has been achieved by reducing the background due to collection of elastically scattered electrons. As in the experiments of Linder and Schmidt, the cross-section rises quite rapidly from the threshold and does not remain very small until an energy of 1·0 eV is reached as in the earlier experiments of Schulz (see Chap. 11, p. 744). The observed rate of rise of the cross-section for excitation of the $v = 1, 2, 3$, and 4 levels is, in cm^2 eV^{-1}, $4 \cdot 3 \times 10^{-17}$, $7 \cdot 2 \times 10^{-19}$, 4×10^{-20}, and 3×10^{-21} respectively. For D_2 the corresponding values for $v = 1$ and 2 are $2 \cdot 0 \times 10^{-17}$ and $3 \cdot 5 \times 10^{-19}$.

Crompton, Gibson, and Robertson† have attempted to relate data obtained from beam and swarm experiments about the $v = 0 \to 1$ vibrational excitation. They find quite large discrepancies at energies near threshold and are unable to reconcile the large rate of rise observed by Burrow and Schulz‡ with either the swarm data or the beam observations of Ehrhardt et al.§

Observations at higher energies are of interest in connection with the relative role of sharp ('Feshbach') resonances and of direct scattering in producing vibrational excitation. Linder and Schmidt‖ measured differential cross-sections for excitation of the $v = 1$ and $v = 4$ levels of $H_2(^1\Sigma_g^+)$ as a function of impact energy at fixed angles of scattering. They found sharp peaks at the resonance energies 11·62 and 11·3 eV (see Table 10 A.1). A comparison was then made between the energy loss spectra near the $v = 4$ excitation at the resonance energy 11·62 eV and near the $v = 1$ excitation at an energy 10·80 eV where resonance effects are unimportant. A striking difference was found in that the former spectra showed a sharp peak at the $v = 1$ loss but with very little accompanying rotational excitation. This was, however, quite strong at 10·80 eV.

The principal cause of this difference is that the 11·62 eV resonance arises by capture of a σ_g electron into the $X\ ^1\Sigma_g^+$ ground state of H_2. The σ_g orbital is predominantly of s character and will not contribute to rotational excitation. Direct excitation of the $v = 1$ level involves electrons in a number of angular momentum states which can contribute to this rotational transition. However, some rotational excitation is expected in the resonance case as the σ_g orbital is not of pure s character. This was established both by observation of the angular distribution of the excitation cross-section and by the observation of weak rotational excitation at the resonance energy.

† CROMPTON, R. W., GIBSON, D. K., and ROBERTSON, A. G., Phys. Rev. A**2** (1970) 1386.
‡ BURROW, P. D. and SCHULZ, D. J., ibid. **187** (1969) 97.
§ loc. cit., p. 3437.   ‖ loc. cit., p. 3423.

A detailed experimental study of the differential cross-sections for excitation of the vibrational levels with $v = 1, 2,$ and 3 by electrons in the energy range 7–81·6 eV and angular range 10–80° has been made by Trajmar, Truhlar, Rice, and Kupperman.†

3.3. Dependence on density of the drift velocity of electrons in normal and para H_2 and D_2 at 77 °K

The drift velocity u of electrons in a gas at a fixed temperature T is a function of the ratio of field strength F to gas concentration N but for fixed F/N should be independent of N. This depends on the assumption that N is not so large that many-body collisions become important. However, the improved precision of drift velocity measurements, thanks largely to Crompton and his group, has revealed a small but real dependence on N in certain cases at pressures p of less than one atmosphere.

The first evidence came from the experiments of Lowke‡ in N_2 at 77 °K which showed a decrease in u of up to 3 per cent as p increased from 50 to 500 torr. Somewhat later, changes of up to 1 per cent in para H_2§ and in D_2‖ and about 0·5 per cent in normal H_2‖ were observed over a range of p from 50 to 500 torr.

According to these results the drift velocity u at a concentration N is related to that, u_0, in the limit of vanishing N by

$$u(N) = u_0/(1+\alpha N) \qquad (7)$$

where α depends on F/N, T, and the nature of the gas. Since the effects are small,
$$u(N) \simeq u_0(1-\alpha N). \qquad (8)$$

Frommhold†† suggested that the term in αN arises through temporary capture of electrons to form autodetaching negative ion states. If the lifetime τ_a of these states is short compared with the mean free time between collisions
$$\alpha N = \nu_a \tau_a,$$
where ν_a is the frequency of formation of negative ion states and is proportional to N.

Stimulated by this suggestion, Crompton and Robertson‡‡ carried out further experiments at 77 °K in which they made precise measurements of drift velocities in (a) para H_2, (b) normal H_2, (c) normal D_2, and

---

† TRAJMAR, S., TRUHLAR, D. G., RICE, J. K., and KUPPERMAN, A., *J. chem. Phys.* **52** (1970) 4516.
‡ LOWKE, J. J., *Aust. J. Phys.* **16** (1963) 115.
§ CROMPTON, R. W., GIBSON, D. K., and MCINTOSH, A. I., ibid. **22** (1969) 715.
‖ CROMPTON, R. W., ELFORD, M. T., and MCINTOSH, A. I., ibid. **21** (1968) 43.
†† FROMMHOLD, L., *Phys. Rev.* **172** (1968) 118.
‡‡ CROMPTON, R. W. and ROBERTSON, A. G., *Aust. J. Phys.* **24** (1971), 543.

(d) helium, over a range of F/N from 8×10^{-20} to 4×10^{-18} V cm^2 for pressures ranging from 50 to 700 torr.

In terms of the characteristic energy eD/μ (see Chap. 2, p. 54) the following results were obtained. For para H_2, α in (7) initially increases rapidly with D/μ and then changes only slowly, while for normal H_2, α is zero at low D/μ and then increases rapidly to a broad maximum. The ratio of the values of α for para and normal H_2 varies with D/μ but is in general less than $\frac{1}{4}$. For D_2, α is large for low D/μ and decreases as D/μ increases. No effect was found for He.

If it is assumed that Frommhold's explanation is correct and the negative ion states result from attachment of an electron to one or more rotationally excited molecules, the results for para H_2 require a resonance associated with the states with $J = 2$, while those for normal H_2 require a further resonance to be associated with $J = 3$; otherwise the α value would be exactly four times that for para H_2. Finally, the observation of an effect for D_2, large at low D/μ, could be taken to reflect the fact that at 77° the mean electron energy is already close to that for exciting the $J = 2$ transition even in the absence of any electric field. A resonance effect associated with the $J = 2$ state would be well developed at low D/μ in contrast to H_2 for which the $J = 2$ threshold is somewhat higher.

Although these results do provide circumstantial evidence of the validity of Frommhold's suggestion, there remains the very great difficulty of understanding how any resonance effects can be associated with H_2 and/or D_2 in their ground electronic and vibrational states simply due to the existence of rotational excitation. It is well known that at distances near the equilibrium separation no low-lying state of H_2^- exists. Even the so-called resonance state lying about 2·3 eV above the ground state of H_2 is not really such a state but only the energy at which the time spent by an electron in a $2p\sigma$ scattering state in the neighbourhood of the molecule is a maximum (see p. 3427). How the existence of one or two quanta of rotation could change this position so dramatically as to produce a resonance state at very low energy defies imagination.

3.4. *Oxygen*

In Chapter 10, Fig. 10.37, p. 723, the results of observations made by Boness and Hasted of the transmission of electrons through O_2 as a function of electron energy between 0·2 and 1·2 eV are illustrated. A fairly regular fine structure is observed which was interpreted tentatively as arising from resonance effects associated with different vibrational levels of a compound state.

The validity of this feature has since been well established experimentally. Linder and Schmidt,† using the same equipment as in their experiments in H_2 described above, were able to resolve separately elastic collisions and collisions leading to excitation of vibrational states with $v = 1, 2, 3,$ and 4. At a fixed angle of scattering and impact energies between 0·2 and 1·7 eV the intensities of these different scattering processes all exhibited a number of sharp resonance effects which occurred at the same energies irrespective of the exit channel. The energy sequence agreed quite well with that observed by Boness and Hasted in their early transmission experiments. This situation is quite different from that in N_2 for which the resonance peaks occur at different energies in the different exit channels.

The origin of this difference is almost certainly that the intermediate O_2^- complex has a lifetime long compared with the vibrational period whereas in N_2^- it is at least comparable with this period. It has long been known from experiments on three-body capture of electrons in O_2 that the potential energy curve of the ground Π_g state of O_2^- lies closely below that of O_2, the best determination of the electron affinity being that from the energy spectrum of photoelectrons from O_2^- (0·44 eV, see p. 3522). It follows that vibrational states of O_2^- which lie above the ground vibrational state of O_2 will be unstable towards autodetachment (see Chap. 13, p. 1016). The intermediate complex can therefore be regarded simply as O_2^- with sufficient excess vibrational excitation.

The abundance and precision of the experiments which have now been made to determine the locations and relative intensities of the low energy resonances in O_2 have made it possible to determine some of the properties of the ground electronic state of O_2^- and of its vibrational levels. A detailed study with this aim was first made by Boness and Schulz‡ using data on transmission,§ on the differential electron scattering,‡ and on the total scattering at large angles,‖ all of which gave results in substantial agreement. This analysis was later extended by Larkin and Hasted,†† taking into account the experiments of Linder and Schmidt referred to above.

Assuming the value of 0·44 eV for the electron affinity of O_2 the

† LINDER, F. and SCHMIDT, H., Z. Naturf. **26** (1971) 1617.
‡ BONESS, M. J. W. and SCHULZ, G. J., Phys. Rev. A**2** (1970) 2182.
§ HASTED, J. B. and AWAN, A. M., J. Phys. B (GB) **2** (1969) 367.
‖ SPENCE, D. and SCHULZ, D. J., Phys. Rev. **188** (1969) 280.
†† LARKIN, I. W. and HASTED, J. B., J. Phys. B (GB) **5** (1972) 95.

individual quantum numbers of the vibrational levels of O_2^- which give rise to the resonance peaks can be identified. Evidence from the trapped electron experiments of Spence and Schultz† suggests that the $v = 3$ vibrational level of O_2 is accidentally close to coincidence with one of the levels of O_2^- and this was used as a check on the assignment. It is then found that the spectroscopic constant ω_e for O_2^- is 1089 cm^{-1} (0·135 eV) with the anharmonicity $\omega_e x_e \simeq 10$ cm^{-1}. To determine the equilibrium separation Larkin and Hasted‡ used the formula (41) of Chapter 12, p. 841, which gives the relative intensity of a particular transition $v \to v'$ via a particular vibrational level n of the collision complex in terms of

TABLE 11 A.2

Resonance energies (eV) and energy-integrated cross-sections (10^{-20} cm^2 eV) for excitation of different vibrational transitions (0–v) from the ground state of O_2 by electron impact

Resonance energy (eV)	0·082	0·207	0·330	0·450	0·569	0·686	0·801	0·914	1·025	1·135	1·242	1·346	1·449
$v = 1$	—	—	25	82	110	100	61	35	17	9	5	—	—
2	—	—	—	—	8·5	25	32	28	19	12	5·8	2·4	1·1
3	—	—	—	—	—	—	1·3	5·5	7·3	7	5·8	3·8	1·8
4	—	—	—	—	—	—	—	—	—	1	1·9	2	1·7

the square of the product of the Franck–Condon overlap factors between v and n and n and v'. Larkin and Hasted determined the equilibrium separation r_e in O_2^- by assuming Morse type potential energy curves for O_2 and O_2^- and adjusting r_e to give the best fit with the relative intensities for the different resonance peaks observed by Linder and Schmidt for the $v = 0 \to 1$ transition. They found $r_e = 1\cdot39$ Å, somewhat larger than that determined from photodetachment studies (see p. 3522).

It appears that the vibrational levels of O_2^- up to $n = 3$ inclusive lie below the ground vibrational level of O_2. Linder and Schmidt, whose absolute energy calibration is correct to ± 20 meV, confirmed that the $v = 3$ level of O_2 coincides closely with a level ($v = 8$) of O_2^-. Further comments on the relation of the conclusions to the rate and interpretation of three-body attachment in O_2 are made on p. 3469.

Returning to vibrational excitation in O_2, Linder and Schmidt have measured cross-sections for different transitions integrated over each separate resonance. These are somewhat smaller than the values

---

† SPENCE, D. and SCHULTZ, D. J., loc. cit. p. 3443.
‡ LARKIN, I. W. and HASTED, J. B., *J. Phys. B (GB)* **5** (1972) 95.

assumed by Hake and Phelps in their analysis of swarm experiments in O_2 (see Chap. 11, p. 788).

Linder and Schmidt also measured the angular distribution of electrons scattered after producing resonance excitation and find only a small change in intensity over the observed angular range from 20 to 110°. They also were able to analyse the rotational structure of the resonances.

3.5. *Nitric oxide*

The transmission data for electrons in NO at energies between 0·4 and 1·5 eV obtained by Boness and Hasted (see Chap. 10, Fig. 10.37, p. 723) show resonance features very similar to those observed for oxygen, although somewhat less sharp. Later observations have confirmed and extended these conclusions. Spence and Schulz† have carried out a detailed study of these resonances both in elastic scattering and in vibrational excitation, using the trapped-electron method (see Chap. 5, p. 284 and Chap. 11, p. 745). Good agreement as to the location of the resonances is obtained with the early transmission measurements (allowing for some uncertainty in absolute energy calibration in the early work) and later work, using either transmission‡ or differential electron scattering.§

The behaviour is generally similar to that observed with O_2 in that the resonance peaks appear at the same energies in the different exit channels but the widths are certainly broader than O_2. It is also found that the resonance widths increase with the electron energy. The spectroscopic constants were found to be

$$\omega_e = 1371 \text{ cm}^{-1} \,(0\cdot170 \text{ eV}), \qquad \omega_e x_e = 0\cdot01 \text{ eV}.$$

With these values the ground vibrational level of NO lies 0·054 eV below that of NO. r_e was estimated as 1·286 Å using Badger's rule which provides a relation between ω_e and r_e involving empirical constants characteristic of the rows of the periodic table to which the individual atoms belong. Photodetachment studies give 1·258 Å (see p. 3522).

In contrast to O_2^- only the ground vibrational level lies below the ground level of NO.

Absolute values were found for the vibrational excitation cross-sections. The strongest peak rises to 10^{-17} cm² for the $v = 0 \to 1$ transition near 0·45 eV while near 0·8 eV the $0 \to 2$ and $0 \to 3$ cross-sections are near 8×10^{-18} and $5\cdot5 \times 10^{-18}$ cm² respectively.

† SPENCE, D. and SCHULZ, G. J. *Phys. Rev.* **A3** (1971) 1968.
‡ BONESS, M. J. W., HASTED, J. B., and LARKIN, I. W., *Proc. R. Soc.* **A305** (1968) 493.
§ EHRHARDT, H. and WILLMANN, K., *Z. Phys.* **204** (1967) 462.

3.6. *Nitrogen*

Spence, Mauer, and Schulz† have made further observations, by the trapped-electron method (see Chap. 11, p. 732) of the total inelastic cross-sections for vibrational and rotational excitation of N_2 by impact with electrons with energies in the range 2·5 to 3·8 eV where resonance structure appears. Using the angular distributions measured by Ehrhardt and Willmann (see Chap. 11, p. 741) correction was made for increase in path length caused by elastic scattering. At 2·5 eV the total inelastic cross-section is found to be $6·0 \times 10^{-16}$ cm², which agrees with that measured by Engelhardt, Phelps, and Fisk in their analysis of swarm data in N_2 (Chap. 11, Fig. 11.31, p. 778). Burrow and Schulz,‡ also using the trapped-electron method, have measured the rate of rise from threshold of the cross-sections for excitation of vibrational levels with $v = 1$ and $v = 2$, obtaining the respective values 2·5 and 0·02 in 10^{-18} cm² eV⁻¹.

Birtwistle and Herzenberg§ have developed further the theory of vibrational excitation of N_2 via a compound state of N_2^- which has a lifetime comparable with the vibrational period. Allowance was made for the variation of the lifetime of the state with nuclear separation. By suitable adjustment of the parameters involved quite good reproduction of the observed variation with electron energy of the cross-sections for excitation of different vibrational transitions was obtained. The physical significance of the individual maxima in these cross-sections is still not very clear.

Comer and Read‖ have measured cross-sections for excitation of vibration in N_2 by electrons with energies in the range 11–13 eV, the energy resolution of their equipment, which worked with crossed electron and molecular beams, being about 0·040 eV. Resonance peaks were observed at 11·48, 11·755, and 12·02 eV with the same relative intensity at each angle of scattering for excitation of the levels with $v = 1, 2,$ and 3 as well as elastic scattering (see p. 3430). Assuming the resonance energies to correspond to different vibrational levels of the N_2^- complex the energy of these levels was found to be given by

$$E = E_0 + a(v+\tfrac{1}{2}) - b(v+\tfrac{1}{2})^2,$$

with $E_0 = 11·345$ eV, $a = 0·270 \pm 0·02$ eV, $b = 0·002 \pm 0·002$ eV. By comparison of the resonance elastic scattering with the background

---

† SPENCE, D., MAUER, J. L., and SCHULZ, G. J., *J. chem. Phys.* **57** (1972) 5516.
‡ loc. cit., p. 3440.
§ BIRTWISTLE, D. T. and HERZENBERG, A., *J. Phys. B (GB)* **4** (1971) 53.
‖ COMER, J. and READ, F. H., ibid. **4** (1971) 1055.

potential scattering (see Chap. 9, p. 600) the width at half-height of the resonance at 11·48 eV was found to be 6×10^{-4} eV.

Pavlovic, Boness, Herzenberg, and Schulz[†] have observed vibrational excitation cross-sections for electrons of energy from 20 to 28 eV. Relative differential cross-sections were measured for excitation of levels with $v = 1, 2$, and 3. They all showed greater variability with angle than those for elastic scattering which fell monotonically with angle between 30 and 90°. Measurement of the variation of the differential cross-sections with impact energy at fixed angles of scattering revealed maxima near 22 eV at all angles and final vibrational quantum numbers. This must be taken as evidence of a compound N_2^- state with energy considerably greater than that of N_2^+.

3.7. Carbon monoxide

Burrow and Schulz,[‡] using the trapped-electron method, find that the cross-sections for excitation of the $v = 1$ and $v = 2$ levels of CO increase from threshold as 120 and $2 \cdot 4 \times 10^{-18}$ cm² eV⁻¹ respectively, considerably larger than for the corresponding levels for N_2.

Comer and Read,[§] applying the same technique as for N_2 described above, observed excitation of the $v = 1$ vibrational level by electrons with energies ranging from 9·5 to 11·2 eV. Resonant production was observed at $10 \cdot 02 \pm 0 \cdot 03$ eV with natural width $0 \cdot 045 \pm 0 \cdot 005$ eV in both inelastic and elastic channels.

3.8. Carbon dioxide

Spence, Mauer, and Schulz[||] have re-measured the cross-sections for inelastic collisions of electrons of 3·8 eV energy in CO_2 and find a value $8 \cdot 3 \times 10^{-16}$ cm², much larger than anticipated (see Chap. 11, pp. 791–2).

[†] Pavlovic, Z., Boness, M. J. W., Herzenberg, A., and Schulz, G. J., *Phys. Rev.* A**6** (1972) 676.
[‡] loc. cit., p. 3440.
[§] Comer, J. and Read, F. H., *J. Phys. B (GB)* **4** (1971) 1678.
[||] loc. cit., p. 3446.

12

COLLISIONS OF ELECTRONS WITH MOLECULES—ELECTRONIC EXCITATION, IONIZATION, AND ATTACHMENT—GENERAL THEORETICAL CONSIDERATIONS AND EXPERIMENTAL METHODS

1. Time-of-flight measurements on metastable atoms and molecules—translation spectroscopy

THE importance of determining the energy distribution of charged fragments resulting from ionization or attachment reactions of electrons with neutral molecules is discussed in Chapter 12, § 3 while in § 7 of the same chapter an account is given of the experimental methods used to measure these energy distributions. It is of equal importance to be able to measure the energies of excited neutral products resulting from electron impact. This may be done by time-of-flight techniques, provided the lifetime of the excited state is as high as 10^{-5} s. In the last few years a number of experiments have been carried out† on these lines to study, not only atoms and molecules in metastable states as usually understood, but also atoms in highly excited (Rydberg) states. The radiative lifetime of an atom in an excited state with total quantum number n is proportional to n^3 and so is between 10^{-5} and 10^{-4} s for n around 20.

The technique involved is to bombard a molecular beam with a pulsed electron beam crossing it at 90°. The long-lived excited fragments pass along with the molecular beam to impinge on a suitable detector. Electronic equipment is arranged to record the signal as a function of time since the end of the exciting pulse. The usual form of detector depends on the emission of electrons from a suitable surface due to collisions of the second kind with the excited atoms or molecules. The emitted electron current is then amplified in a suitable multiplier. For this technique to be effective the excitation energy must exceed the

† LEVENTHAL, M., ROBISCOE, R. T., and LEA, K. R., *Phys. Rev.* **158** (1967) 49; FREUND, R. S. and KLEMPERER, W., *J. chem. Phys.* **47** (1967) 2897; CLAMPITT, R. and NEWTON, A. S., ibid. **50** (1969) 1997; CLAMPITT, R., *Phys. Lett.* **28A** (1969) 581; FREUND, R. S., *J. chem. Phys.* **54** (1971) 3125; BORST, W. L. and ZIPF, E. C., *Phys. Rev.* **A4** (1971) 153.

work function of the surface. A caesium surface, deposited by evaporation *in vacuo*, has a low work function but is difficult to keep clean. It is an advantage to use less sensitive surfaces such as Be–Cu if it is not required to detect fragments with excitation energy less than, say, 4 eV.

Atoms A in highly excited Rydberg states may be distinguished from the usual metastable atoms by taking advantage of the small amount of energy required to release an electron from them. One effective way is to ionize the atoms near the surface of a metal, an effect first observed by Latypov, Kupriyanov, and Tunitskii.† A second‡ is to transfer the electron to a molecule with high electron affinity so as to produce a negative ion, e.g.
$$A + SF_6 \rightarrow A^+ + SF_6^-,$$
a process which will have a high cross-section. Other methods depending on ionization by application of an electric field or in collision with gas molecules may be used.

Having observed the time-of-flight spectrum, excitation functions for fragments associated with particular features of the spectrum may be obtained. Appearance potentials may then be used to help in the identification of the fragments involved.

Applications of these techniques are described on pp. 3463, 3466. Similar techniques may be applied to study long-lived neutral fragments arising from photon interactions with molecules (see p. 3503).

2. Angular distribution of products of molecular dissociation
(see Chap. 12, § 4.2, p. 830)

O'Malley and Taylor§ have discussed in detail the angular distribution of the products of dissociative attachment reactions in terms of a resonance formation.

Under conditions in which the rotational energy spacing is small compared with the kinetic energy of the ions produced, or with the width of the resonance associated with the capture, they find that the differential cross-section for production of ions moving in the direction (θ, ϕ) with respect to the direction of the incident electrons is given by

$$I(\theta, \phi)\, d\omega = \frac{4\pi^3}{k^2} e^{-\rho} \left| \sum_{L=|\mu|}^{\infty} a_{L|\mu|}(k) Y_{L\mu}(\theta, \phi) \right|^2 d\omega, \tag{1}$$

where k is the wave number of the incident electron, the $a_{L|\mu|}$ are matrix

† LATYPOV, Z. Z., KUPRIYANOV, S. E., and TUNITSKII, N. N., *Zh. eksp. teor. Fiz.* **46** (1964) 883 (*Soviet Phys. JETP* **19** (1964) 570).
‡ FREUND, R. S., loc. cit., p. 3448.
§ O'MALLEY, T. F. and TAYLOR, H. S., *Phys. Rev.* **176** (1968) 207. See also O'MALLEY, T. F., ibid. **150** (1966) 14.

elements of an effective interaction energy averaged over the initial and final 'vibrational' wave-functions (the latter corresponding to an unbound state) and $e^{-\rho}$ is a survival factor against autodetachment. The suffix μ is restricted so

$$\mu = \Lambda_r - |\Lambda_i| \tag{2}$$

where Λ_r, Λ_i are the quantum numbers specifying the component of electronic orbital angular momentum along the nuclear axis for the resonance state and for the initial state of the target molecule respectively. There is the further restriction that, according as the electronic transition does or does not involve a change of parity, L must be odd or even respectively.

In general the series converges more rapidly the smaller the electron wave number k. Thus for sufficiently small k the distribution is proportional to $|Y_{L_0\mu_0}(\theta,\phi)|^2$ where L_0 is the lowest allowed value of L and $\mu_0 = |\Lambda_r| - |\Lambda_i|$.

These results do not conflict with the rules given by Dunn, stated on p. 832, which refer only to cases in which the nuclear axis is parallel or perpendicular to the direction of the electron beam.

Application to the experimental study of associative detachment in oxygen is described on p. 3468.

3. Determination of appearance potentials of different products in reactions involving attachment of electrons with molecules

The complications introduced into the determination of the threshold energies for particular reactions arising from electron attachment with molecules have been discussed in Chapter 12, pp. 843–5. These arise largely because of the temperature motion of the gas atoms. As a result of this motion, ions which would be produced with a single energy W_i if the target molecules were at rest, will be produced with a velocity distribution given by†

$$f(W)\,dW = (4\pi w_i)^{\frac{1}{2}}[\exp\{-(w^{\frac{1}{2}}-w_i^{\frac{1}{2}})^2\}-\exp\{-(w^{\frac{1}{2}}+w_i^{\frac{1}{2}})^2\}]\,d\omega, \tag{3}$$

where $w = W/\kappa T$, $w_i = W_i/\kappa T$. In most circumstances $W_i > \kappa T$ and

$$f(W) \simeq (4\pi w_i)^{\frac{1}{2}} \exp\{-(w^{\frac{1}{2}}-w_i^{\frac{1}{2}})^2\}, \tag{4}$$

with a maximum at $w = w_i$. It is then possible to determine w_i directly if the most probable ion energy can be measured.

This may be applied to eliminate temperature effects in determining threshold potentials for dissociative attachment reactions produced by

† CHANTRY, P. J., *Phys. Rev.* **172** (1968) 125.

highly monochromatic electron beams. For such reactions (see Chap. 12, p. 844)
$$W_i = (1-\beta)\{E_e - (D-A)\}, \tag{5}$$
where β is the ratio of the mass of the ion to that of the target molecule, E_e is the kinetic energy of the incident electron, D the dissociation energy of the molecule, and A the electron affinity of the atom or radical to which the electron attaches.

In recent experiments carried out by Chantry[†] a Wien filter is used in conjunction with a mass spectrograph to determine the energy distribution of the negative ions. The analysis is performed by requiring the ions to pass through crossed electric and magnetic fields E and B respectively. For this they must possess a velocity $v = E/B$. The energy resolution obtainable corresponds to a full width at half maximum of about 0·1 eV.

4. New methods for measuring attachment rates

Two new methods for measuring rates of attachment of electrons under swarm conditions have been introduced.

The first due to Grunberg[‡] is a modification of the pulse method described in Chapter 12, § 7.5.2. In this method electrons, released from a cathode by a pulse of radiation, drift under the influence of an electric field along the axis of a tube containing the gas under investigation, towards an anode. The attachment coefficient is then determined from the ratio of the negative ion to the electron current received at the anode. Grunberg shows how this may be done very simply. In practice the transient current flow is measured by the transient potential difference developed across a resistor R. If an integrating circuit is introduced with time constant $RC \gg T_n$, the time taken for negative ions to drift the distance d between the electrodes, the potential difference developed up to the drift time T_e ($\ll T_n$) for the electrons will be given by
$$V(T_e) = \frac{n_0 e}{C \alpha_a d}(1 - e^{-\alpha_a d}), \tag{6}$$
where n_0 is the number of electrons released in the initial pulse and α_a is the attachment coefficient. The factor $(\alpha_a d)^{-1}(1 - e^{-\alpha_a d})$ then represents the fraction of electrons which arrive unattached. At a time $T_1 > T_n$ the negative ions also will have been collected so, at these times, still $< RC$, we will simply have
$$V(T_1) = n_0 e/C \quad (T_1 > T_n), \tag{7}$$

[†] CHANTRY, P. J., *Phys. Rev.* **172** (1968) 125.
[‡] GRUNBERG, R., *Z. Naturf.* **23A** (1968) 1994.

all the charge in the pulse now having passed to the anode. The ratio
$$V(T_e)/V(T_1) = (1-e^{-\alpha_a d})/\alpha_a d, \tag{8}$$
is very simply and accurately measured on an oscilloscope. Thus the appearance of the oscillogram is a sharp spike of height proportional to $V(T_e)$ from which the voltage grows to a flat plateau of height proportional to $V(T_1)$. It is, of course, assumed that conditions are such that neither ionization nor detachment is significant.

With this method it is unnecessary to restrict the duration of the light pulse to a time much less than T_e. This is an advantage if attachment is rapid, as n_0 must then be large to give a signal for $V(T_e)$ well above amplified noise. Attachment rates in O_2 measured by the method are discussed on p. 3469.

The second method involves a novel use of a static afterglow. In Chapter 12, § 7.5.5 the determination of attachment coefficients by measuring the decay time of the electron concentration in an afterglow containing the attaching gas is described. Puckett, Kregel, and Teague[†] concentrate instead on observing with high time-resolution the positive and negative ion currents which diffuse through a small aperture into a mass spectrometer. The relation of the time variation of these currents to the rates of electron and ion loss processes can be described as follows.

Very soon after the cessation of the ionizing pulse which may be, as in the experiments of Puckett et al., a pulse of ionizing radiation, the electrons will be free and loss will occur by ambipolar diffusion to the walls of the cavity in the usual way. During this phase, as usual, a field will build up across the plasma in a sense to retard electron motion towards the walls. Because of this, negative ions will be confined within the plasma. However, as electrons are lost by attachment a situation will be reached when electron loss to the walls no longer controls the distribution of electric field across the plasma. This breaks down and negative ions begin to reach the walls, so beginning a new phase in which both positive and negative ions are lost by diffusion. The onset of this phase will be apparent in a relatively sudden commencement of negative ion reception by the mass analyser. If n^+, n^-, n_e are the concentrations of positive and negative ions and of electrons respectively, we have
$$n^+ = n^- + n_e \tag{9}$$
and, in the first phase,
$$\frac{dn^+}{dt} = -\nu_d n^+, \tag{10}$$
where ν_d is the diffusion frequency. Assuming that only the fundamental

[†] PUCKETT, L. J., KREGEL, M. D., and TEAGUE, M. W., Phys. Rev. A4 (1971) 1659.

diffusion mode is involved $\nu_d = \Lambda^2/D_a$, where Λ is the fundamental diffusion length and D_a the ambipolar diffusion coefficient (see p. 1952).

If detachment is negligible we have, in the same phase,

$$\frac{dn^-}{dt} = \nu_a n_e, \tag{11}$$

where ν_a is the attachment frequency.

It follows from (9), (10), and (11) that, in the first phase,

$$\frac{dn^-}{dt} + \nu_a n^- = \nu_a n^+(0)e^{-\nu_d t}, \tag{12}$$

$n^+(0)$ being the initial concentration of positive ions. This gives

$$n^- = n^+\{\nu_a/(\nu_a-\nu_d)\}\{1-\exp(\nu_d-\nu_a)t\}. \tag{13}$$

At a time T such that all electrons have effectively disappeared, just before loss of negative ions to the walls becomes significant, we have, substituting $n^+ = n^-$ in (13),

$$T(\nu_a-\nu_d) = \ln(\nu_a/\nu_d). \tag{14}$$

Since ν_d may be determined by the usual techniques it follows that if T may be measured ν_a may be obtained.

By way of illustration Puckett et al. show observed time variations of positive and negative ion wall currents in an afterglow in NO at a pressure of 5×10^{-2} torr, resulting from ionization by an 8 ms pulse of Kr resonance radiation. Up to a time of 170 ms in the afterglow positive ion currents of NO^+ and $(NO)_2^+$ were observed (cf. Chap. 20, § 4.7) which varied smoothly with time in the usual way. During this period only much smaller and very variable negative ion currents of NO_2^- were observed but, just after 170 ms, this current suddenly increased by at least two orders of magnitude after which it decayed very smoothly. This was accompanied by smaller, rapid, downward changes in the positive ion currents which thereafter were much less regular, though showing a gradual mean fall with increasing time. It is natural to suppose that for this case T in (14) may be taken as 170 ms.

Application of this method to study attachment in NO and NO, NO_2 mixtures is discussed on p. 3471 (see also p. 3639).

5. Spurious peaks in electron attachment observations

Chantry[†] has drawn attention to the possibility that peaks may be produced, in certain circumstances, in observed rates of dissociative

[†] CHANTRY, P. J., J. chem. Phys. 55 (1971) 1851.

attachment as functions of electron energy due to processes of the type

$$e(E) + XY \rightarrow XY^* + e(E - E_{ex}), \quad (15\,a)$$

$$e(E - E_{ex}) + XY \rightarrow X + Y^-. \quad (15\,b)$$

where $e(E)$ denotes an electron of energy E.

The first excitation process has the effect of reducing the energy of an electron to a value which is close to one E_r at which the dissociative attachment rate in (15 b) is a maximum. The net effect of the two reactions will be to produce negative ions with maximum probability when the electron energy $E = E_{ex} + E_r$. Such a maximum is spurious in the sense that it does not arise in single collisions of electrons with the attaching molecules.

Spurious maxima of this kind can be identified by the energy relationship and by the fact that they vary as the square of the gas pressure.

Further evidence was obtained by Chantry who introduced an electrode between the collision chamber and electron detector in a typical experiment on negative ion production. If E and E_s ($= E - E_x$) are the energies of an incident electron and of an electron after suffering the inelastic collision then, if the potential on the new electrode relative to the collision chamber is greater than $-E_s$ no effect should be observed on the negative ion current i^- collected. However, when it becomes less than $-E_s$ the current of secondary electrons will be reflected back into the collision chamber, thereby increasing i^-. Finally, when it becomes less than $-E$ the primary beam will also be reflected back, thereby increasing i^- still further. Although in practice the rather sudden changes expected in i^- are smoothed out they are nevertheless observable. Their occurrence not only is evidence that a particular peak is spurious but also through the rough determination of E_s gives some indication of the nature of the excitation process involved. An example is referred to on p. 3467.

13

COLLISIONS OF ELECTRONS WITH MOLECULES—ELECTRONIC EXCITATION, IONIZATION, AND ATTACHMENT— APPLICATIONS TO SPECIFIC MOLECULES

1. Hydrogen and deuterium

1.1. *Electronic excitation*

GEIGER and his colleagues have continued to carry out high-resolution energy loss measurements for electrons scattered through very small angles in H_2. The experiments are of such precision that it is necessary, in comparing the relative intensities of excitation of different vibrational levels associated with a given excited electronic state, to take account of the variation of the amplitude for the electronic transition with the nuclear separation.

Miller and Krauss† have calculated the absolute intensities for different vibrational transitions associated with excitation of the B, B', C, D, and D' electronic states by electrons of incident energy 300 eV scattered in the forward direction. They used Born's approximation and the full expression (11) of Chapter 12, p. 828 which allows for the variation of the amplitude for the electronic transition with nuclear separation. Approximate Hartree–Fock functions built up of gaussian-type atomic orbitals were used for the molecular states.

Ford and Browne‡ carried out similar calculations for the B and C states but for electrons of 34 keV energy. Quite elaborate wave-functions were used for the molecular states. Their results agreed well with those of Miller and Krauss and with the experimental results of Geiger and Schmoranzer§ who observed energy loss spectra of 34 keV electrons in H_2, HD, and D_2 with an energy resolution of about 10 meV.

Ford and Browne‡ also carried out similar calculations for excitation of the E and F electronic states which both possess the same parity as the ground state. The transitions are therefore optically forbidden in the dipole approximation but it was found that, associated with angles of scattering greater than 0·50°, they should be of comparable intensity

---

† MILLER, K. J. and KRAUSS, M., *J. chem. Phys.* **47** (1967) 3754.
‡ FORD, L. and BROWNE, J. C., ibid. **55** (1971) 1053.
§ GEIGER, J. and SCHMORANZER, H., *J. mol. Spectrosc.* **32** (1969) 39.

to the corresponding transitions to the B and C states. In carrying out these calculations it was necessary to use very elaborate approximate wave-functions in order to satisfy the orthogonality conditions. A 40 configuration function was finally used. The lowest root of the secular equation gave the ground state and the next lowest the E and F states.

Cartwright and Kuppermann† have applied the Ochkur and Ochkur–Rudge approximations to calculate cross-sections for excitation of the first $(b\,^3\Sigma_u^+)$ and second $(a\,^3\Sigma_g^+)$ triplet states, processes which lead to dissociation of the molecule into normal atoms (see Chap. 13, § 1.2). The theoretical basis for using these approximations down to electron energies not far above the threshold is very uncertain.

A number of different approximate wave-functions for the initial and final states were used. Results obtained using a Weinbaum function for the initial state and two-parameter functions for the excited states gave results which are comparable with the experimental measurements of Corrigan (see Chap. 13, p. 885). It is difficult to judge from the uncertain data near the peak whether the Ochkur or Ochkur–Rudge approximations gives the better results.

A number of experiments have been carried out which measure the intensity of the Ly α line excited by electron-impact dissociation of H_2. One of the difficulties in this work is to distinguish the Ly α radiation from radiation at neighbouring wavelengths due to excitation of molecular bands. In their experiments on the excitation of Ly α by electron impact on H atoms (see Chap. 4, p. 194), Fite and Brackmann used as detector an iodine-filled Geiger counter in front of which was placed a gas filter consisting of a cell with LiF windows through which was passed a stream of O_2. In the course of their experiments they also observed the signal in this detector due to Ly α from molecular dissociation, plus an unknown contribution from molecular ultraviolet emission. They therefore referred to this signal as due to countable molecular ultraviolet radiation (CUV). In a later experiment Kauppila, Teubner, Fite, and Girnius‡ using the known cross-section for production of Ly α from e–H impact (see p. 3368) have remeasured the effective cross-section for production of CUV by 100 eV electrons as $1{\cdot}48 \pm 0{\cdot}05 \times 10^{-17}$ cm^2.

Estimates of the percentage of CUV which can be ascribed to molecular excitation have since been made by Carriere and de Heer§ who analysed

---

† CARTWRIGHT, D. C. and KUPPERMANN, A., *Phys. Rev.* **163** (1967) 86.

‡ KAUPPILA, W. E., TEUBNER, P. T. O., FITE, W. L., and GIRNIUS, R. J., *J. chem. Phys.* **55** (1971) 1670.

§ CARRIERE, J. D. and DE HEER, T. J., Private communication to MUMMA, M. J. and ZIPF, E. C. (see note ‡, next page).

with a monochromator the photons arising from impact excitation by 100 eV electrons which were transmitted through the LiF–O_2 filter. They estimated that about 20 per cent of the radiation arose from molecular bonds. A comparable figure was arrived at by McGowan and Williams.† Assigning 80 per cent of the CUV to Ly α the cross-section for production of Ly α by 100 eV electrons comes out to be $1 \cdot 2 \times 10^{-17}$ cm².

A number of measurements‡ of the variation with electron energy of the cross-section for production of CUV have been made, covering a range from threshold (near 14 eV) to 500 eV. Quite good agreement is obtained between the results obtained by different observers.

The cross-section for production of H(2s) atoms by electron impact from H_2 may be measured by quenching the H(2s) through conversion to H(2p), with subsequent Ly α emission. Vroom and de Heer‡ used an electrostatic field for quenching and this confined their measurement to electron energies greater than 50 eV in order that the quenching field should not unduly disturb the electron motion. Cox and Smith‡ used a radio-frequency field tuned to a suitable frequency to produce transitions between the $^2S_{\frac{1}{2}}$ and $^2P_{\frac{1}{2}}$ states. This is much more readily carried out for D than for H. This is because the hyperfine splittings are so much smaller for D that, when a suitably chosen single radio frequency of 1060 MHz is used, all the hyperfine components can be quenched without raising the power level high enough to disturb the electron motion. This technique also has the advantage of producing unpolarized Ly α radiation as only $2\ S_{\frac{1}{2}} \to 2\ ^2P_{\frac{1}{2}}$ transitions are involved, whereas electrostatic quenching gives rise to radiation 30 per cent polarized and so anisotropically distributed. Cox and Smith‡ therefore carried out their experiments with D_2 instead of H_2 and measured the cross-section for D(2s) production from threshold ($14 \cdot 8 \pm 0 \cdot 3$ eV) to 500 eV. It rises from threshold to a maximum of $3 \cdot 52 \times 10^{-18}$ cm² at $59 \cdot 38$ eV and then falls gradually at increasing energy to $1 \cdot 43 \times 10^{-18}$ cm² at 506 eV. The measurements of Vroom and de Heer converge towards the results of Cox and Smith at high electron energies but at lower energies become progressively larger so that at 50 eV they are nearly 50 per cent above.

The time-of-flight technique (see p. 3448) has been applied by Leventhal, Robiscoe, and Lea,§ by Clampitt and Newton,‖ and by Clampitt††

---

† McGowan, J. W. and Williams, J. F., *6th int. conf. on the physics of electronic and atomic collisions, Cambridge, Mass.* M.I.T. Press, 1969, p. 431.

‡ Dunn, G. H., Geballe, R., and Pretzer, D., *Phys. Rev.* **128** (1962) 2200; Vroom, D. A. and de Heer, F. J., *J. chem. Phys.* **50** (1969) 580; Mumma, M. J. and Zipf, E. C., ibid. **55** (1971) 1661; Cox, D. M. and Smith, S. J., *Phys. Rev.* A**5** (1972) 2428.

§ loc. cit., p. 3448.   ‖ loc. cit., p. 3448.   †† loc. cit., p. 3448.

to study the metastable H atoms produced by impact dissociation of H_2. The first authors also studied D atoms from D_2. These atoms emerge in two distinct groups, a slow and a fast group. The threshold for the former is $\simeq$ 15 eV and they possess a kinetic energy of around 0·3 eV, while for the latter the corresponding quantities are 25 eV and 4·7 eV respectively. Clampitt also was able to observe metastable ($C^3\Pi$) H_2 molecules (see Chap. 13, p. 880).

1.2. *Ionization*

Van Brunt and Kieffer† have repeated the observations of Dunn and Kieffer (see Chap. 13, § 1.5.2) on the angular distribution of protons and deuterons produced by dissociative ionization of H_2 and D_2. In particular they were concerned with extending the observations down to electron energies close to threshold and making allowance for momentum transfer to the heavy particles in the impact. The arrangement of the equipment was essentially similar to that of Dunn and Kieffer but with several improvements incorporated, including particularly a large increase in sensitivity due to the introduction of counting techniques.

Assuming that at threshold the electron is stopped and transfers all its momentum to the heavy particles the difference $\Delta E(\Theta)$ in the kinetic energy of the dissociation products produced with the same energy $\bar{E}$ in the C.M. system but observed at the angles Θ and $\pi - \Theta$ in the laboratory system is given by

$$\Delta E(\Theta) = 4(M_A/M_B)[(m/M_A)(1-M_A/M_{AB})\{\bar{E}^2 - E(D+E_i)\}]^{\frac{1}{2}} \cos\Theta, \quad (1)$$

M_A being the mass of the fragment observed and M_{AB} of the molecule. D and E_i are the respective dissociation and ionization energies of the molecule. E is the impact energy. In obtaining (1) the energy relation

$$\bar{E} = (1-M_A/M_{AB})\{E-(D+E_i)\} \quad (2)$$

has been used.

$\Delta E(\Theta)$ was determined experimentally by applying a small accelerating or retarding potential to the analysing spectrometer relative to the collision chamber, so as to bring the observed energy distribution measured at Θ and at $\pi - \Theta$ into coincidence. It was then verified that, near threshold, the values of ΔE so obtained agree with (1). The same technique was then applied to determine $\Delta E(\Theta)$ at higher impact energies. To obtain the angular distribution of H⁺ ions in the C.M. system for fixed E and $\bar{E}$ the measurements were made with the appropriate correcting potentials at each angle.

† VAN BRUNT, R. J. and KIEFFER, L. J., *Phys. Rev.* **A2** (1970) 1293.

Near threshold the angular distribution in the C.M. system can be regarded as the sum of an isotropic term and one with a minimum at 90°. Its interpretation is not clear, the origin of the isotropic component being difficult to understand. The anisotropic component cannot be interpreted as arising merely from dipolar terms in the transition matrix element but this is not unexpected.

2. Dissociation of H_2^+ (Chap. 13, § 2)

A series of further investigations of the dissociation of H_2^+ by electron impact (see Chap. 13, § 2.3) have been carried out by Peart and Dolder.† This work has been characterized by the attention paid to eliminating secondary effects so as to obtain accurate values of the cross-sections concerned. In particular, the possibility of spurious contributions at the higher energies from slow electrons accumulated in the interaction region is largely excluded by enclosure of the region within a box at a potential so as to repel the electrons. The H_2^+ ions are produced by electron impact from a molecular beam of H_2 and the arrangement is such that they have very little chance of undergoing collisions before reaching the interaction region. Under these conditions, provided the ionizing electrons have sufficient energy the vibrational distribution in the H_2^+ should be as given by the Franck–Condon principle (see p. 942). This has been checked from experiments on photodissociation (see p. 3496) which confirmed this expectation to about 4 per cent.

A new feature of the experiments is the use of a particle multiplier as detector for the molecular fragments. This makes possible coincidence experiments in which the cross-sections for selected processes of dissociation may be separately measured. At the same time much weaker ion beams may be used. This eases the problem of preparing a source to produce ions under clearly defined conditions and also that of distinguishing the wanted from the background signal. The greater sensitivity also makes possible experiments with inclined beams of the same general kind as those described in Chapter 24, § 1.2.4.3 for beams of heavy particles.

The processes concerned are‡

$$H_2^+ + e \to H^+ + H + e \qquad (3\,a)$$

$$\to H^+ + H^+ + 2e, \qquad (3\,b)$$

with cross-sections Q_{disc} and Q_{cont} respectively. In earlier experiments

† PEART, B. and DOLDER, K. T., *J. Phys. B (GB)* **4** (1971) 1496; **5** (1972) 860; **5** (1972) 1554.

‡ The contribution from dissociative recombination is ignored as it will be very small.

(Chap. 13, § 2.3), the total current of protons produced was measured giving a cross-section Q_p where

$$Q_p = Q_{\text{disc}} + 2Q_{\text{cont}}. \tag{4}$$

Peart and Dolder† in their first experiments counted only single protons from the process (3 b) and so measured the dissociation cross-section Q_{diss} where

$$Q_{\text{diss}} = Q_{\text{disc}} + Q_{\text{cont}}. \tag{5}$$

By counting the protons produced in coincidence with (a) H atoms‡ and (b) other protons§ they were able, in subsequent experiments to measure Q_{disc} and Q_{cont} separately.

These measurements were carried out over an energy range extending from 18·4 to 980 eV. Comparison with the observations of Dunn and van Zyl (see Chap. 13, Fig. 13.45), who measured Q_p under comparable conditions, shows remarkable consistency. Q_{cont} is about 10 per cent of Q_{diss} and $Q_p > Q_{\text{diss}} > Q_{\text{disc}}$ by about this amount. The observed values of Q_{disc} agree remarkably well with those calculated by Peek (see Chap. 13, p. 955), exhibiting in particular the logarithmic behaviour as a function of electron velocity given by (37) of Chapter 13, p. 956. Q_{cont} is more difficult to calculate but a semiclassical calculation by Alsmiller (see Chap. 13, p. 955) gives results which are not far from those observed by Peart and Dolder (the observed peak value of Q_{cont} is $0 \cdot 2\pi a_0^2$ at an impact energy of 110 eV).

The measurement of Q_p by Dance, Harrison, and Rundel‖ are a little higher which is probably due to the ion source, which they used, providing H_2^+ with a somewhat smaller population of highly excited vibrational states.

A remarkable feature of the comparison with Peek's theory is that there appears to be quite good agreement down to very low impact energies. Possible reasons for this are given in Chapter 13, p. 956. To investigate this further Peart and Dolder†† carried out an inclined beam experiment which enabled them to measure cross-sections at impact energies between 3·45 and 31·5 eV. There was sufficient overlap with the lower energy range of the earlier experiments to check the validity of the modified technique Evidence was obtained of departure from

---

† PEART, B. and DOLDER, K. T., *J. Phys. B (GB)* **4** (1971) 1496.
‡ PEART, B. and DOLDER, K. T., ibid. **5** (1972) 860.
§ DOLDER, K. T., private communication.
‖ Fig. 13.45 shows also preliminary measurements by Dance, Harrison, Rundel, and Smith at high electron energies which were reduced in subsequent measurements reported in their published paper. DANCE, D. F., HARRISON, M. F. A., RUNDEL, R. D., and SMITH, A. C. H., *Proc. phys. Soc.* **92** (1967) 577.
†† PEART, B. and DOLDER, K. T., *J. Phys. B (GB)* **5** (1972) 1554.

the theory in that the variation of EQ_{diss} with $\log E$, where E is the impact energy, is no longer linear but shows evidence of structure.

Caudano and Delfosse[†] have measured the energy distribution of protons produced by dissociation of H_2^+ ions by electron impact in a crossed beam experiment which was carried out at electron energies from 30 to 500 eV. The H_2^+ ions were produced from a Penning source and accelerated as a beam to an energy of 10 keV. The initial vibrational distribution is therefore not as well defined as in the experiments of Dunn et al. (see Chap. 13, p. 950) and of Dolder et al. (p. 953). The protons studied were those projected in the same direction as that of motion of the H_2^+ ions and their energy distribution was analysed with a 90° cylindrical electrostatic deflector. Analysis of the observed distribution to give that of the protons relative to the centre of mass of the H_2^+ was carried out as in the earlier work by the same authors on dissociation of H_2^+ by impact with neutral atoms (see Chap. 24, p. 2690). When this was done maxima appeared between 0·2 and 0·4 eV probably due to the $1s\sigma_g \rightarrow 2p\pi_u$ transition and one at about 2 eV corresponding to the $1s\sigma_g \rightarrow 2p\sigma_u$ transition. Other structural features were observed.

3. Nitrogen

3.1. *Excitation*

Lassettre, Skerbele, Dillon, and Ross[‡] have measured energy loss spectra for electrons in N_2 with incident energies between 33 and 100 eV and scattering angles up to 16°. They used an electron spectrometer incorporating an electrostatic lens system to compensate for chromatic aberration. Relative intensities were measured of the vibrational distributions for the $b\,^1\Pi_u \leftarrow X\,^1\Sigma_g^+$ and $a\,^1\Pi_g \leftarrow X\,^1\Sigma_g^+$ transitions, as many as 14 transitions being distinguished in the latter case. The distributions were found to be independent of scattering angle and electron energy.

A number of further measurements have been made of the excitation functions and cross-sections for excitation of electronic states of N_2. Chung and Lin[§] have calculated cross-sections for the $a\,^1\Pi_g$, $c'\,^1\Sigma_u^+$, $a''\,^1\Sigma_g^+$, $w\,^1\Delta_u$, $b'\,^1\Sigma_u^+$, and $b\,^1\Pi_u$ singlet states and for the $A\,^3\Sigma_u^+$, $B\,^3\Pi_g$, $C\,^3\Pi_u$, $D\,^3\Sigma_u^+$, $W\,^3\Delta_u$, and $E\,^3\Sigma_g^+$ triplet states of N_2, using the Born approximation to calculate the direct amplitude and the Ochkur–Rudge formulation for the exchange amplitude. The singlet cross-sections were

[†] CAUDANO, R. and DELFOSSE, J. M., *6th int. conf. on the physics of electronic and atomic collisions, Cambridge, Mass.* Abstracts, M.I.T. Press, 1969, p. 818.

[‡] LASSETTRE, E. N., SKERBELE, A., DILLON, M. A., and ROSS, K. J., *J. chem. Phys.* **48** (1968) 5066.

[§] CHUNG, S. and LIN, C. C., *Phys. Rev.* A6 (1972) 988.

calculated over the energy range from threshold to 2000 eV and the triplet from 0 to 40 eV. The calculations were greatly facilitated by expressing the molecular wave-functions in terms of Gaussian-type orbitals.† Cartwright‡ has also carried out calculations for the excitation of the triplet states, including also $B'\,^3\Sigma_g^-$, with the Ochkur–Rudge formulation as in his calculations for H_2 (see p. 3456). The same reservations apply to the use of the Ochkur–Rudge formulation as those mentioned on p. 3456.

For the $a\,^1\Pi_g$ and $b\,^1\Pi_u$ states the excitation cross-sections for energies between 100 and 500 eV derived from the observations of Lassettre and Krasnow and shown in Chapter 13, Fig. 13.54 remain the most reliable data available. Experiments§,‖ on the excitation function of the Lyman–Birge–Hopfield bands give results for $a\,^1\Pi_g$ which agree well. All are about 25 per cent smaller than those calculated by Chung and Lin. Ajello's†† recent measurements between threshold and 200 eV are nearly three times larger over the overlapping energy range.

Borst‡‡ has measured the cross-section for $a\,^1\Pi_g$ production by electrons with energies in the range 9 to 40 eV, using the same technique as that which he used to determine the corresponding cross-sections for $A\,^3\Sigma_u^+$ and $E\,^3\Sigma_g^+$ production (see below). He finds a peak cross-section of $3\cdot 85\times 10^{-17}$ cm^2 at 16 eV which is a little larger than that observed by Ajello.

For $b\,^1\Pi_u$ the calculated cross-sections are in quite good agreement with those shown in Fig. 13.54.

Very few experimental data are available with which to compare the calculations for other singlet states. For $c'\,^1\Sigma_u^+$ the optical excitation function of the $v' = 0 \to v'' = 1$ component of the transition has been measured by Aarts and de Heer‖ and agrees very well in shape with that calculated over the energy range of observation (60–1000 eV). The transition observed by Lassettre and Krasnow, the cross-section for which is shown in Fig. 13.54 as a $^1\Sigma_u^+$ state, probably applies to $b'\,^1\Sigma_u^+$. Unfortunately, owing to configuration mixing it is difficult to calculate the cross-sections but estimates made are not inconsistent with the observed results.

Turning now to the triplet states we find that there is still a wide spread

---

† IDEM, *App. Opt.* **10** (1971) 1790.
‡ CARTWRIGHT, D. C., *Phys. Rev.* **A2** (1970) 1331.
§ HOLLAND, R. F., *J. chem. Phys.* **51** (1969) 3940.
‖ AARTS, J. F. M. and DE HEER, F. J., *Physica,'s Grav.* **52** (1971) 45.
†† AJELLO, J. M., *J. chem. Phys.* **53** (1970) 1156.
‡‡ BORST, W., *Phys. Rev.* **A5** (1972) 648.

in measured values† for the states investigated. For $C\,^3\Pi_u$ measurement of apparent excitation functions for the second positive ($C\,^3\Pi_u$–$B\,^3\Pi_g$) bands have been made by a number of investigators. As cascade effects are likely to be unimportant in this case the data should give quite closely the true excitation functions. There is a wide spread in the results. The data of Jobe, Sharpton, and St. John† agree most nearly in peak cross-sections with those calculated by Chung and Lin but the observed cross-sections fall off rather faster at higher energies.

For the $B\,^3\Pi_g$ state the situation is a little more satisfactory in that the observations of apparent cross-sections (excitation functions of the first positive bands) by Stanton and St. John‡ agree quite well with those of McConkey and Simpson§ though still being in considerable disagreement with Skubenich and Zapesochnyi.† The cross-section calculated by Chung and Lin is considerably smaller at the peak (by a factor of 2) than that observed by Stanton and St. John, corrected for contributions from cascade. In this case Cartwright's calculated values are closer to the observed values.

The observed apparent cross-sections for excitation of the $D\,^3\Sigma_u^+$ state fall off very slowly with electron energy up to 40 eV, in sharp contrast to the calculated.

Data are available on the cross-sections for excitation of the metastable $A\,^3\Sigma_u^+$ and $E\,^3\Sigma_g^+$ states from the measurements of Borst. He observed the total excitation function‖ for metastable N_2 using a Cu–Be–O surface as detector (cf. p. 962). This included contributions from the $a\,^1\Pi_g$, $A\,^3\Sigma_u^+$, and $E\,^3\Sigma_g^+$ states. To disentangle these†† use was made of the measured lifetime of $a\,^1\Pi_g$ (115 μs),‖ the known very long life ($\sim$ 1 s) for $A\,^3\Sigma_u^+$, and the possibility of eliminating the contribution from $E\,^3\Sigma_g^+$ by using a long drift distance and appropriate detector surfaces.

The time-of-flight technique (see p. 3448) has been applied by Clampitt and Newton‡‡ to observe the sharp excitation function of the $E\,^3\Sigma_g^+$ state, using a detecting surface of Cs_2Sb.

For $A\,^3\Sigma_u^+$ the observed cross-section after allowance for cascade, using

---

† JOBE, J. D., SHARPTON, F. A., and ST. JOHN, R. M., *J. opt. Soc. Am.* **57** (1967) 106; BURNS, D. J., SIMPSON, F. R., and McCONKEY, J. W., *J. Phys. B (GB)* **2** (1969) 52; SKUBENICH, V. V. and ZAPESOCHNYI, I. P., 5th int. conf. on the physics of electronic and atomic collisions, Leningrad, p. 570.
‡ STANTON, P. N. and ST. JOHN, R. M., *J. opt. Soc. Am.* **59** (1969) 252.
§ McCONKEY, J. W. and SIMPSON, F. R., *J. Phys. B (GB)* **2** (1969) 923.
‖ BORST, W. L. and ZIPF, E. C., *Phys. Rev.* **A3** (1971) 979.
†† BORST, W., loc. cit., p. 3462.
‡‡ loc. cit., p. 3448.

the data of Stanton and St. John† and of Jobe, Sharpton, and St. John‡ for $C\,^3\Pi_u$ and $B\,^3\Pi_g$ excitation, has a maximum near 11 eV of

$$5\cdot 3 \genfrac{}{}{0pt}{}{+4\cdot 0}{-3\cdot 0} \times 10^{-17}\ \mathrm{cm}^2$$

and falls rapidly with increasing energy, being five times smaller at 25 eV. The theoretical cross-sections have the same general shape but are at least twice as large at the peak and fall off more slowly at higher energies.

Evidence that the excitation cross-section for the $E\,^3\Sigma_u$ state shows a sharp peak around 12 eV was discussed in Chapter 13, pp. 962–3. Borst§ has verified the existence of this peak at 12·2 eV and measured the peak cross-section as $0\cdot 7 \times 10^{-17}$ cm^2. The width, 0·4 eV, of the peak at half-height agrees well with the earlier measurements of Ehrhardt and Willman (see Chap. 13, Fig. 13.51, p. 963). No evidence was obtained of any background contribution to the cross-section.

3.2. *Dissociative excitation*

A number of measurements have been made of apparent cross-sections for collisions in which excited atoms are produced by impact dissociation of N_2. Thus Mumma and Zipf∥ have measured excitation cross-sections over the impact energy range from threshold to 350 eV for (*a*) the $2p^2\,3s\,^4P$ multiplet via the transition to $2p^3\,^4S^0$ radiating at 1200 Å, (*b*) the $2p^2(^1D)3s\,^2D$ via $2p^3\,^2D^0$ at 1243 Å, and (*c*) the $2p^2\,3s\,^2P$ via $2p^3\,^2D^0$ at 1493 Å. The values obtained at 100 eV, for the excitation of the respective ultraviolet multiplets are, in 10^{-18} cm^2, 6·70, 1·45, and 2·53. The measurements of Aarts and de Heer†† give values between 30 and 40 per cent smaller for (*a*) and (*b*) but agree closely for (*c*). Ajello's observations‡‡ give much larger values for (*c*) but agree moderately well for (*a*).

Mumma and Zipf∥ also measured cross-sections for excitation of the multiplet at 1177 Å $(2p^3\,^2D^0\text{–}2p^2\,4s\,^2P)$, and 1164 Å $(2p^3\,^2D^0\text{–}2p^2\,3d\,^2D)$ at 100 eV as $5\cdot 16 \times 10^{-19}$ and $1\cdot 63 \times 10^{-19}$ cm^2 respectively.

3.3. *Excitation of the negative bands of* N_2^+

A number of further measurements of the excitation function for the (0, 0) first negative band of N_2^+ have been made in addition to those discussed in Chapter 13, p. 977 (see Figs. 13.61 and 13.62). Thus Borst

---

† STANTON, P. N. and ST. JOHN, R. M., *J. opt. Soc. Am.* **59** (1969) 252.
‡ loc. cit., p. 3463.   § BORST, W., loc. cit., p. 3462.
∥ MUMMA, M. J. and ZIPF, E. C., *J. chem. Phys.* **55** (1971) 5582.
†† AARTS, J. F. M. and DE HEER, F. J., *Physica, 's Grav.* **52** (1971) 45.
‡‡ AJELLO, J. M., *J. chem. Phys.* **53** (1970) 1156.

and Zipf,† using photon-counting techniques, measured the excitation cross-section over the impact energy range from threshold to 3 keV. They observed at the angle of 54° 44′ to eliminate polarization effects (see p. 3369). At energies above 300 eV their results agree well with other recent observations,‡,§,∥,†† but all are more than twice as large as the earlier observations shown in Chapter 13, Fig. 13.62. The maximum cross-section at 100 eV is $16 \cdot 4 \times 10^{-18}$ cm² according to Borst and Zipf as compared with 21·2, 16·8, 15·6, and $15 \cdot 2 \times 10^{-18}$ cm² according to the other recent experiments of Aarts, de Heer, and Vroom;‡ Srivastava and Mirza;§ Stanton and St. John;∥ and McConkey, Woolsey and Burns†† respectively. All of these are much higher than the value $6 \cdot 2 \times 10^{-18}$ cm² observed in the earlier experiments of Stewart (see Fig. 13.62). At high energies the cross-section observed by Borst and Zipf falls off as $AE^{-1}\ln(BE)$ where E is the impact energy in eV, $A = 1 \cdot 91 \times 10^{-15}$ cm² eV, and $B = 2 \cdot 60 \times 10^{-2}$ eV^{-1}.

Kupriyanov‡‡ has observed highly excited N atoms resulting from impact dissociation of N_2, using the detection techniques referred to on p. 3449.

4. Oxygen

4.1. *Excitation*

The excitation of the $a\,^1\Delta_g$ and $b\,^1\Sigma_g^+$ metastable states of O_2 (see Chap. 13, p. 986 and Fig. 13.69) has been studied by Trajmar, Cartwright, and Williams§§ and by Linder and Schmidt∥∥ in the course of their experiments on vibrational excitation of O_2 (see p. 3443). The former authors observed differential cross-sections for excitation of both states for incident electron energies from 4 to 45 eV and scattering angles from 10 to 90°. For the $b\,^1\Sigma_g^+$ state the differential cross-sections fall very rapidly as the scattering angle decreases below 20°. At larger angles maxima and minima appear. On the other hand for $a\,^1\Delta_g$ the cross-section has a forward peak, a minimum between 50 and 80°, and then increases at larger angles. The integrated cross-sections show maxima of 9×10^{-18} and 2×10^{-18} cm² for $a\,^1\Delta_g$ and $b\,^1\Sigma_g^+$ respectively, at impact energies near 7 eV. These results agree well with those obtained by

---

† Borst, W. L. and Zipf, E. C., *Phys. Rev.* **A1** (1970) 834.
‡ Aarts, J. P. M., de Heer, F. J., and Vroom, D. A., *Physica*, *'s Grav.* **40** (1968) 197.
§ Srivastava, B. N. and Mirza, I. M., *Phys. Rev.* **176** (1968) 137.
∥ Stanton, P. M. and St. John, R. M., *J. opt. Soc. Am.* **59** (1969) 252.
†† McConkey, J. W., Woolsey, J. M., and Burns, D. J., *Planet. Space Sci.* **15** (1967) 1332.
‡‡ Kupriyanov, S. E., *Soviet Phys. JETP* **28** (1969) 240.
§§ loc. cit., p. 3424. ∥∥ loc. cit., p. 3443.

Linder and Schmidt up to energies of 4 eV and are consistent with the assumptions made by Hake and Phelps in their analysis of swarm data in O_2 (see Chap. 13, p. 1017).

4.2. *Dissociative excitation*

Apparent excitation cross-sections for production of the 1304 Å ($^3S^0 \to {}^3P$) resonance radiation of atomic oxygen have been measured by Mumma,[†] Lawrence,[‡] and Ajello.[§] They observe maximum cross-sections at 100 eV of 3·8, 3·05, and $3·0 \times 10^{-18}$ cm^2 respectively. Ajello also observed the emission of the 1356 Å multiplet arising from $^5S^0 \to {}^3P$ transitions due to production of O atoms in $^5S^0$ metastable states.

Detailed studies of the production of these atoms, which have a spontaneous transition probability of $1·68 \times 10^5$ s^{-1}, have been made by Freund[‖] and by Borst and Zipf,[††] using the time-of-flight technique (see p. 3448). Both experiments used a Be–Cu dynode as detector, placed sufficiently far away for most highly excited atoms to have radiated before reaching it. In addition Freund introduced a second detector, rather closer to the source, to observe the highly excited atoms through the reaction

$$A^* + SF_6 \to A^+ + SF_6^-. \qquad (6)$$

On the whole there is quite good agreement between the results obtained in the two experiments. Broadly speaking, the time-of-flight spectrum shows peaks corresponding to a slow ($\simeq 0·6$ eV) and a fast ($\simeq 2·9$ eV) group of metastable atoms. The threshold energy for the slow group is at $14·3 \pm 0·2$ and $14·9 \pm 0·5$ eV according to Borst and Zipf and to Freund respectively. The ionization function rises to a flat maximum near 30 eV. For the first group the threshold is near 21 eV and the maximum excitation cross-section occurs near 100 eV.

It seems very probable that the metastable state concerned is the $^5S^0$ state. Thus the ^1D state has insufficient energy to eject an electron from the Cu–Be surface and while the ^1S may just be able to do this the observed threshold indicates a dissociation limit for the parent O_2 state of 14·3 eV which agrees with the $^5S + {}^3P$ limit. Furthermore, it was verified that the atoms concerned can produce Penning ionization of naphthalene so the energy of the state must be above 8·1 eV.

Freund also verified that the fast group did not arise from highly excited atoms. Thus the relative intensities of the fast and slow groups

[†] MUMMA, M. J., Thesis, Pittsburgh (1970).
[‡] LAWRENCE, G. M., *Phys. Rev.* A**2** (1970) 397.
[§] AJELLO, J. M., *J. chem. Phys.* **55** (1971) 3156.
[‖] FREUND, R. S., ibid. **54** (1971) 3125.
[††] BORST, W. L., and ZIPF, E. C., *Phys. Rev.* A**4** (1971) 153.

were exchanged when the drift distance was reduced by a factor of 3 or more and they were unaffected when SF_6 was admitted to the drift chamber. However, up to four peaks due to highly excited atoms were observed by Freund using the SF_6 detector. Highly excited O atoms arising from dissociation of O_2 by electron impact have also been observed by Kupriyanov.[†]

4.3. Attachment

4.3.1. *Dissociative attachment.* Further experiments have been carried out on the temperature-dependence of the rate of dissociative attachment of electrons to O_2 (see Chap. 13, pp. 1002–4 and Fig. 13.79). The work has been directed particularly towards verifying the explanation by O'Malley of the remarkable effects previously observed. Henderson, Fite, and Brackmann,[‡] apart from carrying out a further careful study of the temperature dependence, produced evidence that it is the vibrational (T_v) and not the rotational (T_r) temperature which is important, They compared results obtained with two furnaces. In one, possessing a large aperture, and operated at low pressure so that effusive flow prevails, T_r and T_v are the same but in the other, operated at a high pressure with a small aperture, T_v remains at the furnace temperature because of slow vibrational relaxation (see Chap. 17, p. 1518) while T_r is greatly reduced because rotational relaxation is rapid. No difference was found between the results obtained with these two furnaces.

Spence and Schulz[§] used the retarding potential difference method (see p. 114–6) with the ion collector system enclosed in an iridium furnace which could be heated to 950 °K. They obtained results for the variation with temperature of the peak attachment cross-section, the onset energy and the energy at peak which agree very well with O'Malley's theoretical results. There is also quite good agreement with the results of Henderson *et al.*, which are much the same as their earlier data reported in Chapter 13, pp. 1002–4. At the highest temperature in their work (1930 °K) they observed a rather larger variation than predicted by O'Malley for the peak cross-section.

Chantry[||] has studied O^- production in O_2 at relatively high pressure (10^{-2} torr) and finds a peak at an electron energy near 15 eV which disappears at low pressures. Using his reflection technique (see p. 3454)

[†] loc. cit., p. 3465.
[‡] HENDERSON, W. R., FITE, W. L., and BRACKMANN, R. T., *Phys. Rev.* **183** (1969) 157.
[§] SPENCE, D. and SCHULZ, G. J., ibid. **188** (1969) 280.
[||] loc. cit., p. 3453.

Chantry found that the electrons which produced the peak had an energy close to 6·5 eV, suggesting that they result from energy loss in primary excitation of the $B\,{}^3\Sigma_u^-$ continuum (see Chap. 13, Fig. 13.69).

Van Brunt and Kieffer[†] have measured the angular distributions of O^- ions resulting from dissociative attachment in O_2 of electrons with energies in the range 5·75–8·40 eV. They used essentially the same apparatus as in their experiments on the dissociative ionization of H_2 and D_2 (see p. 3458). The main aim was to determine the quantum numbers of the state of O_2^- which is involved in the attachment process (see Chap. 12, Fig. 12.6 (a)). According to the analysis of O'Malley and Taylor (see p. 3459) the dominant term in the angular distribution varies as $|Y_{L|\mu|}(\theta,\phi)|^2$, where θ and ϕ are the polar coordinates of the direction of relative motion of ion and atom relative to the direction of the electron beam. μ is given by $|\Lambda_f|-|\Lambda_i|$ where Λ_i, Λ_f are the quantum numbers specifying the axial components of angular momentum of the initial and final molecular electronic states. L is the lowest integer, not less than $|\mu|$, which is even or odd according as the transition does or does not involve a change of parity of the molecular state. For O_2 the initial state is ${}^3\Sigma_g^-$ with $\Lambda_i = 0$. The possible final states are Π_u, Π_g, Δ_u, and Σ_u^-. Σ_u^+ is excluded since a Σ^- to Σ^+ transition is forbidden. For the four respective states the dominant term is therefore expected to be $|Y_{1,1}|^2$, $|Y_{2,1}|^2$, $|Y_{3,2}|^2$, and $|Y_{1,0}|^2$ corresponding to variation as $\sin^2\theta$, $\sin^2 2\theta$, $\sin^2 2\theta \sin^2\theta$, and $\cos^2\theta$ respectively.

Van Brunt and Kieffer found angular distributions which fell off very rapidly near 0 and 180°, thereby eliminating Σ_u^- as a possibility. They are well fitted by the form $(\sin\theta + \alpha \sin\theta \cos^2\theta)^2$ which would be given for a ${}^2\Pi_g$ final state if the next most important term to the dominant one, proportional to $Y_{3,1}$, is included. α is found to vary from 0·353 at an impact energy of 8·40 eV to 0·187 at 5·75 eV. This would be expected, as the relative importance of successive terms in the angular distribution falls off with the electron wave number. Also, theoretical estimates of the relative importance of the second term are consistent with the values of α which are found. Possible combinations arising from other states do not give such good agreement and it seems likely that the upper state can be identified as ${}^2\Pi_u$.

4.3.2. Three-body attachment.

Using the technique described on p. 3451 Grunberg[‡] has measured the ratio α/p of the attachment coefficient to the gas pressure p as a function of F/p, where F is the electric

---

[†] VAN BRUNT, R. J. and KIEFFER, L. J., *Phys. Rev.* A**2** (1970) 1899.
[‡] loc. cit., p. 3451.

field strength, over the range 0·1 to 18 V cm^{-1} torr^{-1} of F/p and pressures from 14·9 to 880 torr. This greatly extends the pressure range of the previous measurements (see Chap. 13, Fig. 13.81). While for F/p less than 1·5 V cm^{-1} torr^{-1}, in which range dissociative attachment is negligible, F/p is found to vary as the pressure for p less than 40 torr, at higher pressures deviations are observed in the sense expected—at sufficiently high pressure saturation occurs and α/p no longer depends on the pressure of the third body. The deviations indicate that the lifetime of the O_2^- state initially formed towards autodetachment is $< 10^{-10}$ s.

Spence and Schulz† have studied the three-body attachment process by a beam method using the same apparatus as in their experiments on associative detachment described above. On p. 3443 we described the evidence which strongly suggests that vibrational excitation in O_2 proceeds through formation of an O_2^- compound state possessing a well-defined set of vibrational energy levels. Three-body attachment could then be expected to occur simply by stabilization of such a compound state through energy withdrawal by the colliding third body before autodetachment has occurred. The dependence of the three-body rate on electron energy should therefore consist of a series of sharp peaks corresponding to the different vibrationally excited states of O_2^- which lie above the ground vibrational state of O_2. No experiments with sufficient energy resolution to observe these peaks had been carried out. However, Spence and Schulz using the retarding potential difference method (see p. 21) were able to work with an electron energy distribution of width between 80 and 100 meV about mean energies from 0·1 eV upwards. Their observations of the effective cross-section as a function of electron energy confirm the resonance description—peaks are found corresponding to the vibrational levels $v = 4$ to 10 of O_2^- as determined from the experiments on vibrational excitation (see p. 3443). Up to energies of 0·6 eV the rate coefficients measured by Pack and Phelps (see Chap. 13, p. 1010) at 300 °K and 500 °K agree very well with the new data with the peaks averaged out. At higher energies the swarm results are larger but exact agreement could not be expected. The absolute values agree well with those calculated by Chapman and Herzenberg‡ on the resonance picture. The temperature dependence of the peak value of the rate constant at 0·09 eV was observed over the

† Spence, D. and Schulz, G. J., *Phys. Rev.* A5 (1972) 724.
‡ Chapman, C. J. and Herzenberg, A. *7th int. conf. on the physics of electronic and atomic collisions, Amsterdam.* Abstracts, North Holland, 1971 p. 1166.

range 300–750 °K, and is consistent with that observed in the swarm experiments at 77, 300, and 500 °K.

5. Carbon monoxide

Lassettre and Skerbele† have carried out a careful measurement of absolute generalized oscillator strengths for excitation of the $A\,^1\Pi$, $B\,^1\Sigma^+$, $C\,^1\Sigma^+$, and $E\,^1\Pi$ states. The absolute magnitudes were obtained using the absolute elastic scattering cross-sections measured by Bromberg with the same technique as in his measurements for helium (see pp. 3376 and 3402). An electron spectrometer of much greater resolution than in earlier experiments of Lassettre and Silverman (see Chap. 13, Fig. 13.95, p. 1023) was used and very clearly resolved vibrational spectra obtained.

For the fourth positive $(A\,^1\Pi \to X\,^1\Sigma^+)$ bands the vibrational intensity distribution was found to be independent of scattering angle for angles $> 1°$ at impact energies of 200 eV or higher and, at lower kinetic energies, remained unchanged at all angles. Very precise generalized oscillator strengths for the transition to the vibrational level $v' = 2$ were obtained and extrapolated to zero momentum transfer to give an optical oscillator strength $f(v' = 2) = 0.0429$. This compares well with the value 0.045 ± 0.005‡ obtained with an earlier spectrometer from observations at 0° in which the calibration was made by comparison with the known value for helium. From the vibrational intensity distribution, averaged over all angles and energies, the corresponding oscillator strengths were obtained for other transitions, giving for the entire $A\,^1\Pi \leftarrow X\,^1\Sigma^+$ transitions an optical oscillator strength 0·195.

The new values for the oscillator strengths are on the average 0·727 times those obtained in the earlier measurements of Lassettre and Silverman. It is considered that this factor largely arises from the pressure measurements which, in the earlier work, did not allow for the streaming error in the McLeod gauge used for absolute pressure calibration. In this connection it is significant that the values of $\mu - 1$, where μ is the refractive index, calculated from the f-values obtained in earlier measurements for N_2, O_2, and H_2O were larger than the observed values by a factor close to $1/0.77$ in each case. Analysis of the vibrational distribution in terms of calculated Franck–Condon factors shows that the dipole moment matrix element depends on the nuclear separation R, increasing linearly by 20 per cent as R decreases from 1·16 to 1·02 Å.

† LASSETTRE, E. N. and SKERBELE, A., *J. chem. Phys.* **54** (1971) 1597.
‡ MEYER, V. D. and LASSETTRE, E. N., ibid. **54** (1971) 1608.

The optical oscillator strengths for transitions to $B\,^1\Sigma^+$, $C\,^1\Sigma^+$, and $E\,^1\Pi$ states with $v' = 0$ were redetermined as 0.0153 ± 0.0014, 0.163 ± 0.015, and 0.094 ± 0.009 respectively.

A number of measurements have recently been made of excitation cross-sections for different band systems of CO and CO$^+$.[†]

Kupriyanov[‡] has observed highly excited carbon and oxygen atoms resulting from dissociative ionization of CO by electron impact.

Chantry[§] has studied dissociative attachment in CO (see Chap. 13, § 5.2) using the technique referred to on p. 3451 in which the most probable energy of the O$^-$ ions produced by associative detachment is measured as a function of the electron energy. Two processes were distinguished, with thresholds at 9.70 ± 0.10 eV and 10.95 ± 0.10 eV, the peak cross-section for the former being 21 times larger. The threshold values agree well with those expected for the processes

$$e + CO \to O^- + C(^3P), \quad 9.62 \text{ eV} \tag{7a}$$

$$\to O^- + C^*(^1D), \quad 10.88 \text{ eV} \tag{7b}$$

assuming the dissociation energy of CO $= 11.09$ eV, the electron affinity of O$^- = 1.465$ eV, and the excitation energy of C(^1D) $= 1.26$ eV.

Stamatovic and Schulz[∥] have carried out experiments with high electron energy resolution (width 0.07 eV), obtained with a trochoidal monochromator and a quadrupole mass-spectrometer in which they were particularly concerned to study the formation of C$^-$ ions (see Chap. 13, Table 13.16). This is difficult because the peak cross-section for C$^-$ production is only about 6×10^{-23} cm^2 and the ion count rate in the experiments was only about 23 s^{-1}. The onset potential for C$^-$ production was found to be 10.20 ± 0.04 eV which is close to that (10.07 ± 0.1) found by Lagergren in earlier experiments. A second minimum in C$^-$ production is observed at about 0.5 eV higher.

6. Nitric oxide

Chantry[§] has studied dissociative attachment in NO (see Chap. 13, p. 1033) using the same technique as described above for CO. He finds only one reaction

$$e + NO \to O^- + N^*(^2D) \tag{8}$$

with threshold at 7.50 ± 20.10 eV close to that calculated (7.42 eV).

[†] SKUBENICH, V., *Optika Spektrosk.* **23** (1967) 540 ($d\,^3\Delta_i$, $b\,^3\Sigma^+$, $B\,^1\Sigma$ of CO, $A\,^2\Pi$, $B\,^2\Sigma^+$ of CO$^+$); AARTS, J. F. M. and DE HEER, E. J., *J. chem. Phys.* **52** (1970) 5354 ($A\,^1\Pi \leftarrow X\,^1\Sigma^+\,(0, 1)$); AJELLO, J. M., ibid. **55** (1971) 3158 ($A\,^1\Pi \to X\,^1\Sigma^+$, $a\,^3\Pi \to X\,^1\Sigma^+$ of CO, $B\,^2\Sigma^+ \to X\,^2\Sigma^+$, $A\,^2\Pi \to X\,^2\Sigma^+$ of CO$^+$).

[‡] loc. cit., p. 3465.   [§] loc. cit., p. 3450.

[∥] STAMATOVIC, A. and SCHULZ, G. J., *J. chem. Phys.* **53** (1970) 2663.

Puckett, Kregel, and Teague,[†] using the afterglow technique described on p. 3452, have measured the rate constant for three-body attachment of thermal electrons to NO as $6 \cdot 8 \pm 3 \cdot 4 \times 10^{-32}$ cm^6 s^{-1} over an NO pressure range from 0·05 to 0·7 torr (see p. 3639). This is to be compared with $1 \cdot 3 \pm 0 \cdot 1 \times 10^{-31}$ cm^6 s^{-1} measured over a pressure range from 1 to 5 torr by Weller and Biondi[‡] in the course of their experiments on recombination in an afterglow in NO (see Chap. 20, p. 2251).

7. Carbon dioxide

Freund and Klemperer[§] have observed time-of-flight spectra of the dissociation products from electron impact in CO_2. A peak was obtained at a velocity of 105 cm s^{-1}. This is too high to ascribe to metastable molecules of CO_2. As the fragments were detected using a tantalum detector with work function 4·1 eV they could not be metastable O or C. This leaves as the only possibility metastable molecules of CO.

Observations of apparent excitation functions for ultraviolet radiation emitted in dissociative excitation of CO_2 have been made by a number of investigators.[||] Cross-sections for excitation of the $\tilde{A}\,^2\Pi_u$ and $\tilde{B}\,^2\Sigma^+$ band systems of CO_2^+ have also been measured.[††]

Spence and Schulz[‡‡] noted that the onset energy for formation of O$^-$ (see Chap. 13, Fig. 13.115) ions with zero kinetic energy from CO_2 shows a marked temperature dependence. This work was later extended to study the variation with temperature, from 300 to 1000 °K, of the onset, shape, and magnitude of the attachment cross-sections. The first onset decreased from 3·85 eV at 300 °K to 3·35 at 950 °K.

8. Attachment in N_2O

The study of the attachment of electrons in N_2O has proved to be of unexpected interest. Early measurements[§§] of the variation of the total cross-section for attachment, at low gas pressures ($\leqslant 10^{-4}$ torr), as a function of electron energy, agree quite well at energies above 2 eV but at energies below 1·5 eV show much greater differences than is usually the case. Furthermore, the threshold for the dissociative attachment

---

[†] loc. cit., p. 3452.
[‡] WELLER, C. S. and BIONDI, M. A., *Phys. Rev.* **172** (1968) 198.
[§] loc. cit., p. 3448.
[||] MUMMA, M. J., ZIPF, E. S., and STONE, E. J., *J. chem. Phys.*; AJELLO, J. M., ibid. **55** (1971) 3169.
[††] MCCONKEY, J. W., BURNS, D. J., and WOOLSEY, J. H., *J. Phys. B (GB)* **1** (1968) 71; NISHIMURA, H., *J. phys. Soc. Japan* **24** (1968) 130.
[‡‡] loc. cit., p. 3467.
[§§] SCHULZ, G. J., *J. chem. Phys.* **34** (1961) 1778; RAPP, D. and BRIGLIA, D. D., ibid. **43** (1965) 1480.

reaction
$$e + N_2O \rightarrow N_2 + O^- \tag{9}$$
should be at an energy less than $D-A$, where D is the dissociation energy of the N_2–O band and A the electron affinity of O. Taking the accepted values 1·67† and 1·46 eV for D and A respectively suggests that the threshold energy should exceed 0·21 eV but the experiments indicated a zero energy onset. A possible explanation suggested by Kaufman† is that attachment takes place to vibrationally excited N_2O molecules. This would mean that the attachment rate depends quite strongly on temperature.

Confirmation of this was obtained by Chantry‡ who studied the temperature dependence of the total attachment cross-section and of the energy distributions of the mass-identified O^- ions produced, using the experimental arrangements described on p. 3451. He found that, for impact energies below 1·5 eV, the cross-section does vary rapidly with temperature over the range 160 to 1040 °K. The variation is indeed great enough to explain the discrepancies between earlier data—a difference of only 20 °K gives rise to readily distinguishable results for the variation of cross-section with energy below 1·5 eV. As the temperature increased above about 400 °K the probability of attachment of very low-energy electrons increased rapidly, the cross-section reaching a sharp peak of 10^{-15} cm^2 at the highest temperature. On the other hand the peak cross-section at 2·5 eV remained unchanged at $8·6 \times 10^{-18}$ cm^2. This suggests that attachment of very slow electrons in hot N_2O could provide the best source of beams of O^- ions.

At a given temperature the kinetic energy distribution of the O^- ions at different electron energies showed the same general behaviour. If attachment occurred to ground-state N_2O and produced ground state N_2, neither possessing vibrational excitation, the most probable kinetic energy would be given by (5) (p. 3451). At the lowest electron energies the peak occurred at a slightly higher energy while at energies above 0·5 eV it occurred at lower energies. Indeed at electron energies $> 1·5$ eV it remained constant at about 0·4 eV. This suggests that, at low energies, attachment occurs to vibrationally excited N_2O and at higher energies leads to production of vibrationally excited N_2.

Confirmation of the strong temperature dependence of the low-energy attachment has been provided from the swarm experiments of Chaney and Christophorou.§

† KAUFMAN, F., *J. chem. Phys.* **46** (1967) 2649.
‡ CHANTRY, P. J., ibid. **51** (1969) 3369.
§ CHANEY, E. L. and CHRISTOPHOROU, L. G., ibid. **51** (1969) 883.

The theoretical interpretation of these results depends on the prediction that, whereas N_2O is known to be linear in its equilibrium state, N_2O^-, being isoelectronic with NO_2, should have a bent equilibrium configuration.† The excitation of bending vibrations in N_2O would therefore be expected to increase markedly the chance of capture of a low-energy electron into a low vibrational level of N_2O^-. If this level remains above the limit for dissociation into N_2 and O^- the net result will always be dissociative attachment. Wentworth, Chen, and Freeman‡ have carried out a semi-quantitative analysis on these lines which has also been shown to be theoretically plausible by Bardsley.§

Warman, Fessenden, and Bakale|| have studied electron loss in afterglows in N_2O and mixtures of N_2O with various gases and vapours, using the technique described in Chapter 12, p. 875. They measured effective attachment rates at pressures (4 to 27 torr) at which three-body attachment occurred and then extrapolated to zero pressure to obtain the two-body coefficients. With pure N_2O and mixtures with N_2 the limiting rate at 300 °K was only 1×10^{-15} cm^3 s^{-1} but with C_2H_6 or C_3H_8 as diluents rates six times larger were obtained. This can be understood if allowance is made for detachment through the reactions

$$O^- + N_2O \rightarrow NO + NO^-, \qquad (10\,a)$$

$$NO^- + N_2O \rightarrow NO + N_2O + e \qquad (10\,b)$$

which would be absent when the hydrocarbons are present due to their acting irreversibly with O^-. The low electron affinity of NO (see p. 3522) makes (10 b) very probable.

Other measurements at pressures for which three-body attachment is important have been made by Phelps and Voshall†† and by Warman and Fessenden.‡‡

Chantry§§ has investigated the processes which lead to production of O^-, NO^-, and N_2O^- ions, at pressures ranging from 10^{-6} to 5×10^{-3} torr, by electron attachment. He has produced evidence which strongly indicates that the latter ions arise from the reactions

$$O^- + N_2O \rightarrow NO^- + NO, \qquad (11\,a)$$

$$NO^- + N_2O \rightarrow N_2O^- + NO. \qquad (11\,b)$$

† FERGUSON, E. E., FEHSENFELD, F. C., and SCHMELTEKOPF, A. L., *J. chem. Phys.* **47** (1967) 3085.
‡ WENTWORTH, W. E., CHEN, E., and FREEMAN, E., ibid. **55** (1971) 2075.
§ BARDSLEY, J. N., ibid. **51** (1969) 3384.
|| WARMAN, J. M., FESSENDEN, R. W., and BAKALE, G., ibid. **57** (1972) 2702.
†† PHELPS, A. V. and VOSHALL, R. E., ibid. **49** (1968) 3246.
‡‡ WARMAN, J. M. and FESSENDEN, R. W., ibid. **49** (1968) 4718.
§§ CHANTRY, P. J. ibid. **51** (1969) 3369.

9. Negative-ion production from halogen-containing molecules
(see Chap. 13, p. 1052)

9.1. *Sulphur hexafluoride*

Fehsenfeld[†] has measured the rate of attachment of electrons to SF_6 at temperatures between 293 and 523 °K in a flowing afterglow (see p. 2021) with helium as buffer gas, the pressure ranging from 0·1 to 1·5 torr. Because of the high attachment rate it was not possible to add sufficiently small measured amounts of pure SF_6. Instead, analysed mixtures of SF_6 with Ar and He were introduced (0·0495 per cent in Ar and 0·0497 per cent in He). The observed increase of SF_6^- on adding the mixture was well correlated with a reduction of electron density measured using Langmuir probes operating in the orbital limited mode (see Chap. 20, p. 2168). The rate constant was found to be $2·2 \times 10^{-7}$ cm^3 s^{-1}, independent of pressure and temperature. Measurements were also made of the relative yield of SF_5^-. This was found to increase from about 5×10^{-5} at room temperature to 4×10^{-2} at 500 °K.

A number of other measurements have also been made. Davis and Nelson[‡] used a modified time-of-flight technique. They found results which were practically independent of the buffer gas, which included He, 1 per cent CH_4+99 per cent Ar, 11 per cent CH_4+89 per cent Ar, H_2, N_2, CO, CO_2, CH_4, CF_4, C_2H_2, C_2H_4, C_2F_6, *cis*-C_4H_8, and C_4H_8. The mean value, close to $1·93 \times 10^{-7}$ cm^3 s^{-1}, agrees quite well with that observed by Fehsenfeld. Somewhat larger values were obtained by Mahan and Young[§] and Young[||] using the afterglow technique described in Chapter 12, p. 875. With helium as buffer they obtained $3·1 \times 10^{-7}$ and $2·7 \times 10^{-7}$ cm^3 s^{-1} respectively.

The evidence strongly suggests that, under the working conditions, the chance of stabilization of the initial SF_6^- complex is virtually unity for all buffer gases. In Fehsenfeld's experiments, if collisions between the complex and the buffer gas atoms occur at a rate determined by orbiting (see Chap. 19, p. 2005), then the time between collisions at the lowest pressure used is about 0·5 μs so the autodetachment lifetime is $\gg 0·5$ μs. An early time-of-flight experiment[††] gave a lifetime of 10 μs.

9.2. *Other hexafluorides*

Negative ion formation has been studied in the hexafluorides of Se,

[†] FEHSENFELD, F. C., *J. chem. Phys.* **53** (1970) 2000.
[‡] DAVIS, F. J. and NELSON, D. R., *Chem. Phys. Lett.* **6** (1970) 277.
[§] MAHAN, B. H. and YOUNG, C. E., *J. chem. Phys.* **44** (1966) 2192.
[||] YOUNG, C. E., *UCRL Rept.* UCRL–17171 (1966).
[††] EDELSON, D., GRIFFITHS, J. E., and MCAFEE, K. B., *J. chem. Phys.* **37** (1962) 917.

Te, Mo, Re, U,[†] and Xe[‡] over the electron energy range from 0 to about 10 eV. Only for Mo and Re were hexafluoride ions formed directly by electron attachment. SeF_6^-, TeF_6^-, and UF_6^- were produced by charge exchange with thermal SF_6^-. XeF_6^- was not observed but XeF^-, XeF_2^-, XeF_3^-, XeF_4^-, F_2^-, and F^- were detected as products of electron impact on XeF_6.

9.3. *Iodine*

Truby[§] has measured the dissociative attachment rate of electrons in I_2 using the afterglow technique described in Chapter 12, p. 875. As in earlier experiments of Biondi (see Chap. 13, § 11.2, p. 1057) he used iodine vapour buffered with helium but to avoid the risk of any reactions with excited helium species the electrons were initially produced by photo-ionization by light in the 1150–1300 Å region, emitted by a low pressure hydrogen flash lamp. The iodine partial pressure was determined by measuring the absorption of 4840 Å radiation. With single pulse operation (see Chap. 20, p. 2163) the attachment coefficient was found to be $1 \cdot 8 \times 10^{-10}$ cm³ s⁻¹ at 295 °K corresponding to an attachment cross-section of $1 \cdot 7 \times 10^{-17}$ cm², considerably smaller than that, $3 \cdot 9 \times 10^{-16}$ cm² (see Chap. 13, Fig. 13.126, p. 1061), observed earlier by Biondi. In the latter experiments repetitive pulses were used and it was verified by Truby that when this procedure was used in his experiment a larger attachment rate was observed. As there is a risk that negative ion concentration may build up when repetitive pulses are used, so vitiating the analysis of the data, greater weight must be attached to the results obtained with single pulse operation.

9.4. *Bromine*

Truby[∥] has applied the same method as used for I_2[§] to measure the attachment rate for electrons in Br_2 and finds a much smaller value: $0 \cdot 82 \times 10^{-12}$ cm³ s⁻¹ at 296 °K.

[†] STOCKDALE, J. A. D., COMPTON, R. N., and SCHWEINLER, H. C., *J. chem. Phys.* **53** (1970) 1502.
[‡] BEGUN, G. M. and COMPTON, R. N., ibid. **51** (1969) 2367.
[§] TRUBY, F. K., *Phys. Rev.* **172** (1968) 24.
[∥] IDEM, ibid., A4 (1971) 114.

14

COLLISIONS INVOLVING ELECTRONS AND PHOTONS—RADIATIVE RECOMBINATION, PHOTOIONIZATION, AND BREMSSTRAHLUNG

SINCE Vol. 2 was written a great deal of progress has been made in the application of photoelectron spectroscopy, particularly to the investigation of molecular structure. Much of this work has made use of soft X-ray sources of ionizing radiation. The volume of new material is such as to make it impossible to summarize it effectively here. References may be made to the extensive reviews by Siegbahn† and by Turner, Baker, Baker, and Brundle.‡ We shall confine attention here to the more basic aspects of the subject, associated with the spectra of simple systems. This includes the study of the angular distribution of the photoelectrons, the photoelectron spectroscopy of radicals, and of autoionizing peaks.

There have been a number of investigations directed towards the production of polarized electrons through photoionization. This work has been confined to the study of alkali metal atoms for which spin–orbit interaction effects had already been recognized. Photoelectrons ejected by circularly polarized light are partially polarized because of this interaction. Another more obvious possibility, already suggested as long ago as 1930 by Fues and Hellmann,§ is to photoionize a polarized atom beam with unpolarized light. Both of these possibilities have been verified experimentally and studied in some detail theoretically.

A considerable amount of new work has been carried out on double and indeed multiphoton ionization through the use of intense laser sources of radiation.

Apart from these major directions, further applications of existing theoretical and experimental techniques have been made, including a

† SIEGBAHN, K., *Phil. Trans. R. Soc.* **268** (1970) 33; publications of Uppsala University Institute of Physics **7** (1971).
‡ TURNER, D. W., BAKER, C., BAKER, A. D., and BRUNDLE, C. R., *Molecular photoelectron spectroscopy* (Wiley–Interscience, London, 1970).
§ FUES, E. and HELLMANN, H., *Physik. Z.* **31** (1930) 465; COOPER, J. and ZARE, R. N. *J. chem. Phys.* **48** (1968) 942.

very comprehensive study of the autoionizing states of the heavier rare-gas atoms and a thorough investigation of the photoionization of H_2^+.

1. Calculation of atomic cross-sections and polarization parameters

1.1. *Some general formulae*

The angular distribution of photoelectrons produced by linearly polarized light in ejecting an electron from an orbital with quantum numbers n, l is given in terms of the differential cross-section

$$I_l(\theta)\,d\omega = (4\pi)^{-1}Q_{nl}^a\{1+\beta P_2(\cos\theta)\}\,d\omega, \qquad (1)$$

where Q_{nl}^a is the total cross-section for ejection of the electron and θ is the angle between the direction of motion of the ejected electron and the polarization of the incident light. Provided the magnetic substates of the initial state are equally populated and the wave functions are represented by antisymmetrized products of spin orbitals, the asymmetry parameter β is given by

$$\beta = \{l(l-1)r_{l-1}^2+(l+1)(l+2)r_{l+1}^2-6l(l+1)r_{l-1}r_{l+1}\cos(\xi_{l+1}-\xi_{l-1})\} \times$$
$$\times [(2l+1)\{lr_{l-1}^2+(l+1)r_{l+1}^2\}]^{-1}. \qquad (2)$$

The $r_{l\pm 1}$ are the radial matrix elements

$$r_{l\pm 1} = \int_0^\infty R_{nl}\,r R_{\kappa l\pm 1}\,dr, \qquad (3)$$

where $r^{-1}R_{nl}$, $r^{-1}R_{\kappa l\pm 1}$ are the radial wave-functions for the initial and final states of the electron. The $\xi_{l\pm 1}$ are phase shifts such that

$$R_{\kappa,l}(r) \sim \sin(\kappa r - \tfrac{1}{2}l\pi - \alpha \ln 2\kappa r + \xi_l), \qquad (4)$$

κ being the wave number of the ejected electron and $\alpha = -e^2m/\kappa\hbar^2$ (see Chap. 6, §§ 3.9, 3.10). In the special case $l = 0$, $\beta = 2$ independent of the energy of the ejected electron.

For unpolarized light the angular factor in (1) is changed to

$$1-\tfrac{1}{2}\beta P_2(\cos\theta),$$

giving a $\sin^2\theta$ distribution for ejection of an electron from an s orbital.

Lipsky† has generalized the formula for β to apply to transitions in which the ion resulting from the photoionization is in an excited state, LS coupling being supposed to apply throughout. The final channel is then defined by the orbital angular momentum of the ejected electron and the final total orbital angular momentum.

† LIPSKY, L., *5th int. conf. on the physics of electronic and atomic collisions, Leningrad.* Abstracts, 1967, p. 617.

The formulae for the polarization effects to be expected in photo-ionization of alkali-metal atoms depend on a parameter x defined by

$$x = (2r_{\frac{3}{2}} + r_{\frac{1}{2}})/(r_{\frac{3}{2}} - r_{\frac{1}{2}}). \quad (5)$$

$r_{\frac{3}{2},\frac{1}{2}}$ are the radial matrix elements

$$r_{\frac{3}{2},\frac{1}{2}} = \int_0^\infty R_s r R_{\kappa\frac{3}{2},\frac{1}{2}} \, dr, \quad (6)$$

where $r^{-1}R_s$ is the wave-function for the outer s orbital of an alkali-metal atom and $r^{-1}R_{\kappa\frac{3}{2},\frac{1}{2}}$ for the continuum p orbitals with $j = \frac{3}{2}, \frac{1}{2}$ respectively. $R_{\kappa\frac{1}{2}}$ and $R_{\kappa\frac{3}{2}}$ will differ due to spin–orbit interaction.

The degree of polarization of photoelectrons from unpolarized atoms, ejected by circularly polarized light, is given by[†]

$$P_1 = (1+2x)/(2+x^2) \quad (7)$$

while that of photoelectrons ejected from atoms with polarization P_a by unpolarized light is

$$P_2 = P_a x^2/(2+x^2). \quad (8)$$

If the light is circularly polarized there will be an asymmetry between the rate of photoionization by left- and right-hand circularly polarized light which is determined by the parameter

$$\delta = (2x-1)/(2+x^2). \quad (9)$$

We now discuss briefly the calculations which have recently been carried out for different atoms including those which give the asymmetry parameter β.

1.2. Helium

Jacobs and Burke[‡] have calculated cross-sections for photoionization of helium leaving the He$^+$ in an excited state with $n = 2$. As the energy difference between the final $2s$ and $2p$ states is too small to resolve experimentally the relative probability of the ion being left in either can only be determined from observations of the angular distribution of the photoelectrons. Jacobs and Burke calculated the total and differential cross-sections required using a 56-term Hylleraas bound-state wave-function for He and a $1s\,2s\,2p$ close-coupling continuum wave-function obtained by Burke and McVicar (see Chap. 9, p. 621). Both length and velocity matrix elements (see Chap. 14, § 2) were used and gave results in good agreement. They find that $\{Q(2s)+Q(2p)\}/Q(1s)$, the ratio of the cross-sections for ionization leaving the He$^+$ in the $n = 2$ state to that

[†] FANO, U., *Phys. Rev.* **178** (1969) 131; **184** (1969) 250.
[‡] JACOBS, V. L. and BURKE, P. G., *J. Phys. B* (*GB*) **5** (1972) L67.

leaving it in the ground state, has a maximum of 0·138 at a photon energy of 77 eV, tending to 0·0482 in the limit of very high energies. Values of $Q(2s)$, $Q(2p)$ and the asymmetry parameter $\beta(2p)$ are given for photoelectron energies ranging from 1·35 to 94·5 eV. Comparison with experimental data is discussed on p. 3484.

Bell and Kingston[†] have calculated photoionization cross-sections for He and for the series of isoelectronic ions. They used a 20-parameter correlated wave-function for the ground state and the Hartree–Fock (exchange) approximation for the continuum function. Both length and velocity matrix elements were calculated. Comparison with more elaborate calculations for He and with experiment indicates that the length matrix element calculated in this way should give satisfactory results. The calculated cross-sections for O^{6+} and Ne^{8+} were compared with observed data on K-shell ionization of O[‡] and Ne,[§] allowance being made for the differences in ionization energies. Quite good agreement was obtained.

Norcross[§] has calculated cross-sections for photoionization of He($2\,^3S$) and He($2\,^1S$) using bound-state wave-functions obtained by Cohen and McEachran[∥] and final-state functions given by the $1s\ 2s\ 2p$ close-coupling expansion.

1.3. Neon, argon, krypton, and xenon

Kennedy and Manson[††] have calculated cross-sections and the asymmetry parameters β for photoionization of the $2s$ and $2p$ subshells of Ne; $2s$, $2p$, and $3s$ of Ar; $3s$, $3p$, $3d$, $4s$, and $4p$ of Kr; and $3s$, $3p$, $3d$, $4s$, $4p$, $4d$, $5s$, and $5p$ of Xe. For the discrete states Hartree–Fock wave-functions were used. These were taken to be the same for both the atom and ion. A number of calculations were also carried out using the Slater-type orbitals as tabulated by Herman and Skillman.[‡‡] The continuum orbital was calculated according to the exchange (Hartree–Fock) approximation. With these approximations autoionization does not occur. Calculations were carried out with both length and velocity matrix elements, and extended up to photon energies as high as 2 keV. The calculated asymmetry parameters were found to depend very little on the assumed

---

[†] BELL, K. L. and KINGSTON, A. E., *J. Phys. B (GB)* **4** (1971) 1308.
[‡] COMPTON, A. H. and ALLISON, S. K., *X-rays in theory and experiment* (Macmillan, 1935); BEARDEN, A. J., *J. appl. Phys.* **37** (1966) 1681; WUILLEUMIER, F., *J. Phys., Paris* **26** (1965) 776.
[§] NORCROSS, D. W., *J. Phys. B (GB)* **4** (1971) 652.
[∥] COHEN, M. and McEACHRAN, R. P., *Proc. phys. Soc.* **92** (1967) 37 and 539.
[††] KENNEDY, D. J. and MANSON, T., *Phys. Rev.* **A5** (1972) 227.
[‡‡] HERMAN and SKILLMAN, S., *Atomic structure calculations* (Prentice-Hall, N.J., 1963).

wave-functions or form of the matrix elements, even though these different approximations lead to quite markedly different cross-sections.

Amus'ya, Cherepkov, and Chernysheva[†] have calculated the total cross-section for photoionization of argon allowing for electron correlation effects in the final state by the random phase approximation of Altick and Glassgold.[‡] The correlation effects are quite significant and substantially improve the agreement with observation which is discussed on p. 3486.

1.4. Alkali-metal atoms

Most attention has been devoted to the calculation of the parameter x defined in (5) which determines the polarization effects though the total photoionization cross-section is usually calculated in passing.

Weisheit[§] determined the wave-functions $R_{\kappa\frac{3}{2},\frac{1}{2}}$ by solving the one-body problem for motion of an electron under the influence of a potential

$$V(r) = V_c(r) - \tfrac{1}{2}\alpha_d\, r^{-4}[1-\exp\{-(r/r_1)^6\}] - \tfrac{1}{2}\lambda r^{-6}[1-\exp\{-(r/r_1)^8\}] +$$
$$+ (c_1 + c_2\, r)\exp(-r/r_0) + V_{so}(r) \quad (10)$$

in atomic units. V_c is the mean interaction with the Hartree–Fock field of the core, α_d is the core polarizability, and r_0 and r_1 are chosen cut-off parameters. λ, c_1, and c_2 are chosen to give good agreement with observed spectra, V_{so} is the spin–orbit interaction taken to be

$$V_{so} = \tfrac{1}{2}\alpha^2 \mathbf{l}\cdot\mathbf{s}(Z-a)/r^3, \quad (11)$$

where α is the fine-structure constant, $\mathbf{l}$ and $\mathbf{s}$ the orbital and spin angular momenta, and a a screening constant determined so as to fit measured fine-structure energy defects of P fine-structure levels.

Allowance must be made in calculating the matrix elements $r_{\frac{3}{2},\frac{1}{2}}$ of (6) for the contribution to the dipole moment operator from the core. This is done approximately by multiplying r in (6) by the factor

$$1 - \alpha_d\, r^{-3}[1-\exp\{-(r/r_c)^3\}], \quad (12)$$

where r_c is an effective core radius. r_c was treated as an adjustable parameter to obtain agreement with the measured energy at which x vanishes. This could not be applied to sodium for which no measured values exist. In this case r_c was adjusted to give agreement between the locations of the observed and calculated minima in the total photoionization cross-section (see Fig. 14.32). The results were found to be quite sensitive to the choice of r_c.

[†] AMUS'YA, M. Y., CHEREPKOV, N. A., and CHERNYSHEVA, L. V., Zh. eksp. teor. Fiz. **60** (1971) 160; Soviet Phys. JETP **33** (1971) 90.
[‡] ALTICK, P. L. and GLASSGOLD, A. E., Phys. Rev. **133** (1964) A632.
[§] WEISHEIT, J. C., ibid. A5 (1972) 1621.

For sodium x is small enough to give appreciable polarization only in a very narrow photon energy range but for the heavier alkali-metal atoms photoelectrons with appreciable polarization should be produced over a useful energy range.

Chang and Kelly† calculated the same quantities for K, Rb, and Cs using relativistic Hartree–Fock equations to determine the discrete and continuum wave-functions. Correlation effects are therefore neglected but the only adjustability in their results comes from a choice between including calculated or observed energies in the formulae.

Discussion of the comparison of these two sets of calculations with experiment is deferred until p. 3488.

1.5. *Atomic nitrogen and oxygen*

Koppel‡ has calculated cross-sections for photoionization from the different terms of the ground configurations of N and O in which one-electron wave-functions for the initial and final states of the active electron are determined by the scaled Fermi–Thomas potential method of Stewart and Rotenburg.§ Thus if the Fermi–Thomas statistical model gives the field of the core as $Ze^2r^{-1}\phi(r/\mu)$ where μ is the usual parameter (see p. 378) and Z is the atomic number, the effective potential in which the electron moves is taken as

$$V = Ze^2r^{-1}\phi(r/\mu\alpha), \qquad (13)$$

where the scaling parameter α is adjusted so as to give the correct binding energy for the electron. There is difficulty in deciding what value α should take for the continuum states as extrapolation of α_n, the value for the nth bound state, to $n \to \infty$ is not practicable. Koppel therefore took α to be the same for both the initial and final states. With this choice length and velocity matrix elements give the same results.

The calculated cross-sections differ from those calculated earlier with Hartree–Fock wave-functions by no more than do the results obtained in the latter calculations using length and velocity matrix elements.

Comparison with experimental results is discussed on p. 3490.

1.6. *Other atoms*

Amus'ya, Cherepkov, and Chernysheva‖ have calculated cross-sections for photoionization of calcium and zinc using Hartree–Fock

† CHANG, J.-J. and KELLY, H. P., *Phys. Rev.* **A5** (1972) 1713.
‡ KOPPEL, J. V., *J. chem. Phys.* **55** (1971) 123.
§ STEWART, J. C. and ROTENBERG, M., *Phys. Rev.* **140** (1965) A1508.
‖ AMUS'YA, M. Y., CHEREPKOV, N. A., and CHERNYSHEVA, L. V., *7th int. conf. on the physics of electronic and atomic collisions, Amsterdam.* Abstracts, North Holland, 1971, p. 196.

wave-functions with and without correction for correlation as in their calculations for argon (see p. 3481). As in the latter case correlation effects are found to be important.

2. Experimental measurements of photoionization—total and differential cross-sections

2.1. *The measurement of cross-sections for unstable species*

Comes, Elzer, and Speier[†], [‡] have developed a crossed-beam technique for the measurement of photoionization cross-sections of unstable species such as atomic oxygen and nitrogen.

The technique consists in crossing a beam of ultraviolet radiation with a molecular beam partially dissociated by an electric discharge. The degree of dissociation α is measured by observing, with a time-of-flight mass spectrometer, the current of molecular ions produced by radiation at a particular wavelength, with and without the discharge turned on. Allowance must be made for the change in temperature of the gas due to the discharge. This was measured by including about 8 per cent of Ar in the gas and observing the proportional change in Ar$^+$ photo-ionization current produced when the discharge is on. Having measured the cross-section $Q(X_2)$ for photoionization of X_2 using an ionization chamber in the usual way, the absolute value of that, $Q(X)$, for X was obtained by comparison. Thus if the same geometrical factors apply we have

$$Q(X) = Q(X_2) \frac{i(X^+)}{i(X_2^+)} \frac{A_m(X_2^+)}{A_m(X^+)} \frac{1-\alpha}{\alpha} \tag{14}$$

where $i(X^+)$, $i(X_2^+)$ are the respective currents of X^+ and X_2^+ ions measured by a secondary electron multiplier and the A_m are the secondary electron yields for the respective ions impinging on the multiplier cathode. Correction must be made for X^+ ions produced by dissociative photoionization of X_2. This may be done by comparing the X^+ and X_2^+ yields from an undissociated beam.

It is important that no appreciable contribution comes through photoionization of metastable X_2 or X present in the molecular beam. The absence of the former may be checked by the absence of a signal when radiation of frequency below the threshold for ionization of X_2 and X but above that for ionization of metastable X_2 is used. Metastable atoms will usually be deactivated quite quickly in collisions with the neutral molecules (see pp. 3586–92).

[†] Comes, F. J., Speier, F., and Elzer, A., *Z. Naturf.* **23A** (1968) 125.
[‡] Comes, F. J., Elzer, A., and Speier, F., ibid. **23A** (1968) 114.

This method has been applied successfully to measure photoionization cross-sections for O† and N‡ atoms (see p. 3490).

We now consider observed results obtained for different atoms and their relation to theory.

2.2. Helium

Krause and Wuilleumier§ measured the ratio $\{Q(2s)+Q(2p)\}/Q(1s)$ where $Q(2s)$, $Q(2p)$, and $Q(1s)$ are cross-sections for photoionization of helium leaving the He$^+$ ion in the $2s$, $2p$, and $1s$ states respectively. They also measured the asymmetry parameter $\beta(2p)$ for the photoelectrons produced, leaving the ion in the $2p$ state. To do this they used photoelectron spectroscopy to distinguish photoelectrons associated with the ion in the states with $n = 1$ and $n = 2$ arising from these. Let $i_1(\theta)$, $i_2(\theta)$ be the respective currents of photoelectrons ejected at an angle θ with respect to the direction of the incident, unpolarized photons. At an angle $\theta_0 = 54° 44'$ $P_2(\cos\theta_0) = 0$ and according to (1) we have

$$i_2(\theta_0)/i_1(\theta_0) = S(\theta_0) = \{Q(2s)+Q(2p)\}/Q(1s). \tag{15}$$

When $\theta = \tfrac{1}{2}\pi$, however, since $P_2(0) = -\tfrac{1}{2}$, $\beta(2s) = \beta(1s) = 2$,

$$i_2(\tfrac{1}{2}\pi)/i_1(\tfrac{1}{2}\pi) = S(\tfrac{1}{2}\pi) = [Q(2s)+6^{-1}\{4+\beta(2p)\}Q(2p)]/Q(1s), \tag{16}$$

so
$$\beta(2p) = 6S'-4+6(S'-1)Q(2s)/Q(2p), \tag{17}$$

where $S' = S(\theta_0)/S(\tfrac{1}{2}\pi)$.

The observed values of $\{Q(2s)+Q(2p)\}/Q(1s)$, extending over a range of photon energy from 110 to 200 eV, agree well with those calculated by Jacobs and Burke (see p. 3479). Measurements of $S(\tfrac{1}{2}\pi)$ agree well with earlier measurements by Samson‖ and are also consistent with the theoretical values. Using theoretical values of $Q(2s)/Q(2p)$ and measured values of S' in (17) gave $\beta(2p)$ in good agreement with theory at 120 eV but too small at 106 eV.

2.3. Neon

Codling, Madden, and Ederer†† have published an account of a comprehensive exploration of the resonance structure in the photoionization continuum of neon, in the photon energy range from 20 to 150 eV, extending their earlier work carried out with the synchrotron source of radiation as described in Chapter 14, pp. 1118–21.

† loc. cit. †, p. 3483.   ‡ loc. cit. ‡, p. 3483.
§ Krause, M. O. and Wuilleumier, F., *J. Phys. B* (*GB*) **5** (1972) L143.
‖ Samson, J. A. R., *Phys. Rev. Lett.* **22** (1969) 693.
†† Codling, K., Madden, R. P., and Ederer, D. L., *Phys. Rev.* **155** (1967) 26.

Discrete structure has been observed (a) between the $^2P^0_{\frac{1}{2}}$ and $^2P^0_{\frac{3}{2}}$ ionization limits near 22 eV, (b) between 44 and 60 eV, and (c) near 80 eV. The first arises from autoionizing levels which are equivalent to those first observed by Beutler† in the spectra of Ar, Kr, and Xe. In the second range the levels arise from excitation of a 2s electron or simultaneous excitation of the two 2p electrons. The weak resonances observed near 80 eV are due to simultaneous excitation of a 2s and a 2p electron. 61 levels in all have been observed and all but 7 have been identified. For the states $2s\,2p^6\,np\,{}^1P^0$, with $n = 3$, 4, and 5 and the state $2p^4(^3P)3s\,3p\,{}^1P^0$, the profile parameters q and Γ (see Chap. 14, p. 1077) have been determined.

The smooth 'background' photoionization cross-section for Ne calculated by Kennedy and Manson (see p. 3480) agrees quite well with that observed (see Chap. 14, p. 1119, Fig. 14.23).

2.4. *Argon, krypton, and xenon*

Very extensive studies have been made of the absorption spectra of argon, krypton, and xenon by Madden, Ederer, and Codling.‡

They used the 180 MeV synchrotron at the National Bureau of Standards as the source of continuum radiation in conjunction with two spectrometers, a 3 m grazing-incidence spectrograph and a 3 m grazing-incidence scanning monochromator using photoelectric detection. The wavelength resolution was about 0·06 Å. Wavelength calibration was carried out by superimposing the absorption spectra of helium and neon.

For argon the absorption spectrum was examined photographically over the wavelength range from 600 to 80 Å. Discrete structure was observed between 466 and 210 Å, the first resonance being due to the lowest possible optically allowed excitation of a 3s electron ($3s^2\,3p^6 \rightarrow 3s\,3p^6\,4p$). At a little shorter wavelength, 427 Å, the lowest energy two-electron excitation was found. No structure was seen in the other wavelength regions.

In the wavelength region between 466 and 320 Å, series of autoionizing levels involving both single- and two-electron excitation (involving two 3p electrons) were observed, the former giving rise to window-type resonances. Between 320 and 210 Å less prominent features due to simultaneous excitation of a 3s and 3p electron were observed, 17 involving single-electron excitation and 133 involving excitation of two 3p electrons. The parameters determined for the first three members of

† BEUTLER, H., *Z. Phys.* **93** (1935) 177.
‡ MADDEN, R. P., EDERER, D. L., and CODLING, K., *Phys. Rev.* **177** (1969) 136.

the $3s\,3p^6\,np$ series and the two most prominent which involve two-electron excitation are given in Table 14 A.1. The notation for the parameters is as in Chapter 9, pp. 608–11 and Chapter 14, § 4.3.

TABLE 14 A.1

Parameters characterizing certain autoionizing levels in argon

		$3s\,3p^6\,(^2S_{\frac{1}{2}})\,np\,{}^1P_1^0$			$3s\,3p^5\,(^3P)\,nl\,n'l'$	
		$(^2S_{\frac{1}{2}})4p\,{}^1P_1$	$(^2S_{\frac{1}{2}})5p\,{}^1P$	$(^2S_{\frac{1}{2}})6p\,{}^1P$	$(^3P)\,4s\,(^2P_{\frac{1}{2}})\,4p$	$(^3P)\,3d\,(^2P_{\frac{3}{2}})\,4p$
E_r (eV)	(a)	26·614	27·996	28·509	29·224	30·847
	(b)	25·81	27·83	28·44	—	—
q	(a)	−0·22	−0·21	−0·17	−0·2	−1·9
Γ (eV)	(a)	0·080	0·0282	0·0126	0·02	0·006
	(b)	0·99	0·12	0·043	—	—
ρ^2	(a)	0·86	0·83	0·85	0·10	0·20
n^*	(a)	2·277	3·310	4·318	—	—
	(b)	1·99	3·12	4·13	—	—

(a) Observed by Madden, Ederer, and Codling.
(b) Calculated by Ormonde, Smith, and Torres (see p. 3418).

The total photoionization cross-sections calculated by Amus'ya *et al.*,† including correlation effects, agree quite well with the observed values‡, § over a photon energy range from threshold to 300 eV. The Hartree–Fock calculations of Kennedy and Manson‖ reproduce the shape of the variation quite well up to 600 eV but there is a very considerable difference between the predictions obtained using length and velocity matrix elements. Below 30 eV the length values agree well with the observations while between 30 and 50 eV the best agreement is obtained with the velocity formulation.

The asymmetry parameter β for photoelectrons ejected from the $3p$ shell has been measured over an energy range from 2 to 7 eV. There is a considerable spread in the observed data††, ‡‡, §§, ‖‖, ††† but the values

† loc. cit., p. 3481.
‡ SAMSON, J. A. R., *Adv. atom. mol. Phys.* **2** (1966) 177.
§ LUKIRSKII, A. P., BRYTOV, I. A., and ZIMKINA, T. M., *Optika Spektrosk.* **17** (1964) 438; *Optics Spectrosc.* **17** (1964) 234.
‖ loc. cit., p. 3480.
†† MORGENSTERN, R., NIEHAUS, A., and RUF, M. W., *7th int. conf. on the physics of electronic and atomic collisions, Amsterdam*. Abstracts, North Holland, 1971, p. 167.
‡‡ VELESOV, F. I. and LOPATIN, Z. N., *Ves. Lening. Univ. ser. Fiz. i Khim.* **64** (1970).
§§ BERKOWITZ, J., EHRHARDT, H., and TEKAAT, T., *Z. Phys.* **200** (1967) 69.
‖‖ SAMSON, J. A. R., *Phil Trans. R. Soc.* **A268** (1970) 141.
††† CARLSON, T. A., MCGUIRE, G. F., JONAS, E., CHENG, K. L., ANDERSON, C. P., LEE, C. C., and PULLEN, B. P., *Proc. int. conf. on electron spectroscopy, California 1971*. North Holland, 1972, p. 207.

calculated by Kennedy and Manson using the Hartree–Fock field (see p. 3480) are consistent with these data.

While for argon there is already a considerable overlap between the different singly- and doubly-excited Rydberg series of autoionizing levels, for krypton and xenon it is much more difficult still to distinguish the separate series which perturb each other appreciably.

Codling and Madden† have observed a total of 153 resonances in the wavelength range 500–337 Å in krypton and 254 in the range 600–375 Å in xenon. Of these 45 in krypton and 56 in xenon have been grouped tentatively into Rydberg series. As for argon they are associated either with excitation of an inner s electron or double excitation of two outer p electrons. Ederer‡ has analysed the profiles for 12 Kr and Xe resonances.

The total photoionization cross-sections calculated by Kennedy and Manson§ compare with the observed values∥,†† in much the same way as for argon, a conclusion which also applies to the comparison between observed‡‡,§§ and calculated asymmetry parameters β.

Cairns, Harrison, and Schoen∥∥ have measured cross-sections for production of Xe^{++} and Xe^{+++} in a single photoionizing collision. They used time-of-flight mass analysis to distinguish the different ions which could be counted individually with an open electron multiplier. At quantum energies between the threshold, 33 eV, and 67 eV the cross-section for Xe^{++} production is about 30 per cent of the total.

2.5. Alkali-metal atoms

2.5.1. *Total photoionization cross-sections.* Marr and Creek††† have made further measurements of photoionization cross-sections for K, Rb, and Cs, paying particular attention to separating out effects due to the molecules K_2, Rb_2, and Cs_2. Their results disagree rather strongly with earlier data (see Chap. 14, §§ 7.4.3–4) at short wavelengths for K and Cs. For Rb good agreement is obtained with the earlier observations of Mohler and Boeckner (see Fig. 14.34), corrected by including improved vapour pressure data, but there are no early observations at wavelengths

---

† CODLING, K. and MADDEN, R. P., *Phys. Rev.* **A4** (1971) 2261.
‡ EDERER, D. L., ibid. 2263.　　　　　　　　　§ loc. cit., p. 3480.
∥ See references ‡, § on p. 3486.
†† EDERER, D. L. and TOMBOULIAN, D. H., *Phys. Rev.* **133** (1964) A1525.
‡‡ See references ††, ‡‡, §§, ∥∥, ††† on p. 3486.
§§ McGOWAN, J. W., VROOM, D. A., and COMEAUX, A. R., *J. chem. Phys.* **51** (1969) 5626.
∥∥ CAIRNS, R. B., HARRISON, H., and SCHOEN, R. J., *6th int. conf. on the physics of electronic and atomic collisions, Cambridge, Mass.* Abstracts, M.I.T. Press, 1969, p. 127.
††† MARR, G. V. and CREEK, D. M., *Proc. R. Soc.* **A304** (1968) 233.

shorter than 2400 Å with which to compare. In their review Marr and Creek consider that for wavelengths shorter than 2000 Å the observed data are unreliable.

2.5.2. *Autoionization in lithium.* Ederer, Lucatorto, and Madden† have observed resonance effects due to autoionization, in the absorption of photons with energy between 50 and 70 eV in lithium. These resonances are associated with autoionizing states with configuration $1s\,2s\,np$ or $1s\,nl\,n'l'$. A number of well-developed series were observed. Comparison of the observed locations of the resonances with the close-coupling calculations of Cooper *et al.* (see p. 3417) reveals quite good agreement for many states.

2.5.3. *Polarization phenomena.* The first observations of polarization phenomena in photoionization (see p. 3479) were made by Kessler and Lorenz‡ and by Baum, Lubell, and Raith.§ The former observed the polarization P of electrons produced by photoionization of a beam of unpolarized caesium atoms by polarized light. This is given by (7). P was measured by Mott scattering (see Chap. 5, § 6), in the wavelength range 2750 to 3150 Å where it is estimated theoretically to be about 85 per cent. Kessler and Lorenz found 65 ± 15 per cent.

Baum, Lubell, and Raith measured the asymmetry δ defined in (9), between the photoelectron yields from ionization of polarized K, Rb, and Cs atoms by right- and left-handed circularly polarized light. A beam of alkali-metal atoms was polarized by passage through a hexapole magnet and ionized by circularly polarized light from a high-pressure xenon lamp. Values of the parameter x defined by (5) were measured over a photon-energy range from threshold passing through the cross-section minimum. Their results, which agree quite well for Cs with those of Kessler and Lorenz, are in good general agreement with the theoretical calculations of Weisheit and of Chang and Kelly, referred to on pp. 3481–2. Although these theoretical results are not very precise and are semi-empirical there seems little doubt that the interpretation in terms of spin–orbit interaction is sound.

The most thorough detailed study directed towards obtaining electrons with a high polarization and relatively high intensity from photo-ionization has been that of Hughes, Long, Lubell, Posner, and Raith∥

† EDERER, D. L., LUCATORTO, T., and MADDEN, R. P., *Phys. Rev. Lett.* **25** (1970) 1537.
‡ KESSLER, J. and LORENZ, J., ibid. **24** (1970) 87.
§ BAUM, G., LUBELL, M. S., and RAITH, W., ibid. **25** (1970) 267.
∥ HUGHES, V. W., LONG, R. L., LUBELL, M. S., POSNER, M., and RAITH, W., *Phys. Rev.* **A5** (1972) 195.

who observed electrons with the predicted, quite high, degree of polarization produced by photoionization of polarized Li and K atoms.

By deflexion in a hexapole magnetic field, atoms with $m_J = \frac{1}{2}$ were selected from an atomic beam and passed out into a longitudinal magnetic field H strong enough to produce a high degree of decoupling of nuclear and electron spin. To produce 90 per cent polarization of the $m_J = \frac{1}{2}$ atoms, fields of 163, 574, 330, and 182 G are required for Li^6, Li^7, K^{39}, and K^{40} respectively. In the actual experiments allowance must also be made for imperfect state selection in the hexapole field. When this was done the polarization expected for the K beam was 57 per cent with $H = 90$ G. With a comparable field for Li an even higher value, near 80 per cent, may be attained.

Hot wire detectors were used to monitor the K and Li beam intensities. For Li it is necessary to use a detector of high work function and iridium was used. It was calibrated by measurement of the signal as a function of wire temperature determined by a pyrometer.

The light source for the experiment with K atoms was a high-pressure d.c. mercury arc and for the continuous experiments with Li a Xe–Hg high-pressure arc with Suprasil envelope. With Li a powerful pulsed source in the form of a condensed high-voltage spark discharge between tungsten electrodes about 1 mm apart, in 7 atmospheres of argon or xenon, was used at a repetition rate of $10\ s^{-1}$, dissipating about 5 J per flash. Of this about 1 mJ was in the form of radiation effective in producing photoionization of the Li beam.

Light from the source was focused by an aluminized spherical mirror with a MgF_2 protective coating. The converging light, the axis of propagation of which was normal to the atom beam, was reflected through 90° within the vacuum system by a plane mirror through which a hole of diameter 0·47 cm was cut to allow the beam to pass through in the ultimate direction of propagation of the light.

Photoelectrons produced in the atomic beam were collimated by the longitudinal magnetic field H. These electrons were longitudinally polarized and for analysis of their polarization by Mott scattering (see Chap. 5, § 6) it is necessary to convert this to transverse polarization. This was done by electrostatic deflection. The Mott scattering technique also requires that the electrons should be accelerated to an energy of about 100 keV. Careful attention had to be paid to the electron-optical design to ensure that a suitable beam emerged for the final scattering analysis which was carried out using foils of aluminium, gold, and Formvar.

Account had to be taken of certain depolarizing effects. That due to spin–orbit interaction (see (11) of p. 3481) is not important as it only affects the polarization over a narrow range of wavelengths near the minimum for K and is very weak in Li. For K the presence of K_2 molecules is serious. Photoelectrons arising from these molecules, for which the threshold is at a lower frequency than for K, are unpolarized because the ground state of K_2 is a $^1\Sigma$ state. Furthermore, the photoionization cross-section for K_2 is considerably larger than for K (see p. 3505). It was found that by heating the oven orifice, so that the atomic beam emerged from unsaturated vapour, molecule formation was rendered unimportant.

A second depolarizing effect, which was important for K, arises from the near-resonance between the 4050 Å Hg arc line and the $4\,^2S$–$5\,^2P$ resonance line of K. This leads to an appreciable population of excited K atoms which may be photoionized, as was established by observing photoelectron production when the ionizing radiation was passed through glass which absorbed all radiations with wavelength < 3200 Å. The production of photoelectrons from $5\,^2P$ atoms was eliminated by passing the ionizing radiation through about 1 cm of a $NiSO_4$ solution in a quartz tank which absorbed the 4050 Å radiation.

With these arrangements photoelectrons with almost the predicted polarization were observed both for K and for Li.

2.6. *Atomic oxygen and nitrogen*

Comes, Speier, and Elzer[†] have measured the photoionization of atomic oxygen from threshold to a wavelength of 450 Å, using the technique described on p. 3483. Apart from four strong autoionization peaks the observed values agree quite well with those calculated by Dalgarno, Henry, and Stewart (see Fig. 14.39 (a)), using the velocity matrix element, and by Koppel (see p. 3482), at least out to 500 Å. The identification of the autoionization peaks is discussed.

Similar measurements were carried out by Comes, Elzer, and Speier[‡] for ionization of atomic nitrogen. The results agree quite well out to 500 Å with those calculated by Henry using the velocity matrix elements which are comparable with the corresponding calculations of Dalgarno *et al.* for O. They are, however, about 30 per cent smaller than those calculated by Koppel (see p. 3482). Resonances are observed below the $N^{+\,5}S$ limit which agree well in position with the first few levels of the $2s\,2p^3(^5S^0)np\,^4P$ series as calculated by Smith and Ormonde (see p. 3418)

[†] loc. cit., p. 3483.   [‡] loc. cit., p. 3483.

and observed by Carroll, Huffman, Larrabee, and Tanaka† in absorption measurements.

2.7. *Other atoms*

Further measurements of photoionization cross-sections and autoionization parameters have been made in calcium. As in all measurements with vaporized metals the determination of the atom concentration is a major source of error. McIlraith and Sandeman‡ have used a method based on comparison of measurements of photoionization current and of anomalous dispersion (see pp. 248–9) which makes possible the elimination of the atom concentration. They find for Ca at 1886 and 1765·1 Å ionization cross-sections of $52\pm2\times10^{-18}$ and $71\pm10\times10^{-18}$ cm² respectively, which are larger by a factor of 2·2 than the later measurements of Carter, Hudson, and Breig§ which depend on knowledge of the vapour pressure.

The latter authors made detailed measurements of the absolute photoabsorption cross-section of Ca between 2028 and 1753 Å and between 1589 and 1424 Å. They have repeated the analysis for the $^1S_0 \rightarrow {}^1P_1^0$ transitions using two-channel quantum defect theory (see Chap. 14, § 7.4.5) and find better agreement with the observed results for the autoionization structure than obtained by Moores (see p. 1133).

Autoionization structure in the absorption spectrum of strontium has been observed and analysed by Garton, Grasdalen, Parkinson, and Reeves.‖

Marr and Austin have carried out similar experiments and analysis for absorption by cadmium†† and zinc†† for which, in the wavelength region from 1320 to 750 Å, the autoionizing levels arise from excitation of a d electron from the otherwise closed d-shell.

3. Cross-sections for photoionization of molecules

3.1. *Introductory theoretical considerations*

A major problem in the calculation of cross-sections for photoionization of molecules is the determination of the continuum wave-function. In Chapter 14, § 4.2, pp. 1074–6 we described the approximation

---

† CARROLL, P. K., HUFFMAN, K. F., LARRABEE, J. C., and TANAKA, Y., *Ap. J.* **146** (1966) 553.
‡ MCILRAITH, T. J. and SANDEMAN, R. J., *J. Phys. B* (*GB*) **5** (1972) L217.
§ CARTER, V. L., HUDSON, R. D., and BREIG, E. L., *Phys. Rev.* **A4** (1971) 821.
‖ GARTON, W. R. S., GRASDALEN, G. L., PARKINSON, W. H., and REEVES, E. M., *J. Phys. B* (*GB*) **1** (1968) 114.
†† MARR, G. V. and AUSTIN, J. M., *Proc. R. Soc.* **A310** (1969) 137; *J. Phys. B* (*GB*) **2** (1969) 107.

developed by Burgess and Seaton for this wave-function for atoms. Referring to (30) of p. 1075 the wave-function for electrons of wave-number k and orbital angular momentum quantum number l is written in the form

$$r^{-1}G_{kl}(r) = r^{-1}[G_l^c(\epsilon,r)\cos\pi\mu_l(\epsilon) + \{1-\exp(-\tau_l r/a_0)\}^{2l+1}H_l^c(\epsilon,r)\sin\pi\mu_l(\epsilon)] \quad (18)$$

where $\epsilon = k^2 a_0^2$. G_l^c and H_l^c are the wave-functions for motion of electrons in the field of a charge e which have the asymptotic form in atomic units

$$\begin{matrix} G_l^c \\ H_l^c \end{matrix} \sim \begin{matrix} \sin \\ \cos \end{matrix} (kr + \tfrac{1}{2}l\pi + k^{-1}\ln 2kr + \eta_l), \quad (19)$$

with $\eta_l = \arg\Gamma(l+1+i\alpha)$. $\mu_l(\epsilon)$ is the extrapolated quantum defect for the appropriate Rydberg series which converges on the particular continuum state. τ_l is a cut-off parameter $\simeq 10/l(l+1)$.

Tuckwell† has generalized (18) to apply to the continuum states of electrons in the two-centre field of a diatomic molecule. The basic functions involved are now solutions of the wave equation for motion of an electron in the field of two centres of charge $+\tfrac{1}{2}e$ at the appropriate distance R apart. In terms of the usual spheroidal coordinates λ, μ, ϕ such a solution can be written

$$L_l^m(c,\lambda)S_l^m(kR,\mu)e^{\pm im\phi} \quad (l \geq m), \quad (20)$$

where the $S_l^m(d,\mu)$ are spheroidal harmonics. The L_l^m satisfy the equation

$$\frac{d}{d\lambda}\left\{(\lambda^2-1)\frac{dL}{d\lambda}\right\} + \left\{\tfrac{1}{4}k^2R^2\lambda^2 - A_l^m(kR) + R\lambda - \frac{m^2}{\lambda^2-1}\right\}L = 0 \quad (21)$$

with A_l^m determined from the solution of the equation in μ.

Solutions exist with asymptotic form

$$(\lambda^2-1)^{-\tfrac{1}{2}}\begin{matrix}\sin\\ \cos\end{matrix}\{\tfrac{1}{2}kR\lambda + k^{-1}\ln(kR\lambda) + \tfrac{1}{2}(l+1)\pi + \zeta_l^m\}. \quad (22)$$

Denoting these solutions by $(\lambda^2-1)^{-\tfrac{1}{2}}G_l^{m,c}$, $H_l^{m,c}$ respectively then a molecular function corresponding to (18) may be built up in which G_l^c and H_l^c are replaced by $G_l^{m,c}$ and $H_l^{m,c}$ respectively.

Extrapolated quantum defects $\mu_l^m(\epsilon)$ can be obtained in principle from the appropriate Rydberg series of molecular terms‡ and an approximate cut-off factor introduced of the form

$$[1-\exp\{-\tau_l^m(kR)(\lambda-1)\}]^{2l+m-1}. \quad (23)$$

This procedure, while satisfactory in principle, is as yet difficult to

† TUCKWELL, H. C., *J. Phys. B (GB)* **3** (1969) 293.
‡ WEINBERG, M., BERRY, R. S., and TULLY, J. C., *J. chem. Phys.* **49** (1968) 122.

apply because of inadequate observational data on molecular Rydberg series. Also the separate series are often more difficult to disentangle than for atoms (see p. 3499).

As far as autoionization effects appearing in molecular photoionization cross-sections quite close to the threshold are concerned (see pp. 3494–5), it seems that they arise from excitation of vibrationally excited Rydberg states. Berry† has shown that, in general, the autoionization rate which arises from the breakdown of the Born–Oppenheimer separation decreases as $|v_i - v_f|$ increases, where v_i and v_f are the initial and final vibrational quantum numbers respectively. Further discussions of these autoionization rates have been contributed by Bardsley‡ and by Nielsen and Berry.§ The latter authors have compared the rates with those for predissociation. They find that the autoionization rates increase rapidly with v_i, the rate of increase of the rate increasing with $|v_i - v_f|$. For $n > 4$, where n is the total quantum number of the initial Rydberg state, the rate varies as n^{-3}. Predissociation was found to be faster only for a few values of n.

We now consider experimental and theoretical results for different molecules.

3.2. *Hydrogen* (H_2)

Chupka and Berkowitz∥ have carried out a thorough experimental and analytical study of the relative absorption and photoionization cross-sections of normal and para H_2 in the wavelength range 745–810 Å, at 300 and 78 °K. The resolution they obtained, 0·04 Å, is considerably higher than in earlier work (see Chap. 14, p. 1148).

The light source was the Hopfield continuum of helium, generated by a d.c. condensed discharge and dispersed by a 1 m vacuum u.v. monochromator. Ions produced by photoionization were detected and mass analysed by a 60° magnetic sector mass spectrometer.

When working with an ionization chamber cooled with liquid N_2 special precautions were taken to avoid effects due to condensation of insulating material on electrode surfaces. Thus the potentials were applied to grids of fine gold or platinum wires which covered only a small part of the surface and could be warmed with an electric current, a possibility which in fact was not necessary.

A great amount of structural detail due to autoionization was

† BERRY, R. S., *J. chem. Phys.* **45** (1966) 1228.
‡ BARDSLEY, J. N., *Chem. Phys. Lett.* **1** (1967) 229.
§ NIELSEN, S. E. and BERRY, R. S., ibid. **2** (1968) 503.
∥ CHUPKA, W. A. and BERKOWITZ, J., *J. chem. Phys.* **51** (1969) 4244.

observed. In the neighbourhood of the threshold the most prominent absorption transitions are from the $X\,^1\Sigma_g^+$ ground state with $v = 0$ to the $3p\pi\,^1\Pi_u$ ($v' = 6$), $4p\pi\,^1\Pi_u$ ($v' = 3$), $5p\pi\,^1\Pi_u$ ($v' = 2$), and $4p\sigma\,^1\Sigma_u^+$ ($v' = 4$) levels. Of these, all of the rotational lines associated with the transition to $5p\pi\,^1\Pi_u$ ($v' = 2$) are observed to autoionize but none of those associated with $4p\pi\,^1\Pi_u$ ($v' = 3$). For the other two transitions some of the rotational lines autoionize and some do not. From analysis of these latter bands close limits were imposed on the ionization potential of H_2. These depended on the use of selection rules for rotational transitions in autoionization. If K is the quantum number specifying the total orbital angular momentum (nuclear rotation+electron orbital motion) then for transitions to $^1\Sigma_u^+$ states, $\Delta K = \pm 1, \pm 3,...$ and to $^1\Pi_u$, $\Delta K = 0, \pm 2,...$ for the Q branch lines and $\pm 1, \pm 3,....$ for the P and R branches. These limitations are based on symmetry requirements and hence are rigorous. In addition, because the time for electron transitions is small compared with the period of nuclear rotation it is likely that only the minimum changes ΔK will be observed.

On this basis Chupka and Berkowitz† find from analysis of the $X\,^1\Sigma_g^+ \to 3p\pi\,^1\Pi_u$ ($v' = 6$) system that the ionization potential lies between 124416·2 and 124393·5 cm^{-1}, the upper limit being less certain. A recent very precise determination has been made by Herzberg‡ based on determination of the convergence limits of the $R(0)$ lines of the Rydberg series $np\sigma\,^1\Sigma_u^+ - X\,^1\Sigma_g^+$ derived from absorption measurements in para H_2, and of the $Q(1)$ lines of the series $np\pi\,^1\Pi_u - X\,^1\Sigma_g^+$ observed in absorption in normal H_2. He obtains $124418·4\pm 0·4$ cm^{-1}.

Analysis of the $X\,^1\Sigma_g^+ \to 4p\sigma\,^1\Sigma_u^+$ ($v' = 4$) band observed by Chupka and Berkowitz gives less definite results and there is evidence that autoionization occurs with $\Delta K = -3$ for $K = 4$ and $K = 3$ levels of $4p\sigma\,^1\Sigma_u^+$ ($v' = 4$).

An attempt to observe the threshold for direct ionization of H_2 was made using para H_2 at 78 °K but ionization was observed 23 cm^{-1} below the threshold even when special precautions were taken to remove traces of o-H_2. Possible sources of this ionization are ionization of highly excited Rydberg states either by collisions or by the electric field in the ionization chamber.

From the ratio of the measured photoionization and absorption cross-sections at the peaks the relative probability that the excited level decays by autoionization or by predissociation can be obtained.

† CHUPKA, W. A. and BERKOWITZ, J., loc. cit., p. 3493.
‡ HERZBERG, G., *Phys. Rev. Lett.* **23** (1969) 1081.

The results are somewhat lower (in some cases by a factor of 2) than those obtained by Comes and Wellern† who made no mass analysis of the ions and worked at higher pressures.

The autoionization rates for a few lines, whose widths were larger than instrumental, were measured. These were all associated with $np\sigma$ transitions and are given in Table 14 A.2.

TABLE 14 A.2

Measured autoionization rates (in 10^{11} s^{-1}) of certain levels of H_2

Wavelength (Å)	Level designation	Observed rate
800·86	$8\,p\sigma\,(v = 1)$	4
787·90	$8\,p\sigma\,(v = 2)$	13
776·21	$8\,p\sigma\,(v = 3)$	27
762·98	$9\,p\sigma\,(v = 4)$	27
755·06	$8\,p\sigma\,(v = 5)$	20
753·34	$9\,p\sigma\,(v = 5)$	58

These results show a strong correlation between the rate and the lowest possible change Δv of vibrational quantum number in the autoionizing transition. The smaller the value of Δv the faster the rate. A number of other levels with widths below the instrumental value can all ionize with Δv small. Further confirmation of this is provided by the measurements made by Berkowitz and Chupka‡ of the energy distributions of photoelectrons ejected from some of the autoionizing states. In these cases the transition with the minimum value of Δv is the most probable. This behaviour is different from that observed with other molecules (see pp. 1158–64) but agrees with theoretical expectation for autoionizing transitions arising from breakdown of the Born–Oppenheimer approximation (see p. 3493). The order of magnitude of the autoionizing rate also agrees with theory.

Cross-sections for photoionization of the E, B, C, a, and c excited states of H_2 have been calculated by Cohn§ using the Coulomb approximation method (see p. 3492). For the unbound states the phase shifts required were taken from the calculations of Temkin and Vasavada (see p. 3425) instead of from extrapolated quantum defects.

† COMES, F. J. and WELLERN, H. O., *Z. Naturf.* **23A** (1968) 881.
‡ BERKOWITZ, J. and CHUPKA, W. A., *J. chem. Phys.* **51** (1969) 2341.
§ COHN, A., ibid. **57** (1972) 2456.

3.3. H_2^+ and D_2^+

Von Busch and Dunn† have extended and refined their measurements of the photodissociation cross-section of H_2^+ and D_2^+ (see Chap. 14, p. 1153) so that accurate results are now available over a wavelength range from 2472 to 13 613 Å. The design of the experiment is similar to that used by Dunn and von Zyl (Chap. 13, p. 948) to measure cross-sections for dissociation of H_2^+ by electron impact.

The interpretation of the results depends on the nature of the ion source as it determines the vibrational distribution of the H_2^+ (or D_2^+) ions. An electron bombardment source was used with the electron energy of the order 100 eV, large compared with the difference between the excitation thresholds for production of H_2^+ ions in different vibrational states. This was operated at low pressure and a temperature of 100 °C at which the target molecules are all in their ground vibrational state.

The extracted beam of H_2^+ ions was accelerated to 8 keV. Before entering the interaction region it passed through a lens which prevented passage of particles with half the beam energy. This excluded protons formed by break-up collisions with residual gas, an otherwise very serious source of background signals.

A Hg–Xe lamp was used as light source between 2000 and 5800 Å. This was replaced by a pure Xe lamp as a source between 4600 and 10 500 Å. The light beam was chopped at about 28 Hz to facilitate phase-sensitive detection and then passed through a fast grating monochromator of Ebert type. The optical system provided a 1:1 image of the arc at the intersection with the H_2^+ beam, the cone angle being 34°. After passing through the interaction region the light was collected in a cone covered with platinum black to prevent appreciable reflection.

The intensity of the light beam was measured with a power calorimeter consisting of a cup connected by a rod to a metal block as heat sink. The temperature gradient along the rod produced by absorption of the light in the cup is proportional to the power input. Calibration was carried out by feeding a known amount of power into the cup by passing a current through a coil wound round it.

Protons resulting from photodissociation in the interaction region have an energy less than 2·5 eV in the C.M. system and so pass out of the region with almost the same speed and direction as the parent H_2^+ ions. The mixed beam passed into a 45° parallel-plate electrostatic analyser and the separated ions collected in Faraday cups.

The beam-overlap factor (see (36), p. 952) was measured by driving

† VON BUSCH, F. and DUNN, G. H., *Phys. Rev.* **A5** (1972) 1726.

two perpendicular slits through the interaction region, one of the slits carrying a photocell behind it.

Special care was taken to search for and remove systematic errors. It was verified that the signal was proportional to the H_2^+ current and the overlap factor and that it was independent of the beam velocity and the chopping frequency. The possibility of modulation of the background signal by release of adsorbed gas by the light beam was examined in detail and evidence found that it was unimportant. Finally, a small change in the signal with gas pressure in the ion source was found. Because of this the observed signals were extrapolated to zero pressure, involving an increase of 0·8 per cent above the highest measured signal.

The resulting cross-section was compared with that calculated with accurate H_2^+ wave-functions, assuming the initial vibrational distribution of the H_2^+ ions is that obtained if the Franck–Condon principle may be strictly applied to the process of ionization of H_2 molecules in their ground vibrational state by electrons of energy 100 eV or so (see Chap. 13, p. 942 and Fig. 13.39, and also p. 3459). In other words it was assumed that, for this process, (77) of Chapter 14 is applicable. Quite good agreement was found but the differences were nevertheless outside expected experimental error. The reason for this was carefully studied. Effects due to rotational excitation and to autoionization did not seem able to account for it but it was found that almost exact agreement could be obtained if, in calculating the initial vibrational distribution, the dipole moment was allowed to vary with nuclear separation R as

$$1+0\cdot56(R-0\cdot99)^2 \tag{24}$$

where, in atomic units, $1 \leqslant R \leqslant 2$.

This modification of the initial vibrational distribution of the H_2^+ has much less effect on the dissociation by electron impact (see p. 3459).

3.4. He_2^+

Beaty and Browne[†] have calculated cross-sections for photodissociation of He_2^+ through excitation of the repulsive $^2\Sigma_g^+$ and $^2\Pi_g$ states from the ground $^2\Sigma_u^+$ state. Potential energy curves used for the $^2\Sigma_u^+$ and $^2\Sigma_g^+$ states were those given by Gupta and Matsen[‡] and for the $^2\Pi_g$ state by Browne.[§] Full quantum calculations were carried out for initial vibrational states with $v = 0$ to 7. The maximum cross-section is $6\cdot4 \times 10^{-18}$ cm^2 occurring at a wavelength of 2000 Å.

[†] BEATY, E. C. and BROWNE, J. C., *7th int. conf. on the physics of electronic and atomic collisions, Amsterdam*. Abstracts, North Holland, 1971, p. 190.
[‡] GUPTA, B. K. and MATSEN, F. A., *J. chem. Phys.* **47** (1967) 4860.
[§] BROWNE, J. C., ibid. **45** (1966) 2707.

3.5. Oxygen (O_2)

The cross-section for dissociative photoionization of O_2

$$O_2 + h\nu \to O^+ + O + e, \tag{25}$$

has been measured by Comes, Speier, and Elzer[†] using the technique described on p. 3483 which they developed particularly for the measurement of the photoionization of atomic oxygen and nitrogen. They find a maximum cross-section of about $4 \cdot 2 \times 10^{-18}$ cm² for a wavelength close to 525 Å. Some uncertainty attaches to the measurements because a considerable fraction of the O^+ ions are formed with a kinetic energy greater than thermal.[‡] There is therefore some doubt as to whether the transmission of the time-of-flight mass spectrometer used in the experiments remains the same for these energetic ions as for the thermal ions used in the calibration.

Bahr, Blake, Carver, Gardner, and Kumar[§] have measured photoelectron spectra at wavelengths close to several autoionizing states of O_2 as well as at neighbouring wavelengths.

They used the dispersed radiation from a one-metre vacuum monochromator to study the wavelength range from 775 to 947 Å with a resolution of 1·6 Å. The observed spectra corresponding to the autoionization resonances show greater extension of the vibrational structure of the O_2^+ produced towards lower photoelectron energies. Interpreting this in terms of Franck–Condon transitions, values of the equilibrium nuclear separation were determined for these autoionizing states.

The asymmetry parameter β (see p. 3478) determining the angular distribution of photoelectrons has been measured for O_2 by Morgenstern, Niehaus, and Ruf[||] at wavelengths of 584, 736, and 744 Å. For transitions leaving the O_2^+ in the $X\ ^2\Pi_g$ state after ionization by 736 Å radiation, β was found to vary with the final vibrational quantum number (see p. 3499 for N_2).

Jonathan, Morris, Ross, and Smith[††] have observed high-resolution photoelectron spectra from photoionization of $O_2(^1\Delta_g)$ metastable molecules by HeI 584 Å radiation. The $O_2(^1\Delta_g)$ molecules were produced from pure O_2 by an electrodeless discharge, any atomic oxygen being removed by passage over mercuric oxide placed downstream from the

---

[†] loc. cit., p. 3483.
[‡] DOOLITTLE, P. H., SCHOEN, R. I., and SCHUBERT, K. E., *J. chem. Phys.* **49** (1968) 5108.
[§] BAHR, J. L., BLAKE, A. J., CARVER, J. H., GARDNER, J. L., and KUMAR, V., *J. quant. Spectrosc. radiat. Transfer* **11** (1971) 1853.   [||] loc. cit., p. 3486.
[††] JONATHAN, N., SMITH, D. J., and ROSS, K. J., *Chem. Phys. Lett.* **53** (1970) 3758; with MORRIS, A., *J. chem. Phys.* **54** (1971) 4954.

discharge before the photoionization region. Three new bands were observed in the photoelectron spectrum which disappeared when a copper wire was inserted between the discharge and photoionization regions. $^2\Pi_g$, $^2\Pi_u$, $^2\Phi_u$, and $^2\Delta_g$ states of O_2^+ could result from photoionization of a valence shell electron of $O_2(^1\Delta_g)$. The first was positively identified but the identifications of the other two bands, which correspond to adiabatic ionization potentials of 18·81 eV and 16·45 eV, are still uncertain.

3.6. Nitrogen (N_2)

Calculations of cross-sections for photoionization of N_2 have been carried out by Schneider and Berry† and by Tuckwell.‡ The former authors obtained the phase-shifts ζ_l^m for the motion of the ejected electron from the quantum defects of two Rydberg series (see p. 3492). They then used these phase-shifts to fix the parameters in a simple model pseudopotential representing the interaction of the active electron with the core.

Tuckwell calculated the cross-section for photoionization of N_2, by radiation of frequency extending from the threshold value to 500 Å, leaving the N_2^+ ion in either the $X\,^2\Sigma_g^+$, $A\,^2\Pi_u$, or $B\,^2\Sigma_u^+$ states. For the bound state wave-function, combinations of Slater-type orbitals expressed in spheroidal coordinates were used. An attempt was made to use the procedure described on p. 3492 with phase-shifts for the continuum wave determined by extrapolation of quantum defects observed for the appropriate Rydberg series. However, the problem of disentangling the different series makes this procedure unreliable. Most calculations were carried out taking as continuum states the appropriate solution for motion in the field of two fixed centres of charge $+\tfrac{1}{2}e$.

Comparison with experimental data for the total photoionization cross-section (see Chap. 14, § 8.5) shows quite good agreement for wavelengths from the threshold to 661 Å but the calculated values become considerably too large at shorter wavelengths. The relative magnitudes of the cross-section leaving the ion in the different states considered are in qualitative agreement. The calculations of Schneider and Berry are only comparable with the observations quite close to the threshold.

The asymmetry parameter β has been measured by Morgenstern et al.,§ and behaves similarly to that for O_2 (p. 3498).

Bahr, Blake, Carver, Gardner, and Kumar|| have observed photoelectron spectra in N_2 at six well-defined resonances between 775·7 and

† SCHNEIDER, B. and BERRY, R. S., Phys. Rev. 182 (1969) 141.
‡ TUCKWELL, H. C., J. Phys. B (GB) 3 (1970) 293.　　§ loc. cit., p. 3498.
|| BAHR, J. L., BLAKE, A. J., CARVER, J. H., GARDNER, J. L., and KUMAR, V., J quant. Spectros. radiat. Transfer 11 (1971) 1839.

744·5 Å using the same technique as for O_2 (see p. 3498). Similar features were found and estimates made of the equilibrium nuclear separations.

Cook, Klingberg, and McNeal[†] have observed the photoionization spectrum in the wavelength range 795–840 Å of vibrationally excited N_2 produced in an electric discharge. With the discharge on, 50 new bands were observed most of which can be ascribed to transitions between $N_2(X\,^1\Sigma_g^+, v = 1$ to $5)$ and autoionizing states converging either to $N_2^+(A\,^2\Pi_u)$ or vibrationally excited levels of $N_2^+(X\,^2\Sigma_g^+)$.

3.7. CO and NO

Bahr et al.[‡] have observed photoelectron spectra in CO and NO at a number of well-defined resonances between 796·8 and 838·5 Å and between 771·3 and 914·6 Å respectively. Similar features were found as for O_2 and N_2.

3.8. Halogen molecules F_2, Cl_2, Br_2, I_2

3.8.1. F_2 and HF. A considerable amount of attention has been devoted to experimental study of the photoionization of F_2 with the aim of obtaining definite information about the dissociation energy $D(F_2)$ of the molecule.

From a study[§] of the equilibrium in partly dissociated fluorine gas $D(F_2)$ was found to be $1·592\pm0·006$ eV while, from the vacuum ultra-violet spectrum of F_2, Iczkowski and Margrave[‖] derived a value $1·63\pm0·1$ eV, in marked disagreement with a later value $1·44\pm0·07$ eV derived by Stricker and Krauss.[††] Evidence in favour of an even lower value was obtained by Dibeler, Walker, and McCulloh[‡‡] from the appearance potentials for dissociative photoionization of F_2, determined from measurements made with similar apparatus to that used in the experiments referred to on p. 1148. They derived $D(F_2) = 1·34\pm0·03$ eV. Furthermore, the same authors[§§] determined the thresholds for the production of H+ ions through photoionization of HF from which they derived $D(HF) = 5·74\pm0·03$ eV, appreciably lower than that, $5·86\pm0·01$ eV, obtained spectroscopically by Johns and Barrow.[‖‖] $D(HF)$ is related to $D(F_2)$ by

$$D(HF) = \tfrac{1}{2}D(F_2)+\tfrac{1}{2}D(H_2)-\Delta H_{fo}(HF), \qquad (26)$$

[†] Cook, G. R., Klingberg, R. A., and McNeal, R. J., *7th int. conf. on the physics of electronic and atomic collisions, Amsterdam.* Abstracts, North Holland, 1971, p. 157.
[‡] loc. cit., p. 3498.
[§] Stampfer, J. G. and Barrow, R. F., *Trans. Faraday Soc.* **54** (1958) 1592.
[‖] Iczkowski, R. P. and Margrave, J. L., *J. chem. Phys.* **30** (1959) 403.
[††] Stricker, W. and Krauss, L., *Z. Naturf.* **23A** (1968) 486.
[‡‡] Dibeler, V. H., Walker, J. A., and McCulloh, K. F., *J. chem. Phys.* **50** (1969) 4592. [§§] ibid. **51** (1969) 4230.
[‖‖] Johns, J. W. C. and Barrow, R. F., *Proc. R. Soc.* **A251** (1959) 504.

where ΔH_{fo}, the heat of formation of HF from its elements, is $-64\cdot8\pm0\cdot3$ kcal/mole ($-2\cdot81$ eV). With D(HF) as determined by Dibeler et al. $D(\text{F}_2)$ comes to $1\cdot38\pm0\cdot03$ eV, consistent with the value they derived directly.

Berkowitz, Chupka, Guyon, Holloway, and Spohr† therefore carried out a further experimental study in which special attention was paid to the criteria for locating the thresholds, allowing for finite slit widths and thermal spread. They used similar apparatus to that for their experiments on absorption and photoionization of H_2 (see p. 3493) the wavelength resolution being $0\cdot4$ Å.

The relative photoionization cross-sections were measured at 75 and 300 °K, from threshold to 620 Å, for production not only of F_2^+ and F^+ ions but also F^- arising from the polar photodissociation (cf. Chap. 14, § 8.12),
$$\text{F}_2 + h\nu \rightarrow \text{F}^+ + \text{F}^-. \tag{27}$$

If the ionization cross-section rises linearly from threshold the effect of a finite slit width is to add a 'tail' to an otherwise linear rise. Extrapolation of the linear portion to zero then gives the correct threshold. On the other hand if the ionization at and near threshold is represented by a step function the finite slit width converts it into a sigmoid curve. In this case extrapolation of the tangent at the point of inflexion to the zero energy axis yields an energy smaller than the true threshold by $\frac{1}{2}\delta$ where δ is the half-width of the slit function (assumed triangular). In the experiments of Berkowitz et al. $\delta = 0\cdot4$ Å corresponding to 8 meV. On the other hand Dibeler et al. chose as threshold that energy at which the first measurable signal is obtained. For F_2^+ production a linear extrapolation of the results of Berkowitz et al. gave $15\cdot684$ and $15\cdot680$ eV as the ionization energy derived from the observations at 78 °K and 300 °K respectively. It was considered that a correction $\frac{1}{2}\delta$ should be included giving $15\cdot684\pm0\cdot006$ eV as the average value. This agrees well with results obtained from photoelectron spectra (see p. 1177) and from Rydberg series extrapolation.‡

Autoionization structure is observed near threshold. This fades out at about 770 Å but new structures appear below 740 Å. Data from photoelectron spectra (see Chap. 14, p. 1177) show that at least four vibrational levels ($v = 0$ to 3) of $\text{F}_2^+(^2\Pi_g)$ are appreciably populated and that the spin–orbit splitting in the $^2\Pi_g$ state is 337 ± 40 cm^{-1}. The first results suggest that, near threshold, the autoionization process is similar

† BERKOWITZ, J., CHUPKA, W. A., GUYON, P. M., HOLLOWAY, J. H., and SPOHR, R., J. chem. Phys. **54** (1971) 5165.
‡ ICZKOWSKI, R. P. and MARGRAVE, J. L., loc. cit., p. 3500.

to that in H_2, involving transfer of vibrational energy to the electron in the excited Rydberg orbital. On this basis the energies of the auto-ionizing levels were calculated using the ionization energy as above, the vibrational constant for the $^2\Pi_g$ state of F_2^+ and spectroscopic information on quantum defects. Good agreement was obtained with the observed locations of the peaks.

The cross-section for production of F^- shows two strong peaks near 791 and 792 Å as well as other smaller peaks. As pointed out by Dibeler et al.† the transition

$$F_2 + h\nu \to F^+(^3P) + F^-(^1S) \qquad (28)$$

involves a change of spin and should be very improbable (see Chap. 18, p. 1756). However, it is difficult to interpret the structure observed other than in terms of Rydberg states which converge to the same $^2\Pi_g$ state of F_2^+. Berkowitz et al. suggest that the potential energy curves of these states cross some curve which dissociates to $F^+(^3P) + F^-(^1S)$. They produce reasons why this may be a $^3\Sigma_u$ curve which is crossed near a turning point of the Rydberg curve. Under these otherwise near-adiabatic conditions there may be an appreciable probability of a jump to the $^3\Sigma_u$ curve even though the interaction is very weak.

The cross-section for F^+ production shows exactly the same structure in the region 770–800 Å as does that for F^-. With the experimental arrangement used the sensitivity of the mass spectrometer is much higher for positive than for negative ions. To allow for this the negative ion current was normalized to agree with that of F^+ in the wavelength range below the threshold for dissociative ionization

$$F_2 + h\nu \to F^+ + F + e. \qquad (29)$$

Having normalized the negative ion current in this way it was then subtracted from the total F^+ current above this threshold. When this was done, the remaining current, representing the relative cross-section for dissociative ionization, was found to exhibit a clear threshold at 652·50 Å (19·001 eV), allowance being made for the slit function as before. With allowance for thermal rotational energy of about 0·007 eV this may be taken as 19·01 eV which is to be compared with 18·76 eV derived from their observations by Dibeler et al. who did not allow for the contribution from the reaction (28). Since the ionization energy of atomic fluorine is 17·42 eV the new threshold energy for (29) gives $D(F_2) = 1\cdot59$ eV which agrees within estimated experimental error with the original data. For further confirmation the threshold for (28),

† loc. cit., p. 3500.

assuming the electron affinity of F, should be at 797·1 Å very close to an observed threshold at 796 Å.

An elaborate theoretical calculation† of $D(F_2)$ including allowance for correlation gives 1·55 eV in close agreement with the higher value.

Berkowitz et al. also re-examined the photoionization of HF and DF. They find adiabatic ionization energies of $16·00\pm0·010$ eV for HF and $16·03\pm0·010$ eV for DF. The former value is a little lower than that, 16·05 eV, derived from photoelectron spectroscopy but higher than that obtained by Dibeler et al. (15·92 eV). The threshold energies for the reactions

$$HF+h\nu \to H^++F, \qquad (30\,a)$$

$$DF+h\nu \to D^++F \qquad (30\,b)$$

were found to be $19·42\pm0·01$ eV and $19·489\pm0·005$ eV respectively for (30 a) and (30 b). The value for (30 a) is higher than that of Dibeler et al. by 0·105 eV, the difference arising from the procedure adopted to derive the threshold value from the data. The new values give $D(HF) = 5·845$ eV, agreeing well with that derived by Johns and Barrow and consistent through (26) with $D(F_2) = 1·59$ eV.

Finally, Chupka and Berkowitz‡ have measured the kinetic energies of ions produced in the polar photodissociation processes

$$HF+h\nu \to H^++F^-, \qquad (31\,a)$$

$$F_2+h\nu \to F^++F^-, \qquad (31\,b)$$

at a number of wavelengths near threshold. Account was taken of the thermal energy of the neutral molecules as in the analysis of data on dissociative ionization by electron impact (see pp. 843–5). The results obtained for the threshold frequencies for production of H⁺ ions of different kinetic energies from (31 a) are completely consistent with $D(HF) = 5·85$ eV. For the F⁺ ions from (31 b) the extrapolation is complicated by the fact that the ions may be in the $^3P_0$, $^3P_1$, or $^3P_2$ states but the results are consistent with $D(F_2) = 1·59$ eV.

3.8.2. Cl_2. The technique of translational spectroscopy (see p. 3448) has been applied to the photodissociation of Cl_2.§, ‖ In the experiments of Busch, Mahoney, Morse, and Wilson a molecular beam of Cl_2 was crossed at 90° by an intense pulse of polarized light from the second harmonic of a ruby laser. The duration of a pulse was about 20 ns and

† DAS, G. and WAHL, A. C., *Phys. Rev. Lett.* **24** (1970) 440.
‡ CHUPKA, W. A. and BERKOWITZ, J., *J. chem. Phys.* **54** (1971) 5126.
§ BUSCH, G. E., MAHONEY, R. T., MORSE, R. I., and WILSON, K. R., ibid. **51** (1969) 449.
‖ DIESEN, R. W., WAHR, J. C., and ADLER, S. E., ibid. **50** (1969) 3635.

it produced about 2×10^{18} photons cm^{-2}. Cl atoms recoiling in a direction perpendicular to the plane of the intersecting beams were observed by a universal detector (see p. 1345 (e)) so a time-of-flight energy analysis could be performed.

The flux of the product atoms was found to show a peak when the direction of the electric vector was perpendicular to the direction of recoil of the atoms. This shows that the transition dipole moment is perpendicular to the nuclear axis, consistent with the interpretation of the process as a transition from the $^1\Sigma_u^+$ ground state to a $^1\Pi_u$ state rather than to the neighbouring $^3\Pi_u$ state. This is in agreement with the observed translational energy of the Cl atoms which supports the interpretation

$$\text{Cl}_2(^1\Sigma_u^+) \rightarrow \text{Cl}_2(^1\Pi_u) \rightarrow \text{Cl}(^2P_{\frac{3}{2}}) + \text{Cl}(^2P_{\frac{3}{2}}) \tag{32 a}$$

and rules out

$$\text{Cl}_2(^1\Sigma_u^+) \rightarrow \text{Cl}_2(^3\Pi_u) \rightarrow \text{Cl}(^2P_{\frac{3}{2}}) + \text{Cl}(^2P_{\frac{1}{2}}). \tag{32 b}$$

3.8.3. Br_2 *and* I_2. A considerable amount of data on absorption and photoionization in Br_2 and I_2 has been obtained recently.

The vacuum ultraviolet spectrum of Br_2 has been photographed by Venkateswarlu† over the wavelength range 1700–1170 Å. From the convergence limit of five Rydberg series the ionization energy is determined as $10 \cdot 56 \pm 0 \cdot 01$ eV ($85\,165 \pm 80$ cm^{-1}). This agrees well with the value, $10 \cdot 55$ eV, obtained from the photoionization threshold‡ but is a little higher than that measured from the photoelectron spectrum§ ($10 \cdot 51$ eV), possibly because the latter value is an effective average of ionization energies of molecules with vibrational quantum numbers 0, 1, and 2.

Carver and Gardner‖ measured the absorption and photoionization cross-section of Br_2 in the range 1180 to 600 Å using a double ion chamber (see p. 1087). They also measured photoelectron spectra at 584 Å and 1216 Å, obtaining values for the differences between the energies for ionization which leave the Br_2^+ ion in the $^2\Pi_{\frac{1}{2},g}$, $^2\Pi_{\frac{3}{2},g}$, $^2\Pi_u$, and $^2\Sigma_g^+$ states. These differ appreciably from the earlier results of Frost, McDowell, and Vroom§ (see p. 1177). Relative yields of Br_2^+ ions in the different final states were determined.

For I_2 the absorption and photoionization cross-sections have been measured between 2300 and 1050 Å by Myer and Samson,†† 584 to 1245 Å

---

† VENKATESWARLU, P., *Can. J. Phys.* **47** (1969) 2525.
‡ WATANABE, K., *J. chem. Phys.* **26** (1956) 542.
§ FROST, D. C., McDOWELL, C. A., and VROOM, D. A., ibid. **46** (1967) 4255.
‖ CARVER, J. H. and GARDNER, J. L., *J. quant. Spectros. radiat. Transfer* **12** (1972) 207.
†† MYER, J. A. and SAMSON, J. A. R., *J. chem. Phys.* **52** (1970) 716.

by Carver and Gardner,† and between 600 and 1000 Å by Yoshino, Ida, Wakiya, and Suzuki.‡ Photoelectron spectra have also been measured by Carver and Gardner† at 584 and 1216 Å. Myer and Samson observed two thresholds in the photoionization yield curve which they associated with direct production of I_2^+ and with polar dissociation. From the latter, using the ionization potential of I_2 derived by Venkateswarlu§ as in his work on Br_2, the electron affinity of I was obtained as $3·073\pm0·014$ eV which is close to values derived in other ways (see Chap. 15, Table 15.5, p. 1272). As for Br_2 Gardner and Carver derived the differences between the ionization energies required to leave the I_2^+ ion in the different final states as well as determining the relative yields of ions in these states. Agreement of the ionization energy differences with the earlier results of Frost et al.|| is closer than for Br_2.

3.9. HCN

Berkowitz, Chupka, and Walter†† have recently investigated the photoionization of HCN and have, in particular, determined relative cross-sections for production of H^+ by dissociative ionization and of CN^- by polar dissociation. From the difference in the threshold values for the two processes the electron affinity of CN was determined as $3·82\pm0·03$ eV.

3.10. Alkali-metal dimers

Creek and Marr‡‡ (see p. 3487) have measured cross-sections for photo-ionization of K_2, Rb_2, and Cs_2 in the range 3000–1100 Å. In conjunction with earlier data§§ (see p. 1097) the results show that, near the threshold for atomic photoionization, the cross-sections are around 10^{-18} cm² for Na_2, K_2, and Rb_2, rising to several times this value at longer wavelengths.

4. Multiphoton ionization of atoms and molecules

In Chapter 14 we discussed double-photon ionization both theoretically by standard perturbation methods and experimentally for ionization of caesium. The subject has now advanced considerably and multiphoton processes, involving as many as fifteen or so photons in

† loc. cit., p. 2503.
‡ YOSHINO, M., IDA, A., WAKIYA, K., and SUZUKI, H., J. phys. Soc. Japan 27 (1969) 976.
§ VENKATESWARLU, P., Bull. Am. phys. Soc. 14 (1969) 622.
|| FROST, D. C., McDOWELL, C. A., and VROOM, D. A., J. chem. Phys. 26 (1967) 4255.
†† BERKOWITZ, J., CHUPKA, W. A., and WALTER, T. A., ibid. 50 (1969) 1497.
‡‡ CREEK, D. M. and MARR, G. V., J. quant. Spectros. radiat. Transfer 8 (1968) 1431.
§§ HUDSON, R. D., J. chem. Phys. 43 (1965) 1790; HUDSON, R. D. and CARTER, V. L., Phys. Rev. A139 (1965) 1426.

some cases, have been studied experimentally using focused high-power Q-switched laser beams. When so many photons are involved perturbation theory becomes cumbersome and semi-classical methods would seem more appropriate. We shall begin by summarizing the theoretical developments and then discuss them in relation to experimental results.

4.1. *Theories of multiphoton ionization*

Some time ago Bebb and Gold† discussed multiphoton ionization in terms of perturbation theory. According to this theory the rate of photo-ionization per atom due to absorption of N photons, in which the electron is emitted in the direction θ, ϕ, within the solid angle $d\Omega$ is

$$w^{(N)}(\theta,\phi)\,d\Omega = \frac{m}{\hbar}(\alpha F\nu)^N |M_{\text{if}}^{(N)}|^2 k\,d\Omega, \tag{33}$$

where α is the fine structure constant, ν the frequency of the radiation, F the photon flux, k the wave-number of the ejected electron, and $M_{\text{if}}^{(N)}$ the Nth order perturbation matrix element. This is given by

$$M_{\text{if}}^{(N)} = \sum_{\alpha\beta\gamma\ldots} \left\{ \frac{z_{i\alpha} z_{\alpha\beta} \ldots z_{\omega f}}{(\nu_{i\alpha}+\nu)(\nu_{\alpha\beta}+2\nu)\ldots(\nu_{\psi\omega}+(N-1)\nu)} \right\}, \tag{34}$$

where $\alpha, \beta, \gamma, \ldots$ distinguish separate atomic states of energy $E_\alpha, E_\beta, \ldots$, etc. $\nu_{\alpha\beta}$ is then written for $h^{-1}(E_\alpha - E_\beta)$. $z_{\alpha\beta}$ is the dipole matrix element in the z-direction for an $\alpha \to \beta$ transition.

The problem of carrying out the summation in (34) is difficult. No allowance has been made for damping so that at certain frequencies singularities will arise due to vanishing denominators in (34). To avoid this a term $-\frac{1}{2}i\gamma$ was added to each factor in the denominator in (34), γ being taken as the combined width of the initial and final states, following Weisskopf and Wigner.‡ Bebb and Gold then carried out the summations involved in (34) by introducing average frequencies. Usually there will be some intermediate atomic state, λ say, for which $\nu_{i\lambda} + P\nu = 0$, $1 \leqslant P \leqslant N$. A major contribution is then likely to arise from terms involving this denominator. Bebb and Gold therefore write

$$M_{\text{if}}^{(N)} = \sum_\lambda \frac{z_{\lambda f}^{N-P} z_{i\lambda}^P}{(\nu_{i\lambda}+P\nu-\frac{1}{2}i\gamma) \prod_{S=1,\neq P}^{N} \{\bar{\nu}_\lambda(S)+S\nu-\frac{1}{2}i\gamma\}}, \tag{35}$$

where $z_{\alpha\beta}^S = \langle\alpha|z^S|\beta\rangle$ and $\bar{\nu}_\lambda(S)$ is a mean frequency which depends on the state λ. As a further approximation a mean $\bar{\nu}_\lambda(S)$ was replaced by $\bar{\bar{\nu}}$

---

† BEBB, H. B. and GOLD, A., *Phys. Rev.* **143** (1966) 1.
‡ WEISSKOPF, V. F. and WIGNER, E. P., *Z. Phys.* **63** (1930) 54; **65** (1930) 18.

assumed independent of λ. From detailed calculations for atomic hydrogen it was found that $\bar{\nu}$ may be taken as the value of $\nu_{i\alpha}$ where α is the first excited state. This was assumed in applications to ionization of other atoms. Even with these approximations numerical calculations are quite formidable.

Bebb and Gold† calculated ionization rates for H as functions of photon energy for multiphoton processes with $N = 2$ to 12. In all cases resonance peaks occur at certain energies due to the vanishing of the real parts of certain denominators in the expressions for $M_{if}^{(N)}$. Results were also obtained for the rare gases with $N = 14, 13, 9, 8$, and 7 for He, Ne, Ar, Kr, and Xe respectively.

TABLE 14 A.3

Calculated values of two-photon ionization cross-sections per unit light intensity, in 10^{-33} cm^4 W^{-1}, for atomic hydrogen in the ground state

Wavelength (Å)	(a)	(b)	(c)	(d)
1020	—	6·75	5·52	5·522
1100	1·0	0·4	0·4049	0·4049
1200	66	63·03	58·03	58·03
1300	11	12·76	12·83	12·83
1400	7	8·450	8·453	8·453
1600	6·8	9·154	9·143	9·143
1700	8·4	10·25	10·24	10·23

(a) BEBB, H. B. and GOLD, A., loc. cit., p. 3506.
(b) CHANG, F. T. and TANG, C. L., *Phys. Rev.* **185** (1969) 42.
(c) KLARSFELD, S., *Nuov. Cim. Lett.* **2** (1969) 548.
(d) GONTIER, Y. and TRAHIN, M., loc. cit., below.

More elaborate procedures for carrying out the summations for hydrogen atoms have been introduced by Gontier and Trahin.‡ Table 14 A.3 gives a comparison between two-photon ionization rates in terms of the cross-section per unit light intensity (see Chap. 14, p. 1231) calculated by different authors who used different methods for carrying out the state sum in (34).

It is important to have some idea of the range of validity of perturbation theory. The expression (33) may be written

$$w^{(N)} = \text{const} \times (0.85 \times 10^{-33} Fh\nu)^N |M_{if}^{(N)}|^2 k \tag{36}$$

with the quantum energy $h\nu$ expressed in eV, F in cm^{-2} s^{-1}, and the matrix elements $z_{\alpha\beta}$ in atomic units. If $M_{if}^{(N)}$ varies slowly with N we

† loc. cit., p. 3506.
‡ GONTIER, Y. and TRAHIN, M., *Phys. Rev.* **A4** (1971) 1896.

could expect the perturbation treatment to be valid if $F < 10^{33}\,\text{cm}^{-2}\,\text{s}^{-1}$. However, owing to the effect of resonance denominators the assumption of gradual variation of the matrix element with N may be unjustified in certain conditions. Thus, although the photon flux in typical experiments is less than $10^{33}\,\text{cm}^{-2}\,\text{s}^{-1}$ we cannot assume that the perturbation theory is valid in all cases.

A different approach to the problem, which throws some light on the conditions of validity of the perturbation theory, is the semi-classical theory first introduced by Keldysh[†] and then expanded and refined by a number of other investigators.[‡] Keldysh proceeded from the limit of very low frequencies in which case the theory given in the early work of Oppenheimer[§] is applicable, ionization taking place by leakage of an electron through the potential barrier which is present when the atom is subjected to the action of a uniform electric field $\mathscr{E}$. The rate of ionization of an H atom is given by

$$w(\nu = 0) = \frac{e^2}{2ha_0}\frac{\mathscr{E}_\text{H}}{\mathscr{E}} \exp\left\{-\frac{2}{3}\frac{\mathscr{E}_\text{H}}{\mathscr{E}}\right\}, \qquad (37)$$

where $\mathscr{E}_\text{H} = e^2/a_0^2$, provided $\mathscr{E}/\mathscr{E}_\text{H} \ll 1$. This formula in the static limit remains adiabatically valid for an alternating electric field of frequency ν provided the time taken for the electron to tunnel through the barrier is short compared with the period of oscillation of the field. The width of the barrier is $E_\text{H}/e\mathscr{E}$ where E_H is the ionization energy $e^2/2a_0$ and the average electron speed is of order $(2E_\text{H}/m)^{\frac{1}{2}}$, m being the electron mass. The tunnelling time τ is therefore given by

$$\tau = \nu_\text{t}^{-1} = (\tfrac{1}{2}E_\text{H}\,m)^{\frac{1}{2}}/e\mathscr{E}, \qquad (38)$$

and the situation remains quasi-adiabatic provided

$$\nu/\nu_\text{t} \ll 1. \qquad (39)$$

In terms of multiphoton ionization it may be shown that, under these conditions, a major contribution comes from processes in which very large numbers of low-frequency photons are concerned. In other words we are very far away from conditions in which perturbation methods are applicable. On the other hand when $\nu/\nu_\text{t} \gg 1$ but $\nu \ll E_\text{H}/h$ the major contribution comes from the N-photon process in which N is the smallest integer greater than $E_\text{H}/h\nu$ and the ionization rate varies as $\mathscr{E}^{2N}$ or, since the photon flux F is given by $(c\mathscr{E}^2/8\pi h\nu)$, as F^N. Allowance may be

[†] KELDYSH, L. V., *JETP (USSR)* **47** (1964) 1945 (*Soviet Phys. JETP* **20** (1965) 1307).
[‡] PERELOMOV, A. M., POPOV, V. S., and TERENT'EV, M. V., *JETP (USSR)* **50** (1966) 1393 (*Soviet Phys. JETP* **23** (1966) 924); *JETP (USSR)* **51** (1966) 309 (*Soviet Phys. JETP* **24** (1967) 207); *JETP (USSR)* **52** (1967) 514 (*Soviet Phys. JETP* **25** (1967) 336).
[§] OPPENHEIMER, J. R., *Phys. Rev.* **31** (1928) 66.

made for resonance effects† in this treatment. It suggests that the perturbation theory should give good results when

$$v_t/v = e\mathscr{E}/(\tfrac{1}{2}E_\mathrm{H} m)^{\frac{1}{2}}v = (16\pi \hbar e^2 F/E_\mathrm{H} mvc)^{\frac{1}{2}} \ll 1. \tag{40}$$

Putting in numerical values for hydrogen this requires that

$$5 \times 10^{-31} F/\epsilon \ll 1, \tag{41}$$

where F is in cm^{-2} s^{-1} and ϵ is the photon energy in eV. In general, for an atom of ionization energy E_i the condition will be

$$5(E_\mathrm{i}/E_\mathrm{H})10^{-31} F/\epsilon \ll 1. \tag{42}$$

This is a little more severe than the condition we arrived at from consideration of the perturbation theory.

Reiss‡ has also developed a semi-classical theory which is valid under the conditions
$$h\nu/E_\mathrm{i} \ll 1, \qquad h\nu_t/E_\mathrm{i} \ll 1. \tag{43}$$

A further matter of interest concerns the relative effectiveness of circular and plane-polarized radiation in producing multiphoton ionization. Klarsfeld and Maquet,§ using perturbation theory, find that the maximum ratio of the ionization rate by circularly as compared with linearly polarized light is $(2N-1)!!/N!$ which increases quite rapidly with N. In any actual case the ratio may be greater or less than unity.

4.2. *Experimental study of multiphoton ionization*

The first measurements of the rate of multiphoton ionization of an atom were made by Voronov and Delone‖ for xenon, using ruby laser radiation of quantum energy 1·79 eV. Since that time a number of other measurements have been made using the same radiation from a Q-switched laser or from a neodymium glass laser (quantum energy 1·16 eV). In some cases measurements have also been made using the first harmonic of the laser radiation.

The principal difficulty in obtaining quantitative results is the measurement of the effective photon flux and the interaction volume. It is usually a good approximation to write the flux F in the form

$$F(x, y, z, t) = F_0 \psi(x, y, z) \phi(t), \tag{44}$$

† KOTOVA, L. P. and TERENT'EV, M. V., *JETP (USSR)* **52** (1967) 732 (*Soviet Phys. JETP* **25** (1967) 481).
‡ REISS, H. R., *Phys. Rev.* **A1** (1970) 803; **6** (1972) 817; **D4** (1971) 3533. See also FAISAL, F. H. M., *J. Phys. B (GB)* **5** (1972) L196.
§ KLARSFELD, S. and MAQUET, A., *Phys. Rev. Lett.* **29** (1972) 79. See also FAISAL, F. H. M., *J. Phys. B (GB)* **5** (1972) L233.
‖ VORONOV, G. S. and DELONE, N. B., *JETP (USSR)* **50** (1965) 78 (*Soviet Phys. JETP* **23** (1966) 54).

where x, y are cartesian coordinates in a plane normal to that of propagation of the light from the laser and z is measured along that direction, the origin being at the focus. F_0 is such that the total number of photons $\mathcal{N}$ passing through the focal region during a laser pulse is given by

$$\mathcal{N} = F_0 S \tau, \qquad (45)$$

with $$S = \iint \psi(x, y, 0)\, dx dy, \qquad \tau = \int \phi(t)\, dt. \qquad (46)$$

If the flux distribution does not change with intensity then, provided F_0 is not too large, the ionization rate W per atom will be proportional to F_0^k so that
$$W = \delta F_0^k. \qquad (47)$$
The number of ions n_i formed per pulse is then given in terms of the atom concentration by
$$n_i = n \delta V_k \tau_k F_0^k, \qquad (48)$$
where V_k, τ_k are the effective volume and irradiation times respectively for a k-photon process. Thus

$$V_k = \int \psi^k(x, y, z)\, dx dy dz, \qquad (49\,\mathrm{a})$$

$$\tau_k = \int \phi^k(t)\, dt. \qquad (49\,\mathrm{b})$$

It is clear that accurate measurement of ψ and ϕ is very important because they appear to the kth power in these formulae. There is no special problem in measuring ϕ with a photocell of resolution less than 1 ns. In typical experiments ψ has been measured by attenuating the laser beam and then forming an image of the focal spot on a photographic film using a microscope objective. The intensity distribution over the spot may be measured photoelectrically. This can be repeated to give measurements for different values of z by displacing the microscope along the z-axis. The total number of photons emitted per pulse can be determined calorimetrically.

In the early experiments of Voronov, Delone (G. A.), and Delone (N. B.) in krypton† the gas pressure was 10^{-4} torr. The time τ was measured both for the total flux and the flux very close to $z = 0$ and found to be 35·4 and 37·0 ns respectively, consistent with the assumption of the form (44) for F. The effective cross-sectional area at $z = 0$ was $2 \cdot 85 \times 10^{-6}$ cm^2 and the total number $\mathcal{N}$ of photons per pulse $10^{17 \cdot 8}$. With these values the flux F_0 is $6 \cdot 3 \times 10^{30}$ cm^{-2} s^{-1} corresponding to an alternating electric field of amplitude $1 \cdot 15 \times 10^7$ V cm^{-1}. From measurements of the variation of n_i with $\mathcal{N}$ the exponent k in (47) was obtained as 6·3. With this value $\tau_k = 13 \cdot 7$ ns, $V_k = 2 \times 10^{-8}$ cm^3.

† VORONOV, G. S., DELONE, G. A., and DELONE, N. B., *JETP (USSR)* 51 (1966) 1660 (*Soviet Phys. JETP* 24 (1967) 1122).

According to the perturbation theory the rate of ionization per atom should be proportional to F^N, where N is the smallest integer greater than $E_i/h\nu$, E_i being the ionization energy and ν the photon frequency. For most of the experiments carried out in the rare gases the results show a variation as F^k, where F is the appropriate mean flux, but with $k < N$. Thus for xenon, Voronov and Delone[†] in their first experiments found $k = 6.2$ whereas $N = 7$, for krypton Voronov, Delone, and Delone[‡] found $k = 6.3$ as compared with $N = 8$ and Agostini et al.,[§] who used Nd laser radiation, found $k = 9$ as compared with $N = 12$. The latter authors also found for argon, neon, and helium, values of $k = 10, 13$, and 18 as compared with $N = 14, 19$, and 21.

The reason for these discrepancies is not completely clear. They were first ascribed to the failure of the perturbation treatment to take account of the disturbance of atomic energy levels by the intense electromagnetic field. As the field increases level-broadening increases so that resonance effects due to near equality of an integral multiple of the photon energy with a (modified) atomic energy become more likely. However, at the flux densities available, it would not be expected, from the considerations outlined on pp. 3508–9 that there would be sufficient level-broadening for these effects to be important.

Chin, Isenor, and Young[||] pointed out that, in typical experiments, the number of ions produced per laser pulse is not small compared with the number of target atoms present within the effective reaction volume. Under these circumstances of at least partial saturation, the slope of a plot of $\log(n_i/\tau)$ against $\log F$ where n_i is the total number of ions produced per pulse, τ the pulse width at half-height, and F the effective photon flux, will no longer be a straight line of slope N. As F increases the yield will be less than that given by this line, in at least qualitative agreement with observation. Chin et al. observed ionization of Hg and Xe using radiation from a Q-switched ruby laser. For Xe at low power ($F < 10^{29.7}$ cm^{-2} s^{-1}) they found n_i/τ to vary as F^k with $k = 7.4 \pm 0.8$, consistent with the expected value of 7. The best fit for all values of F used in the experiments (up to 10^{30} cm^{-2} s^{-1}) gave $k = 5.8 \pm 0.4$ as compared with 6.23 ± 0.14 found by Voronov and Delone,[†] who worked with a maximum peak flux of about $10^{30.25}$ cm^{-2} s^{-1}. The absolute value of the ionization rate δF^k, derived by Chin et al. from their low power data,

[†] loc. cit., p. 3509.
[‡] VORONOV, G. S., DELONE, G. A., and DELONE, N. B., loc. cit., p. 3510.
[§] AGOSTINI, P., BARJOT, G., BONNAL, J. F., MAINFRAY, G., MANUS, C., and MORELLEC, J., *J. quant. Electronics* **4** (1968) 667
[||] CHIN, S. L., ISENOR, N. R., and YOUNG, M., *Phys. Rev.* **188** (1969) 7.

gave δ as $10^{-203 \cdot 7 \pm 2 \cdot 8}$ cm^{2k} s^{k-1} consistent with the value calculated by Bebb and Gold† ($10^{-205 \cdot 5}$ cm^{2k} s^{k-1}) and derived by Voronov and Delone‡ from their observations ($10^{-206 \cdot 3 \pm 3 \cdot 2}$ cm^{2k} s^{k-1}).

For Hg, Chin et al.§ again found evidence of saturation at high powers but for $F < 10^{29 \cdot 4}$ cm^{-2} s^{-1} a linear log plot was obtained with $k = 6 \cdot 3 \pm 0 \cdot 7$ consistent with the expected value of 6. In this case δ comes out to be $10^{-171 \cdot 5 \pm 2 \cdot 3}$ cm^{2k} s^{k-1}.

Chin and Isenor‖ have developed the theory of the saturation effect in more detail but insufficient information is available with which to check it thoroughly. For a laser pulse length τ of 10 ns they found, on the basis of ionization rates calculated by perturbation theory, that saturation effects would become important at photon fluxes near $10^{27 \cdot 5}$, 10^{29}, and $10^{29 \cdot 8}$ cm^{-2} s^{-1} for Cs, K, and Hg respectively.

A number of experiments have been carried out on the ionization of Cs and K which require relatively few photons per atom. Held, Mainfray, and Morellec†† observed the ionization of K with Nd-glass laser radiation at comparatively low power (flux between 6×10^{27} and 6×10^{28} cm^{-2} s^{-1}). They found a linear log–log plot giving $k = 3 \cdot 9 \pm 0 \cdot 2$ consistent with the expected value of 4. When allowance was made for the fact that the laser radiation included a great number of modes so that the ionization probability would be increased by a factor close to 4! above that given by a single mode, the absolute ionization rate was consistent with that calculated from perturbation theory.

Evans and Thonemann‡‡ investigated ionization of caesium by ruby laser light over a flux range from $10^{27 \cdot 6}$ to 3×10^{29} cm^{-2} s^{-1}. Below $10^{27 \cdot 9}$ cm^{-2} s^{-1} a linear relation between $\log n_i$ and $\log F$ was obtained with the expected slope of 3. Above this flux the yield fell off as if saturation were occurring and indeed this is at about the value at which the effect should become apparent according to the estimates of Chin and Isenor. However, with further increase of flux the yield reached a maximum near $F = 5 \times 10^{28}$ cm^{-2} s^{-1}, falling to a minimum near 7×10^{28} cm^{-2} s^{-1} and exhibiting other structural features as it rose with the flux up to the highest flux observed. Evans and Thonemann interpret the maximum as arising from near-resonance between the excitation energy of the $9D$ states of Cs and that of two laser photons which becomes detuned by line broadening due to the electric field. Other effects

† loc. cit., p. 3506.   ‡ loc. cit., p. 3509.
§ loc. cit., p. 3511.
‖ CHIN, S. L. and ISENOR, N. R., Can. J. Phys. 48 (1970) 1445.
†† HELD, B., MAINFRAY, G., and MORELLEC, J., Phys. Lett. A39 (1972) 57.
‡‡ EVANS, R. G. and THONEMANN, P. C., ibid. A39 (1972) 133.

probably arise from the existence of more than one region where the light intensity is a local maximum.

Fox and his collaborators have investigated especially the relative effectiveness of circularly polarized as against linear polarized light in the triple† and double photoionization‡ of Cs due to the first and second harmonics of ruby laser radiation. They found the circularly polarized radiation to be 1·28 and 2·15 times as effective in the respective cases.

† Fox, R. A., Kogan, R. M., and Robinson, E. J., *Phys. Rev. Lett.* **26** (1971) 1416.
‡ Kogan, R. M., Fox, R. A., Beunham, G. T., and Robinson, E. J., *Bull. Am. phys. Soc.* **16** (1971) 1411.

15
COLLISIONS INVOLVING ELECTRONS AND PHOTONS—PHOTODETACHMENT AND RADIATIVE ATTACHMENT

THE most important developments of technique for the study of photodetachment have followed the introduction of laser sources for the detaching radiation. On the one hand, by using a laser beam at a fixed frequency it has been possible to measure very precisely the energy spectrum and angular distribution of the detached electrons. From the former it is possible to determine the energies associated with particular vibrational transitions, when the target is a negative molecular ion, and hence to obtain the electron affinity as distinct from the vertical detachment energy. Observations of the angular distribution make it possible to obtain information about the nature of the final state, a matter of some interest again in dealing with molecular ions.

By using a tunable dye laser as source it is possible to determine with high resolution the fine structure in a photodetachment cross-section near threshold. Another tunable source consisting of an optical parametric oscillator pumped with a Chromatix Model 1000 E Nd YAG laser has been used in connection with an ion cyclotron spectrometer to determine relative cross-sections for photodetachment, and from the threshold frequency, the electron affinity of the ion concerned.

For many applications it is necessary to know photodetachment cross-sections for negative molecular ions in thermal equilibrium with the surrounding gas whereas in the crossed-beam experiments the ions will have a distribution of vibrational states which is far from thermal. This difficulty may be overcome in principle if measurements may be made of the energy spectrum of the ejected electrons, but even in such experiments the relative intensity of transitions from the ground or low-lying vibrational states may be extremely weak and hard to measure accurately if the ion beam possesses a high degree of vibrational excitation. In an attempt to overcome these difficulties Woo, Branscomb, and Beaty[†] carried out experiments in which photodetachment occurred from ions which were drifting in a buffer gas at pressure p under the

[†] Woo, S. B., BRANSCOMB, L. M., and BEATY, E. C., *J. geophys. Res.* **74** (1969) 2933.

action of an electric field F such that F/p is not large enough for the mean energy of the ions to exceed appreciably the thermal value. In this way the ions undergo about 10^6 collisions before reaching the region within which photodetachment occurs, and so should be in thermal vibrational equilibrium.

A drift tube of the type used by Beaty and Patterson for mobility and ionic reaction measurements (see Chap. 19, p. 1976) was adapted for these experiments which were first applied to O_2^-. The light source was a high-pressure xenon arc chosen because its spectrum resembles that of sunlight. To avoid the necessity of determining the absolute total intensity of the light as a function of position in the interaction region measurements were made relative to O^- for which the photodetachment cross-section is well known as a function of frequency (see Chap. 15, § 5.2). This involved measurements for each ionic species of the total number of ions and electrons per pulse and the drift velocity. The ions were identified by their mobility, a procedure which leads to ambiguity because O_4^- ions which may arise under some conditions have nearly the same mobility as O_2^- (see p. 3634).

We give a brief account of the experimental arrangements involved in this new work before describing new results obtained for different negative ions in these ways as well as by the application of techniques developed earlier which are described in Chapter 15.

1. New experimental techniques

1.1. *The measurement of kinetic energies and angular distributions of electrons resulting from photodetachment*

The first experiments to measure the energy spectra of electrons resulting from photodetachment were carried out by Brehm, Gusinow, and Hall† and applied to measure the energy of the metastable He^- ion in the $1s\,2s\,2p\,^4P$ state.

The general arrangement of the apparatus, used also in later experiments, consists of a hot cathode arc discharge source of negative ions, rather similar to that in the early experiments of Seman and Branscomb (Chap. 15, p. 1266). The ions are focused by a series of Einzel and quadrupole lenses and mass-analysed by passage through a Wien filter from which they emerge with an energy of 700 eV. They are focused to a diameter less than 1 mm through the intercavity focused beam (focal spot about $\frac{1}{4}$ mm diam.) from an argon ion laser operated at 4880 Å or occasionally at 5145 Å.

† BREHM, B., GUSINOW, M. A., and HALL, J. L., *Phys. Rev. Lett.* **19** (1967) 737.

Electrons produced by photodetachment which are emitted into a solid angle of about 10^{-2} sr, in a direction normal to both intersecting beams, pass out of the field-free interaction chamber into a hemispherical electron analyser and are detected by a particle multiplier. It is possible for calibration, including the elimination of contact potential effects, to accumulate data giving the electron spectra from two different ions, the ion under study being changed every 8·3 ms.

Discrimination against background is carried out by subtraction, the laser being pulsed at 120 Hz.

The angular distribution of the electrons relative to the direction of polarization of the laser light is measured by inserting an accurate half-wave plate between the laser discharge and interaction chamber to change the direction of polarization as desired.

Counting rates varied from 10^3 s^{-1} for the H$^-$ photodetachment peak to about 1 s^{-1} for the weaker peak in C$^-$.

The technique provides a powerful new means for studying photodetachment from molecular ions as it is possible to distinguish clearly electrons arising from different vibrational transitions

$$XY^-(v'') \rightarrow XY(v').$$

The accuracy in location of the peaks associated with these transitions and in determining their relative intensities is so high that it is possible not only to determine the true electron affinity to high precision but also to determine the molecular parameters r_e and B_e for the electronic states of the negative ion.

Applications of this kind have been made to NO$^-$[†] and O$_2^-$ and are discussed in § 3.1.

1.2. *Use of tunable lasers*

In their experiments on the fine structure in the photodetachment cross-section of S$^-$ near threshold Lineberger and Woodward[‡] used a mass-analysed, focused beam of ions at an energy of 1 keV, crossed at 90° by the radiation from a Rhodamine 6G dye laser. The laser pulses are of 0·3 μs duration, of energy not more than 1 mJ and tunable from 6200 to 5700 Å with an echelle grating.

The negative ions are separated electrostatically from the neutral atoms which are produced by photodetachment. The ion beam is monitored by collection in a Faraday cup and the neutral atom beam

[†] SIEGEL, M. W., CELOTTA, R. J., HALL, J. L., LEVINE, J., and BENNETT, R. A., *Phys. Rev.* **A6** (1972) 607. See also CELOTTA, R. J., BENNETT, R. A., HALL, J. L., SIEGEL, M. W., and LEVINE, J., ibid., p. 631.

[‡] LINEBERGER, W. C. and WOODWARD, B. W., *Phys. Rev. Lett.* **25** (1970) 424.

by a 15-stage electron multiplier. A constant fraction of the laser cavity power is separated by a beam splitter and monitored by a thermopile.

The first use of an ion cyclotron-resonance spectrometer to measure electron affinities through observation of threshold frequencies for photodetachment was made by Brauman and Smyth.† Negative ions formed by electron capture were trapped within an analyser cell, as in the experiments on ionic reactions described on p. 3617, and detected by monitoring the radio-frequency power absorbed from an oscillator tuned to the cyclotron frequency of a particular ion.

In one type of experiment‡ the instrument was operated in a steady state, with the ions formed continuously in the source region and allowed to drift under the action of crossed electric and magnetic fields to enter the analyser region. By irradiating this region with light from the laser at an average power level of 30 mW as many as half of the PH_2^- ions present suffered photodetachment.

In a second type of experiment‡ a pulsed ion cyclotron-resonance spectrometer was used. Negative ions were formed by a 30 ms pulse of the electron beam and trapped within the cell for a time around 200 ms, during which they could be irradiated with a number of laser pulses. Resonance signals with and without irradiation made it possible to study photodetachment with particular reference to the determination of threshold frequencies.

With this equipment it is possible to study a variety of identifiable negative ions in quite small concentrations.

2. Results for specific negative atomic ions (see Chap. 15, §§ 5 and 6)

2.1. H⁻

Hall and Siegel,§ using the apparatus described on p. 3515, have measured the angular distribution of electrons resulting from photodetachment from H⁻ by the 4880 Å radiation from an argon ion laser. As only an s–p transition is involved the theoretical distribution is simply given by $\cos^2\theta$, where θ is the angle between the direction of motion of the electrons and the polarization of the laser beam. The observed results are consistent with this.

2.2. O⁻

Similar observations of the angular distribution of the ejected photoelectrons as for H⁻ have been made by Hall and Siegel.§ In this case

† BRAUMAN, J. I. and SMYTH, K. C., *J. Am. chem. Soc.* **91** (1969) 7778.
‡ SMYTH, K. C., McIVER, R. T., BRAUMAN, J. I., and WALLACE, R. W., *J. chem. Phys.* **54** (1971) 2758.   § HALL, J. L. and SIEGEL, M. W., ibid. **48** (1968) 943.

p–s and p–d transitions are involved and the distribution would be expected to deviate somewhat from the isotropic form which would be expected if the p–d transition rate were negligible. Detailed calculations of the expected distribution have been carried out by Cooper and Zare† using the same approximations as those made in the calculation of total photoionization cross-sections by Robinson and Geltman (see p. 1266). The results, which show only comparatively small departures from a $\sin^2\theta$ form ($\beta = -1$), agree very well with the observations of Hall and Siegel. The observed value for the asymmetry parameter β (see (1), p. 3478) is $-0\cdot885$ as compared with $-0\cdot95$ calculated.

2.3. C⁻

Hall and Siegel‡ have also measured the angular distribution of electrons photodetached from C⁻ by polarized laser light at 4880 Å and 5145 Å respectively. The asymmetry parameter β (see (1) p. 3478) was measured as $-0\cdot715$ and $-0\cdot805$ for the respective wavelengths as compared with $-0\cdot65$ and $-0\cdot73$ calculated by Cooper and Zare.†

In the course of these experiments, from measurements of the electron energy compared with corresponding measurements for O⁻, the electron affinity of carbon was found to be $0\cdot195\pm0\cdot010$ eV less than that of O. Taking the latter to be $1\cdot465\pm0\cdot005$ eV (see p. 1263) this gives $1\cdot270\pm0\cdot010$ eV for carbon (cf. Table 15.4, p. 1267).

2.4. S⁻

Lineberger and Woodward,§ applying the technique described on p. 3516, have observed the relative cross-sections for photodetachment from S⁻ over an energy range from 2 to $2\cdot18$ eV. A number of apparent thresholds were observed of which four could be identified unambiguously from the known fine-structure splitting of the ground ^3P state of neutral S. Following from these identifications the data given in Table 15 A.1 were obtained, including a very accurate value for the electron affinity of S.

The separation (397 cm⁻¹) between transitions (a) and (b) and between (c) and (d) is almost equal to that between the $^3P_2$ and $^3P_1$ level of S, $396\cdot8$ cm⁻¹, while that between (a) and (c) and between (b) and (d), 482 cm⁻¹, can be taken to be that between the $^2P_{\frac{1}{2}}$ and $^2P_{\frac{3}{2}}$ levels of S⁻.

The cross-section, involving a p–s transition, should rise above a

---

† COOPER, J. and ZARE, R. N., *J. chem. Phys.* **48** (1968) 942.
‡ loc. cit., p. 3517. § loc. cit., p. 3516.

threshold as $\Delta E^{\frac{1}{2}}$, where ΔE is the energy of the photon above the threshold value. This was confirmed by the observed results.

The relative strengths of the different transitions in Table 15 A.1 are not simply in the ratio of the statistical weights of the states concerned. It is suggested by Lineberger and Woodward that this occurs because the final state, arising from attachment of an s electron to an atom in a $^3P_1$ state is a linear combination of $^2P_{\frac{3}{2}}$ and $^2P_{\frac{1}{2}}$ states.

TABLE 15 A.1

Observed data on transitions in S⁻ leading to photodetachment

Transition	Threshold		Relative intensity	Remarks
	cm⁻¹	eV		
(a) $^2P_{\frac{1}{2}} \to {}^3P_2$	16 273	2·0175	1	
(b) $^2P_{\frac{1}{2}} \to {}^3P_1$	16 670	2·0667	2·3	
(c) $^2P_{\frac{3}{2}} \to {}^3P_2$	16 755	2·0772±0·0005	7·6	Electron affinity
(d) $^2P_{\frac{3}{2}} \to {}^3P_1$	17 152	2·1264	3·9	

2.5. He⁻

The $1s\,2s\,2p\,^4P$ term of He⁻, which lies below the $2\,^3S$ level of He, is metastable towards autodetachment (see Chap. 24, p. 2850). In particular the $^4P_{\frac{5}{2}}$ level has a lifetime of about 0·35 ms.[†] It is therefore possible to observe photodetachment from a beam of ions in this state in the usual way—ions of kinetic energy 1 keV travel a distance of 50 m before suffering appreciable autodetachment.

Brehm, Gusinow, and Hall[‡] have applied the technique described on p. 3515 to measure the energy of the electrons produced by photodetachment from a beam of He⁻ ions. The ions were produced, together with H⁻ and D⁻ ions, by double charge exchange (see Chap. 24, p. 2851) from the corresponding positive ions on passage, at an energy of 2·5 keV, through an oven containing potassium vapour at a pressure of a few times 10⁻³ torr. In this way a well-defined beam of ions with a quite small energy spread is produced. These ions were then decelerated to 700 eV before entering the interaction chamber.

In determining the energy of the photoelectrons in the centre of mass system allowance must be made for the effect of recoil momentum gained by the atom. This may be calculated from the kinetic energy of the ions, or calibrated by comparison with the observed shift between the electron energies observed with H⁻ and D⁻, or from the observed splitting

[†] BLAU, L. M., NOVICK, R., and WEINFLASH, D., *Phys. Rev. Lett.* **24** (1970) 1268.
[‡] loc. cit., p. 3515.

between the energies resulting when the He atom is left in $2\,^3S$ or $2\,^3P$ states. Calibration to avoid effects of potential difference is carried out by comparison with results for H⁻ or D⁻ for which the electron affinity is accurately known.

The final value obtained for the electron affinity, relative to He($2\,^3S$) is $0\cdot080\pm0\cdot002$ eV. This is quite close to the value $0\cdot072$ eV calculated by Weiss.†

The probability of the He atom being left in the $2\,^3S$ state under the experimental conditions, in which the electrons collected were ejected in a direction perpendicular to both intersecting beams, was found to be about 7 times larger than for it to be left in a $2\,^3P$ state.

2.6. Cl⁻

Mück and Popp‡ have applied the technique described in Chapter 15, § 4 to study the radiation emitted from the axis of a low-current arc in chlorine including the affinity continuum of Cl. From a quantitative study on similar lines to that described on pp. 1253–4 they obtained an electron affinity of $3\cdot616$ eV which is very close to the values derived from shock-wave absorption and emission studies (see Table 15.5). The photodetachment cross-section derived by detailed balance has a maximum of about $1\cdot2\times10^{-17}$ cm² at a photon energy of $4\cdot25$ eV.

3. Results for specific negative molecular ions

3.1. NO⁻ and O_2^-

In applying the technique described in § 1.1 to NO⁻ and O_2^- Celotta et al.§ produced both ions by direct extraction from a glow discharge source containing N_2O at a pressure of $0\cdot05$ torr. This yielded a beam of about 2×10^{-8} Å of NO⁻ and O_2^- and about 6×10^{-8} Å of O⁻ which could be used for calibration purposes.

As in the corresponding experiments for He⁻ allowance must be made for the recoil momentum gained by the atom in the detaching process and for any departure of the direction of motion of the electrons before collection from being normal to the ion beam. This was done as for He⁻ by comparison with measurements for H⁻ and D⁻.

Great care was taken in the calibration of the hemispherical electron energy analyser. The difference in energy of the photoelectrons from S⁻ and O⁻ was measured and compared with the difference in the electron affinities as measured by Lineberger and Woodward for S⁻ (see

---

† Weiss, A. W., private communication.
‡ Mück, G. and Popp, H. P., *Z. Naturf.* **23** (1968) 1213.
§ loc. cit., p. 3516.

p. 3518) and by Branscomb et al. for O^-. The observed energy difference was $3\cdot0\pm0\cdot5$ per cent too small. An error of about the same magnitude was found in the observed energy difference for transitions from a single O_2^- vibrational state to the $v' = 1$ levels of the ground $X\,^3\Sigma_g$ and excited $a\,^1\Delta_u$ electronic states of O_2. Assumption of the same error throughout the energy range involved in the experiments led to completely consistent interpretation of the data for both NO^- and O_2^-.

Photoelectron spectra including 7 clearly defined peaks were obtained for NO^- while for O_2^- at least 10 peaks were observed, some of which arose from transitions to vibrational levels of the excited $a\,^1\Delta_u$ state of O_2.

The major problem is the identification of the different transitions involved. In both cases, for transitions between the same electronic states the peaks are nearly equally spaced in energy. This suggests that they correspond either to transitions in which the initial or final vibrational quantum number v'' or v' remains the same. More precisely it is found that the spacing decreases smoothly with the electron energy. This shows that it is the initial number v'' which is constant.

To determine this constant value of v'' account is taken of the fact that it is only for transitions from $v'' = 0$ that the 'envelope' of the vibrational transition probabilities associated with an upper electronic state shows a single maximum. This is so for both NO^- and O_2^- and hence v'' may be taken as 0 for both. Further evidence for this comes from the fact that in no measurements using NO^- beams produced under different source conditions was a different spectrum obtained. For O_2^- some weak transitions from $v'' = 1$ were observed.

Finally, to determine the values of v' associated with the different peaks three different procedures were adopted. The first involved very precise measurement of peak spacing and comparison with that between the different vibrational levels of the neutral molecule. With the analyser correcting factor of $1\cdot03$ introduced, very good agreement was obtained with a particular assignment of v'. The second method makes use of the Franck–Condon intensity distribution. This gives a smooth variation of intensity with v'. For NO^- extrapolation of the observed intensity envelope to smaller v' would give peaks discernible above the background. The absence of such peaks can only mean that $v' = 0$ for the last observed peak on the high energy side. For O_2^- more detailed calculations of the distribution were made, based on an assumed Morse potential for the negative ion in which the parameters ω_e'' and $\omega_e x_e''$ were taken from the experiments of Boness and Schulz (see p. 3443). r_e was varied together with the assignment of v' until a good fit was obtained with the observed peak heights. Finally, measurements were made of shifts in the peaks produced by replacing O^{16} by the O^{18} isotope. From these, knowing ω'_e, $\omega_e x'_e$, ω''_e, and $\omega_e x''_e$ it is possible to determine v' unambiguously. All these methods gave completely consistent results.

Having determined the vibrational assignments the vertical detachment energies may be obtained. Values of 0·0364 and 0·4392 eV are found for the (0, 0) transition but these must be corrected for rotational excitation and for spin–orbit multiplicity. From the temperature of the source, determined from the relative population of the $v' = 0$ and 1 vibrational levels of O_2^-, the rotational distributions in the negative ions may be obtained. Using the known statistical weights and selection rules the effect of transitions between the various fine structure and rotational levels on the peak location can be determined and appropriate corrections made to determine the true electron affinity. These were found to be $-0·012$ eV for NO^- and $+0·001$ eV for O_2^-, giving the final electron affinities

$$E_A(NO) = 0·024 \text{ eV}, \qquad E_A(O_2) = 0·44 \text{ eV}.$$

This value for NO is in agreement with the inference from the threshold determination† for the charge transfer reaction of NO and I^- that $E_A(NO) \geqslant 0·09 \pm 0·1$ eV. For O_2 the result agrees well with that determined by Pack and Phelps (see Chap. 19, § 4.2.2.3, p. 2096).

The molecular constants determined for NO^- are

$$\omega_e = 1470 \pm 200 \text{ cm}^{-1} \; (0·182 \pm 0·023 \text{ eV}), \qquad r_e = 1·258 \pm 0·010 \text{ Å},$$
$$B_e = 1·427 \pm 0·02 \text{ cm}^{-1}.$$

Spence and Schulz, using the trapped electron technique (see p. 284), obtained $\omega_e = 0·170 \pm 0·020$ eV which agrees very well.

For O_2^- the constants obtained are

$$r_e = 1·341 \pm 0·010 \text{ Å}, \qquad B_e = 1·17 \pm 0·02 \text{ cm}^{-1}.$$

The value for r_e is somewhat smaller than that determined from the position of low energy resonances in electron scattering from O_2 (see p. 3444).

The measurements of Woo, Branscomb, and Beaty,‡ on the photodetachment of electrons from drifting ions in oxygen using the procedure described on p. 3515, gave a photodetachment rate in sunlight, of ions with mobility close to that of O_2^-, of $0·3 \pm 0·1$ s^{-1}. The identity of the ions is not certain as O_4^-, which has much the same mobility (see p. 3634), could well have been formed.

† BERKOWITZ, J., CHUPKA, W. A., and GUTMAN, D., J. chem. Phys. 55 (1971) 2733.
‡ loc. cit., p. 3514.

3.2. SH⁻

Steiner† has published a complete account of his investigations, a preliminary account of which is given in Chapter 15, p. 1277. The electron affinity is found to be $2 \cdot 319 \pm 0 \cdot 010$ eV. The structural parameters, determined as for OH⁻ (see Chap. 15, § 6.1), are the same within experimental error as for the neutral radical, namely r_e (equilibrium separation) $1 \cdot 35 \pm 0 \cdot 02$ Å, $\omega_e = 2700 \pm 300$ cm⁻¹, $B_e = 9 \cdot 46 \pm 0 \cdot 32$ cm⁻¹.

3.3. Other molecular ions

Sinnott and Beaty,‡ using a technique similar to that of Woo, Branscomb, and Beaty except that the xenon arc source was replaced by a dye laser with a spectral line width of 1 Å, have made preliminary measurements of the photodetachment cross-section of O_3^-.

Warneck§ has measured photodetachment cross-sections for NO_2^- by the conventional crossed-beam technique, absolute values being obtained by comparison with H⁻. Although it is difficult to distinguish transitions from different vibrational states Warneck produced evidence that the electron affinity of NO_2 is $3 \cdot 1 \pm 0 \cdot 05$ eV (see, however, p. 3636).

Using the ion cyclotron-resonance technique Smyth et al.∥ observed photodetachment from PH_2^- and NH_2^- obtaining electron affinities of 1·26 and 0·74 eV respectively.

Burt†† has applied the technique referred to on p. 3515 to measure rates for photodetachment of electrons by sunlight from CO_3^-, $CO_3^- \cdot H_2O$, and O_4^- ions.

4. Multiphoton photodetachment

Faisal‡‡ has applied the method of Reiss (see p. 3509) to calculate cross-sections for photodetachment in multiphoton collisions. Two-photon rates for I⁻, Br⁻, and F⁻ at ruby laser frequency are explicitly calculated and the values obtained agree quite well with those calculated by Geltman using perturbation theory (see Chap. 15, p. 1294). For Cl⁻ the three-photon rate was calculated as

$$W^{(3)} = 0 \cdot 12 \times 10^{-80} F^3 \text{ s}^{-1}$$

where F is the photon flux in cm⁻² s⁻¹.

---

† STEINER, B., J. chem. Phys. **49** (1968) 5097.
‡ SINNOTT, G. and BEATY, E. C., 7th int. conf. on the physics of electronic and atomic collisions, Amsterdam. Abstracts, North Holland, 1971, p. 176.
§ WARNECK, P., Chem. Phys. Lett. **3** (1969) 532.   ∥ loc. cit., p. 3517.
†† BURT J. A., J. chem. Phys. **57** (1973) 4649.
‡‡ FAISAL, F. H. M., J. Phys. B (GB) **5** (1972) L258.

16

COLLISIONS BETWEEN ATOMS AND MOLECULES UNDER GAS-KINETIC COLLISIONS—ELASTIC COLLISIONS

1. Introductory remarks and summary

THE first aim in this work has been to establish the principles involved in calculating differential and total cross-sections for elastic collisions between gas atoms and/or molecules with gas-kinetic velocities. This initial phase of the programme, which is largely complete, was described in Vol. 3. The next phase, that of applying the principles to determine the interaction energy between atoms and/or molecules at separations such that this energy is comparable with or less than thermal energy, which had already commenced at the time of writing Vol. 3, has advanced considerably. This is because of the improvements in energy and angular resolution which have been achieved experimentally. Also the first experiments naturally used alkali-metal atoms as one of the colliding partners because of the ease of detection by thermal ionization. Since Vol. 3 was written a number of experiments have been carried out with high resolution in which detection by electron impact ionization and mass analysis has been used.

Whereas the experiments described in Vol. 3 gave little information about atomic interactions apart from the parameters ϵ (the depth of the attractive well) and r_m (the separation at which the interaction is a minimum) (see pp. 1299 and 1334) the latest experiments are beginning to give information about the shape of the interaction as well. In fact it is now worth while to attempt to use observed data to derive the deflexion function ϑ (see p. 1309 (27)) and hence the interaction energy $V(r)$ by inversion. This has been successfully applied already for Na–Hg collisions.

To summarize, the main new developments and results may be listed as follows.

(a) The development of practical techniques for deriving interaction energies by inversion and the successful application to Na–Hg collisions.

(b) The employment of nozzle sources for molecular beams (see p. 1343) which reduce the velocity spread and increase the intensity.

(c) Observation of interference effects due to symmetry in collisions between atoms of the same kind. These effects are observed in the variation of the total cross-section with relative velocity and of the differential cross-section with angle at large angles.

(d) Much improved data on atomic and/or molecular interactions, showing clear departures from the simple empirical Lennard–Jones (12, 6) form.

(e) In particular, new results for He–He indicate that no bound state exists for the He–He system.

In addition, the technique of measuring cross-sections for scattering of state-selected beams has been developed so that data on spin-exchange collisions between alkali-metal atoms and on the dependence of cross-sections for collisions of molecules with atoms on the initial rotational state are now available. However, the latter falls essentially within the subject matter of Chapter 17 as rotational transitions on collision are involved and the former within that of Chapter 18. They will therefore be discussed under those headings.

We shall now enlarge briefly on the subjects falling within Chapter 16 as outlined above.

2. The determination of the interaction energy by inversion, using experimental data

On p. 1309 we defined the classical angle of deflexion

$$\vartheta = \pi - 2 \int_{r_0}^{\infty} \frac{\mathrm{d}r}{r\phi(r)}, \tag{1}$$

where
$$\phi(r) = \left\{ \frac{r^2}{p^2} - 1 - \frac{r^2}{p^2 E} V(r) \right\}^{\frac{1}{2}}, \tag{2}$$

p being the impact parameter, E the relative kinetic energy of impact, and $V(r)$ the interaction energy at separation r. r_0 is the outermost zero of $\phi(r)$. If p is a single-valued function of ϑ and $0 \leqslant \vartheta \leqslant \pi$ then we may equate ϑ with the angle of scattering θ in the C.M. system. Under these conditions, which prevail if the interaction energy is monotonic and repulsive, we may, from the observed differential cross-section $I(\theta)\,\mathrm{d}\omega$, derive the interaction energy at distances beyond the classical closest distance of approach by Firsov's method described in Chapter 22,

p. 2384. Thus we have (see (76), p. 2385)

$$\{p(\theta)\}^2 = 2\int_\theta^\pi I(\theta)\sin\theta\, d\theta. \tag{3}$$

Firsov then shows that, if we consider r as a function of ψ where

$$\psi = r^2\{1 - V(r)/E\}, \tag{4}$$

then

$$r(\psi_0) = \psi_0^{\frac{1}{2}} \exp\left\{\pi^{-1} \int_{\psi_0^{\frac{1}{2}}}^\infty \theta(p)(p^2-\psi_0)^{-\frac{1}{2}}\, dp\right\}. \tag{5}$$

For the collisions we are concerned with p is not a single-valued function of ϑ and the condition $0 \leqslant \vartheta \leqslant \pi$ does not hold for all p as may be seen from Figs. 16.5(a) and 16.6. In fact, for the case in which the rainbow angle $< \pi$, we distinguish the following regions in the variation of θ with p

$$p \to 0, \quad \theta \to \pi, \qquad\qquad \vartheta = \theta, \tag{6a}$$
$$0 < p \leqslant p_g, \quad 0 < \theta < \pi, \qquad \vartheta = \theta, \tag{6b}$$
$$p_g < p < p_r, \quad 0 < \theta < \theta_r, \qquad \vartheta = -\theta, \tag{6c}$$
$$p > p_r, \quad \theta < \theta_r, \qquad\qquad \vartheta = -\theta, \tag{6d}$$
$$p \to \infty, \quad \theta \to 0, \qquad\qquad \vartheta = -\theta. \tag{6e}$$

θ_r is the rainbow angle at which $|d\theta/dp| = 0$. In the region (6b) for $\theta > \theta_r$ the relation between p and θ is single valued and we may use (3–5) but for smaller angles it is not so and (4, 5) are invalid. If, however, in some way we could still determine θ as a function of p from the experimental data, the formulae could still be applied.

Buck and Pauly[†] have devised a practical means for doing this when at least three rainbow maxima (see p. 1321 and Fig. 16.9) can be clearly located. This is based on an improved formula derived by Berry[‡] which gives for the contribution to the differential cross-section for $\theta < \theta_r$ arising from impact parameters p_a, $p_b > p_g$ (i.e. that for which $\vartheta < 0$, see Fig. 16.5), with rapid oscillations averaged out,

$$I(\theta) = \pi\{I_a^{\frac{1}{2}} + I_b^{\frac{1}{2}}\}^2 |z|^{\frac{1}{2}} \operatorname{Ai}^2(-|z|) + \pi\{I_a^{\frac{1}{2}} - I_b^{\frac{1}{2}}\}^2 |z|^{-\frac{1}{2}} \operatorname{Ai}'^2(-|z|) \tag{7}$$

where, in the notation of p. 1313, I_a and I_b have the classical values $(p_{a,b}/\sin\theta)|dp_{a,b}/d\theta|$, Ai and Ai' are the Airy function (see Fig. 16.8) and its first derivative respectively, and

$$4/3 z^{\frac{3}{2}} = 2\eta(p_a) - 2\eta(p_b) + k\theta(p_b - p_a). \tag{8}$$

[†] Buck, U. and Pauly, H., J. chem. Phys. **51** (1969) 1662; Buck, U., ibid. **54** (1971) 1923.
[‡] Berry, M. V., Proc. phys. Soc. **89** (1966) 479.

$\eta(p_{a,b})$ are the phase-shifts given by the semi-classical approximation (see Chap. 16, § 5.3) and k is the wave number of relative motion. As the contribution from the third impact parameter p_c (see p. 1320) will vary smoothly with θ the rainbow maxima are given by the zeros of Ai' and the minima by those of Ai.

To use (7) Buck and Pauly break up the region in which $\vartheta < 0$ into subregions as follows.

(i) Around the rainbow minimum. Here we may write as in (81), p. 1321,
$$\vartheta(p_{a,b}) = -\theta_r + q(p_{a,b} - p_r)^2, \qquad (9)$$
giving
$$z = k^{\frac{2}{3}} q^{-\frac{1}{3}} (\vartheta_r - \theta). \qquad (10)$$

(ii) In the neighbouring region
$$\vartheta(p_a) = -c_1 p_a^{-c_2} \qquad (11)$$
$$\vartheta(p_b) = -a_1(p_b - p_g) \qquad (12)$$
giving, with $\alpha = 1/c_2$,
$$z = (\tfrac{3}{4})^{\frac{2}{3}} \{ 2\eta_g + kp_g \theta + \tfrac{1}{2} k a_1^{-1} \theta^2 - kc_1^\alpha (1-\alpha)^{-1} \theta^{1-\alpha} \}^{\frac{2}{3}}, \qquad (13)$$
where η_g is the phase-shift at the glory angle where $p = p_g$.

(iii) In the asymptotic region $\theta \to 0$, in which the scattering comes mainly from the van der Waals interaction $-C/r^6$, we have
$$c_1 = (15/16)\pi(C/E), \qquad c_2 = 6. \qquad (14)$$

q, θ_r, p_r, η_g, p_g, a_1, c_1, c_2, and C may be taken as parameters to be determined. Of these η_g may be determined if the glory oscillations in the total cross-section (see p. 1328) are observed. A theoretical value may be taken for C while continuity conditions must be satisfied across the boundaries between subregions. This leaves 5 parameters to be determined by least-squares fitting to the observations.

The only gap left is the deflexion function for $\vartheta > 0$ and $< \vartheta_r$. This may be filled by interpolation.

Additional information may be obtained if the rapid oscillations are resolved since these determine the reduced wave number kr_m.

A check was carried out for a simulated case in which the scattering was calculated for a Lennard-Jones potential and the potential then recovered by applying the above procedure.

An alternative approach based on the use of semiclassical action functions and their expansion in terms of the reduced angle variable τ ($= E\theta$) (see Chap. 22, p. 2388), which depends on use of the observed amplitudes of the supernumerary rainbows as well as the location of the extrema of both the rainbows and the rapid oscillations, has been given by Pritchard.† A further alternative has been given by Miller‡ while Klingbeil§ has suggested a somewhat different procedure involving first a phase-shift analysis of the observed differential cross-sections and then use of the relation (73), p. 1320.

TABLE 16 A.1

The interaction energy between Na *and* Hg *atoms*

$r_m = (4.72 \pm 0.5\%) \times 10^{-8}$ cm, $\quad \epsilon = (8.79 \pm 3\%) \times 10^{-14}$ erg.

r/r_m	V/E	r/r_m	V/E	r/r_m	V/E
0.67	2.34	0.86	−0.48	1.22	−0.597
0.68	2.09	0.88	−0.63	1.26	−0.507
0.70	1.70	0.90	−0.748	1.30	−0.423
0.72	1.34	0.94	−0.917	1.40	−0.252
0.74	1.00	0.98	−0.989	1.50	−0.157
0.76	0.68	1.02	−0.996	1.60	−0.102
0.78	0.39	1.06	−0.940	1.70	−0.058
0.80	0.13	1.10	−0.871	1.80	−0.035
0.82	−0.10	1.14	−0.781	1.90	−0.021
0.84	−0.31	1.18	−0.688	2.00	−0.015

2.1. *Application of inversion procedure to* Na–Hg *interaction*

Buck and Pauly‖ carried out high-resolution measurements of the differential cross-section for Na–Hg collisions. By using nozzle beams sufficient intensities were obtained to make possible the use of a velocity selector on the incident Na beam as well as velocity analysis of the mercury beam by a second selector. The relative velocity was resolved to better than 2 per cent and could be measured to within 1·2 per cent.

Measurements were made for relative energies ranging from 28.7×10^{-14} to 40.2×10^{-14} erg and up to eight rainbow maxima could be clearly distinguished and located to $\pm 0.2°$, except for angles less than 4° ($\pm 0.3°$) and for the primary rainbow ($\pm 0.3°$). The superimposed rapid oscillations were also visible.

Using these data in conjunction with the calculated value of the van der Waals constant C (4.76×10^{-58} erg cm^6)†† and observed locations of glory extrema in the velocity dependence of the total cross-section,‡‡ Buck and Pauly, applying their method outlined above, obtained consistent values for the interaction $V(r)$ between Na and

† PRITCHARD, D. E., *Phys. Rev.* **A1** (1970) 1120.
‡ MILLER, W. H., *J. chem. Phys.* **51** (1969) 3631.
§ KLINGBEIL, R., *7th int. conf. on the physics of electronic and atomic collisions, Amsterdam.* Abstracts, North Holland. 1971, p. 529.
‖ BUCK, U. and PAULY, H., *J. chem. Phys.* **54** (1971) 1929.
†† STWALLEY, W. C. and KRAMER, H. L., ibid. **49** (1968) 5555.
‡‡ BUCK, U., KÖHLER, K. A., and PAULY, H., *Z. Phys.* **244** (1971) 180.

Hg atoms from measurements at different impact energies. Their results are given in Table 16 A.1. As compared with a (12, 6) potential with the same ϵ and r_m, the derived potential has a wider potential well and tends to zero more rapidly as r increases.

Experiments† on K–Hg and Cs–Hg collisions have also been carried out with the same apparatus and analysed in the same way. The shape of the interaction is effectively the same for all these collision pairs in the region of the potential well but rises less steeply in the repulsive regions for Na–Hg than for Cs–Hg.

3. The interactions between alkali-metal and rare-gas atoms

The interactions between alkali-metal atoms and rare-gas atoms (see Chap. 16, Tables 16.13 and 16.14) have been determined to greater precision, information being obtained about the shape as well as the parameters ϵ and r_m. This has been achieved by analysing the more accurate data available on differential angular and glory scattering in terms of more elaborate potentials than the $(n, 6)$, $(12, \alpha, 6)$, and $(\exp \alpha, 6)$ forms referred to in Chapter 16, § 6.1.

Buck and Pauly‡ used the form

$$V(r) = \frac{\kappa_1 \epsilon}{m^2 - \kappa_1} \left\{ \left(\frac{r_m}{r}\right)^n - \frac{n^2}{\kappa_1}\left(\frac{r_m}{r}\right)^{\kappa_1/n} \right\} \quad \begin{array}{l} (r \leqslant r_m, \\ n^2 > \kappa_1), \end{array} \quad (15\,\mathrm{a})$$

$$= \frac{36\epsilon}{\kappa_2 - 36} \left\{ \left(\frac{r_m}{r}\right)^{\kappa_2/6} - \frac{\kappa_2}{36}\left(\frac{r_m}{r}\right)^6 \right\} \quad \begin{array}{l} (r \geqslant r_m, \\ \kappa_2 > 36), \end{array} \quad (15\,\mathrm{b})$$

in which ϵ and r_m have the same significance as before while κ_1 and κ_2 are the curvatures of $V(R)$ on the left- and right-hand sides of the potential minimum. This is referred to as an $(n, \kappa_1, \kappa_2, 6)$ potential. Düren, Raabe, and Schlier§ based their analysis on modification of the (12, 6) potential in the neighbourhood of two points r_m and r_1 by writing

$$V(r) = \epsilon f(x) - \epsilon \{f(x)+1\} \left[\Gamma_0 \exp\left\{-\left(\frac{x-1}{\gamma_0}\right)^2\right\} + \Gamma_1 \exp\left\{-\left(\frac{x-x_1}{\gamma_1}\right)^2\right\} \right], \quad (16)$$

with $\qquad x = r/r_m, \qquad f(x) = x^{-12} - 2x^{-6}.$

This not only leaves the significance of ϵ and r_m unaltered but also retains the expression $2\epsilon r_m^6$ for the van der Waals constant.

With these more elaborate potentials the adjustable parameters are sufficient to obtain good fits to the experimental data both on differential

† Buck, U., Kick, M., and Pauly, H., *7th int. conf. on the physics of electronic and atomic collisions, Amsterdam.* Abstracts, North Holland, 1971, p. 543.
‡ Buck, U. and Pauly, H., *Z. Phys.* **208** (1968) 390.
|| Düren, R., Raabe, G. P., and Schlier, Ch., ibid. **214** (1968) 410.

and total cross-sections. Table 16 A.2 compares the values found for ϵ and r_m using the respective interactions. Values obtained by use of the total cross-section (glory undulations) only are bracketed. The remaining values were obtained using both differential and total cross-section data.

TABLE 16 A.2

Parameters ϵ and r_m derived for interactions between alkali-metal atoms and rare-gas atoms using the interactions (15) and (16)

	Ar		Kr		Xe	
	(15)	(16)	(15)	(16)	(15)	(16)
Li						
r_m (Å)	[4·95]	—	[4·87]	[4·65]	[4·90]	—
ϵ (10^{-14} erg)	[0·85]	—	[1·37]	[1·27]	[2·11]	—
Na	5·01	[4·78]	4·96	4·73	5·06	4·91
	0·89	[0·83]	1·39	1·37	2·08	1·99
K	[5·34]	5·05	5·24	[4·84]	5·25	—
	[0·84]	0·86	1·42	[1·45]	2·20	—
Rb	—	—	[5·29]	—	—	—
	—	—	[1·45]	—	—	—
Cs	[5·50]	—	[5·44]	—	[5·47]	—
	[0·90]	—	[1·47]	—	[2·18]	—

Reasonably good agreement is obtained between the two sets of results. As far as the shape of the interaction is concerned the potential well is found to be somewhat broader and the short range repulsion somewhat less steep than for the (12, 6) interaction. This is consistent with the earlier analysis which obtained a better fit with an (8, 6) interaction. The deviations from the (12, 6) form are opposite to those found for interactions between rare-gas atoms (see p. 3537).

High-resolution data on differential cross-sections for collisions of Li atoms with Kr and Xe[†] have been obtained over an impact energy range from 0·03 to 0·3 eV. For Li–Xe at a relative velocity of $2·19 \times 10^5$ cm s^{-1}, 10 maxima were observed in a range of laboratory scattering angles from 1 to 18°. The data were analysed in terms of (12, 6) and Düren–Schlier interactions (see (16)).

Absolute measurements[‡] of the total cross-section for Li–Ar collisions have been made over the relative velocity range $1·4 \times 10^5$ cm s^{-1} to an estimated accuracy of better than 2 per cent. Analysis of the data is

---

[†] AUERBACH, D., DETZ, C., REED, K., and WHARTON, L., *7th int. conf. on the physics of electronic and atomic collisions, Amsterdam*. Abstracts, North Holland, 1971, p. 541.

[‡] URY, C. B. and WHARTON, L., *J. chem. Phys.* **56** (1971) 5832.

estimated to yield an interaction correct to 5 per cent over the range of separations from 6 to 8 Å. At 7 Å it is $-1\cdot 97\times 10^{-15}$ erg.

4. Interactions between alkali-metal atoms

In Chapter 16, § 12.2, the measurements made by Buck and Pauly† on total cross-sections for collisions between alkali-metal atoms are discussed. The derived values of the van der Waals constant, given in Table 16.24, are about one-half of the calculated. Further light on this discrepancy is afforded from experiments on the scattering of spin selected beams and is discussed in the notes on Chapter 18 on p. 3603.

5. Interactions between rare-gas atoms

5.1. *The effect of symmetry on the scattering of atoms by like atoms* (see Chap. 16, pp. 1308 and 1405)

In the collisions between like atoms which obey the Bose–Einstein statistics the differential cross-section is given by

$$I(\theta)\,d\omega = \tfrac{1}{2}(|f(\theta)+f(\pi-\theta)|^2)\,d\omega, \tag{17}$$

where
$$f(\theta) = \frac{1}{2ik}\sum(2l+1)(e^{2i\eta_l}-1)P_l(\cos\theta), \tag{18}$$

η_l being the usual phase shift and k the wave number of relative motion. The resulting total cross-section

$$Q = \frac{8\pi}{k^2}\sum_{l\,\text{even}}(2l+1)\sin^2\eta_l, \tag{19}$$

so that no contribution comes from odd-order phase shifts. We may write (19) in the form

$$Q = \frac{4\pi}{k^2}\{\Sigma_l(2l+1)\sin^2\eta_l\}+\frac{2\pi}{k^2}\{\Sigma_l(2l+1)\cos l\pi-\Sigma_l(2l+1)\cos(2\eta_l-l\pi)\} \tag{20}$$

$$= Q_0+\Delta_s Q, \tag{21}$$

where Q_0 is the cross-section obtained if symmetry is ignored and $\Delta_s Q$ is the modification due to symmetry which is given by the second term of (20). Under semiclassical conditions we first convert the sums over l to integrals and then apply the method of stationary phase just as is done in Chapter 16, p. 1317 for Q_0. No contribution comes in this way

† BUCK, U. and PAULY, H., *Z. Phys.* **185** (1965) 155.

from $\Sigma(2l+1)\cos l\pi$ as no stationary point occurs in the integrand. We then have

$$\Delta_s Q \simeq -\frac{2\pi}{k^2} \int 2l \cos(2\eta_l - l\pi) \, dl, \tag{22}$$

and the point of stationary phase is given by

$$\frac{d\eta_l}{dl} = \tfrac{1}{2}\pi. \tag{23}$$

Since
$$\frac{d\eta_l}{dl} = \tfrac{1}{2}\vartheta(l) \tag{24}$$

where ϑ is the deflexion function (see Chap. 16 (27)), we have

$$\vartheta(l) = \pi, \tag{25}$$

which is satisfied by $l = 0$ (see Fig. 16.5, p. 1313). Near $l = 0$ we may write

$$\eta_l = \eta_0 + \tfrac{1}{2}\pi l - \lambda l^2. \tag{26}$$

λ may be written in terms of the interaction parameters ϵ and r_m, the quantum parameter $A = kr_m$, where k is the wave number of relative motion, and the reduced energy $K \, (= E/\epsilon)$ in the form

$$\lambda = f(E/\epsilon)/A. \tag{27}$$

The function f depends on the shape of the interaction. For collisions between rigid spheres $f = \tfrac{1}{2}$ while for a (12, 6) interaction†

$$f = \tfrac{1}{2}(E/\epsilon)^{1/12}\pi^{\frac{1}{2}}\Gamma(\tfrac{13}{12})/\Gamma(\tfrac{7}{12}). \tag{28}$$

We now have

$$\Delta Q_s = -\frac{2\pi}{k^2} \int_0^\infty 2l \cos(2\eta_0 - 2\alpha l^2/A) \, dl$$

$$= -\frac{\pi r_m^2}{Af}\left\{1 + \frac{f}{A}\left(\frac{\pi A}{f}\right)^{\frac{1}{2}} + \frac{\pi f}{2A}\right\}^{\frac{1}{2}} \cos(2\eta_0 - \tfrac{1}{2}\pi + \phi), \tag{29}$$

with
$$\cot \phi = \{1 + 2(A/\pi f)^{\frac{1}{2}}\}. \tag{30}$$

ΔQ_s is thus of undulatory form with extrema determined by the variation with relative velocity of the zero-order phase-shift (cf. the theory of the glory undulations, Chap. 16, § 6.3).

5.2. *The* He–He *interaction*

A number of measurements of total cross-section for He–He collisions have been carried out which are of sufficient velocity resolution to observe the undulations due to the symmetry. The first to give positive

---

† KIHARA, T., *Proc. phys. math. Soc. Japan* **25** (1943) 73.

results were those of Dondi, Scoles, Torello, and Pauly.† They used a nozzle source cooled to liquid N_2 temperature to produce the beam which was then velocity-selected, with a 5 per cent resolution, before entering the scattering chamber cooled to 2 °K. The intensity of the beam was monitored with a bolometer detector.‡ Measurements were made over a relative velocity range from 650 to 1200 m s^{-1} and showed a clear minimum in the total cross-section at a velocity of 850 m s^{-1} and maximum at 1050 m s^{-1}. Analysis in terms of the theory outlined above gave $r_m = 3 \cdot 0$ Å. The accuracy of these observations was improved in later experiments§ and the velocity range extended to 1700 m s^{-1} so that two further extrema were resolved. The data were analysed in terms of a potential of the form (15). The best fit was obtained with $n = 11$, $\kappa_1 = 66$, $\kappa_2 = 72$, $\epsilon = 0 \cdot 840$ meV $= 1 \cdot 344 \times 10^{-15}$ erg, and $r_m = 2 \cdot 96$ Å.

Even at the lowest relative velocities the impact energy in these experiments is more than five times the depth of the attractive well so the derived interaction is likely to be accurate only in the initial part of the repulsive region.

Bennewitz, Busse, and Dohmann∥ extended the observations down to relative velocities as low as 160 m s^{-1} and so were able to obtain more precise information about the attractive interaction. Some values are tabulated in Table 16 A.3. The parameters ϵ and r_m are found to be 0·90 meV (1·44 × 10^{-15} erg) and 2·98 Å respectively. In contrast with

TABLE 16 A.3

The He–He *interaction derived by Bennewitz et al.*

r (Å)	2·1	2·3	2·5	2·7	2·9	3·1	3·3	3·5	3·7	3·9	4·1	4·3	4·5
$V(r)$ (meV)	31·0	11·6	2·2	−0·16	−0·86	−0·84	−0·68	−0·50	−0·35	−0·25	−0·19	−0·14	−0·11

the interaction of alkali atoms with rare-gas atoms and with Hg atoms the observed interaction is sharper than the (12,6) interaction with the same values of ϵ and r_m. It is also not quite strong enough to support a bound state for the He$_2$ system.

In later work Bennewitz, Busse, Dohmann, and Schrader†† observed the velocity variation of the total cross-section for collisions between

† Dondi, M. G., Scoles, G., Torello, F., and Pauly, H., *J. chem. Phys.* **51** (1969) 392.
‡ Cavallini, M., Gallinaro, G., and Scoles, G., *Z. Naturf.* **22A** (1967) 413.
§ Cantini, P., Dondi, M. G., Scoles, G., and Torello, F., *J. chem. Phys.* **56** (1972) 1946.
∥ Bennewitz, H. G., Busse, H., and Dohmann, H. D., *Chem. Phys. Lett.* **8** (1971) 235.
†† Bennewitz, H. G., Busse, H., Dohmann, H. D., and Schrader, W., *7th int. conf. on the physics of electronic and atomic collisions*, Amsterdam. Abstracts, North Holland, 1971, p. 651.

^{3}He atoms which obey the Fermi–Dirac statistics so that in place of (19),

$$Q = \frac{\pi}{k^2}\left\{3\sum_{l\,\text{odd}}\sin^2\eta_l + \sum_{l\,\text{even}}\sin^2\eta_l\right\}. \tag{31}$$

With the interaction given in Table 16 A.3 a minimum is expected at a velocity of about 250 m s^{-1} and clear evidence of this minimum was obtained in the measurements.

At the opposite extreme Gegenbach, Hahn, Toennies, and Welz† have measured the total cross-section over a relative energy range from 0·01 to 2·8 eV. Over this range they find that an interaction which gives a good fit to the data is given by

$$\begin{aligned}V &= A\exp(-\alpha r - \beta r^\nu) \quad (r < r_e),\\ &= 4\epsilon_m[\exp\{2\gamma(1-x)\} - \exp\{\zeta\gamma(1-x)\}] \quad (r_e < r < r_0),\end{aligned} \tag{32}$$

with $A = 191{\cdot}47$ eV, $\alpha = 3{\cdot}848$ Å^{-1}, $\nu = 8{\cdot}59$, $\beta = 9{\cdot}74\times 10^{-4}$ Å$^{-\nu}$, $r_e = 2$ Å, $r_0 = 2{\cdot}682$ Å, $\gamma = 6{\cdot}443$, $\zeta = 1{\cdot}13$, $\epsilon_m = 0{\cdot}74$ meV and $x = r/r_0$. For $r > r_0$ the interaction is taken to be that given in Table 16 A.3.

The first measurements of differential cross-sections with sufficiently high resolution to observe the symmetry undulations at large angles of scattering were those of Siska, Parson, Schafer, and Lee,‡ the crossed beam being produced from nozzle sources with a Mach number of 15–25. Observations were carried out over the entire angular range for relative kinetic energies of 10·1 and 3·74×10^{-14} erg. In the former case 5 maxima and in the latter 3 maxima due to symmetry interference were observed over the range 0–90° of C.M. angle. Similar measurements were later made by Farrar and Lee§ and a collision energy as low as 0·8×10^{-14} erg obtained by cooling both He beams in liquid H$_2$. These data have been analysed in terms of an interaction of the form

$$V(r) = \epsilon f(x), \quad x = r/r_m, \quad c_s = C_s/\epsilon r_m^s, \tag{33}$$

with

$$\begin{aligned}f(x) &= A\exp\{-\alpha(x-1)\} \quad (0 < x < x_1),\\ &= \exp\{a + x - x_1\}[a_2 + (x-x_2)\{a_3 + (x-x_1)a_4\}] \quad (x_1 < x < x_2),\\ &= \exp\{-2\beta(x-1)\} - 2\exp\{-\beta(x-1)\} \quad (x_2 \leqslant x \leqslant x_3),\\ &= b_1 + (x-x_3)[b_2 + (x-x_4)\{b_3 + (x-x_3)b_4\}] \quad (x_3 \leqslant x \leqslant x_4),\\ &= -c_6 x^{-6} - c_8 x^{-8} - c_{10} x^{-10} \quad (x_4 \leqslant x < \infty). \end{aligned} \tag{34 a}$$

---

† Gegenbach, R., Hahn, Ch., Toennies, J. P., and Welz, W. *7th int. conf. on the physics of electronic and atomic collisions, Amsterdam.* Abstracts, North Holland, 1971, p. 653.

‡ Siska, P. E., Parson, J. M., Schafer, T. P., and Lee, Y. T., *J. chem. Phys.* **55** (1971) 5762.

§ Farrar, J. M. and Lee, Y. T. ibid. **56** (1972) 5801.

The short-range repulsion parameters were taken as $A = 0\cdot 684$, $\alpha = 12\cdot 53$ as suggested by Matsumoto, Bender, and Davidson† (see p. 2356) and the long range attraction as given by Starkschall and Gordon‡ (c_6), Davison (c_8),§ and Dalgarno and Stewart (c_{10});∥ namely $c_6 = 1\cdot 250$, $c_8 = 0\cdot 388$, and $c_{10} = 0\cdot 160$. x_1 was chosen to be $0\cdot 5$. The best fit of the data, over the whole energy and angular range covered, was obtained by taking

$$\epsilon = 0\cdot 152 \pm 0\cdot 003 \times 10^{-14} \text{ erg } (11\cdot 0 \pm 0\cdot 2 \text{ °K}), \quad r_\mathrm{m} = 2\cdot 963 \text{ Å},$$
$$\beta = 6\cdot 19 \pm 0\cdot 18, \quad x_2 = 0\cdot 8377, \quad x_3 = 1\cdot 1446, \quad x_4 = 1\cdot 50,$$
$$a_1 = 5\cdot 973, \quad a_2 = 15\cdot 635, \quad a_3 = -9\cdot 380, \quad a_4 = 94\cdot 56,$$
$$b_1 = -0\cdot 6500, \quad b_2 = 1\cdot 439, \quad b_3 = -4\cdot 306, \quad b_4 = 5\cdot 300.$$

(34 b)

This interaction agrees well with that given in Table 16 A.3. In particular the attraction is not strong enough to give rise to a bound state. It agrees well with an interaction determined by Beck†† which is derived from analysis of data on the transport properties and second virial coefficients of He (see Chap. 16, § 10) and is of the form

$$V(r) = A\exp(-\alpha r - \beta r^6) - 0\cdot 869(r^2+a^2)^{-3}\{1+(2\cdot 709+3a^2)/(a^2+r^2)\}$$
$$(r > 0\cdot 5 \text{ Å}) \quad (35)$$

with

$$a = 0\cdot 675 \text{ Å}, \quad \alpha = 4\cdot 390 \text{ Å}^{-1}, \quad \beta = 3\cdot 746 \times 10^{-4} \text{ Å}^{-6},$$
$$A = 398\cdot 7 \text{ eV},$$

for which $\epsilon = 0\cdot 143 \times 10^{-14}$ erg, $r_\mathrm{m} = 2\cdot 969$ Å. However, the interaction (34) is appreciably different for $r < 2$ Å.

Theoretical calculations have been carried out based on Hartree–Fock wave-functions and including correlation energy to give values of ϵ and r_m quite close to those of the interaction (35). Thus McLaughlin and Schaefer‡‡ find $\epsilon = 0\cdot 166 \times 10^{-14}$ erg and $r_\mathrm{m} = 2\cdot 944$ Å, and Bertoncini and Wahl§§ $\epsilon = 0\cdot 145 \times 10^{-14}$ erg and $r_\mathrm{m} = 2\cdot 92$ Å.

The consistency of the interaction (34) with data obtained from experiments at higher energies will be discussed in the notes on Chapter

† MATSUMOTO, G. H., BENDER, C. F., and DAVIDSON, E. R., *J. chem. Phys.* **46** (1969) 402.
‡ STARKSCHALL, G. and GORDON, R. G., ibid. **54** (1971) 663.
§ DAVISON, W. D., *Proc. phys. Soc.* **87** (1966) 133.
∥ DALGARNO, A. and STEWART, A. L., *Proc. R. Soc.* **A238** (1956) 269.
†† BECK, D. E., *Mol. Phys.* **14** (1968) 311. See also BRUCH, L. W. and MCGEE, I. J. *J. chem. Phys.* **52** (1970) 5884.
‡‡ MCLAUGHLIN, D. R. and SCHAEFER, H. F., *Chem. Phys. Lett.* **12** (1971) 244.
§§ BERTONCINI, P. and WAHL, A. C., *Phys. Rev. Lett.* **25** (1970) 991.

22. It seems clear, however, that for $r > 2$ Å (34) must be quite close to the true interaction.

5.3. *Other rare gases*

Measurements of differential cross-sections for Ne–Ne[†] and Ar–Ar[‡],[§] and Kr–Kr[||] collisions have been made over a wide angular range with sufficient resolution to observe the extrema due to the nuclear symmetry as well as the rainbow extrema.

For Ne–Ne eleven maxima were observed at C.M. angles between 15° and 90° at locations which agree well with those expected on the assumption of a (12, 6) interaction with parameters $\epsilon = 0.494 \times 10^{-14}$ erg, $r_\mathrm{m} = 3.09 \times 10^{-8}$ cm derived from transport properties of Ne.

For Ar–Ar Parson *et al.*[‡] carried out measurements of differential cross-sections for C.M. angles between 4° and 90° using a nozzle source which provided atoms in the crossed beams with a mean impact energy of 1.00×10^{-13} erg. Between 12 and 14 symmetry oscillations were observed as well as the rainbow structure which agreed with that observed in early experiments by Scoles and his collaborators.[††] They analysed their data in terms of an interaction of the form (34) except that x_1 and x_2 were taken as zero. Account was also taken of data on second virial coefficients. c_6 was given the semi-empirical value $(6.48 \times 10^{-11}$ erg Å$^6)$ found by Starkschall and Gordon, x_4 was taken as 1·4 and x_3 chosen so that $f(x_1) = -0.7$. The best fit consistent with the rainbow and symmetry oscillations as well as the second virial coefficient data was obtained with

$$\epsilon = 1.943 \times 10^{-14} \text{ erg}, \qquad r_\mathrm{m} = 3.760 \text{ Å}, \qquad \beta = 6.279,$$
$$c_8 = 4.75 \times 10^{-10} \text{ erg Å}^8, \qquad b_1 = -0.7, \qquad b_2 = 1.8337,$$
$$b_3 = -4.5740, \qquad b_4 = 4.3667, \qquad x_3 = 1.12636,$$

and
$$x_4 = 1.4. \tag{36}$$

The resulting interaction agrees quite well with that derived by Barber, Fisher, and Watts[‡‡] which fits a large number of properties of argon in different states.

[†] SISKA, P. E., PARSON, J. M., SCHAFER, T. P., TULLY, F. P., WONG, Y. C., and LEE, Y. T., *Phys. Rev. Lett.* **25** (1970) 271.
[‡] PARSON, J. M., SISKA, P. E., and LEE, Y. T., *J. chem. Phys.* **56** (1972) 1511.
[§] MUELLER, C. R., SHEEN, S., SEARCY, J., and WENDELL, K., *7th int. conf. on the physics of electronic and atomic collisions, Amsterdam.* Abstracts, North Holland, 1971, p. 536.
[||] SCHAFER, T. P., SISKA, P. E., and LEE, Y. T., ibid., p. 546.
[††] CAVALLINI, M., GALLINARO, G., MENEGHETTI, L., SCOLES, G., and VALBUSA, U., *Chem. Phys. Lett.* **7** (1970) 203.
[‡‡] BARBER, J. A., FISHER, R. A., and WATTS, R. O., *Mol. Phys.* **21** (1971) 657.

Mueller *et al.* also derived interactions from their measurements which were carried out at mean impact energies of 9·58 and 32·2×10⁻¹⁴ ergs. The potential they found which gives the best fit to their results and to the virial coefficient data has a value of r_m somewhat larger and a value of ϵ a little larger than those derived by Parson *et al.*

Schafer *et al.* analysed their Kr–Kr data using again an interaction of the form (34) and obtained a good fit using the parameters

$$\epsilon = 2\cdot 64 \times 10^{-14}\ \mathrm{erg}, \qquad r_\mathrm{m} = 4\cdot 10\ \mathrm{\AA}, \qquad \beta = 7\cdot 1,$$
$$c_6 = 1\cdot 27 \times 10^{-10}\ \mathrm{erg\ \AA^6}, \qquad c_8 = 3\cdot 26 \times 10^{-10}\ \mathrm{erg\ \AA^8},$$
$$x_1 = 1\cdot 084, \quad \text{and} \quad x_2 = 1\cdot 500. \tag{37}$$

In contrast to interactions between alkali-metal and rare-gas atoms (see p. 3530) the (12, 6) potential gives too soft a repulsion and too wide an attractive well.

High resolution differential cross-sections have also been measured for Ne–Ar,[†] Ne–Kr,[†] Ne–Xe,[†] Ar–Kr,[‡] and Ar–Xe[‡] collisions. Lempert, Corrigan, and Wilson,[§] using a time of flight method of velocity analysis, have observed glory effects in Ne–Kr and Ne–Xe collisions.

In terms of $(n, 6)$ interactions a better fit with the differential cross-sections for the Ne collisions is obtained with $n = 20$ than $n = 12$, just as for Ar–Ar collisions. The derived parameters ϵ, r_m give values of ϵr_m which are consistent with those derived from the observations of the velocity variation of the total cross-section.[§]

For Ar–Kr and Ar–Xe Parson *et al.*[‡] analysed their data in terms of a potential of the form (34) and obtained good agreement with a suitable choice of parameters.

6. Viscosity of atomic hydrogen (see p. 1424)

New measurements have been made by Cheng and Blackshear[||] which yield values of the viscosity about 15 per cent lower than those measured earlier by Browning and Fox[††] (see Fig. 16.62, p. 1428). The latter authors assumed that the surface slip conditions were the same for both H and H_2. In the later work it was possible to work either under continuum or slip-flow conditions while the concentration of H in the H–H_2 mixtures whose viscosity was measured could be varied over a wider

[†] Parson, J. M., Schafer, T. P., Tully, F. P., Siska, P. E., Wong, C., and Lee, Y. T., *J. chem. Phys.* **53** (1970) 2123.
[‡] Idem, **53** (1970) 3755.
[§] Lempert, G. D., Corrigan, S. J. B., and Wilson, J. F., *Chem. Phys. Lett.* **8** (1971) 67.
[||] Cheng, D. Y. and Blackshear, P. L., *J. chem. Phys.* **56** (1972) 213.
[††] Browning, R. and Fox, J. W., *Proc. R. Soc.* **A278** (1964) 274.

range (20–70 per cent). This made it possible to determine the viscosity η_1 of H as well as the fictitious viscosity η_{12} (see p. 1427) by an iterative procedure directly from the observations.

Allison[†] has recalculated η_1 using the Kolos–Wolniewicz[‡] H–H interaction (see p. 1421) and finds good agreement with the new results.

7. Collisions involving molecules

Considerable interest is attached to accurate determination of He–H_2 interactions as they are not only important in applications (see Chap. 17, p. 1576) but are also within the scope of theoretical prediction. Absolute measurements of total cross-sections for He–H_2[§] and He–D_2[||] collisions have been made.

For He–H_2 Gegenbach et al. analysed their data with interactions of the same form as they used for He–He (see p. 3534) except that β was taken as zero. The best choice of parameters was found to be

$$A = 280 \text{ eV}, \quad \alpha = 3\cdot648 \text{ Å}, \quad r_e = 2\cdot17 \text{ Å}, \quad r_0 = 3\cdot08 \text{ Å},$$
$$\gamma = 5\cdot59, \quad \zeta = 1, \quad \epsilon_m = 1\cdot15 \text{ meV}, \quad \text{and} \quad \epsilon = 1\cdot03 \text{ meV}.$$

This agrees well for $r < 2\cdot32$ Å with the average over all molecular orientations of the interaction calculated by Gordon and Secrest.[††]

Among other measurements with molecular targets there are those of glory effects in collisions of He, HD, and D_2 with CH_4, N_2, O_2, NO, CO, and CO_2[‡‡] and of rainbow extrema in the scattering of D_2 and H_2 by O_2.[§§]

[†] ALLISON, A. C., J. chem. Phys. **56** (1972) 6266.
[‡] KOLOS, W. and WOLNIEWICZ, L., ibid. **43** (1965) 2429.
[§] GEGENBACH, R., HAHN, C., TOENNIES, J. P., and WELZ, W., loc. cit., p. 3534.
[||] LILINFELD, H. V., LANG, N. C., PARKS, E. K., and KINSEY, J. L., 7th int. conf. on the physics of electronic and atomic collisions, Amsterdam. Abstracts, North Holland, 1971, p. 656.
[††] GORDON, H. D. and SECREST, D., J. chem. Phys. **52** (1970) 120.
[‡‡] BUTZ, H. P., FELTGEN, R., PAULY, H., and VEHMEYER, H., 7th int. conf. on the physics of electronic and atomic collisions, Amsterdam. Abstracts, North Holland, 1971, p. 557.
[§§] GORDON, R. J., COZZIOLA, M., and KUPPERMANN, A., ibid., p. 552.

17

COLLISIONS BETWEEN ATOMS AND MOLECULES UNDER GAS-KINETIC CONDITIONS—COLLISIONS INVOLVING EXCITATION OF VIBRATION AND ROTATION—REACTIVE COLLISIONS BETWEEN MOLECULAR BEAMS

1. Introduction

SINCE this chapter was written the most important new advances have been in the study of reactive scattering. Further measurements on vibrational and rotational relaxation phenomena concerning different molecules have been made on the lines already described in §§ 1–5.

Laser techniques are being used to an increasing extent, for the production of molecules in particular vibrational states and in other ways. Much new work has been carried out on vibration–vibration transfer, which is an important process in the design of certain lasers. Perhaps the most important developments in this field are likely to come from scattering experiments using molecular beams whose rotational or vibrational state has been selected in some way as in the experiments of Bennewitz, Kramer, Paul, and Toennies described on pp. 1595–1605. Further experiments on these and similar lines are in progress. A novel procedure to investigate rotational excitation has been introduced by Beck and Förster† in which energy loss by the atomic beam is observed by making a velocity analysis before and after scattering.

The study of reactive scattering has expanded rapidly. The introduction of nozzle sources has not only provided beams with greater intensity but also with relatively narrow velocity spread. In addition the atoms from a nozzle source containing a light gas with an admixture of a heavier component will all have the same mean velocity which will be that of the predominant, light, component. Thus by seeding the inflow of light gas to the source with a heavy gas the atoms or molecules of this latter gas can be accelerated to relatively high energies. This provides a means of studying a wide range of endothermic reactions.

† BECK, D. and FÖRSTER, H., *Z. Phys.* **240** (1970) 136.

While a great deal of the work has still involved the reactions of alkali-metal beams this has covered a wider range of reactants with consequent increase in empirical knowledge of the possibilities of different reactive processes. Reactions studied include those leading to production of positive and negative ions. The range of experiments still using hot-wire detectors has been extended by the use of beams of alkali-metal dimers. Experiments have also been carried out to investigate orientation effects on reactions of CH_3I[†] and CF_3I[‡] with alkali-metal atoms.

On the theoretical side, apart from a number of papers which discuss the dynamics of collisions by classical or semiclassical methods, a thorough analysis of data on scattering of K, Rb, and Cs from CCl_4, CH_3I, and $SnCl_4$ in terms of the optical model discussed on pp. 1624–34 has been carried out by Harris and Wilson.[§]

Whereas at the time of writing of Vol. 3 very few experiments had been carried out on reactive scattering in which it was necessary to use electron-beam ionization and mass analysis to identify and monitor the flux of selected products, a rapidly increasing number of such experiments have been carried out in the last few years. These have included reactions of H and D atoms with halogen molecules as well as of halogen atoms with various molecules.

2. Vibrational relaxation (see Chap. 17, §§ 2–4)

Progress has been made in measuring vibrational relaxation times in H_2 and D_2 at ordinary temperatures and in the determination of rates of transfer of vibration between diatomic molecules. The use of lasers to produce vibrational excitation has made much of this work possible.

The decay of infrared fluorescence produced by chemical laser excitation of hydrogen halides has been used to study the vibrational relaxation of pure halides as well as mixtures of halides with other atoms and molecules.

Ducuing and his collaborators[‖,††] have made use of the fact that a Q-switched ruby laser beam will produce vibrational excitation in H_2 through Raman scattering. In the first experiments[‖,††] the relaxation of this excited population was followed by observing the intensity of Raman anti-Stokes scattering of a probing ruby laser beam.

An alternative method[††] used in later experiments is based on the

[†] BEUHLER, R. J. and BERNSTEIN, R. B., *Chem. Phys. Lett.* **2** (1968) 166.
[‡] BROOKS, P. R., *J. chem. Phys.* **50** (1969) 5031.
[§] HARRIS, R. M. and WILSON, J. F., ibid. **54** (1971) 2088.
[‖] DE MARTINI, F. and DUCUING, J., *Phys. Rev. Lett.* **17** (1966) 117; DUCUING, J. and DE MARTINI, F., *J. Chim. phys.* **64** (1967) 209.
[††] DUCUING, J., JOFFRIN, C., and COFFINET, J. P., *Optics Communs* **2** (1970) 245.

change with time of the refractive index of the vibrationally excited volume of gas. This was monitored by observing the scattering of light from a He–Ne laser at 6328 Å. Provided that heat flow from the excited volume is negligible during the relaxation and the time for a sound pulse to propagate across the volume is small compared with the relaxation time τ the change $\Delta n(t)$ in refractive index is proportional to $N(t)$, the number of vibrational quanta transferred to translation since the initial excitation at $t = 0$. If $\Delta n(t) \ll 1$ the intensity of scattering is proportional to $\{N(t)\}^2$ and hence to $\{1-\exp(t/\tau)\}^2$.

With this latter technique τ was found to be $3 \cdot 1 \pm 0 \cdot 30 \times 10^{-4}$ s atm corresponding to a mean deactivation cross-section of $1 \cdot 6 \times 10^{-22}$ cm^2. Earlier measurements based on the intensity of anti-Stokes scattering gave a value more than 3 times larger. The reason for this discrepancy is not clear.

Infrared emission from vibrationally excited CO in the region immediately behind a shock wave has been used to observe vibrational relaxation of gases seeded with CO.

2.1. Observed vibrational relaxation of H_2, D_2, and HD

Cross-sections for deactivation of vibration of D_2 and HD in collision with each other and with H_2 have been derived from observations made by Hopkins and Chen† of the relaxation of mixtures of the hydrogen isotopes with HCl excited by a chemical laser.

Results obtained are given in Table 17 A.1 together with those for H_2–H_2 collisions obtained with the stimulated Raman scattering technique referred to above.

Linear extrapolation on a $(\log \tau, T^{-\frac{1}{3}})$ diagram of the high-temperature data obtained with shock tubes (see Chap. 17, p. 1518) gives much smaller cross-sections. The fact that the self-deactivation cross-section for HD is higher than for H_2 and D_2 suggests that transfer of vibration to rotation may be important for the unsymmetrical case.

Calvert‡ has calculated cross-sections for vibration–translation (VT) and vibration–vibration (VV) transfer in H_2, D_2, and their mixtures, up to energies of 2 eV, and has derived relaxation times for temperatures between 300 and 2000 °K. He used the distorted-wave method (Chap. 8, p. 525 and Chap. 17, p. 1550) in three dimensions and, following Bauer, assumed that the interaction between the molecules is given by the sum of identical Morse interactions between individual atoms with the atomic

† HOPKINS, B. M. and CHEN, H.-L., *J. chem. Phys.* **57** (1972) 3161.
‡ CALVERT, J. B., ibid. **56** (1972) 5071.

centres collinear. This leads to a spherically symmetric interaction. Allowance for steric effects is then made by reducing the calculated cross-sections by a factor representing the average of $\cos^2\theta$, where θ is the orientation angle of the molecular axis, over all angles for each

TABLE 17 A.1

Observed cross-sections for vibrational deactivation at 300 °K of hydrogen isotopes

Excited molecule	Collision partner	Cross-section 10^{-22} cm²	Probability P per collision	Vibrational excitation energy (cm⁻¹)
D_2	D_2	1·4	$5·3 \times 10^{-8}$	2990
D_2	H_2	14·0	$5·1 \times 10^{-7}$	2990
D_2	HD	8·2	$3·0 \times 10^{-7}$	2990
HD	HD	6·9	$2·6 \times 10^{-7}$	3627
H_2	H_2	5·4	$2·0 \times 10^{-7}$	4159
		1·6	$5·9 \times 10^{-8}$	4159

molecule involved in a vibrational transition. Remarkably good agreement is obtained with the observations which refer only to VT transfer, both at 300° and at shock tube temperatures. This provides evidence that vibration–rotation (VR) transfer is not important in these cases.

For VV transfer in H_2–D_2 collisions at 300° the calculated cross-section is $4·9 \times 10^{-18}$ cm².

3. Vibration–vibration transfer

Hancock and Smith[†] have measured relaxation rates for highly excited CO† molecules (with vibrational quantum number v between 4 and 13) formed in the reaction sequence

$$O + CS_2 \to SO + CS \tag{1a}$$

$$O + CS \to CO^\dagger + S + 75 \text{ kcal/mol} (3·2 \text{ eV}). \tag{1b}$$

The decay of vibrational excitation with a particular pressure of added foreign quenching gas was measured from observation of the spectrum of the infrared luminescence. It is then possible from a knowledge of the spontaneous emission coefficients and estimates of the effects of wall loss and self-quenching effects in the resulting mixture, which are kept as small as possible, to determine the individual rates for deactivating collisions with the foreign gas molecules by carrying out observations at different foreign gas pressures. When the foreign gas is monatomic the

† HANCOCK, G. and SMITH, I. W. M., *Chem. Phys. Lett.* **8** (1971) 41.

relaxation process involves VT transfer but if it is diatomic VV transfer will generally dominate because the internal energy change is smaller in this case. In these experiments CO itself may be introduced as an added foreign gas.

Results obtained for quenching by CO, OCS, O_2, and He are given in Table 17 A.2.

3.1. $CO^\dagger$ collisions with other diatomic molecules

Shock tube measurements of vibrational relaxation of $CO^\dagger$ ($v = 1$) by N_2,†,‡,§ O_2,† D_2,† and H_2† have been carried out in which the relaxation behind the shock wave in a suitable mixture is monitored by observation of infrared emission from CO and in some cases H_2O. From these data VV transfer rates between $CO^\dagger$ ($v = 1$) and the different molecules in their ground vibrational states have been derived. For N_2 there is a considerable spread between the results of different observers. On a plot of $\log \tau$, where τ is the relaxation time, against $T^{-\frac{1}{3}}$, where T is the temperature in °K (see Chap. 17, p. 1524), the data are consistent with a linear increase of $\log \tau$ as $T^{-\frac{1}{3}}$ increases, for $T > 2000$ °K, but, for smaller T, $\log \tau$ varies very little down to the lowest temperature of observation (950 °K). Over this latter range the VV transfer probability per collision is around 2×10^{-4}. At 2000 °K Sato et al. find this probability to be 6×10^{-5}, 4×10^{-5}, and 10^{-5} for O_2, D_2, and H_2 respectively.

3.2. VV collisions between other diatomic molecules

Other measurements of VV transfer rates involving diatomic molecules include a number of observations of transfer from halogen halides to other halogen halides and to HD, D_2, N_2, and CO.‖ In these experiments the initial vibrational excitation was produced by a chemical laser source and the relaxation monitored from the decay of the resonance infrared fluorescence. The probability per collision that HCl ($v = 1$) will transfer vibrational excitation is found to be 7×10^{-2}, 6.0×10^{-2}, 1.1×10^{-4}, 8.0×10^{-4}, 3.4×10^{-4}, and 5.0×10^{-4} in collision with D_2, HBr, N_2, HI, CO, and DCl respectively. These cases are in order of increasing internal energy change ΔE, in cm^{-1}, 108, 327, 555, 656, 743, and 795 respectively. It is noteworthy that the probabilities do not correlate well with ΔE.

† SATO, Y., TSUCHIYA, S., and KURATANI, K., J. chem. Phys. **50** (1969) 1911.
‡ MCLAREN, T. I. and APPLETON, J. P., 8th int. shock tube symposium, London, No. 27 (1971).
§ VON ROSENBERG, C. W., BRAY, K. N. C., and PRATT, N. H., J. chem. Phys. **56** (1972) 3230.
‖ CHEN, H.-L. and MOORE, C. B., J. chem. Phys. **54** (1971) 4080, HCl–HBr, HCl–N_2, CO, DCl, and D_2–HCl; CHEN, H., ibid. **55** (1971) 5551, HBr–HCl, HI.

TABLE 17 A.2

Probability per collision for deactivation of $CO^{\dagger}$ (v) *to* $CO^{\dagger}$ ($v-1$) *in collisions with different gas atoms and molecules at* 300 °K

	4	5	6	7	8	9	10	11	12	13
Frequency of transition $v \rightarrow v-1$ in cm^{-1}	2064	2038	2011·5	1985	1959	1934	1907	1883	1856	1830
	Probability									
Quenching gas										
CO Frequency of $1 \rightarrow 0$ transition 2143 cm^{-1}	$6{\cdot}7\times10^{-3}$	$4{\cdot}3\times10^{-3}$	$2{\cdot}1\times10^{-3}$	$1{\cdot}1_5\times10^{-3}$	$5{\cdot}9\times10^{-4}$	$3{\cdot}8\times10^{-4}$	$2{\cdot}0\times10^{-4}$	$1{\cdot}5\times10^{-4}$	$1{\cdot}1\times10^{-4}$	
OCS Frequency of infrared active ν_3 mode, 2062 cm^{-1}	0·114	0·15	0·14	0·090	0·036	0·018	0·010	$7{\cdot}8\times10^{-3}$	$4{\cdot}8\times10^{-3}$	$2{\cdot}5\times10^{-3}$
O_2 Frequency of $1 \rightarrow 0$ transition 1556 cm^{-1}	—	—	—	—	—	—	—	—	$7{\cdot}1\times10^{-5}$	$1{\cdot}5\times10^{-4}$
He	—	—	—	—	—	$4{\cdot}2\times10^{-7}$	$7{\cdot}1\times10^{-7}$	$9{\cdot}1\times10^{-7}$	$1{\cdot}4\times10^{-6}$	$1{\cdot}7\times10^{-6}$

3.3. VV collisions between CO_2 and other molecules

Vibrational relaxation of CO_2 has continued to attract much attention particularly in connection with laser devices.† The transfer process

$$CO_2^{\dagger}(\nu_3) + {}^{14}N_2 \rightleftharpoons CO_2 + {}^{14}N_2\ (v=1) \qquad (2)$$

is of additional interest because the internal energy change is very small (18 cm^{-1}).

A number of recent experiments have been carried out to measure the rate of this process. These include shock-tube experiments,‡ in which the vibrational excitation of CO_2 in CO_2–N_2 mixtures is monitored by infrared emission, and experiments at lower temperatures (300–1000 °K)§ in which the ν_3 (001) mode of CO_2 (see Chap. 17, Fig. 17.26) is excited by a Q-switched CO_2–N_2 laser and the relaxation followed by observing the infrared emission. Although there is a big spread in the results at temperatures below 1600 °K the general trend of the variation of the relaxation rate with temperature is similar to that for $CO^{\dagger}$–N_2 transfer collisions described above. Below 1000 °K the relaxation time varies roughly as $T^{\frac{1}{2}}$ but at higher temperatures it is roughly proportional to $T^{-\frac{3}{2}}$.

3.4. Theory of VV transfer processes

The theory of VV transfer processes between diatomic molecules has been discussed in Chapter 17, § 4.2. This theory regarded the process as determined by the short-range exponential interaction used with considerable success to provide a theory of VT transfer (see Chap. 17, § 4.4). However, it was pointed out by Sharma and Brau‖ that long-range forces may become important when the internal energy change ΔE is small, particularly at low temperatures. They investigated the contribution from these forces to the cross-section for VV transfer from $CO_2^{\dagger}$ (001) to N_2 by an impact parameter method in which the trajectories of relative motion of the molecules were assumed to be linear and the interaction to be represented by the first term of a multipole expansion, that arising from the instantaneous dipole moment of the CO_2 and the quadrupole moment of N_2. Quite good agreement was found with the observations at temperatures below 1000 °K in which temperature range the relaxation time increases with the temperature.

---

† TAYLOR, R. L. and BITTERMAN, S., *Rev. mod. Phys.* **41** (1969) 26.
‡ TAYLOR, R. L. and BITTERMAN, S., *Bull. Am. Phys. Soc.* **13** (1968) 1591.
§ MOORE, C. B., WOOD, R. E., HU, B., and YARDLEY, J. T., *J. chem. Phys.* **46** (1967) 4222 (300 °K); ROSSER, W. A., WOOD, A. D., and GERRY, E. T., ibid. **50** (1969) 4996.
‖ SHARMA, R. D. and BRAU, C. M., ibid. **50** (1969) 924.

In a later paper Sharma† applied a similar analysis to transfer between $^{12}\mathrm{C}^{16}\mathrm{O}_2$ and the different isotopic molecules. Dipole–dipole coupling is found to give quite good agreement to the room-temperature data for transfer to $^{12}\mathrm{C}^{16}\mathrm{O}_2$ and to $^{16}\mathrm{O}^{12}\mathrm{C}^{18}\mathrm{O}$ but for $^{13}\mathrm{C}^{16}\mathrm{O}_2$ dipole–octopole coupling is more important.

Jeffers and Kelley‡ took into account both short- and long-range forces in a theoretical analysis of the experiments of Hancock and Smith described above on the rates of the reaction

$$\mathrm{CO}^\dagger\,(v) + \mathrm{CO}\,(0) \to \mathrm{CO}^\dagger\,(v-1) + \mathrm{CO}^\dagger\,(1). \tag{3}$$

They obtained good agreement with the observed results at 300 °K (see Table 17 A.2). For $v > 9$ the short-range interaction is dominant but for $v < 8$ by far the major contribution comes from the long-range dipole–dipole interaction. According to these calculations, although the transition between the regions moves to smaller v as the temperature decreases, the short range forces eventually dominate even for small v as the temperature increases.

Further detailed calculations on VV transfer between diatomic molecules have been carried out with a two-dimensional model in which accurate numerical solution of the classical equations of motion for the coplanar collision of two diatomic molecules is carried out. The interaction was taken to be the sum of Morse functions between each non-bonded pair of atoms. These functions include three parameters, the attractive well depth ϵ, the equilibrium distance r_m, and a range parameter α. ϵ and r_m were fixed from empirical data available on other molecular properties and α so that calculated rates for VT transfer were consistent with observed data.§ For $\mathrm{N}_2^\dagger\,(v=1)$ to $\mathrm{O}_2\,(v=0)$ transfer, with an energy discrepancy of 775 cm^{-1}, quite good agreement is obtained∥ with the available experimental data†† which cover the temperature range from 300 to 4500 °K. In this case the energy discrepancy is large and the dominant contribution comes from the short range repulsion. The theory described in Chapter 17, § 4.2 gives the variation with temperature correctly but overestimates the transfer rate. Reasonably good agreement is obtained with the classical method for transfer from $\mathrm{N}_2^\dagger\,(v=1)$ to $\mathrm{HCl}\,(v=0)$∥ and to $\mathrm{CO}\,(v=0)$.§

† SHARMA, R. D., *Phys. Rev.* **177** (1969) 102.
‡ JEFFERS, W. Q. and KELLEY, J. D., *J. chem. Phys.* **55** (1971) 4433.
§ BEREND, G. C. and BENSON, S. W., ibid. **51** (1969) 1480.
∥ BEREND, G. C., THOMMARSON, R. L., and BENSON, S. W., ibid. **57** (1972) 3601.
†† BAUER, H. J. and ROESSLER, H., *Molecular relaxation processes*, Academic Press, New York, 1966, p. 245; WHITE, D. R., *J. chem. Phys.* **49** (1968) 5472; BRESHEARS, W. D. and BIRD, P. F., ibid. **48** (1968) 4768; KAMIMOTO, G. and MATSUI, H., *AIAA J.* **7** (1970) 2358.

4. Transfer between vibration and rotation on collision

In Chapter 17, p. 1530 attention was drawn to the observed dependence of the relaxation time as a function of fundamental vibration frequency at 300 °K on the number of hydrogen atoms in the molecule, as noted by Lambert and Salter. It was suggested that the velocity of a rotating H atom, in a complex atom, relative to a colliding vibrator may well be considerably higher than that of relative translation. Under these circumstances transfer from vibration to rotation (VR transfer) may be more effective than VT. Experiments in which H is replaced by D support this interpretation.

The experiments of Chen and Moore[†] on vibrational relaxation in pure HCl and mixtures with DCl, following on laser excitation at 300 °K, show similar features to the earlier experiments with complex molecules described in Chapter 17, p. 1530. The observed cross-sections for HCl–HCl, DCl–DCl, and DCl–HCl vibrational deactivation are $4 \cdot 2 \pm 0 \cdot 4$, $1 \cdot 3^{+0 \cdot 2}_{-0 \cdot 4}$, and $2 \cdot 9 \pm 0 \cdot 3 \times 10^{-19}$ cm^2 respectively. If the main process was one of VT transfer the cross-sections for HCl–HCl would be expected to be smaller than for DCl–DCl because the fundamental vibrational frequency is greater. The observed results, including that for DCl–HCl, are consistent with the assumption that in the collisions most of the vibrational energy is converted into rotational energy of the same molecule. Data taken in shock-tube experiments[‡] show the same features.

5. Rotation–translation (RT) transfer

Beck and Förster[§] have carried out experiments on the scattering of a K-atom beam by a crossed thermal beam of CO_2 molecules in which a velocity analysis of the K atoms was made before and after scattering. Observations were carried out for centre of mass scattering angles ranging from 6° to 28° and at impact energies of 16·8 and $20 \cdot 2 \times 10^{-14}$ erg (109 and 150 κ). The energy loss suffered by the K atoms corresponded to excitation of high rotational states of CO_2 in which the change of rotational quantum number was between 6 and 22. It is likely that these inelastic collisions proceed through formation of an intermediate complex (see p. 3554).

Rotational relaxation in normal and para-H_2 has been measured by Yoder[||] using a cryogenic shock tube, the relaxation being followed by

[†] CHEN, H.-L. and MOORE, C. B., *J. chem. Phys.* **54** (1971) 4072.
[‡] BRESHEARS, W. D. and BIRD, P. F., ibid. **50** (1969) 333.
[§] BECK, D. and FÖRSTER, H., *Z. Phys.* **240** (1970) 136.
[||] YODER, M. J., *J. chem. Phys.* **56** (1972) 3226.

a schlieren method (see Chap. 17, p. 1484) using a laser light source. Observations made over a temperature range of 140–450 °K give a probability of deactivation per collision of $2·7 \times 10^{-3}$ and $2·3 \times 10^{-3}$ at 200° and 350° respectively, consistent with earlier results obtained by other methods (see Chap. 17, Table 17.9).

Polanyi and Woodall[†] have analysed data on rotational relaxation derived from observations of chemiluminescence. They have shown that, in many cases, the relaxation of an initial non-thermal distribution of rotational states proceeds according to the relation

$$P(J \to J - \Delta J) = N \exp(-C \Delta E), \qquad (4)$$

where N and C are constants, $P(J \to J - \Delta J)$ being the probability of a change of rotational quantum number from J to $J - \Delta J$ and ΔE the internal energy change involved.

Further calculation of cross-sections for rotational excitation of H_2 by H impact have been carried out, both by close coupling[‡] and distorted wave[§] methods (see Chap. 17, p. 1576). The results depend quite strongly on the interaction assumed.

6. Further theoretical discussion of inelastic collisions

The theory of collinear collisions between atoms and molecules (see p. 1540) has been analysed in detail so that comparison may be made between distorted wave, classical, and exact results for assumed exponential and Morse interactions. It has been shown[||] that in these cases the probabilities given by the distorted wave method can be corrected by multiplying by the energy independent factor

$$(1 - 3M/4)^2,$$

where $M = M_1 M_3 M_2^{-1}(M_1 + M_2 + M_3)^{-1}$, M_1, M_2, and M_3 being the respective masses of the incident atom, the nearer, and the further molecular atom respectively.

A complete three-dimensional close-coupling treatment in which both vibrational and rotational transitions are allowed for is being applied by Burke and his collaborators.[††]

[†] POLANYI, J. C. and WOODALL, K. B., *J. chem. Phys.* **56** (1972) 1563.
[‡] HAYES, E. F., WELLS, C. A., and KOURI, D. J., *Phys. Rev.* **A4** (1971) 1017; WOLKEN, G.. MILLER, W. H., and KARPLUS, M., *J. chem. Phys.* **56** (1972) 4930.
[§] TANG, K. T., *Phys. Rev.* **187** (1969) 122.
[||] ROBERTS, R. E., *J. chem. Phys.* **45** (1966) 1437 and **55** (1971) 100; ROBERTS, R. E. and DIESTLER, D. J., ibid. **57** (1973) 2998. See also HEIDRICH, F. E., WILSON, K. R., and RAPP, D., ibid. **54** (1971) 3885.
[††] BURKE, P. G., SCRUTTON, D., TAIT, J. H., and TAYLOR, A. J., *J. Phys. B (GB)* **2** (1969) 1155. See also EASTES, W. and SECREST, D., *J. chem. Phys.* **56** (1972) 640.

The classical two-dimensional calculations of Berend and Benson† have already been referred to above. Three-dimensional semi-classical approximations have been developed by Wartell and Cross‡ and by Burke and Taylor.§

7. Scattering of state-selected molecular beams

In Chapter 17, § 5.5.2.1 we described experiments carried out on the scattering of beams of the polar molecule TlF in selected rotational states. These included measurements of the dependence of the elastic cross-section for collisions with rare gas atoms on the initial rotational state as well as of cross-sections for rotational excitation. In these experiments state selection was carried out using an inhomogeneous electric field. This is not applicable to non-polar molecules but Moerkerken, Prior, and Reuss∥ have developed a magnetic technique which they were able to apply to select ortho-H_2 molecules in chosen rotational states.

For ortho-H_2 the magnetic moment in a strong magnetic field is given by

$$\mu(m_I, m_J) = (5 \cdot 58 m_I + 0 \cdot 88 m_J) \text{ nuclear magnetons,} \qquad (5)$$

where m_I and m_J are the respective components of nuclear spin and rotational angular momentum along the magnetic field. By use of inhomogeneous magnetic fields it is practicable to select molecules with chosen m_I but not m_J. To select the latter, advantage may be taken of magnetic resonance to produce transitions between the Zeeman substates. For a particular m_J and m_I there is a resonance frequency which will produce a transition in which $\Delta m_J = 0$, $\Delta m_I \neq 0$. If the inhomogeneous fields are adjusted so that m_I is selected, the transmitted flux of molecules will drop markedly at this frequency, the reduction being proportional to the number of beam molecules in the particular m_J state. By carrying out measurements of the strength of the reduced signal with and without scattering gas present, at different resonance frequencies, the dependence of the total scattering cross-section on m_J could be determined.

Preliminary application of this technique has yielded values for the ratio

$$A = \frac{Q_0}{Q_\pm}$$

† loc. cit., p. 3546.
‡ WARTELL, M. A. and CROSS, R. J., J. chem. Phys. **55** (1971) 4983.
§ BURKE, P. G. and TAYLOR, A. J., 7th int. conf. on the physics of electronic and atomic collisions, Amsterdam. Abstracts, North Holland, 1971, p. 9.
∥ MOERKERKEN, H., PRIOR, M., and REUSS, J., Physica, 's Grav. **50** (1970) 499.

of the cross-sections Q_0 when $m_J = 0$ to those $Q_\pm$ for $m_J = \pm 1$, of 1·009, 0·978, and 1·012 for H_2 colliding with Ar, Xe, and N_2 respectively.

Reuss and Stolte[†] have carried out calculations, by the distorted wave method, so as to relate the ratio A to the anisotropic part of the H_2–atom interaction.

A similar magnetic resonance technique is being applied by Brink and Wightman[‡] to the scattering of state-selected O_2 and OH.

Experiments are also being carried out[§] with NO beams, in the $^2\Pi_{\frac{3}{2}}$, $J = m_J = \frac{3}{2}$ state. This is produced by passing a beam, at a temperature of 180 °K, through an inhomogeneous magnetic field, an electrostatic 6-pole field, and a velocity selector. Molecules in the required state are sharply focused on the detector slit by using the positive linear Stark effect of NO while other molecules form a roughly uniform background over a relatively wide region. By varying the magnetic field the sharply focused beam may be swept across the detector while the background signal due to molecules in other states varies little. The state-selected beam passes into the scattering chamber in which a magnetic field may be produced to align the molecules in any direction in the plane of collision. For NO–CCl_4 collisions preliminary results gave

$$\frac{Q_\| - Q_\perp}{Q_\|} = (-0\cdot4 \pm 0\cdot1) \text{ per cent,}$$

where $Q_\|$, $Q_\perp$ are the total cross-sections for collisions in which the NO axis is respectively parallel and perpendicular to the most probable direction of relative motion.

8. Collisions involving change of orientation only

In Chapter 17, § 5.8 the relation of the nuclear spin-lattice relaxation time $\mathscr{T}$ in H_2, and in mixtures of H_2 with other gases, to the rates of collisions in the gas which cause reorientation of the rotational angular momentum of the molecule is discussed.

Observations of $\mathscr{T}$ for H_2–He mixtures have been extended to temperatures of 300 °K by Riehl, Kinsey, Waugh, and Rugheimer,[∥] so that the analysis of Riehl, Kinsey, and Waugh (Chap. 17, p. 1620) could be extended to obtain more definite information about the anisotropic

---

[†] REUSS, J. and STOLTE, S., *Physica, 's Grav.* **42** (1969) 111.

[‡] BRINK, G. O. and WIGHTMAN, C., *7th int. conf. on the physics of electronic and atomic collisions, Amsterdam.* Abstracts, North Holland, 1971, p. 661.

[§] STOLTE, S., REUSS, J., and SALWAY, H. L., ibid., p. 658.

[∥] RIEHL, J. W., KINSEY, J. L., WAUGH, J. S., and RUGHEIMER, J. H., *J. chem. Phys.* **49** (1968) 5276.

component of the He–H$_2$ interaction which causes reorientation on collision.

Above 150 °K it appears to be necessary to take into account collisions in which excitation as well as reorientation occurs. At lower temperatures a reasonably good fit to the data was obtained with an interaction, in eV,

$$V(r,\theta) = e^{-3\cdot 85 r}\{583\cdot 43 + 200\cdot 12 P_2(\cos\theta)\} + \\ + r^{-6}\{4\cdot 535 + 1\cdot 264 P_2(\cos\theta)\}, \qquad (6)$$

r being in Å.

9. Reactive scattering

9.1. *Reactions involving alkali-metal atoms*

Difficulties have arisen in the use of the subtraction method for distinguishing between scattered alkali-metal atoms and alkali halide product molecules (see p. 1622). This method depends on the difference between the signals produced from two hot filaments both of which respond to alkali-metal atoms but differ greatly in response to the halides according to the surface contamination. However, many compounds such as chlorides, fluorides, and oxides poison the surface of one or both filaments. There are also problems in normalizing the two sets of detector readings.

Gordon, Herm, and Herschbach[†] have extended their earlier measurements (p. 1623) in which the alkali-metal atoms could be deflected away from the detector by an inhomogeneous magnetic field. They checked data for many reactions which had been previously obtained by the subtraction method. The same method has also made possible the study of reactive collisions of Li atoms[‡] with various halide and other molecules.

Electric resonance spectroscopy has been used[§] to investigate the distribution of vibrational states in CsF resulting from reactive collisions of Cs with halogen-containing molecules. The Stark spectrum arises from transitions between components with different M_J of particular vibration–rotation states (v, J) of the molecule. Rotational states are well resolved for J ranging from 1 to 4. For each J value Stark lines

[†] GORDON, R. J., HERM, R. R., and HERSCHBACH, D. R., *J. chem. Phys.* **49** (1968) 2684.
[‡] PARISH, D. D. and HERM, R. R., *7th int. conf. on the physics of electronic and atomic collisions, Amsterdam.* Abstracts, North Holland, 1971, p. 29.
[§] BENNEWITZ, H. G., HAERTEN, R., KLAIS, O., and MÜLLER, G., ibid., p. 28. FREUND, S. M., FISK, G. A., HERSCHBACH, D. R., and KLEMPERER, W., *J. chem. Phys.* **54** (1971) 2510.

associated with each vibrational level from $v = 0$ to 4 may be identified by comparison with the Stark spectrum of a thermal CsF beam. Results obtained in this way are discussed on p. 3556.

Grice, Mosch, Safron, and Toennies† have developed a technique for observation of the rotational distribution of reaction products when these are polar molecules. The reaction products are collimated to pass through a velocity selector and directed along the axis of an electric quadrupole. It may be shown that the variation of the intensity at the detector with voltage on the quadrupole is the transform of the rotational state distribution. Applied to RbBr formed by the Rb–Br$_2$ reaction, this distribution was found to be thermal about a temperature of 2500 ± 300 °K, independent of C.M. scattering angle and velocity over the range of the experiments. These results agree well with earlier observations‡ of average rotational energies.

Systematic observations have been made of the reactive scattering of alkali-metal atoms with both homo- and heteronuclear halogen molecules so that regular variations in behaviour between the different alkali-metal and halogen atoms may be discussed.

For K, Rb, and Cs with Cl$_2$,§ Br$_2$,∥ and I$_2$∥ the following features are confirmed.

(a) The reaction cross-sections are all very large ($\geqslant 1\cdot5\times10^{-14}$ cm^2).

(b) The alkali halide is largely scattered forward relative to the incident alkali beam, the projection angle θ in the C.M. system being $< 60°$ though about 5–15 per cent of the intensity is projected into the range $90° < \theta < 180°$.

(c) The sharpness of the forward peaking is almost independent of the alkali-metal atom but tends to decrease in the sequence Cl$_2$, Br$_2$, I$_2$.

(d) The angular distribution of the elastically scattered alkali-metal atoms falls much more rapidly at large angles (in the C.M. system) than for collisions involving non-reactive atoms of comparable size. For Cl$_2$ the decrease is most marked with K, less so for Rb and Cs. For Br$_2$ and for I$_2$ the shape of the distribution is much the same for all three alkali-metal atoms but does vary between the halogens.

† GRICE, R., MOSCH, J., SAFRON, S. A., and TOENNIES, J. P., *J. chem. Phys.* **53** (1970) 3376.
‡ MALTZ, C., Ph.D. thesis, Harvard Univ., 1969. See also Chapter 17, p. 1640.
§ GRICE, R. and EMPEDOCLES, P. B., *J. chem. Phys.* **48** (1968) 5352.
∥ BIRELY, J. H., HERM, R. R., WILSON, K. R., and HERSCHBACH, D. R., ibid. **47** (1967) 993.

(e) Most of the energy released in the reactions appears as internal excitation of the product.

The same features were found in reactions with ICl and IBr.† In addition, evidence from kinematic analysis of the scattering indicates that in these cases the principal product is MCl or MBr not MI, M being the alkali-metal atom.

Parrish and Herm‡ have studied the reactions of Li with Cl_2, ICl, and Br_2 using the magnetic deflection technique to distinguish between Li and lithium halides. While the reaction cross-sections are large and much of the energy released appears as internal excitation, the angular distribution is much less sharply peaked in the forward direction.

Systematic observations of the reactions of alkali-metal atoms with alkyl halides (Chap. 17, § 6.4) have confirmed that they behave quite differently. For example Kwei, Norris, and Herschbach§ have studied a number of reactions of K atoms with RI molecules, R being an alkyl group. The following general features emerge, consistent with the earlier results referred to in Chapter 17, pp. 1634–44.

(a) The reaction cross-sections are around 3×10^{-15} cm², ‖ varying little with the alkyl group.

(b) The angular distribution of the KI product has a peak in the backward direction which becomes less pronounced as the size of the alkyl group increases.

(c) The angular distribution of the elastically scattered K atoms is similar to that for scattering by nonreactive molecules of comparable size, and depends little on the alkyl group.

(d) There is evidence that an increasing fraction of the energy released in the reaction appears as internal excitation as the size of the alkyl group increases. For $K+CH_3I$ the fraction is about 0·5.

Wilson and Herschbach,†† in an attempt to follow the transition between the two extreme types of stripping and rebound reactions, studied the reactive scattering of K, Rb, and Cs atoms by two series of

† KWEI, G. H. and HERSCHBACH, D. R., *J. chem. Phys.* **51** (1969) 1742.
‡ loc. cit., p. 3551.
§ KWEI, G. H., NORRIS, J. A., and HERSCHBACH, D. R., *J. chem. Phys.* **52** (1969) 1317.
‖ EDELSTEIN, S. A. and DAVIDOVITS, P. (*7th int. conf. on the physics of electronic and atomic collisions, Amsterdam.* Abstracts, North Holland, 1971, p. 31) have measured reactive cross-sections for Na, K, Rb, and Cs collisions with I_2 by using photo-dissociation of the alkali iodide in the presence of a known concentration of I_2 molecules. The rate of decay of the alkali-metal concentration after passage of a dissociating light pulse was observed by resonance absorption. Reaction cross-sections of 0·968, 1·67, and 1·95 × 10⁻¹⁴ cm⁻² were obtained for Na, K, Rb, and Cs respectively.
†† WILSON, K. R. and HERSCHBACH, D. R., *J. chem. Phys.* **49** (1968) 2676.

polyhalide molecules which in order of decreasing reaction cross-section are

(A) PBr_3, PCl_3, BBr_3.

(B) CBr_4, $SnCl_4$, CCl_4, $CHCl_3$, $SiCl_4$.

For PBr_3, PCl_3, CBr_4, and $SnCl_4$ the reaction cross-section is greater than $1\cdot5 \times 10^{-14}$ cm² and the angular distribution of the reaction product is strongly peaked in the forward direction. As the cross-section decreases the distribution changes. For CCl_4 it exhibits a maximum (in the C.M. system) at an angle of 50° for Cs, 70° for Rb, and 90° for K while for $CHCl_3$, with a smaller reaction cross-section, most of the product is projected into the backward hemisphere in the C.M. system. Again the relative elastic scattering at large angles decreases as the reactivity increases. Nevertheless the total scattering (elastic+reactive) is very much the same. These measurements, in which the filament difference method was used, were checked[†] using magnetic deflection.

Examples of crossed beam reactions of alkali-metal atoms which proceed through the formation of complexes with lifetimes long compared with their rotational period were first provided by Miller, Safron, and Herschbach.[‡] They observed the reactive scattering of K and Cs atoms by RbCl and the corresponding inverse reactions. If a complex is formed with a lifetime long compared with its rotation period the distribution of wide-angle elastic and of reactive scattering is determined solely by geometrical considerations arising from the angular momentum relations. If

$$L_m = (k^2 Q_c/\pi)^{\frac{1}{2}} \gg J_m, \qquad (7)$$

where J_m is the most probable value of the quantum number specifying the initial internal angular momentum of the system, Q_c the cross-section for complex formation, and k the wave number of initial relative motion, the angular distribution, according to the statistical model for the complex, is a function only of a parameter

$$\chi = L_m/K, \qquad (8)$$

where
$$K = (\kappa T_r I_r/\hbar^2)^{\frac{1}{2}}. \qquad (9)$$

T_r is the rotational temperature of the complex and $I_r = I_\perp I_\|/(I_\perp - I_\|)$, $I_\|$, $I_\perp$ being the moments of inertia of the complex about the axis of symmetry, i.e. along the direction of initial relative velocity and about a perpendicular axis respectively.

[†] GORDON, R. J., HERM, R. R., and HERSCHBACH, D. R., loc. cit., p. 3551.
[‡] MILLER, W. B., SAFRON, S. A., and HERSCHBACH, D. R., *Discuss. Faraday Soc.* **44** (1967) 108.

In all cases the angular distribution of reactive scattering in the C.M. system is symmetrical about 90°. Whereas for a prolate complex it exhibits peaks in the forward and backward directions which become more pronounced as χ increases, for an oblate complex the distribution shows a peak at 90° except under certain special circumstances when the peak shifts to an intermediate position. For elastic scattering, allowance must be made for the small-angle scattering due to interactions at long range. The formation of the complex at impact parameters $< L_m/k$ is then manifest in the appearance of a peak at large angles although the distribution as a whole will not be symmetrical about 90°.

Observations of the angular distribution in the laboratory system of the reaction products were consistent with the C.M. distribution typical of a prolate complex, as described above. The angular distribution of elastic scattering also showed a characteristic peak in the backward direction. It was possible to obtain the ratio Γ_r/Γ_e of the chance that the complex will break up yielding reaction products or merely through emission of the initial reactants.

Kwei, Lees, and Silver† have investigated reactive scattering between Li atoms and potassium halide molecules. The observed angular distributions of K atoms for KF are characteristic of those expected for the stripping model but for KBr the distribution includes a strong component characteristic of complex formation. It is nevertheless not symmetrical about 90° in the C.M. system showing that the lifetime of the complex is not long but comparable with the rotational period.

Evidence for the formation of collision complexes involving four atoms has been obtained by Miller, Safron, and Herschbach‡ from observations of the angular distribution of scattering of alkali halides from crossed molecular beams of CsCl with KCl and with KI. On the other hand Foreman, Kendall, and Grice§ find that the reactive scattering of potassium dimers K_2 by halogen molecules is essentially similar to that of alkali-metal atoms, the reaction cross-section for

$$K_2 + Br_2 \rightarrow KBr + K + Br \qquad (10)$$

being large and the angular distribution of the KBr and K in the C.M. system being strongly peaked in the forward direction.

At about the same time as Miller *et al.* reported their results for alkali–alkali halide collisions Ham and Kinsey∥ also reported experimental

† Kwei, G. H., Lees, A. B., and Silver, J. A., *J. chem. Phys.* **55** (1971) 456.
‡ Miller, W. B., Safron, S. A., and Herschbach, D. R., ibid. **56** (1972) 3581.
§ Foreman, P. B., Kendall, G. M., and Grice, R., *Molec. Phys.* **23** (1972) 127; Whitehead, J. C., Hardin, D. R., and Grice, R., ibid. **25** (1973) 515.
∥ Ham, D. O. and Kinsey, J. L., *J. chem. Phys.* **48** (1968) 939.

results on the scattering of velocity-selected K and Cs atoms by SO_2, CO_2, and NO_2 molecules. In these cases no reactive scattering is observed because the reactions are all too endothermic. The angular distributions were found to be of characteristic form expected when complex formation is probable. At angles in the C.M. system greater than 80°, such that only a negligible contribution comes from direct scattering which is concentrated towards small angles, the observed distributions agree well with that expected on the basis of the statistical theory discussed on p. 3554.

The reactive scattering of Cs from SF_6 is of special interest because experimental data are available on the angular and energy distribution,[†] mean rotational energy,[‡] and vibrational energy distribution[†,§] of the CsF product. The angular distribution shows that a long-lived prolate complex is formed and that the translational, vibrational, and rotational temperatures are approximately equal at $1 \cdot 0 - 1 \cdot 2 \times 10^3$ K. This implies that about 80 per cent of the energy released by the reaction goes into internal energy of the SF_5 product. This is consistent with a statistical model[||] in which there is equipartition of the available energy in the C.M. system among the vibrational modes and one rotational mode unconstrained by the conservation of angular momentum.

Experiments have been carried out in which alkali-metal atom beams react with oriented beams of CH_3I molecules. A molecule such as CH_3I with a permanent dipole moment μ acquires a first-order Stark energy in an electric field of strength F given by

$$E = -\mu \langle \cos \theta \rangle F, \tag{11}$$

where in terms of the usual rotational quantum numbers J, K, and M with the axis of quantization along the field

$$\langle \cos \theta \rangle = KM/J(J+1). \tag{12}$$

Because of this it is possible to focus beams of these molecules, which possess a particular value of $\langle \cos \theta \rangle$, by passage through a six-pole electric field. After focusing the beam may be passed through a homogeneous electric field which aligns it with respect to the most probable direction of relative motion of the colliding systems.

Results obtained in reactive scattering using beams oriented in this

---

[†] FEREND, S. M., FISK, G. A., HERSCHBACH, D. R., and KLEMPERER, W., loc. cit., p. 3551.
[‡] MALTZ, C., Ph.D. thesis, Harvard Univ., Cambridge, Mass., 1970.
[§] BENNEWITZ, H. G., HAUTER, R., KLAIS, O., and MILLER, G., loc. cit., p. 3551.
[||] SAFRON, S. A., Ph.D. thesis, Harvard Univ., Cambridge, Mass., 1969.

way were first reported in 1966.† In later experiments by Beuhler and Bernstein‡ on the Rb–CH$_3$I reaction, in which the CH$_3$I beam was velocity-selected, the intensity of the RbI product was measured as a function of laboratory scattering angle for different alignments of the CH$_3$I with respect to the relative velocity vector. In the C.M. system the differential cross-section for reactive scattering in the backwards direction is four times larger with the collision alignment RbICH$_3$ than with RbCH$_3$I. The cross-section does, however, remain finite even for the unfavourable alignment.

Gersh and Bernstein§ have observed the variation with relative impact energy of the reaction cross-section for K–CH$_3$I collisions. Measurements were made over an energy range from 0·1 to 1 eV by using a nozzle source for the CH$_3$I beam which could be seeded with different gases. By monitoring the flux densities of both beams and determining the total flux of KI product by integration over spherical coordinates (in the laboratory system) both in and out of the plane of scattering, relative reaction cross-section measurements were made to an accuracy of about 10 per cent. The cross-section was found to rise to a maximum at 0·18 eV after which it decreased steadily. The existence of the maximum would not have been predicted from an optical model (see p. 3559 below).

Results obtained in the study of reactive collisions which give rise to positive and negative ions involving alkali-metal atoms from sputter sources are discussed on p. 3676.

By using HCl beams excited by an HCl chemical laser‖ the reaction rate of K atoms with HCl molecules in the first vibrational state has been shown to be greater by orders of magnitude than that for the slightly endoergic reaction with HCl in the ground vibrational state.

Production of Na($3\,^2P$) and K($4\,^2P$) excited atoms has been observed in the reactions of Cl atoms with Na$_2$ and K$_2$,†† the production cross-section being between 10^{-15} and 10^{-14} cm^2.

9.2. Optical model analysis for reactive scattering (Chap. 17, § 6.2)

The differential cross-section for elastic scattering when the phase shifts η_l are taken to be complex may be written in the form

$$I_\mathrm{e}(\theta)\,\mathrm{d}\omega = |f_\mathrm{e}(\theta)|^2\,\mathrm{d}\omega, \tag{13}$$

† BROOKS, P. R. and JONES, E. M., *J. chem. Phys.* **45** (1966) 3449; BEUHLER, R. J., BERNSTEIN, R. B., and KRAMER, K. H., *J. Am. chem. Soc.* **88** (1966) 5331.
‡ BEUHLER, R. J. and BERNSTEIN, R. B., *J. chem. Phys.* **51** (1969) 5305.
§ GERSH, M. E. and BERNSTEIN, R. B., ibid. **56** (1972) 6131.
‖ ODIORNE, T. J., BROOKS, P. R., and KASPER, J. V. V., ibid. **55** (1971), 1980.
†† STRUVE, W. S., KITAGAWA, T., and HERSCHBACH, D. R., ibid. **54** (1971) 2759.

where
$$f_e(\theta) = (2ik)^{-1} \sum_{l=0}^{\infty} (2l+1)[\exp(2i\lambda_l)\{1-\mathscr{P}(l)\}^{\frac{1}{2}}-1]P_l(\cos\theta). \quad (14)$$

λ_l is the real part of the phase shift and $\mathscr{P}(l) = 1-e^{-4\mu_l}$ is the probability of an inelastic collision when the relative angular momentum quantum number is l. In Fig. 17.72, p. 1627 typical distributions are illustrated which were calculated on the black-sphere model

$$\mathscr{P}(l) = 0 \quad (l > l_c), \quad = 1 \quad (l < l_c). \quad (15)$$

Harris and Wilson† have carried out a detailed analysis of observed data on elastic scattering of K, Rb, and Cs atoms by CCl_4, $SnCl_4$, and CH_3I molecules. For the first two of these molecules they assumed that

TABLE 17 A.3

Parameters in optical model calculations of scattering of K atoms by CCl_4

Impact energy	ϵ	r_m	l_c/kr_m	d/kr_m
kcal/mole	kcal/mole	Å		
1·48	0·59	2·5	0·90	0·103
1·95	0·59	4·0	0·95	0·141
2·45	0·59	4·0	0·97	0·125
3·20	0·62	5·0	0·94	0·083
4·30	0·62	5·0	0·93	0·060

the interaction is spherically symmetrical and took for $\mathscr{P}(l)$ the form first used by Bernstein and Levine‡

$$\mathscr{P}(l) = \mathscr{P}_0[1+\exp\{(l-l_c)/d\}]^{-1}. \quad (16)$$

For scattering by CCl_4, which is only weakly reactive, the real phase shifts λ_l were assumed to be those given by a real (12, 6) interaction with parameters ϵ and r_m; $f_e(\theta)$ was calculated for a range of choices of these parameters and of l_c and d. For this purpose a semi-classical formula for $f_e(\theta)$ was used which is the extension to complex phase shifts of that derived by Berry (see (7) of p. 3526) for scattering by real potentials and in which the rapid oscillations are averaged out.

Quite good agreement was obtained with observed results, obtained by the same authors, when the parameters were suitably chosen. Values found for K–CCl_4 are given in Table 17 A.3.

It was found that the opacity function, expressed as a function of y/r_m where y is the distance of closest approach, was nearly independent of

† loc. cit., p. 3540.
‡ BERNSTEIN, R. B. and LEVINE, R. D., *J. chem. Phys.* **49** (1968) 3872.

the impact energy. In going from Cs to Rb to K, $\mathscr{P}(y/r_m)$ became less diffuse.

For collisions with $SnCl_4$, which is strongly reactive, an interaction was chosen which took account of the transition from a weak van der Waals attraction to a strong attraction between ions, occurring at a crossing between potential energy surfaces for covalent (A+BC) and ionic (A$^+$+BC$^-$) states (see p. 2660). In the neighbourhood of such a crossing the interaction is approximately given by[†]

$$V(R) = -c_6 R^{-6} - A e^{-\alpha R}, \qquad (17)$$

where the parameters A and α are determined by the ionization potential of A and the electron affinity of BC. With such an interaction, orbiting (see p. 1310) occurs at a particular impact parameter p_0 for each impact energy. The observed behaviour for $SnCl_4$ was well represented by taking for $\mathscr{P}(l)$ the form (16) with l_c given by p_0/kr_m.

Finally, for collisions with CH_3I the interaction was taken to be anisotropic, the (12, 6) form being modified to

$$c_{12} r^{-12} - c_6 r^{-6}\{1 + a P_2(\cos \gamma)\}, \qquad (18)$$

where γ is the angle the line joining the centres of mass of the atom and molecule makes with the molecular axis. Cross-sections were calculated semi-classically as a function of γ assuming a fixed orientation during the collision, and then averaged over γ. The opacity functions giving the best fit to the experimental data are considerably more diffuse than for CCl_4. Values of the reaction cross-section agree quite well with those determined from reactive scattering experiments.[‡]

9.3. *Reactive scattering between systems not containing alkali-metal atoms*

The investigation of such systems requires the use of the universal ionization detector (see Chap. 16, p. 1345). At the time of writing of Vol. 3 only preliminary experiments of this kind had been carried out. A number of important results are now available, largely concerned with reactions of hydrogen, deuterium, and halogen atoms with other molecules.

9.3.1. $D + H_2 \rightarrow HD + H$. This reaction has been investigated by Geddes, Krause, and Fite.[§] The D beam was produced from a tungsten furnace operating at temperatures up to 3000 °K while the H_2 beam was

---

[†] ANDERSON, R. W., Thesis, Harvard Univ., Cambridge, Mass., 1968. GRICE, K. D., ibid. 1967.

[‡] KWEI, G. H., NORRIS, J. A., and HERSCHBACH, D. R., *J. chem. Phys.* **52** (1970) 1317; KWEI, G. H., Thesis, Univ. of California, Berkeley, 1967.

[§] GEDDES, J., KRAUSE, H. F., and FITE, W. L., *J. chem. Phys.* **56** (1972) 3298.

cooled to 77 °K. By use of phase-sensitive detection based on modulation of the D beam the intensity of production and mean velocity of the HD product could be measured as a function of laboratory angle of projection. The relative differential cross-section in C.M. coordinates deduced from these observations was calculated from observations of the non-reactive scattering of D.

In the C.M. system the HD distribution is strongly peaked in the backward direction showing that no long-lived complex is formed in the collision. The rate constant for the reaction is estimated to be $1\cdot2 \pm 0\cdot6 \times 10^{-11}$ cm^2 s^{-1}, corresponding to a mean reaction energy of 0·48 eV, 0·15 eV above the activation energy.† All of this results from a complex analysis of the primary data and the study of the reaction is still at a quite early stage partly because of the difficulty of measuring the low rate constants against the noise background.

9.3.2. *Reactions of deuterium atoms with halogen atoms* (see Chap. 17, § 6.5). Angular and velocity distribution of deuterium halides projected from reactions of crossed beams of D atoms with Cl$_2$, Br$_2$, I$_2$, ICl, and IBr molecules have been measured by McDonald, Le Breton, Lee, and Herschbach.‡ Experiments were carried out with two sources of D atoms, a hot tungsten oven at 2800 °K and a microwave discharge source at 350 °K. The velocity distribution of the halogen beams was confined by use of a nozzle source while that of the deuterium halide reaction product was measured with a time-of-flight velocity analyser.

The results for the reaction of hot D atoms with Br$_2$ do not agree with those obtained in earlier experiments by Datz and Schmidt described in Chapter 17, § 6.5. The latter authors found a maximum in the laboratory angular distribution near 70° but McDonald *et al.* find a flat maximum at 120°. For reactions with I$_2$ and Cl$_2$ respectively the maximum is at smaller and larger angles respectively. In the C.M. system the angular distribution has a peak beyond 150° for Cl$_2$, near 120–140° for Br$_2$, and near 100–110° for I$_2$. The mean energy appearing in relative translation by the reaction is approximately 45, 34, and 26 per cent of the total energy released for the respective reactions with Cl$_2$, Br$_2$, and I$_2$.

Grosser and Haberland§ have also carried out measurements of the angular distributions of reaction products from interaction of H and of D atoms at a temperature of 3000 °K with Cl$_2$ and Br$_2$ molecules, obtaining results which agree quite closely with those of McDonald *et al.*

† KUPPERMANN, A. and WHITE, J. M., *J. chem. Phys.* **44** (1966) 4352.
‡ MCDONALD, J. D., LE BRETON, P. R., LEE, Y. T., and HERSCHBACH, D. R., ibid. **56** (1972) 769.
§ GROSSER, J. and HABERLAND, H., *Chem. Phys. Lett.* **7** (1970) 442.

For the D atoms at a temperature near 300 °K no reaction was detectable with Cl_2, little with Br_2, but comparable to that for the hot atoms with I_2. In this case the peak in the angular distribution of reaction product moves to between 115 and 125°.

The angular distribution of the halide product suggests that the preferred reaction configuration changes from collinear to a strongly bent form in going from Cl to Br to I, while the energy distribution suggests that repulsion between the halogen atoms after the reaction is a dominating feature.

9.3.3. *Reactions of halogen atoms.* Exchange reactions between halogen atoms and halogen molecules have been studied by Lee, McDonald, Le Breton, and Herschbach† and by Blais and Cross.‡ The striking thing about these reactions is that, while the reaction cross-sections are small, 5–20×10^{-16} cm², the angular distribution is not of the rebound type but shows a peak in the forward direction. In contrast to the rebound reactions (see p. 1644) the short range interaction responsible for the halogen exchange reactions must be attractive. There is evidence of complex formation but in most cases its lifetime is comparable with or even somewhat smaller than the rotational period.

9.3.4. *Reactions involving alkaline earths.* A number of experiments§ are in progress to investigate reactive scattering of atoms of the alkaline earths with halide and other molecules.

9.3.5. *Endothermic reactions.* The use of seeded nozzle beams has made it possible to begin the study of endothermic reactions. Thus the dissociation of alkali halides by energetic Xe atoms has been observed.

† LEE, Y. T., MCDONALD, J. D., LE BRETON, P. R., and HERSCHBACH, D. R., *J. chem. Phys.* **49** (1968) 2447; LEE, Y. T., LE BRETON, P. R., MCDONALD, J. D., and HERSCHBACH, D. R., ibid. **51** (1969) 455.

‡ BLAIS, N. C. and CROSS, J. B., ibid. **52** (1970) 3580; CROSS, J. B. and BLAIS, N. C., ibid. **55** (1971) 3970.

§ OTTINGER, CH. and ZARE, R. N., *Chem. Phys. Lett.* **5** (1970) 243; JONAH, C. D. and ZARE, R. N., ibid. **9** (1971) 65; JONAH, C. D., ZARE, R. N., and OTTINGER, CH., *J. chem. Phys.* **56** (1973) 263; BATALLI-COSMOVICI, C. and MICHEL, K. W., *Chem. Phys. Lett.* **11** (1971) 245; FRICKE, J., KIM, B., and FITE, W. L., *7th int. conf. on the physics of electronic and atomic collisions, Amsterdam.* Abstracts, North Holland, 1971, p. 37.

18

COLLISIONS BETWEEN ATOMS (OR MOLECULES) UNDER GAS-KINETIC CONDITIONS—COLLISIONS INVOLVING ELECTRONIC TRANSITIONS

1. Introduction

SINCE Vol. 3 was written the application of delayed coincidence counting techniques to the measurement of the lifetimes of excited states has developed considerably, particularly for the study of quenching cross-sections. New results for these cross-sections have been obtained by this and other methods.

Further new data have been obtained in cross-sections for collisions leading to disorientation and disalignment. In particular the Zeeman scanning technique has been applied to derive these cross-sections from measurements of the depolarization of potassium resonance radiation.

Again, new results are available concerning cross-sections for excitation of fine structure transitions between levels of the resonance P states of alkali-metal atoms. Of special interest are the measurements recently made on the transfer of coherence between fine structure levels on collision. Crossed beam techniques have been applied to study the excitation of transitions between the $6\,^3P_2$ and $6\,^3P_1$ level of mercury atoms.

Further experiments have confirmed that transfer from $n\,^1P$ to nF and from nF to $^{3,1}D$ levels of helium is responsible for pressure dependent effects observed in the excitation of the D levels of helium by electron impact, and new data are available about transfer rates.

A great deal of progress has been made in studying collisions involving metastable atoms. Total cross-sections for Penning and associative ionization of atoms and molecules by $\text{He}(2\,^3S)$ and $\text{He}(2\,^1S)$ metastable atoms have been measured by flowing afterglow and other techniques while energy distributions of electrons ejected in these collisions have been measured with high energy resolution. Results have even been obtained for ionization of atomic hydrogen.

Further progress in the measurement and interpretation of observations on diffusion and charge transfer of $\text{He}(2\,^3S)$ in helium has been made although some difficulties still remain.

A great deal of new results, many of which have been obtained with the use of photon counting techniques, are now available on cross-sections for quenching of O(^1S) and O(^1D) atoms. Considerable progress has also been made in correlating laboratory data with results obtained, using rocket-borne equipment, on the spectral intensity distribution of the atmospheric day airglow.

The first detailed measurements of spin exchange cross-sections for collisions between alkali-metal atoms, using crossed atomic beams, have been made and provide new data of value for the determination of the interactions between alkali-metal atoms.

A detailed study of spin-reversal collisions between polarized Rb atoms and rare-gas atoms has been made which shows that spin reversal not only occurs through spin–orbit interaction in sudden collisions but also in 'sticking' collisions in which an Rb atom is temporarily bound to a rare gas atom.

Much of the new experimental work referred to above has been accompanied by related theoretical work which is summarized in the last section of the notes on this chapter. In particular we draw attention to the work on the interpretation of cross-sections for fine structure transitions, on the analysis of the new data on spin exchange cross-sections to obtain information about the long range forces between alkali-metal atoms (see p. 3531), and on the analysis of the new results on spin-reversal collisions of polarized Rb atoms with Kr atoms.

2. Quenching of resonance radiation

In addition to the techniques for measuring cross-sections for quenching of resonance radiation which are described in Chapter 18, §§ 2.1–2.4, a delayed coincidence technique† has also been introduced successfully. This is essentially similar in principle, for example, to the delayed coincidence technique used to study the lifetimes of free positrons in gases (see Chap. 26, pp. 3130–5). The mixture of fluorescing gas or vapour and admixed quenching gas is excited with a short pulse of radiation and the subsequent emission of quanta of fluorescent light recorded in a multichannel photon-counting system, each channel corresponding to a particular time interval since the termination of the exciting pulse. In this way a decay curve of the population of the excited

† HERON, S., McWHIRTER, R. W. P., and RHODERICK, E. H., Proc. R. Soc. A**234** (1956) 565; BENNETT, W. R., KINDLMANN, P. J., and MERCER, G. N., Applied Optics (supplement **2**, Chemical lasers), 1965, p. 34; BOLLINGER, L. M. and THOMAS, G. E., Rev. scient. Instrum. **32** (1961) 1044; KIBBLE, B. P., COPLEY, G., and KRAUSE, L., Phys. Rev. **159** (1967) 11.

state responsible for the fluorescence may be obtained. From this the lifetime of the state may be derived. By carrying out such measurements at different pressures of the quenching gas the quenching cross-section may be obtained in the usual way.

New measurements made by different techniques, for quenching of $Na(3\,^2P)$ and $K(4\,^2P)$, which do not appear in Table 18.4, p. 1693 of Chapter 18, are summarized in Table 18 A.1.

TABLE 18 A.1

Observed cross-sections (in 10^{-16} cm^2) for deactivation of $Na(3\,^2P)$ and $K(4\,^2P)$ by impact with various molecules

$Na(3\,^2P)$																
H_2			HD		D_2		N_2				O_2	CO		CO_2		H_2O
(a)	(b)	(c)	(b)	(c)	(b)	(c)	(a)	(b)	(c)	(d)	(a)	(a)	(c)	(a)	(e)	(a) (d)
8	16·2	12·2	11·6	11·5	9·8	10·2	21	40·3	30·7	22	34	41	41	50	65→40	2·2 2·2

$K(4\,^2P)$									
(a)	(f)	(f)	(f)	(a)	(f)	(a)	(a)	(a)	(a)
3·4	9·4	11·9	8·0	19	34	49	44	66	2·6

(a) HOOYMAYERS, H. P. and LIJNSE, P. L., *J. quant. Spectros. radiat. Transfer* **9** (1969) 995. Flame technique.
(b) KIBBLE, B. P., COPLEY, G., and KRAUSE, L., *Phys. Rev.* **159** (1967) 11. Delayed coincidence technique.
(c) BÄSTLEIN, C., BAUMGARTNER, G., and BROSA, B., *Z. Phys.* **218** (1969) 319. Demtröder's method (p. 1662).
(d) LIJNSE, P. L. and ELSENAAR, R. J., *J. quant. Spectros. radiat. Transfer* **12** (1972) 1115.
(e) EARL, B. L., HERM, R. B., LIN, S. M., and MIMS, C. A., *J. chem. Phys.* **56** (1972) 867. Photodissociation technique (p. 1675). The range of values given refers to relative impact velocities from 1·3 to $2·2 \times 10^5$ cm s^{-1}.
(f) COPLEY, G. and KRAUSE, L., *Can. J. Phys.* **47** (1969) 533. Delayed coincidence technique.

The new values given by Hooymayers and Lijnse replace earlier values of Hooymayers and Alkemade given in Table 18.4 under (d) for Na and (b) for K.

Further experiments to detect quenching of Na by rare-gas atoms have been carried out. Copley, Kibble, and Krause† have applied the delayed coincidence technique and find cross-sections $< 10^{-18}$ cm^2 for all the gases at 400 °K. These are the lowest limits which have yet been placed upon them.

The photodissociation technique has been applied by Brus‡ to measure quenching cross-sections for $Na(3\,^2P)$ by Br_2 and by I_2. For the latter cross-sections of 190, 165, and 141×10^{-16} cm^2 were obtained for sodium atom velocities of 1·23, 1·68, 2·09, and 2·47 cm s^{-1}.

† COPLEY, G., KIBBLE, B. P., and KRAUSE, L., *Phys. Rev.* **163** (1967) 34.
‡ BRUS, L. E., *J. chem. Phys.* **52** (1970) 1716.

Cross-sections have been measured somewhat less accurately for quenching of the separated doublet lines of K,[†] Rb,[‡] and Cs.[§]

A number of new measurements have been made of cross-sections for quenching of $Hg(6\,^3P_1)$ (see Chap. 18, § 2.5). These are summarized in Table 18 A.2.

TABLE 18 A.2

Observed cross-sections (in 10^{-16} cm^2) for quenching of $Hg(6\,^3P_1)$ atoms by collision with various molecules

H_2		D_2	N_2		CO		CO_2	
(a)	(b)	(b)	(a)	(b)	(a)	(b)	(a)	(b)
15·7	24·6	22·7	3·0	0·73	16·3	21·7	20·4	10·2

(a) JENKINS, D. R. Private communication to L. Krause.
(b) PITRE, J., DEECH, J. S., and KRAUSE, L., *7th int. conf. on the physics of electronic and atomic collisions, Amsterdam*. Abstracts, North Holland, 1971, p. 671. Delayed coincidence technique.

As in all other cases the flame experiments (a) refer to temperatures near 1400 °K while the experiments (b) refer to room temperature.

3. Collisions which change the orientation and alignment of the total electronic angular momentum

3.1. *Disorientation and disalignment of $Hg(6\,^3P_1)$ by foreign gas atoms*

Faroux[||] has made a thorough experimental study of the relaxation of orientation and alignment of the $(6\,^3P_1)$ levels of different effectively pure isotopes of mercury using the technique described in Chapter 18, p. 1721.

For the even isotopes, possessing no nuclear spin, the values for the disorientation Q_{or} and disalignment Q_{al} cross-sections measured at room temperature differ little from those given in Table 18.8. Important new information has been obtained, however, about the variation of these cross-sections with temperature. It was found that, whereas for collisions with the heavier gases Q_{or} and Q_{al} vary quite closely as $T^{-1/5}$ over the observed temperature range for 20 to 500 °K, for collisions with helium and neon there is very little dependence on temperature over the same range. The theoretical significance of these results is discussed on p. 3598.

For ^{199}Hg, with nuclear spin quantum number $\frac{1}{2}$, detailed measurements of the cross-sections $\bar{Q}_{he}$, $^{\frac{3}{2}}\bar{Q}_{al}$, $^{\frac{3}{2}}\bar{Q}_{or}$, $^{\frac{1}{2}}\bar{Q}_{or}$, and $^{\frac{1}{2},\frac{3}{2}}\bar{Q}_{or}$, defined in

[†] McGILLIS, D. A. and KRAUSE, L., *Can. J. Phys.* **46** (1968) 25.
[‡] HRYCYSHYN, E. S. and KRAUSE, L., ibid. **48** (1970) 2761; BELLISIO, J. A., DAVIDOVITS, P., and KINDLMANN, P. J., *J. chem. Phys.* **48** (1968) 2376.
[§] McGILLIS, D. A. and KRAUSE, L., *Can. J. Phys.* **46** (1968) 1051.
[||] FAROUX, J.-P., Thèse, Paris, 1969.

Chapter 18, § 3.7, p. 1719, were made. Again, in most cases, these differ only a little from the earlier measurements given in Table 18.8. The new values are given in Table 18 A.18, in comparison with calculated values with which they agree very well.

For ^{201}Hg, with nuclear spin quantum number $\frac{3}{2}$, disorientation and disalignment cross-sections which arise may be denoted as $^{\frac{3}{2}}\bar{Q}_{\mathrm{al}}$, $^{\frac{3}{2}}\bar{Q}_{\mathrm{al}}$, $^{\frac{5}{2},\frac{3}{2}}\bar{Q}_{\mathrm{al}}$, $^{\frac{5}{2}}\bar{Q}_{\mathrm{or}}$, $^{\frac{3}{2}}\bar{Q}_{\mathrm{or}}$, $^{\frac{1}{2}}\bar{Q}_{\mathrm{or}}$, $^{1,\frac{3}{2}}\bar{Q}_{\mathrm{or}}$, $^{\frac{3}{2},\frac{5}{2}}\bar{Q}_{\mathrm{or}}$, and $^{\frac{5}{2},\frac{1}{2}}\bar{Q}_{\mathrm{or}}$, while there are three hyperfine transfer cross-sections. The measurement of these cross-sections is more difficult than for ^{199}Hg but Faroux has obtained results for collisions with helium which are discussed in relation to the theory on p. 3599.

3.2. *Disalignment of metastable states of* ^{208}Pb

Gibbs[†] has measured cross-sections for disalignment of levels of the metastable $^3P_1$ and $^3P_2$ states of ^{208}Pb. Aligned atoms in these states were produced by radiative transitions from the $6p7s\ ^3P_1$ excited state (see Chap. 18, p. 1723) in turn produced, with polarization, by absorption of resonance radiation (λ 2833 Å) incident along an external magnetic field. The populations of aligned $^3P_1$ and $^3P_2$ atoms were measured by optical absorption. To obtain the disalignment cross-sections the alignment was destroyed by r.f. resonance and the time taken to recover after removal of the r.f. measured as a function of the pressure of added gas. These pressures could be chosen so that disalignment of the excited $^3P_1$ state was negligible, the radiative lifetime of this state being so much shorter than for the metastable states.

Results obtained are given in Table 18 A.3.

TABLE 18 A.3

Observed disalignment cross-sections (in 10^{-16} cm^2) for metastable states of ^{208}Pb

Quenching gas	He	Ne	Ar	Kr	Xe	H$_2$	N$_2$
$^3P_1$	45	49	70	89	127	52	118
$^3P_2$	63	62	88	118	175	71	133

3.2. *Disalignment of excited states of* Ne

In Chapter 18, p. 1725 we described the experiments of Decomps and Dumont on the disalignment of the $2p_4$ levels (in the Paschen notation, see Fig. 18.41) of the $2p^53p$ configuration of Ne, using laser excitation.

[†] GIBBS, H. M., *6th int. conf. on the physics of electronic and atomic collisions*, Cambridge, Mass. Abstracts, MIT Press, 1969, p. 718.

Carrington and Corney† observed that if a discharge is run in an elongated tube in neon or a neon–helium mixture at a pressure of 1–10 torr, the radiation emitted at right angles to the axis to the tube through transitions from the various $2p$ levels ($2p_2$–$2p_{10}$) to the $1s$ levels (again in the Paschen notation) of the $2p^53s$ configuration is as much as one per cent polarized. This presumably arises because of the propagation down the discharge of the light through successive emission and absorption because of the large population of the metastable $1s_3$ and $1s_5$ levels, and of the $1s_2$ and $1s_4$ levels through imprisonment of resonance radiation from these levels to the ground level. The polarization is a consequence of the anisotropy of the resulting radiation field and, although small, may be observed, using phase-sensitive detection, as a function of the magnetic field strength applied along the direction of observation: i.e. the Hanle effect may be obtained for each of the $2p$ levels giving the sum of the disalignment and quenching cross-sections Q_{al} and Q_q respectively (see p. 1702). Results were obtained for collisions with Ne and He at three temperatures, 85, 315, and 870 °K and are given in Table 18 A.4.

The variation of the cross-section with temperature is by no means the same for the different levels. In the $2p_2$ and $2p_6$ levels, for which the cross-sections are less than 40×10^{-16} cm², there is a marked increase with temperature whereas for the $2p_8$ and $2p_9$ levels, with cross-sections above 70×10^{-16} cm², there is very little dependence on temperature. The theoretical interpretation of this behaviour is discussed on p. 3600.

3.3. *Depolarization of potassium resonance radiation*

In Chapter 18, § 3.10 we described measurements made of cross-sections for disorientation and disalignment of Rb($5\,^2P_{\frac{3}{2}}$) and for the disorientation of Rb($5\,^2P_{\frac{1}{2}}$) and Cs($6\,^2P_{\frac{1}{2}}$) by collisions with rare-gas atoms. Measurements have since been carried out also for disorientation of K($4\,^2P_{\frac{1}{2}}$)‡ and disorientation and disalignment of K($4\,^2P_{\frac{3}{2}}$).§

The method used employs the Zeeman scanning technique which was developed at an early stage‖ to investigate the hyperfine structure of mercury resonance radiation.†† In the application to study impact depolarization of K resonance radiation a potassium resonance lamp is placed in a constant magnetic field of around 5 kG. The resonance

† CARRINGTON, C. G. and CORNEY, A., *J. Phys. B (GB)* **4** (1971) 869.
‡ BERDOWSKI, W. and KRAUSE, L., *Phys. Rev.* **165** (1968) 158.
§ BERDOWSKI, W., SHINER, T., and KRAUSE, L., ibid. **A4** (1971) 984.
‖ MROZOWSKI, S., *Bull. Acad. Polon. Sci.* **A9–10** (1930) 464.
†† BUHL, O., *Z. Phys.* **109** (1938) 180; **110** (1938) 395; BITTER, F., DAVIS, S., RICHTER, R., and YOUNG, J., *Phys. Rev.* **96** (1954) 1531.

TABLE 18 A.4

Observed cross-sections $Q_{al}+Q_q$ (in 10^{-16} cm²) for destruction of alignment of the different 2p levels of neon

Level	Temperature	85 °K	315 °K	870 °K
$2p_2$	Ne	12·8	27·3 (29†)	34·7
	He	7·28	19·7 (14†)	47·5
$2p_4$	Ne	—	92·9 (74‡, 157§)	66·3
	He	—	85·0 (72‡, 63§)	70·4
$2p_5$	Ne	60·9	53·0	45·9
	He	39·1	41·4	49·6
$2p_6$	Ne	14·7	25·5	36·2
	He	4·68	23·4	38·9
$2p_7$	Ne	60·6	52·0	43·9
	He	36·9	37·7	51·7
$2p_8$	Ne	81·6	85·2	72·2
	He	73·7	79·2	79·9
$2p_9$	Ne	111·8	110·7	98·5
	He	109·8	108·2	93·5
$2p_{10}$	Ne	—	2·06	—
	He	—	3·20	—

Apart from the bracketed values, results at 85 and 315 °K are due to Carrington and Corney‖ and at 870 °K to Carrington, Corney, and Durrant.†† The bracketed values have been obtained by laser techniques.

radiation, filtered to transmit only one component of the resonance doublet, is then incident on a fluorescence cell bathed in a magnetic field, the magnitude of which can be varied from 1 to 10 kG.

The light emitted from the lamp in a direction parallel to the constant field is separated into σ^+ and σ^- components by a circular polarizer. Pure $\sigma^{\pm}$ radiation will excite fluorescence according to the selection rules $\Delta m_J = \pm 1$ respectively. The exciting radiation is incident on the fluorescence cell in a direction normal to the variable magnetic field. As this field varies, resonance coincidences will occur at particular fields between Zeeman components in emission from the lamp and absorption by the cell. At each coincidence the fluorescent intensity will exhibit a peak which may be identified and which corresponds to emission of completely polarized radiation, in the absence of foreign gases and of self-depolarization. It is then possible to measure the depolarization of each component of fluorescent radiation due to the presence of different

† FOURNIER, E., DUCLOY, M., DECOMPS, B., and DUMONT, M., *Compt. Rend. Acad. Paris*, **268**B (1969) 1495.
‡ DECOMPS, B. and DUMONT, M., *J. Phys. Paris* **29** (1968) 443.
§ HANSCH, T., ODENWALD, R., and TOSCHEK, P., *Z. Phys.* **209** (1968) 478.
‖ loc. cit., p. 3567.
†† CARRINGTON, C. G., CORNEY, A., and DURRANT, A. V., *J. Phys. B (GB)* **5** (1972) 1001.

pressures of a foreign gas. From a detailed analysis of the results for different Zeeman components the disorientation and disalignment cross-sections may be obtained.

The results obtained are given in Table 18 A.5. Comparison with theory is discussed on p. 3601.

TABLE 18 A.5

Observed disorientation (Q_{or}) and disalignment (Q_{al}) cross-sections (in 10^{-16} cm^2) for collisions of rare-gas atoms with $K(4\,{}^2P_{\frac{1}{2}})$ and $K(4\,{}^2P_{\frac{3}{2}})$ atoms

	$K(4\,{}^2P_{\frac{1}{2}})$				
	He	Ne	Ar	Kr	Xe
Q_{or}	46	39	52	80	107
	$K(4\,{}^2P_{\frac{3}{2}})$				
Q_{or}	86	86	164	248	251
Q_{al}	127	120	240	301	336

4. Sensitized fluorescence

4.1. *Sensitized fluorescence and fine-structure collisions*

4.1.1. *Collision induced transitions between ${}^2P_{\frac{1}{2}}$ and ${}^2P_{\frac{3}{2}}$ states of alkali-metal atoms* (see Chap. 18, § 4.1.1). A number of further measurements have been made of cross-sections for excitation of transitions between fine structure levels of the resonance doublets of alkali-metal atoms.

Referring to the data given in Table 18.9 of Chapter 18, new results have been obtained for Na. In the notation of Table 18.9, Pitre and Krause[†] have measured $\bar{Q}_{12}$ for collisions with He, Ne, Ar, Kr, and Xe, in 10^{-16} cm^2, as 86, 67, 110, 85, and 90 respectively with $\bar{Q}_{21}$ as given from detailed balancing. $\bar{Q}_{12}$ for collisions with Na has been measured as 532×10^{-16} cm^2.

Gallagher,[‡] in the course of a detailed investigation of the variation of the cross-sections for Rb and Cs with temperature, obtained evidence that a number of the previously measured cross-sections $\bar{Q}_{21}$ for rare-gas collisions, given in Table 18.9, were incorrect. In all these cases the observed ratio $\bar{Q}_{21}/\bar{Q}_{12}$ did not agree with that given from detailed balancing. Table 18 A.6 summarizes the results obtained by Gallagher.

A number of measurements have been made of these cross-sections for collisions with molecules. These are summarized in Table 18 A.7.

[†] PITRE, J. and KRAUSE, L., *Can. J. Phys.* **45** (1967) 2671.
[‡] GALLAGHER, A., *Phys. Rev.* **172** (1968) 88.

TABLE 18 A.6

Cross-sections (in 10^{-16} cm²) for excitation of fine structure transitions in Rb and Cs by collision with rare gases, measured by Gallagher at different temperatures

	He	Ne	Ar	Kr	Xe
Rb ($5\,^2P_{\frac{1}{2}} \to 5\,^2P_{\frac{3}{2}}$)					
$T = 300\,°$K	~ 0.06	$\sim 10^{-3}$	5×10^{-4}	3×10^{-4}	2×10^{-4}
$T = 900\,°$K	1	1.5×10^{-2}	10^{-2}	6×10^{-3}	10^{-2}
Rb ($5\,^2P_{\frac{3}{2}} \to 5\,^2P_{\frac{1}{2}}$)					
$T = 300\,°$K	~ 0.1	$\sim 1.2 \times 10^{-3}$	8×10^{-4}	5×10^{-4}	4×10^{-4}
$T = 900\,°$K	0.7	1.4×10^{-2}	8×10^{-3}	4×10^{-3}	7×10^{-3}
Cs ($6\,^2P_{\frac{1}{2}} \to 6\,^2P_{\frac{3}{2}}$)					
$T = 300\,°$K	$\sim 4 \times 10^{-5}$	10^{-6}	—	—	—
$T = 900\,°$K	3×10^{-3}	6×10^{-5}	—	—	—
Cs ($6\,^2P_{\frac{3}{2}} \to 6\,^2P_{\frac{1}{2}}$)					
$T = 300\,°$K	$\sim 3 \times 10^{-4}$	7×10^{-6}	—	—	—
$T = 900\,°$K	4×10^{-3}	6.5×10^{-5}	—	—	—

TABLE 18 A.7

Observed cross-sections (in 10^{-16} cm²) for excitation of fine structure transitions between the resonance doublets of alkali-metal atoms, by collision with molecules

	N_2	H_2	HD	D_2	CH_4	CD_4
Na ($3\,^2P_{\frac{1}{2}} \to 3\,^2P_{\frac{3}{2}}$)	144(a)	80(a)	84(a)	98(a)	148(b)	151(b)
K ($4\,^2P_{\frac{1}{2}} \to 4\,^2P_{\frac{3}{2}}$)	100(c)	76(c)	74(c)	72(c)	—	—
Rb ($5\,^2P_{\frac{1}{2}} \to 5\,^2P_{\frac{3}{2}}$)	16(d)	11(d)	18(d)	22	30(d)	28(d)
Cs ($6\,^2P_{\frac{1}{2}} \to 6\,^2P_{\frac{3}{2}}$)	4.7(e)	6.7(e)	4.8(e)	4.2(e)	—	—

(a) STUPAVSKY, M. and KRAUSE, L., *Can. J. Phys.* **46** (1968) 2127.
(b) IDEM, ibid. **47** (1969) 1249.
(c) MCGILLIS, D. A. and KRAUSE, L., ibid. **46** (1968) 25.
(d) HRYCYSHYN, E. S. and KRAUSE, L., ibid. **48** (1970) 2761.
(e) MCGILLIS, D. A. and KRAUSE, L., ibid. **46** (1968) 1051.

One striking feature of these results is that, for Rb and Cs, the cross-sections are very much larger than for collisions with the rare gases (see Table 18 A.6).

Discussion of the results in relation to theory is given on p. 3601.

4.1.2. *Transfer of coherence between fine structure levels on collision.* Elbel and Schneider† have measured mean cross-sections $\bar{Q}_{tr}$ for transfer of polarization from the $3\,^2P_{\frac{1}{2}}$ to the $3\,^2P_{\frac{3}{2}}$ fine structure level in sodium

† ELBEL, M. and SCHNEIDER, W., *Z. Phys.* **241** (1971) 244.

on collision with rare-gas atoms. This may be defined in terms of the equation

$$\frac{d\langle J_z\rangle_{\frac{3}{2}}}{dt} = n\bar{v}\bar{Q}_{tr}\langle J_z\rangle_{\frac{1}{2}} - (\tau^{-1} + n\bar{v}\bar{Q}_{rel})\langle J_z\rangle_{\frac{3}{2}}, \tag{1}$$

where $\langle J_z\rangle_{\frac{3}{2},\frac{1}{2}}$ are the vector polarizations of the $J = \frac{3}{2}$ and $J = \frac{1}{2}$ states respectively. n is the concentration of the rare-gas atoms, $\bar{v}$ the mean relative velocity of collision, τ the lifetime of the $\frac{3}{2}$ state towards spontaneous decay, and $\bar{Q}_{rel}$ is the mean cross-section for collisions which destroy the polarization of the $\frac{3}{2}$ state.

To measure $\bar{Q}_{tr}$ we have, in equilibrium,

$$\langle J_z\rangle_{\frac{3}{2}}/\langle J_z\rangle_{\frac{1}{2}} = \frac{\tau n\bar{v}\bar{Q}_{tr}}{1+\tau n\bar{v}\bar{Q}_{rel}}. \tag{2}$$

If sodium vapour with admixed rare gas is irradiated with polarized $D_1 \sigma^+$ radiation and the relative intensities of the $D_1\sigma^+$, $D_1\sigma^-$, $D_2\sigma^+$, and $D_2\sigma^-$ fluorescent radiation measured it may be shown that

$$R = \frac{I(D_2\sigma^+) - I(D_2\sigma^-)}{I(D_1\sigma^+) - I(D_1\sigma^-)} = \frac{1}{2}\frac{\langle J_z\rangle_{\frac{3}{2}}}{\langle J_z\rangle_{\frac{1}{2}}}, \tag{3}$$

$I(D_2\sigma^+)$ being the intensity of the fluorescent $D_2\sigma^+$ radiation, etc.

The slope of a plot of R against the pressure p of the rare gas at $p = 0$ is given by

$$(dR/dp)_{p=0} = \frac{1}{2}\frac{\bar{v}\bar{Q}_{tr}\tau}{\kappa T}. \tag{4}$$

In this way Elbel and Schneider[†] obtained for He, Ne, Ar, Kr, and Xe values of $\bar{Q}_{tr}$, in 10^{-16} cm², 22, 24, 26, 20, and 23 respectively.

4.1.3. *Excitation of transitions between the $6\,^3P_2$ and $6\,^3P_1$ levels of mercury.* Doemeny, van Itallie, and Martin[‡] have observed the quenching of Hg($6\,^3P_2$) in an atomic beam. A mercury beam was subject to bombardment by electrons with 7·5 eV energy so that only excited atoms in $6\,^3P_2$, $6\,^3P_1$, and $6\,^3P_0$ states were produced. After a flight time from the bombardment region of 10^{-3} s, long compared with the lifetime of $6\,^3P_1$ (10^{-7} s), only $6\,^3P_2$ and $6\,^3P_0$ excited atoms were present. The beam was then crossed with a beam of molecules M and fluorescent radiation arising from the process

$$\mathrm{Hg}(6\,^3P_2) + \mathrm{M} \to \mathrm{Hg}(6\,^3P_1) + \mathrm{M}, \tag{5}$$

$$\mathrm{Hg}(6\,^3P_1) \to \mathrm{Hg}(5\,^1S) + h\nu, \tag{6}$$

observed by a cooled photomultiplier detector with a 2537 Å interference filter. By modulating the metastable atom beam at 20 Hz,

[†] ELBEL, M. and SCHNEIDER, W., loc. cit.
[‡] DOEMENY, L. J., VAN ITALLIE, F. J., and MARTIN, R. M., *Chem. Phys. Lett.* **4** (1969) 302.

phase-sensitive detection could be used. Cross-sections for (5), with M = CO, N_2, H_2, D_2, CH_4, and CD_4, were observed to be in the ratio

$$1:0\cdot71:0\cdot42:0\cdot10:0\cdot62:0\cdot21.$$

4.2. Sensitized fluorescence and collisions involving transfer of excitation (see Chap. 18, § 4.2)

A number of further experiments have been carried out on sensitized fluorescence involving excitation transfer. Czajkowski, Skardis, and Krause[†] have made a further study of mercury-sensitized sodium fluorescence. Their results do not agree with those of Kraulinya and of Rautian and Khaikin given in Table 18.11 of Chapter 18. They find the largest cross-sections (in 10^{-16} cm²) for population of the following levels of Na in order 38·5 (9S), 31·36 (8D), 9·35 (8P), 8·06 (9D and 9P). The energy differences between these levels and that of Hg($6\,^3P_1$) are, in eV, 0·02, 0·04, $-0\cdot01$, 0·09, and 0·06 respectively. Although part of the discrepancies between the different experiments can be ascribed to resonance trapping and to difficulties in measuring vapour pressures, the reacting system is very complicated and difficulties of interpretation may be greater than realized.

Mercury-sensitized fluorescence of Cd,[‡],[§],[∥], Zn,[§],[∥] and In[††] has also been investigated recently.

4.3. Excitation transfer in helium (see Chap. 18, § 4.2.3)

Jobe and St. John[‡‡] have carried out an extensive experimental study of the effect of excitation transfer on the apparent cross-sections Q'_j (see Chap. 4, § 1.1) for excitation of the various states j of helium by electrons of 100 eV energy and a pressure of 4×10^{-3} to 7×10^{-2} torr. Q'_j was measured at $6\cdot3 \times 10^{-2}$ torr pressure for no less than the following 63 levels: $n\,^1S, n = 3\text{–}10$; $n\,^3S, n = 3\text{–}10$; $n\,^1P, n = 2\text{–}10$; $n\,^3P, n = 2\text{–}10$; $n\,^1D, n = 3\text{–}13$; $n\,^3D, n = 3\text{–}13$; nF (multiplicity unresolved), $n = 4\text{–}10$. Observations were made over a spectrum range from 3000 to 21 000 Å. Corrections were made for overlap of lines arising from highly excited states. From the directly-measured apparent cross-sections Q'_{jk} for excitation of the $k\text{–}j$ transition, the Q'_j were obtained using published[§§]

[†] CZAJKOWSKI, M., SKARDIS, G., and KRAUSE, L., *Can. J. Phys.* **51** (1973) 334.
[‡] GOUGH, W., *Proc. phys. Soc.* **90** (1967) 287.
[§] MOROZOV, E. N. and SOSINSKII, M. L., *Optics Spectrosc.* **26** (1968) 282.
[∥] KRAULINYA, E. K. and ARMAN, M. G., ibid. **26** (1969) 285.
[††] KRAULINYA, E. K. and YANSON, M. L., ibid. **29** (170) 239.
[‡‡] JOBE, J. D. and ST. JOHN, R. M., *Phys. Rev.* **A5** (1972) 295.
[§§] WIESE, W. L., SMITH, M. W., and GLENNON, B. M., *Natl. Std. Ref. Data Sci. Natl. Bur. Std. (USA)* NSRDS–NBS 4 (1966). NILES, F. E., U.S. Army Ballistic Research

values of the radiative probabilities for S, P, and D levels and hydrogenic values for F levels.

Detailed analysis of the data on lines similar to those sketched out in Chapter 4, § 1.1 and Chapter 18, § 4.2.3.2 made it possible to determine the contribution from excitation transfer to the apparent cross-sections Q'_j for different levels, at the pressure concerned. An essential part of the analysis was based on observations of the cross-sections for excitation of the ^1P levels over a wide pressure range so that full allowance could be made for resonance trapping.

The results strongly support the suggestion discussed in Chapter 18, § 4.2.3.2 that the transfer processes $n\,^1P \to nF$ and $nF \to {}^{3,1}D$ play an important role. For $n = 4$ transfer cross-sections of $2\cdot3\pm0\cdot6\times10^{-14}$ and $4\cdot7\pm1\cdot5\times10^{-14}$ cm² are obtained for these respective processes. The direct transfer $3\,^1P \to 3\,^1D$ is found to have a much smaller cross-section, $1\cdot2\times10^{-15}$ cm².

TABLE 18 A.8

Contribution of excitation transfer (in 10^{-20} cm²) to apparent excitation cross-sections for various He levels at $6\cdot3\times10^{-2}$ torr pressure, excited by 100 eV electrons

n	$n\,^1P$	nD	nF
4	−55	28	35
5	−47	25	20
6	−33	14	6·2
7	−21	7·8	3·2
8	−14	3·7	1·2
9	−9	2·0	0·9
10	−7	0·8	0·5

Transfer processes contribute little to excitation of ^1S, ^3S, and ^3P levels. Results for other levels are given in Table 18 A.8. It will be seen that the net effect is to reduce the apparent cross-sections for excitation of $n\,^1P$ and to enhance those for nD and nF levels.

Wellenstein and Robertson† have extended and developed the technique of Tater and Robertson (see Chap. 18, p. 1759) who observed the change in population of certain helium states with $n = 3$ in a discharge when irradiated by selected radiation from a second discharge. If the state whose population is directly perturbed is designated as a and if b

Proving Ground, Aberdeen, Md. Report No. 1354 (1967). GABRIEL, A. H. and HEDDLE, D. W. O., *Proc. R. Soc.* A258 (1960) 124. See Chapter 4, Table 4.2, p. 203.
† WELLENSTEIN, H. F. and ROBERTSON, W. W., *J. chem. Phys.* 56 (1972) 1072.

is a second state whose population is affected by the perturbation of a, then the changes $\Delta[\mathrm{He}_{a,b}]$ of the respective populations are related by

$$\Delta[\mathrm{He}_a]\{A_{ab}+k^n_{ab}[\mathrm{He}]+k^e_{ab}\,n_e\}$$
$$= \Delta[\mathrm{He}_b]\Big\{\sum_j A_{bj}+k^t_b[\mathrm{He}]+k^{e,t}_b\,n_e\Big\}, \qquad (7)$$

n_e being the electron concentration, k^n_{ab}, k^e_{ab} the rate constants for transfer from a to b on impact with a normal helium atom and with an electron respectively, k^t_b the rate constant for total loss from the state b in a collision with a normal atom (this may involve associative ionization as well as excitation transfer or deactivation) and $k^{e,t}_b$ that for loss by collision with an electron. The A_{bj} are the radiative transition probabilities.

TABLE 18 A.9

Cross-sections (in 10^{-16} cm^2) for excitation transfer between the states of helium with $n = 3$ by collision with normal atoms, observed by Wellenstein and Robertson

	$S \to P$	$P \to S$	$S \to D$	$D \to S$	$P \to D$	$D \to P$
Singlet	—	4·5±0·03	—	< 0·04	32±1	10·6±0·7
Triplet	—	2·9±0·03	—	< 0·01	0·067±0·005	0·62±0·05

The ratio $\Delta[\mathrm{He}_a]/\Delta[\mathrm{He}_b]$ may be measured spectroscopically as a function of electron density n_e and helium concentration. Extrapolation to zero n_e gives data from which k^n_{ab} and k^t_b can be determined, using known values of the transition probabilities. Once these rates have been obtained, k^e_{ab} and $k^{e,t}_b$ can be derived.

In practice the pumping light was modulated at 91·5 Hz and phase sensitive detection used to determine the changes in intensity of radiation produced in the second discharge. By means of filter combinations light in the 3889, 5016, 5876, and 6678 Å lines was isolated for use as pumping radiation to modulate the population of the $3\,^3P$, $3\,^1P$, $3\,^3D$, and $3\,^1D$ states respectively. Results obtained are given for the derived transfer cross-sections by impact with normal atoms, in Table 18 A.9. Results for electrons are similar for singlets but much larger for triplets.

For the $P \to D$ transfer the result is not inconsistent with that estimated by Jobe and St. John as described above.

Total cross-sections for loss from a given state on impact with neutral atoms, derived from values of k^t_{ab} obtained from the observed data by curve fitting, are given in Table 18 A.10. Earlier results obtained by

Bennett, Kindlmann, and Mercer† are also given and, except for $3\,^1P$, the agreement is quite good.

Wellenstein and Robertson‡ were able to determine the contribution made to the total cross-sections from associative ionization,

$$\mathrm{He} + \mathrm{He} \to \mathrm{He}_2^+ + \mathrm{e}, \tag{8}$$

by observing the modulation of the concentration [He_2^+] of the molecular ions in the second discharge as well as that of the various 3-quantum

TABLE 18 A.10

Total cross-sections for loss from a particular excited state of helium due to impact with a normal helium atom, and partial cross-sections for loss by associative ionization, in 10^{-16} cm^2

	$3\,^3S$	$3\,^1S$	$3\,^3P$	$3\,^1P$	$3\,^3D$	$3\,^1D$
Total						
Wellenstein and Robertson§	< 8	< 3	5±2	40±10	< 10	42±10
Bennett et al.†	2·1±0·5	1·4±0·5	5·5±0·5	19±5	5±1	44±9
Associative ionization	< 0·01	< 0·1	1·6±0·1	3·1±1·0	4·5±0·5	20±4

excited states. Under the experimental conditions, production of He_2^+ arose from associative ionization, and destruction by diffusion to the walls. Knowing the ambipolar diffusion coefficient (see Chap. 19, p. 2002) the associative ionization cross-sections could be derived. Results are given in Table 18 A.10. If the appropriate transfer cross-sections given in Table 18 A.9 are added to these the resulting values are consistent with the total loss cross-sections of Table 18 A.10.

Abrams and Wolga‖ have measured transfer cross-sections between $n = 4$ states using light from a helium laser to perturb the excited state populations in a helium discharge. Otherwise the principle involved is similar to that of the experiments of Wellenstein and Robertson. They find cross-sections for $4\,^3P \to 4\,^3D$ and $4\,^3P \to 4\,^3S$ transfer of 0.99 ± 0.24 and $0.35\pm0.03 \times 10^{-15}$ cm^2 respectively. In contrast, for transfer from $4\,^3P$ to $4\,^1S$, $4\,^1P$, or $4\,^1D$ the cross-section is less than 2×10^{-17} cm^2. It was also found that the cross-sections for transfer from $4\,^3F$ to $4\,^1D$ was as much as 0·6 of that for transfer to $4\,^3D$, in agreement with the

---

† BENNETT, W. R., KINDLMANN, P. J., and MERCER, G. N., *Applied Optics*, Supplement 2, 1965, 34.
‡ WELLENSTEIN, H. F. and ROBERTSON, W. W., *J. chem. Phys.* **56** (1972) 1077.
§ WELLENSTEIN, H. F. and ROBERTSON, W. W., ibid. **56** (1972) 1411.
‖ ABRAMS, R. L. and WOLGA, G. J., *Phys. Rev. Lett.* **19** (1967) 1411.

suggestion of St. John and Fowler that LS coupling is not valid for the F states (see Chap. 18, p. 1759).

5. Collisions involving metastable atoms

5.1. *Ionization by metastable atoms* (Chap. 18, § 5.3)

Metastable atoms A* colliding with atoms B may produce ionization through the processes

$$A^* + B \rightarrow A + B^+ + e, \tag{9}$$
$$\rightarrow AB^+ + e, \tag{10}$$

usually referred to as Penning and associative ionization respectively. If B is replaced by a diatomic molecule CD, additional possibilities arise including dissociative ionization in which the heavy particle products are A, C, and D$^+$ or A, C$^+$, and D, as well as cases in which the products are AC and D$^+$ or AD and C$^+$.

In § 5.3 of Chapter 18 we discussed measurements, made up to the time of writing of Vol. 3, of total cross-sections for ionization of various atoms and molecules by metastable $2\,^3S$ and $2\,^1S$ helium atoms, and of the energy distribution of the ejected electrons which result. This work has since been extended very considerably. Relative cross-sections have been obtained for production of different ionized products but the relative effectiveness of producing ionization by $2\,^3S$ and $2\,^1S$ atoms still remains uncertain. Other targets such as H and alkali-metal atoms have also been studied, while much higher energy resolution has elevated the study of electron energy distributions to become a new branch of precision electron spectroscopy. Data have also been obtained using metastable Ne and Ar. These experimental advances have been accompanied by increased understanding of the phenomena involved, so information can be obtained about the interaction between the initial and final systems. Attempts at a detailed theory are summarized on p. 3600.

The flowing afterglow technique (see Chap. 18, p. 1799) may be readily adapted for the measurement of cross-sections for destruction of He($2\,^3S$) metastable atoms. For He($2\,^1S$) the problem is much more difficult because of the rapid rate of deactivation of He($2\,^1S$) to He($2\,^3S$) by electron impact (Chap. 4, p. 267). Schmeltekopf and Fehsenfeld[†] overcame this by operating the electron gun, which produces the metastable helium atoms, at an energy below the ionization threshold. Alternatively SF$_6$ was added as an electron scavenger (see p. 3374). There still remained the problem of choosing a suitable detector. For He($2\,^3S$) optical

† SCHMELTEKOPF, A. L. and FEHSENFELD, F. C., *J. chem. Phys.* **53** (1970) 3173.

absorption may be used but with He($2\,^1S$) reduction of the concentration by reaction through Penning ionization by 10 per cent generates sufficient electrons to reduce it still further through superelastic collisions by as much as 5 per cent. Instead a technique involving selective excitation of neon, which does not undergo Penning ionization, by transfer of excitation from metastable He atoms, was developed. Thus He($2\,^1S$) selectively excites the 5689·8 Å line of Ne so that the concentration of He($2\,^1S$) can be monitored by introducing neon into the afterglow just beyond the reaction region and observing the intensity of the 5689·8 Å line. A similar procedure can be used for He($2\,^3S$) with the 7032·4 Å line. It was verified with this detector that the variation of the He($2\,^1S$) intensity with bombarding electron energy with and without SF_6 present was as expected. Measurements of cross-sections for destruction of He($2\,^3S$) and He($2\,^1S$) by impact with Ne, Ar, Kr, Xe, H_2, D_2, N_2, CO, NO, O_2, CO_2, NO_2, NH_3, SF_6, CH_4, C_2H_6, C_3H_8, and C_4H_{10} were made. Results for He($2\,^3S$) with Ar, Kr, H_2, N_2, and CO are given in Table 18 A.11. The ratios of the cross-sections for destruction of He($2\,^1S$) and of He($2\,^3S$) by Ar, Kr, and Xe are given in Table 18 A.12.

Bolden, Hemsworth, Shaw, and Twiddy† have also adapted the flowing afterglow technique to measure cross-sections for destruction of He($2\,^3S$) metastable atoms. They determined the concentration of He($2\,^3S$) down the flow tube by a modulated light absorption technique (see Chap. 4, p. 264) and observed the decrease as a function of distance down the tube from a point at which the reactant gas was injected. Results obtained, which agree well with those of Schmeltekopf and Fehsenfeld, are given in Table 18 A.11. With the exception of H_2 both sets of data also agree with those found by Benton et al. (see p. 1792) in earlier work using time-resolved static afterglows.

The flowing afterglow technique was also applied‡ to measure the cross-section for Penning ionization of atomic hydrogen. The chief problem was the determination of the fractional concentration of atoms in the partly dissociated H–H_2 mixture introduced as reactant. Titration methods were not applicable because of the fast flow rates required. Instead, Ar^+ ions, which do not react with H but do with H_2 according to the reaction

$$Ar^+ + H_2 \rightarrow ArH^+ + H, \tag{11}$$

were used to determine the H_2 fraction by comparing the decay rate of

† BOLDEN, R. C., HEMSWORTH, R. S., SHAW, M. J., and TWIDDY, N. D., J. Phys. B (GB) **3** (1970) 61.
‡ SHAW, M. J., BOLDEN, R. C., HEMSWORTH, R. S., and TWIDDY, N. D., Chem. Phys. Lett. **8** (1971) 148.

injected Ar⁺ ions with distance below the injection point with and without the dissociation of the H_2. However, the rate of (11) depends on the vibrational excitation of the H_2 and this will change when the gas is partly dissociated. This effect was disentangled by taking account of the fact that the dissociation fraction was more sensitive to the flow rate of the buffer He gas accompanying the incoming H_2 than the change in rate constant of (11) through vibrational excitation. The observed cross-section for Penning ionization of H was $22\pm6\times10^{-16}$ cm². Comparison with theory is discussed on p. 3601.

A number of investigations have been carried out in which a metastable atom beam is passed through the target gas and the ions produced collected and mass-analysed.

Sinda and Pesnelle[†] used metastable helium and neon atom beams in which the distribution between the different metastable states was not known. Absolute cross-sections were obtained assuming an electron yield per metastable atom from the detecting surface of 0·27 for He and 0·23 for Ne.

Hotop, Niehaus, and Schmeltekopf[‡] were able to separate the contributions from $He(2\,^3S)$ and $He(2\,^1S)$ by carrying out measurements with and without irradiation of the metastable atom beam by light from a He discharge of such intensity as to eliminate the $2\,^1S$ atoms through the reaction

$$h\nu(2\,^1P-2\,^1S)+He(2\,^1S) \to He(2\,^1P) \to He(1\,^1S)+h\nu(2\,^1P-1\,^1S). \quad (12)$$

This technique has been applied by Dunning and Smith (A. C. H.)[§] to determine the ratio of the cross-sections for ionization of a number of gases by $He(2\,^3S)$ and $He(2\,^1S)$ atoms. In this work advantage was taken of measurements made[‖] on the secondary electron yield from the detecting surface for both types of metastable atoms.

Tang, Marcus, and Muschlitz[††] investigated the dependence of the ionization cross-sections on the velocity of the metastable atoms by using a velocity-selected metastable neon beam. Assuming equal detection efficiency for the $^3P_2$ and $^3P_0$ atoms, the ratio of their concentrations in the beam was found by using an inhomogeneous magnetic deflecting field. Absolute values were obtained on the assumption of an electron yield of 0·27 per metastable atom.

† SINDA, T. and PESNELLE, A., *7th int. conf. on the physics of electronic and atomic collisions*, Amsterdam. Abstracts, North Holland, 1971, p. 1107.
‡ HOTOP, H., NIEHAUS, A., and SCHMELTEKOPF, A. L., *Z. Phys.* **229** (1969) 1.
§ DUNNING, F. B. and SMITH, A. C. H., *J. Phys. B (GB)* **3** (1970) L60.
‖ DUNNING, F. B. and SMITH, A. C. H., *Phys. Lett.* **32a** (1970) 287.
†† TANG, S. Y., MARCUS, A. B., and MUSCHLITZ, E. E., *J. chem. Phys.* **56** (1972) 566.

In Table 18 A.11 comparison is made between total cross-sections for ionization by 2^3S atoms measured by different investigators.

Values for (a) and (e) are obtained on the assumption that the cross-sections for ionization by singlet and triplet metastable atoms are the

TABLE 18 A.11

Observed total cross-sections (in 10^{-16} cm^2) for ionization of different target gases by impact of $\text{He}(2^3S)$ *atoms*

Target atom or molecule	(a)	(b)	(c)	(d)	(e)
Ar	7·6±0·5	6·6±1·3	5·34	7±1·4	7·2±0·5
Kr	9±1·8	10±2	7·73	8±1·6	2·3±0·3
H$_2$	1·6±0·3	6±2·2	1·47	1·5±0·3	1±0·2
N$_2$	7±1	6·4±3·2	5·21	5±1	2·2±0·5
CO	7±1·4		7·35	8±1·6	5·8±0·9

(a) SCHOLETTE, W. P. and MUSCHLITZ, E. E., *J. chem. Phys.* **36** (1962) 3368 (see Chap. 18, p. 1816).
(b) BENTON, E. E., FERGUSON, E. E., MATSEN, F. A., and ROBERTSON, W. W., *Phys. Rev.* **128** (1962) 206 (see Chap. 18, p. 1792).
(c) SCHMELTEKOPF, A. L. and FEHSENFELD, F. C., loc. cit., p. 3576.
(d) BOLDEN, R. C., HEMSWORTH, R. S., SHAW, M. V., and TWIDDY, N. D., loc. cit., p. 3577.
(e) SINDA, T. and PESNELLE, A., loc. cit., p. 3578.

same and that the yield of electrons per metastable atom on impact with the detecting surface is 0·28. It will be seen that there is remarkably good agreement between most of the results and especially between the two sets of measurements with the flowing afterglow technique.

The situation is by no means so satisfactory for destruction of $\text{He}(2^1S)$. Table 18 A.12 summarizes results obtained for the ratio of the cross-sections for destruction of $\text{He}(2^1S)$ and $\text{He}(2^3S)$ as well as for the relative probabilities of Penning and of associative ionization.

It will be seen that for the ratio of the total destruction cross-sections of $\text{He}(2^1S)$ and $\text{He}(2^3S)$ there is a marked discrepancy between the results of the flowing afterglow experiments of Schmeltekopf and Fehsenfeld and those obtained in the other two sets of experiments. The same tendency is clearly apparent also for destruction by O_2, N_2, CO, NO, and H_2 for which Dunning and Smith find ratios 1·31, 1·02, 0·93, 0·93, and 0·67 contrasted with 2·75, 2·40, 3·22, 1·79, and 1·53 obtained in the flowing afterglow measurements. The reason for this discrepancy is still obscure at the time of writing.

5.1.1. *Energy distributions of ejected electrons.* In Chapter 18, § 5.3.3 we described the experiments carried out by Čermak and their application to determine the velocity distributions of the electrons ejected from rare-gas atoms and from N_2, CO, and NO molecules in Penning ionization by He($2\,^3S$) and He($2\,^1S$) atoms. These experiments have been repeated and extended by Niehaus and his collaborators but with increased energy resolution so that the full width at half maximum of an energy 'line' is

TABLE 18 A.12

Relative cross-sections for Penning and associative ionization of rare-gas atoms by impact with He($2\,^1S$) and He($2\,^3S$) metastable atoms with mean kinetic energy corresponding to a temperature of 320 °K. $Q_s(p)$, $Q_t(p)$ are the cross-sections for Penning ionization; $Q_s(a)$, $Q_t(a)$ the corresponding cross-sections for associative ionization

	$\dfrac{Q_s(a)}{Q_s(p)}$	$\dfrac{Q_t(a)}{Q_t(p)}$	$\dfrac{Q_t(p)}{Q_s(p)}$		$\dfrac{Q_s(a)+Q_s(p)}{Q_t(a)+Q_t(p)}$	
	(a)	(a)	(a)	(a)	(b)	(c)
Ar	0·212	0·155	0·91	1·15	1·10±0·25	3·13
Kr	0·130	0·170	0·65	1·48	1·26±0·30	3·64
Xe	0·022	0·110	0·58	1·6	1·27±0·30	3·72

(a) HOTOP, H., NIEHAUS, A., and SCHMELTEKOPF, A. L., loc. cit., p. 3578.
(b) DUNNING, F. B. and SMITH, A. C. H., loc. cit., p. 3578.
(c) SCHMELTEKOPF, A. L. and FEHSENFELD, F. C., loc. cit., p. 3576.

not greater than 0·03 eV. The energy scale was calibrated by observing the energy distribution of the photoelectrons arising from the ionization of the target atoms by He resonance radiation (584 Å). The width of this distribution was less than 20 meV.

It is to be expected that the processes leading to ionization of an atom A essentially involve an electronic transition from a state of the diatomic system He*A to one of HeA⁺. This will take place so rapidly that no appreciable change of nuclear separation will occur during it. In other words the Franck–Condon principle will apply. Evidence in favour of this was obtained by Čermak as described in Chapter 18, § 5.3.3.

Let $V_i(R)$, $V_f(R)$ be the initial and final interaction energies between He*–A and He–A⁺ respectively when the nuclear separation is R, measured from the same energy zero. Transitions between V_i and V_f can occur for all $R \geqslant R_t$ where R_t is the closest distance of approach of He* and A. The energy of the ejected electron due to a transition when the nuclear separation is R will be

$$E_e(R) = V_i(R) - V_f(R). \tag{13}$$

The kinetic energy of nuclear motion will be unchanged from its value $E_k - V_i(R)$ at the instant of transition, E_k being the initial relative impact energy. Hence the final state will be bound or unbound, leading to associative or Penning ionization respectively, according as

$$E_k - V_i(R) \lessgtr -V_f(R) + V_f(\infty) - V_i(\infty), \qquad (14\,\text{a})$$

or
$$E_e \gtrless V_i(\infty) - V_f(\infty) + E_k. \qquad (14\,\text{b})$$

In the absence of associative ionization the highest electron energy will be $V_i(\infty) - V_f(\infty) + E_k$. Although in practice the interpretation of the observations is complicated by the fact that the initial kinetic energy E_k will be distributed thermally about a mean value $\bar{E}_k$, if the measured electron energy distributions extend further beyond $V_i(\infty) - V_f(\infty) + \bar{E}_k$ than would be expected for a thermal distribution of E_k there is strong evidence that associative ionization is occurring.

Results obtained by Hotop and Niehaus† for ionization of Ar, Kr, and Xe show that the energy distributions of electrons produced in ionization by $2\,^3S$ atoms are broader than those for $2\,^1S$. The maxima occur at energies exceeding $V_i(\infty) - V_f(\infty)$ by 46, 25, and 5 meV for Ar, Kr, and Xe respectively as compared with the mean thermal impact energy $\bar{E}_k$ of 60 meV. This shows that, at least for Ar and Kr, the most probable transitions occur at separations R for which $V_i(R) \geqslant 0$. The distributions extend beyond $V_i(\infty) - V_f(\infty) + \bar{E}_k$ by some meV, compatible with the occurrence of associative ionization as given in Table 18 A.12.

Measurements made at two different helium beam temperatures (100 °K and 500 °K) showed no marked variation in the energy distribution for He($2\,^1S$) but a considerable broadening at the higher temperature for He($2\,^3S$). This difference was confirmed in later experiments‡ in which the helium metastable beam was velocity-selected.

In later experiments Hotop and Niehaus§ extended their measurements to the energy distribution arising from ionization of alkali-metal atoms as well as of atomic hydrogen.|| In the latter case the atomic H beam was produced by dissociating hydrogen in an r.f. discharge. For these cases information can be derived about the minimum value of V_i as a function of R, i.e. about the attractive well-depth for the interaction between the metastable and alkali-metal atoms.

† Hotop, H. and Niehaus, A., *Z. Phys.* **228** (1969) 68.
‡ Hotop, H., Illenberger, E., Morgner, H., and Niehaus, A., *7th int. conf. on the physics of electronic and atomic collisions*, Amsterdam. Abstracts, North Holland, 1971, p. 1101.
§ Hotop, H. and Niehaus, A., *Z. Phys.* **238** (1970) 452.
|| Hotop, H., Illenberger, E., Morgner, H., and Niehaus, A., *Chem. Phys. Lett.* **10** (1971) 493.

The minimum energy of an ejected electron is given by

$$E_{\min} = V_i(R_m) - V_f(R_m), \qquad (15)$$

where R_m is the distance at which $V_i(R) - V_f(R)$ is a minimum. In all the cases concerned the minimum in V_i occurs at a separation R_i which is much greater than that R_f for V_f so we may take $V_f(R_m) \simeq V_f(\infty)$, giving

$$E_{\min} = V_i(R_i) - V_f(\infty). \qquad (16)$$

From this the well-depth of V_i may be obtained.

Again the maximum electron energy will be given closely by

$$E_{\max} = V_i(R_{cl}) - V_f(R_{cl}), \qquad (17)$$

where R_{cl} is the closest distance of approach of the metastable and target atoms, so that $V_i(R_{cl}) = E_k + V_i(\infty)$. Hence

$$E_{\max} = E_k + V_i(\infty) - V_f(R_{cl}). \qquad (18)$$

This determines $V_f(R_{cl})$ so that, if V_f is known, as is approximately so for He*–H collisions, R_{cl} may be obtained.

TABLE 18 A.13

Value of ϵ_i and R_{cl} for interactions of He($2\,^3S$) *and* He($2\,^1S$) *atoms with* H *atoms*

	He(2^3S)	He(2^1S)
ϵ_i (eV)	$2\cdot44\pm0\cdot2$	$0\cdot46\pm0\cdot05$
R (a_0)	$2\cdot15\pm0\cdot1$	$3\cdot4\pm0\cdot2$

In this way, using interactions V_f calculated by Miller and Schaefer,[†] Hotop *et al.*[‡] obtained the values for $V_i(R_i)$ ($= \epsilon_i$) and R_{cl} for the He($2\,^3S$) and He($2\,^1S$) interactions with H given in Table 18 A.13.

The ratio of the ionization cross-sections $\{Q_s(a) + Q_s(p)\}/\{Q_t(a) + Q_t(p)\}$ was also determined as $1\cdot5\pm50$ per cent. Theoretical analysis of these processes including comparison with the observed data is discussed on p. 3601.

Making due allowance for the thermal spread in E_k, values for $V_i(R_i)$ were obtained as 735 ± 50, 590 ± 50, and 80 ± 50 meV for He($2\,^3S$) and 300 ± 50, 230 ± 50, and 480 ± 50 meV for He($2\,^1S$), interacting with Na, K, and Hg respectively.

It was also found that the ratio of the total cross-section for ionization by He($2\,^1S$) to that by He($2\,^3S$) was $1\cdot20\pm0\cdot10$ for Na and $1\cdot14\pm0\cdot10$ for K.

[†] MILLER, W. H. and SCHAEFER, H. F., *J. chem. Phys.* **53** (1970) 1421.
[‡] loc. cit., p. 3581, ref. ‖.

Finally, Hotop and Niehaus showed that the form of the velocity distribution for He($2\,^3S$)–Na is given quite well if the formula (see p. 1880)

$$Q_t = 2\pi \int P(p)p \, \mathrm{d}p, \qquad (19)$$

where
$$P(p) = \int_{-\infty}^{\infty} W(R) \, \mathrm{d}t, \qquad (20)$$

is used with suitable choice of the functions involved. p is the impact parameter for a given trajectory of relative motion, $W(R)$ is the probability of transition from V_i to V_f at separation R and (see (343) of Chap. 18, p. 1880)

$$\frac{\mathrm{d}R}{\mathrm{d}t} = \{v^2 - p^2 v^2 R^{-2} - 2V_i(R)/M\}^{\frac{1}{2}} \qquad (21)$$

where v is the initial relative velocity and M the reduced mass. For V_i a (12, 6) potential was taken and

$$W(R) = \exp(-\alpha R/R_i). \qquad (22)$$

A good fit was obtained with $\alpha = 6$.

Ionization by He($2\,^3S$) can only take place effectively through electron exchange but for He($2\,^1S$) it may also occur directly. This would account for a difference between the cross-sections Q_t and Q_s should this be confirmed (see Table 18 A.12). Further remarks on the theory are deferred to p. 3601.

5.2. *Dissociative excitation by metastable atoms*

In addition to Penning and associative ionization, metastable atoms may also produce dissociation of a molecule into atoms one or both of which may be excited, i.e.

$$\text{He*}(2\,^3S) + \text{AB} \to \text{A*} + \text{B}. \qquad (23)$$

In a flowing afterglow tube into which diatomic molecular gases are injected, radiation is observed† from excited atoms resulting from dissociation of these molecules. By making absolute intensity measurements of this radiation the loss rate of metastable atoms due to production of the excited atoms in collisions of the type (23) may be determined. For CO the rate of (23), which is exothermic for molecules in the ground vibrational state, was found to be $2 \cdot 1 \times 10^{-11}$ cm^3 s^{-1} as compared with $9 \cdot 8 \times 10^{-11}$ cm^3 s^{-1} for all reactions. The reaction (23) is endothermic for N$_2$ molecules which possess no vibrational excitation and at 300 °K the measured rate was as low as $2 \cdot 9 \times 10^{-13}$ cm^3 s^{-1}. However, by

† HURT, W. B. and GRABLE, W. C., *Bull. Am. phys. Soc.* **16** (1971) 219.

increasing the vibrational temperature to about 4000 °K by microwave heating, the rate increased to $3 \cdot 5 \times 10^{-12}$ cm^3 s^{-1}.

5.3. *Total and differential cross-sections for collisions of metastable atoms with other atoms and molecules*

5.3.1. *Measurement of total cross-sections* (see Chap. 18, § 5.1.1). The variation of the total cross-section for collisions of metastable argon atoms in the $^3P_0$ and $^3P_2$ states with rare-gas atoms has been observed by Nenner.[†] Argon from a nozzle source seeded with hydrogen was excited by a beam of electrons of 40 eV energy. The metastable H atoms were deflected more by the electron impact so they missed the detector. A time-of-flight method (see p. 3448) was used to measure the relative total cross-sections as a function of argon atom velocity. For Ar–Xe, because of the low R.M.S. velocity of the Xe, the resolution in relative velocity was great enough to show up the glory oscillations.

5.4. *Diffusion coefficient and charge transfer for* He(2 ^3S) *metastable atoms in helium*

Considerable progress has been made in determining the long range interactions between He($2\,^3S$) and normal He atoms. Both the two interactions, $^3\Sigma_u^+$ and $^3\Sigma_g^+$, behave in much the same way at large nuclear separations. Asymptotically they both tend to the attractive van der Waals form but at somewhat smaller but still quite large separations they both become repulsive (see Chap. 18, § 8.1.3). At still smaller separations the $^3\Sigma_u^+$ becomes quite strongly attractive with a potential well of depth around 1 eV near $2a_0$. The height of the potential barrier outside this well is around 0·1 eV or so. It seems likely from recent theoretical calculations that the $^3\Sigma_g^+$ curve behaves in qualitatively the same way, with a higher potential barrier and shallower well. Because of the existence of the potential barrier the cross-section for the resonant excitation transfer

$$\mathrm{He}(2\,^3S) + \mathrm{He}(1\,^1S) \to \mathrm{He}(1\,^1S) + \mathrm{He}(2\,^3S) \tag{24}$$

is very small until the relative kinetic energy exceeds the barrier height.

Accurate calculation of the potential energy curves is difficult. Fitzsimmons, Lane, and Walters[‡] have approached the problem of determining the interactions by measuring the variation with temperature of the diffusion coefficient of He($2\,^3S$) in normal helium. The technique

---

[†] NENNER, T., *7th int. conf. on the physics of electronic and atomic collisions, Amsterdam.* Abstracts, North Holland, 1971, p. 564.

[‡] FITZSIMMONS, W. A., LANE, N. F., and WALTERS, G. K., *Phys. Rev.* **174** (1968) 193.

employed is essentially the same as the afterglow and absorption technique used by Phelps and Pack for measurements at room temperature (see Chap. 18, p. 1781 and Chap. 4, § 3.1.1). By immersing the afterglow cavity in a cryogenic liquid inside a Dewar flask, a temperature range from 1 to 300 °K was covered. Checks were made to verify that the discharge pulses did not heat the gas sample appreciably.

The measured value of the diffusion coefficient at 300°, 480 ± 25 cm^2 s^{-1} at 1 torr, agreed well with that observed by Phelps and Pack (470 ± 25 cm^2 s^{-1}). At 1° it had fallen by a factor of 48.

The data were then analysed in terms of assumed long range interactions of the form

$$V(^3\Sigma_u^+) = \alpha r^2 e^{-\beta r} - Cr^{-6}, \tag{25}$$

$$V(^3\Sigma_g^+) = V(^3\Sigma_u^+) + \gamma e^{-\delta r}, \quad r \geqslant r_0 = 6a_0. \tag{26}$$

For $r < r_0$ the interactions have little influence on the diffusion coefficients in the observed temperature range so simple analytical forms were used to extend the interactions to these values of r.

Also, the van der Waals term has little influence except at the very lowest temperatures so, for most of the analysis, C was taken as zero. The parameters γ and δ were chosen to agree with $V(^3\Sigma_g^+) - V(^3\Sigma_u^+)$ calculated† using 5-term variational trial functions over the range of r from 5 to $6a_0$. In any case the diffusion coefficients were found to be insensitive to these parameters. The remaining parameters α and β were varied to give the best fit with the observations, it being found convenient to work in terms of the value of $V(^3\Sigma_u^+)$ at $r = 6a_0$, instead of α.

The best fit was obtained with $\beta = 1{\cdot}6a_0^{-1}$ and a value of $V(^3\Sigma_u^+)$ at $r = 6a_0$ of 0·002 a.u. The corresponding interactions are considerably smaller, for $r \geqslant 6a_0$, than any of those calculated. Below 2 °K there is evidence that the calculated diffusion cross-section is a little too large, possibly due to failure to include any van der Waals attraction. A value between 0 and 20 a.u. for the van der Waals constant C in (25) would improve the agreement at these very low temperatures.

Fitzsimmons et al. used the interactions they derived to calculate total cross-sections for collisions of He($2\,^3S$) atoms in helium and obtained very good agreement with the observations of Rothe, Neynaber, and Trujillo (see Fig. 18.60, p. 1771).

Finally the same interactions were used to calculate the cross-section for the excitation transfer reaction (24). Although this depends strongly

† MATSEN, F. A. and SCOTT, D. R., *Quantum theory of atoms, molecules and the solid state*, ed. Löwdin, P.-O., Academic Press, N.Y. 1966.

on the difference between the interactions and this was not determined from the analysis of the diffusion data, remarkably good agreement was obtained with the observations of Colegrove, Scherrer, and Walters (Chap. 18, § 5.4.2 and Fig. 18.96, p. 1829). However, it has since been pointed out by Dupont-Roc, Leduc, and Laloë† that, when proper account is taken of the hyperfine coupling, the cross-sections observed by Colegrove et al. need to be multiplied by a factor $\frac{9}{4}$. Dupont-Roc et al. carried out further measurements of line widths at 300 °K both of the $F = \frac{3}{2}$ and $\frac{1}{2}$ hyperfine levels of $He(2\,^3S)$. For the former the results agreed well with the observations of Colegrove et al. The width of the $F = \frac{1}{2}$ level was found to be $0·574 \pm 0·012$ times that of the $\frac{3}{2}$ level in agreement with the value $\frac{4}{7}$ predicted by Dupont-Roc et al. and confirming the need to include the factor $\frac{9}{4}$ in deriving cross-sections from measured widths of $F = \frac{3}{2}$ levels. Rosner and Pipkin‡ have extended the measurements to cover the temperature range from 15 to 115 °K. They overlap well with the earlier data of Colegrove et al. above 60° and are more accurate.

In view of these later results it seems that the difference potential $\gamma e^{-\delta r}$ used by Fitzsimmons et al. is not as accurate as seemed earlier. This probably has little effect on the analysis of the data on diffusion coefficients but offers the opportunity for improved determination of γ and δ.

5.5. *Reactions involving metastable O atoms* (Chap. 18, § 5.5)

5.5.1. *Deactivation of $O(^1S)$.* Since Vol. 3 was written further experiments have been carried out to measure cross-sections for deactivation of $O(^1S)$ metastable atoms in collisions with other atoms and molecules. Because of advances in detecting technique it has become possible to study the quenching of the green line at 5577 Å, resulting from the $O(^1S-^1D)$ transition, by the same general procedure as for the quenching of, for example, resonance radiation.

Thus in experiments carried out by Filseth, Stuhl, and Welge§ $O(^1S)$ atoms were produced by photodissociation of CO_2 by radiation of wavelength less than 1250 Å. The decay time of the $O\,^1S$ atoms was measured as a function of the pressure of admixed foreign gas by the delayed coincidence technique (see p. 3563), recording the emission of 5577 Å quanta. From these observations the quenching cross-section of the foreign gas could be determined.

† Dupont-Roc, J., Leduc, M., and Laloë, F., *Phys. Rev. Lett.* **27** (1971) 467.
‡ Rosner, S. D. and Pipkin, F. M., *Phys. Rev.* **A5** (1972) 1909.
§ Filseth, S. V., Stuhl, F., and Welge, K. H., *J. chem. Phys.* **52** (1970) 239.

In the particular experiments the light source was a pulsed discharge in nitrogen at atmospheric pressure, the flash duration being 3 μs and the repetition frequency about 50 Hz. A short wavelength cut-off was provided in the form of an LiF window. Fluorescence was detected by a cooled EMI 9558 QB photomultiplier placed behind a narrow band interference filter. To avoid accumulation of dissociation products the gas mixture under study flowed continuously through the reaction chamber.

Similar experiments have been carried out by Black, Slanger, St. John, and Young† who used photodissociation of N_2O by a pulsed xenon lamp emitting 1470 Å radiation with a flash duration of about 100 μs and a repetition rate of 20 Hz.

Results obtained for the quenching rates coefficients are given in Table 18 A.14 together with values found in the earlier experiments of Young and Black‡ described in Chapter 18, p. 1832 and in experiments by Stuhl and Welge.§ It will be seen that the agreement between the different results is rarely better than within a factor of 2 and in many cases rather worse. However, for the important cases of O_2 and N_2 the situation is better than in other cases. The great range of values of the rate coefficients for different quenching molecules is remarkable and is not understood.

The increased value of the quenching coefficient for CO_2 means a corresponding increase in that for O, measured relative to that for CO_2 in the experiment of Young and Black discussed in Chapter 18, p. 1832. The revised value for O (see Chap. 18, Table 18.21) now becomes 2×10^{-12} in place of 1.8×10^{-13} cm^3 s^{-1}. This in turn means a corresponding increase in the value of the rate coefficient for the Chapman reaction

$$O+O+O \rightarrow O_2+O(^1S) \tag{27}$$

measured also by Young and Black (see p. 1838), so that it becomes 1.5×10^{-33} cm^3 s^{-1}.

Dandekar and Turtle∥ have analysed data which they recorded in rocket flights of the brightness of the 5577 Å emission in the night airglow. They assumed that production takes place through (27) and loss by quenching in collisions with O_2, N_2, and O for which they assumed rates of 2.1×10^{-13}, 1×10^{-17}, and 1.8×10^{-13} cm s^{-1}. The N_2 and O_2

---

† Black, G., Slanger, T. G., St. John, G. A., and Young, R. A., *Can. J. Chem.* **47** (1969) 1872.
‡ Young, R. A. and Black, G., *J. chem. Phys.* **44** (1966) 3741.
§ Stuhl, F. and Welge, K. H., *Can. J. Chem.* **47** (1969), 1870.
∥ Dandekar, B. S. and Turtle, J. P., *Planet. Space Sci.* **19** (1971) 949.

TABLE 18 A.14

Observed rate coefficient in $cm^3 \, s^{-1}$ for quenching of $O(^1S)$ atoms at 300 °K by different foreign gases

	O_2	N_2	H_2	CO	NO	CO_2	H_2O	N_2O	NO_2
(a)	3.6×10^{-13}	$< 2 \times 10^{-16}$	2.8×10^{-16}	4.9×10^{-15}	8.0×10^{-11}	3.6×10^{-13}	7×10^{-11}	1.1×10^{-11}	5.0×10^{-10}
(b)	5×10^{-13}	$< 3 \times 10^{-15}$	1×10^{-15}	—	5.5×10^{-10}	3×10^{-13}	—	1.6×10^{-11}	—
(c)	1.0×10^{-13}	—	3.0×10^{-15}	—	—	2.5×10^{-14}	—	5.9×10^{-15}	—
(d)	3.2×10^{-13}	$< 2 \times 10^{-16}$	1.5×10^{-15}	9.4×10^{-14}	—	4.6×10^{-13}	—	—	—

(a) Filseth, S. V., Stuhl, F., and Welge, K. H., *J. chem. Phys.* **52** (1970) 239.
(b) Black, G., Slanger, T. G., St. John, G. A., and Young, R. A., *Can. J. Chem.* **47** (1969) 1872.
(c) Young, R. A. and Black, G., *J. chem. Phys.* **44** (1966) 3741.
(d) Stuhl, F. and Welge, K. H., *Can. J. Chem.* **47** (1969) 1870.

concentration as functions of height were taken from the 1965 CIRA model atmosphere and the profile of O concentration required to give the observed variation of 5577 Å emission derived. This was compared with profiles given from the CIRA model and derived from the measured incident solar flux in the Schumann–Runge region which dissociates O_2. Reasonably good agreement was found over the altitude range from 92 to 110 km.

5.5.2 *Quenching of* $O(^1D)$. In a remarkable experiment Noxon† was able to measure cross-sections for quenching of $O(^1D)$ by O_2, N_2, and CO_2 by an optical technique exactly similar in principle to experiments on the quenching of resonance radiation and of $O(^1S)$. $O(^1D)$ atoms were produced by photodissociation of O_2 by 1470 Å radiation and their concentration monitored by direct observation of the 6300 Å red line due to the transition $^1D \to {}^3P$. As the transition probability for this line is only 10^{-2} s^{-1} the photon flux is very small. With a dissociating flux of 10^{14} photons cm^{-2} s^{-1} in a beam about 1 cm diameter, the surface brightness of red line emission is only about 10^4 photons cm^{-2} sterad^{-1} s^{-1}, about 500 times weaker than in the night airglow. With good geometry this produces a current in the photomultiplier of only a few photoelectrons s^{-1}. In addition it is necessary to screen off all but one or two of the 10^{15} photons s^{-1} which are produced by the u.v. lamp source of the dissociating radiation. Nevertheless Noxon was able to observe the 6300 Å line from O_2 buffered with helium so as to render negligible diffusion loss of $O(^1D)$ atoms to the walls of the reaction vessel.

With helium-buffered O_2, containing various quenching gases, the intensity of the 6300 Å line is given by

$$I(6300) = AF[O_2]/(k_1[O_2] + \sum k_i[M_i]), \qquad (28)$$

where [X] denotes the concentration of the species X. $F[O_2]$ is the rate of production of $O\,^1D$, A is the radiative transition probability of the red line, while k_1 and k_i are the rate coefficients for deactivation of $O\,^1D$ by O_2 and M_i respectively. F may be determined from the measured flux of dissociating radiation and the known absorption cross-section. Hence, if no other species M_i is present, k_1 may be determined. The change in $I(6300)$ produced by addition of a further quenching gas, say N_2, enables $k(N_2)$ to be obtained and so on.

In addition to observing the 6300 Å line, Noxon also observed the 7600 Å band emitted in the transition $O_2(b\,^1\Sigma_g - X\,^3\Sigma_g)$.

The excited $O_2(b\,^1\Sigma_g)$ molecules would be expected to be formed in

† Noxon, J. F., *J. chem. Phys.* **52** (1970) 1852.

the process of quenching of $O(^1D)$ by O_2

$$O(^1D)+O_2(^3\Sigma) \to O(^3P)+O_2(^1\Sigma). \qquad (29)$$

If ϵ is the probability that quenching of $O(^1D)$ leads to $O_2(^1\Sigma)$ in the $v'=0$ vibrational level we have

$$\text{rate of production of } O_2(^1\Sigma) = \epsilon k_1[O(^1D)][O_2], \qquad (30\text{a})$$

$$\text{rate of loss of } O_2(^1\Sigma) = [O_2(^1\Sigma)]\{A'+k_2[O_2]+\sum k_i'[M_i]+L\}, \qquad (30\text{b})$$

where A' is the radiative transition probability, k_2 and k_i' are quenching rates for $O_2(^1\Sigma)$, and L is the loss rate to the walls. $[O(^1D)]$ is given by $I(6300)/A$ so from (30) we have, in equilibrium,

$$[O_2(^1\Sigma)] = F[O_2]^2\epsilon\tau k_1/\{k_1[O_2]+\sum k_i M_i\}, \qquad (31)$$

where $\qquad \tau^{-1} = A'+k_2[O_2]+\sum k_i'M_i+L. \qquad (32)$

Noxon verified that the intensity of the 7600 Å band varies with $[O_2]$ in pure O_2 and in helium-buffered O_2 according to (31), if it is assumed that τ is independent of $[O_2]$ at O_2 pressures below 1 torr. Observational evidence in favour of this was also obtained, though not directly. It seems likely that the quenching process is indeed (29) in which case observations of the intensity of the 7600 Å line may, under certain circumstances be used to determine the ratio of the quenching rate coefficients for $O(^1D)$ of different added gases. Thus, if addition of N_2 reduces $[O_2(^1\Sigma)]$ without apparently affecting τ then the effect can only arise from quenching of $O(^1D)$ and the ratio $k_1(O_2)/k(N_2)$ can be obtained from (31). In this case the reciprocal of the intensity of emission of the 7600 Å band should vary as $1+\{k(N_2)/k_1\}\{[N_2]/[O_2]\}$. If this is verified $k(N_2)$ may be obtained.

In earlier experiments, Young, Black, and Slanger† used an ingenious technique for measuring the quenching rates which is, however, more complex in principle than that used by Noxon. The same procedure of ultraviolet photolysis was used to produce the $O(^1D)$ atoms but they determined the $O(^1D)$ quenching rate through the observation of the $O(^3P)$ production under different circumstances.

With a gas such as H_2 or N_2O present which reacts with $O(^1D)$ to produce products which do not include $O(^3P)$ the yield of the latter atoms per quantum absorbed is just unity. If a quenching gas is added in sufficient quantity the yield increases up to twice this amount. Thus if k_R is the rate of a reaction of $O(^1D)$ with a reactant R which does not produce $O(^3P)$ and k_Q that for quenching by a component Q, the yield

† YOUNG, R. A., BLACK, G., and SLANGER, T. G., J. chem. Phys. 49 (1968) 4758.

of O(^3P) per quantum absorbed would be given by $1+f$ where
$$f = k_Q[Q]/\{k_Q[Q]+k_R[R]\}. \tag{33}$$
This assumes that a buffer gas is present so that wall loss is negligible.

The O(^3P) concentration was determined by titration with NO (see Chap. 18, p. 1832). The choice of suitable reactant gases is difficult. Thus H_2 is not affected by the 1470 Å radiation but its presence leads to the possibility of reactions involving H and OH for which the reaction rate constants for reaction with O(^1D) are not known and must be derived in the experiments. N_2O, on the other hand, is dissociated by the u.v. radiation and allowance must be made for this. A valuable check is that the quenching rates for different gases should be the same no matter which reactant is used.

Young et al. did find reasonably close agreement and their results are consistent with those of Noxon for O_2 and N_2. Thus, in 10^{-11} cm^3 s^{-1}, Noxon obtains $k(O_2) = 6$ and $k(N_2) = 9$ as compared with 4 and 5 respectively determined by Young et al. These values are consistent with estimates based on analysis of day and night airglow data.

For CO_2, however, there is a very marked discrepancy. While Young et al. find a large rate, $2\cdot3\times 10^{-10}$ cm^3 s^{-1}, Noxon finds only 3×10^{-12}. Lowenstein[†] has carried out a further measurement based on observation of the quenching of the 7600 Å emission. O_2 was dissociated by a pulse of 1470 Å radiation and the intensity of 7600 Å emission at the end of the pulse determined for different [O_2] and [CO_2]. Assuming equilibrium between production and loss of O(^1D), but no effective loss of $O_2(^1\Sigma)$ during the pulse, the ratio $k_Q(CO_2)/k_Q(O_2)$ may be obtained. A ratio $1\cdot8\pm0\cdot4$ was found which agrees qualitatively with Young et al. The ratio $k_Q(N_2)/k_Q(O_2)$ was also measured as 0·64 which is not inconsistent with the results of both the other experiments. The reason for the discrepancy for CO_2 is not clear.

In the course of his experiments Noxon also measured rates of quenching of $O_2(^1\Sigma)$ by O_2, N_2, and CO_2 as $1\cdot5\times 10^{-16}$, $2\cdot0\times 10^{-15}$, and $3\cdot0\times 10^{-13}$ cm^3 s^{-1} consistent with results obtained in earlier experiments by Stuhl and Welge[‡] for all three gases. The N_2 results are also consistent with the analysis of day airglow data.[§]

5.6. Quenching of N(^2D)

As measured in the day airglow N(^2D) atoms are the metastable species which are in the second highest concentration in the upper

[†] LOWENSTEIN, M., J. chem. Phys. 54 (1971) 2282.
[‡] STUHL, F. and WELGE, K., Can. J. Chem. 47 (1969) 1870.
[§] WALLACE, L. and HUNTEN, D. M., J. geophys. Res. 73 (1968) 4813.

atmosphere during daytime. The radiative lifetime is as long as 26 hours, so measurement of quenching rates is very difficult. Black, Slanger, St. John, and Young[†] have obtained some estimates of rate coefficients recently by noting that in the photodissociation of N_2O at 1470 Å (see p. 3591) the $B\,^2\Pi$–$X\,^2\Pi$ bands of NO are observed in emission. The only possible mechanisms of production seem to be

$$N(^2D)+N_2O \rightarrow N_2+NO(B\,^2\Pi)+1\cdot52\text{ eV}, \qquad (34)$$

$$O(^1S)+N_2O \rightarrow NO(X\,^2\Pi)+NO(B\,^2\Pi)+0\cdot08\text{ eV}. \qquad (35)$$

Since β bands with 0·6 eV vibrational excitation are observed it seems unlikely that the second mechanism is significant. Moreover, the quenching rates for the species responsible for $NO(B\,^2\Pi)$ production have little relation to those for $O(^1S)$. This being so they should apply to $N(^2D)$.

The observed rates for quenching by O_2, N_2, H_2, CO, NO, CO_2, and N_2O are found to be, in 10^{-12} cm^3 s^{-1}, 7, 6×10^{-3}, 5, 6, 180, 0·6, and 3 respectively.

6. Spin-exchange collisions—measurement of differential cross-sections

Pritchard, Carter, Chu, and Kleppner[‡] have measured differential cross-sections for spin-exchange collisions between alkali-metal atoms (Na–Cs, Na–Rb, Na–K, K–Cs, and K–Rb) in the energy range 0·1–0·2 eV.[§] The technique consisted in crossing a spin-selected primary beam with an unpolarized beam and analysing the polarization of the scattered beam so as to determine the differential cross-section for scattering into each of the final spin states.

Supersonic-flow oven sources were used for the alkali-metal beams. State selection was carried out in an inhomogeneous magnetic field, after passage through which the primary beam passed into the scattering region to collide with the target beam. Scattered atoms passed through a slotted-disc velocity-selector before entering a second magnetic analyser. They were finally detected by surface ionization, followed by a mass spectrometer and magnetic strip electron multiplier. To reduce the noise level as low as possible an iridium filament was used as an ionizing surface. It was flashed near breaking point for about 1 hour

[†] BLACK, G., SLANGER, T. C., ST. JOHN, G. A., and YOUNG, R. A., *J. chem. Phys.* **51** (1969) 116.
[‡] PRITCHARD, D. E., CARTER, G. M., CHU, F. Y., and KLEPPNER, D., *Phys. Rev.* **A2** (1970) 1922.
[§] See also BECK, D., HENKEL, U., and SCHULTZ, A., *Phys. Lett.* **27A** (1968) 277.

and then aged for several days at the operating temperature. With this treatment the noise level was kept as low as 1 ion s^{-1} mm^{-2} of detector area.

Two arrangements were used, one in which the incident, scattered, and target beams were in the same plane and the second in which the target beam was perpendicular to the plane of the incident and scattered beams. In both cases, different angles of scattering in the laboratory system were obtained by rotating the direction of the primary beam.

With the most satisfactory nozzle source the velocity distribution of the target beam had a full width at half height of 30 per cent, the most probable velocity being close to the expected value $(5\kappa T/M)^{\frac{1}{2}}$, where T is the temperature and M the mass of the target atoms.

Direct measurements were made of differential cross-sections per unit solid angle in the laboratory system, I'_d and I'_{ex} for scattering without and with change of state respectively, from which the differential cross-section for all elastic collisions I'_0 and probability of state change P', again in the laboratory system,

$$I'_0 = I'_d + I'_{ex} \tag{36}$$

and
$$P' = I'_{ex}/I'_0, \tag{37}$$

may be obtained.

The angular resolution was 2×10^{-3} rad at full width and the velocity resolution measured by the full width at half height was between 5 and 10 per cent. The polarized primary beam intensity was $0 \cdot 3 \times 10^{12}$ atoms and the target beam density 10^{12} cm^{-3}.

P' is not equal to the probability P_{ex} of electron spin change because account must be taken of the hyperfine structure. It may be shown that

$$P_{ex} = P'(4I^2+4I+1)/(4I^2+2I+1), \tag{38}$$

where I is the nuclear spin quantum number. No correction is necessary to the differential cross-section for all elastic collisions so $P_{ex}/P' = I_{ex}/I'_{ex}$.

The corrected cross-sections were converted to the centre of mass system. This presented no difficulty because the spread in relative velocities was limited by the velocity selector in the scattered beam and by the nozzle source for the target beam.

Primary and secondary rainbow maxima were observed in the differential cross-sections I_0. The probability of spin exchange at different energies varied in much the same way with angle, rising from zero at small angles to a maximum close to 0·4 at a value of the reduced scattering angle $\tau = E\theta$ (where E is the relative collision energy and θ the scattering angle in the C.M. system, see Chap. 22, p. 2386) near $1\cdot0\times10^{-3}$ a.u. At larger τ further oscillatory behaviour is evident. The theoretical analysis of these results is discussed on p. 3603.

7. Spin-reversal collisions (Chap. 18, §7)

In Table 18.25 cross-sections for collisions involving disorientation of polarized Rb atoms in collision with rare-gas atoms and with various molecules, measured in experiments carried out up to the time of writing of Vol. 3, are given. Since that time more accurate measurements have been made and account taken of the nuclear spin in analysing the relaxation process.†

For an atom with nuclear and electron spin quantum numbers I and $\tfrac{1}{2}$ respectively there are $2(2I+1)$ basic states and so there are $4I+1$ numbers defining the population fraction in each. Corresponding to these there are $4I+1$ independent physical observables O_i each of which relaxes under sudden collision with a single time constant τ_i. Any physical observable can be expressed as a linear combination of the O_i. The z-component of nuclear spin (quantum number I_z) and the scalar product $\mathbf{S}.\mathbf{I}$ of electronic and nuclear spin are both basic observables but the z-component of electronic spin is not. It can, however, be written as a combination of two basic observables as

$$S_z = O_e + [2/\{(2I+1)^2 - 2\}]I_z. \tag{39}$$

Because of this S_z relaxes with two time constants τ_e, τ_n which are those for relaxation of O_e and of I_z ($= O_n$) respectively. If $\omega_H \tau_c \ll 1$ where ω_H is the angular frequency corresponding to the hyperfine splitting and τ_c is a characteristic collision time (of order 10^{-12} s) then we find

$$\tau_e^{-1} = \tau^{-1}, \tag{40}$$

$$\tau_n^{-1} = \{2/(2I+1)^2\}\tau^{-1}, \tag{41}$$

where τ is the relaxation time for $\mathbf{S}.\mathbf{I}$ and τ_e that for S_z if there were no nuclear spin. S_z should therefore relax with two time constants, τ and a longer time $\{(2I+1)^2/2\}\tau$.

In earlier experiments which observed the relaxation of S_z no attempt was made to analyse the data in terms of two time constants. The single time constant determined usually applied rather more to the longer time than to τ.‡

Experiments have since been carried out in which full account has been taken of the existence of two time constants for relaxation under

---

† MASNOU-SEEUWS, F. and BOUCHIAT, M. A., *J. Phys. (Fr.)* **28** (1967) 406.
‡ See, for example, FRANZ, F. A., *Phys. Rev.* **139** (1965) A603.

conditions in which diffusion loss to the walls is negligible. Aymar, Bouchiat, and Brossel† applied the technique described in Chapter 18, p. 1845 to optically pumped Rb atoms in a state of specified M_J and observed the relaxation in the presence of helium. They carried out measurements both with ^{85}Rb ($I = \frac{5}{2}$) and ^{87}Rb ($I = \frac{3}{2}$) and with ^{4}He and ^{3}He isotopes. The two time constants were separately determined in each case and the corresponding cross-sections found were independent of the helium isotope and given in 10^{-24} cm² by

$$Q_\text{n}(^{87}\text{Rb–He}) = 0.47, \quad Q_\text{n}(^{85}\text{Rb–He}) = 0.27, \quad Q_\text{e}(\text{Rb–He}) = 8.2. \tag{42}$$

According to the theory $Q_\text{n}(^{87}\text{Rb})/Q_\text{n}(^{85}\text{Rb}) = 2.25$ as compared with the observed ratio 1·73. However, larger unexplained discrepancies occurred for the ratio Q_e/Q_n which were predicted to be 18 and 8 for ^{85}Rb and ^{87}Rb respectively as compared with observed ratios of 30·5 and 18. It will be noted that the value given in Table 18.25, 0.62×10^{-24} cm², is much closer to Q_n than Q_e. In the experiments of Franz, in which great care was taken to avoid impurities, the disorientation cross-section was measured as 0.33×10^{-24} cm², close to $Q_\text{n}(^{85}\text{Rb–He})$ which is obtained from the longer time constant for the most abundant isotope ^{85}Rb.

Experiments carried out by Bouchiat, Brossel, and Pottier‡ to study relaxation due to collisions with the heavier rare gases Ar, Kr, and Xe revealed a number of unexpected features. It was found that τ_n^{-1} does not vary linearly with the gas pressure p and depends strongly on the static magnetic field H so that, approximately,

$$\tau_\text{n}^{-1} = \tau_\text{n}^{-1}(\infty) + A\{1 + (H/H_0)^2\}^{-1}. \tag{43}$$

$\tau_\text{n}^{-1}(\infty)$ is the limiting value for high magnetic fields and is a linear function of p. A and H_0 are also functions of p. H_0 is of order 10 G so that its presence indicates a characteristic time $\tau' \simeq 1/\gamma_\text{s} H_0 \simeq 10^{-8}$ s. These unexpected features can be understood when account is taken of the existence of ephemeral molecules such as RbKr. Because of the long-range van der Waals forces several stationary states exist but with binding energy comparable with gas kinetic energies. Molecules formed through three-body collisions will therefore often break up on the next collision. At a Kr pressure of 1 torr the time between collisions is about 6×10^{-8} s so that the lifetime of a molecule under these conditions will be of this order which is comparable with τ'. To examine the effect of

† AYMAR, M., BOUCHIAT, M. A., and BROSSEL, J., *J. Phys. (Fr.)* **30** (1969) 619.
‡ BOUCHIAT, M. A., BROSSEL, J., and POTTIER, L. C., *J. chem. Phys.* **56** (1972) 3703.

the formation of these short-lived molecules we consider a number of special cases. We ignore at first the nuclear spin.

(a) At low pressures. In zero field, if **L** is the relative angular momentum,
$$\mathbf{J} \; (= \mathbf{S} + \mathbf{L}) \tag{44}$$
is conserved. At low pressures the lifetime τ of the molecule is long enough for **S** to precess through several revolutions around **J**. The disorientation probability per sticking collision is then $\tfrac{2}{3}$ (this allows for the fact that when **J** is along J_z no variation of S_z occurs). Hence we have, for the relaxation time,
$$\tau_s^{-1}(\text{low } p, H = 0) = \tfrac{2}{3}\tau_f^{-1}, \tag{45}$$
where τ_f^{-1} is the number of sticking collisions per second, which varies as p^2 because molecules can only be formed in three-body collisions.

As the field is increased, **S** is strongly coupled to **H** and no disorientation occurs. Hence
$$\tau_s^{-1}(\text{low } p, \text{high } H) \to 0. \tag{46}$$

It follows that the plot of τ_s^{-1} against H at low pressures follows the form given by the second term of (43) with $A = \tfrac{2}{3}\tau_f^{-1}$. The width will be given by the value of H for which the coupling $\gamma_s \mathbf{S}.\mathbf{H}$ with the magnetic field is comparable with that of $\bar{\gamma}\mathbf{S}.\mathbf{L}$ with the orbital motion, i.e.
$$H_0 \simeq \bar{\gamma}\bar{L}/\gamma_s, \tag{47}$$
where $\bar{L}$ is the average of L over the thermal distribution of molecules and $\bar{\gamma}$ is a mean interaction parameter.

(b) At high pressures. Under these conditions τ_f is very short and so **S** precesses only through a small angle during the lifetime of a molecule.

In zero magnetic field the average precession angle will be $\bar{\gamma}\bar{L}\tau$. After n random sticking collisions the mean net angle of precession will be $\bar{\gamma}\bar{L}\tau n^{\frac{1}{2}}$. Before disorientation can be said to occur n will need to be comparable with $1/\bar{\gamma}^2\bar{L}^2\tau^2$. As one sticking collision takes place every τ_f seconds the relaxation time must be of order $n\tau_f = \tau_f/\bar{\gamma}^2\bar{L}^2\tau^2$. We therefore have, including the orientational factor $\tfrac{2}{3}$ as before,
$$A(\text{high } p, H = 0) = \tfrac{2}{3}\tau_f^{-1}\bar{\gamma}^2\bar{L}^2\tau^2. \tag{48}$$

As τ and τ_f are proportional to p^{-1} and p^{-2} respectively, A is independent of p under these conditions.

With finite magnetic fields we would expect a factor to be introduced proportional to the Fourier transform of the correlation function of the random interaction, presumably $\exp(-|t|/\tau)$, taken at the spin frequency $\gamma_s H$. This gives
$$\tau_s^{-1} = \tfrac{2}{3}\tau_f^{-1}\bar{\gamma}^2\bar{L}^2\tau^2\{1+(\gamma_s H/\tau)^2\}^{-1}, \tag{49}$$

which is of the same form as the second term on the right-hand side of (43) with A as in (48).

These considerations apply to τ_n when nuclear spin is taken into account provided τ is long compared with the hyperfine period $1/\omega_H$, as it is for $p \ll 250$ torr, and provided $\bar{\gamma}$ and γ_s are replaced by $\bar{\gamma}/(2I+1)$ and $\gamma_s/(2I+1)$ respectively.

With Kr as the relaxing gas, measurements may be made over a sufficiently wide pressure range to observe all the predicted effects. For Ar only high pressure rates are observable against background loss by diffusion to the walls, while for Xe measurement may only be made at low pressures because at high pressures the rates are too large to be measured. For Rb–Kr, analysis of the data at low and intermediate magnetic fields, which are consistent with the above analysis, gives

$$\tau = (5 \cdot 69 \pm 0 \cdot 17) \times 10^{-8} \text{ s}/p \text{ (in torr)},$$

$$\tau_f = (1 \cdot 06 \pm 0 \cdot 05) \times 10^{-2}/p^2 \text{ (in torr}^2), \qquad \bar{\gamma}\bar{L}/\gamma_s = 9 \cdot 59 \pm 0 \cdot 28 \text{ G}.$$

Less complete analysis gives for Rb–Ar,

$$\tau = (4 \cdot 96 \pm 0 \cdot 28) \times 10^{-8} \text{ s}/p \text{ (in torr)}, \quad A(\text{high } p, H = 0) = 2 \cdot 20 \pm 0 \cdot 16 \text{ s}^{-1}$$

and for Rb–Xe

$$\tau_f = (4 \cdot 29 \pm 0 \cdot 23) \times 10^{-3}/p^2 \text{ (in torr}^2), \qquad \bar{\gamma}\bar{L}/\gamma_s = 38 \cdot 1 \pm 1 \cdot 6 \text{ G}.$$

Further notes on the theoretical interpretation of these results are given on p. 3605.

TABLE 18 A.15

Measured relaxation rates, τ_n^*, τ_e for disorientation of Rb by collision with rare-gas atoms

	Ar	Kr	Xe
τ_e^{-1} (sec^{-1} torr^{-1})	$0 \cdot 95 \pm 0 \cdot 03$	$33 \cdot 2 \pm 1 \cdot 5$	185 ± 10
τ_n^{*-1} (sec^{-1} torr^{-1})			
^{85}Rb	$0 \cdot 0525 \pm 0 \cdot 0017$	$1 \cdot 85 \pm 0 \cdot 09$	$10 \cdot 3 \pm 0 \cdot 5$
^{87}Rb	$0 \cdot 118 \pm 0 \cdot 004$	$4 \cdot 15 \pm 0 \cdot 19$	$23 \cdot 2 \pm 1 \cdot 2$
Q_e (10^{-16} cm^2)	$(6 \cdot 08 \pm 0 \cdot 20) \times 10^{-6}$	$(2 \cdot 66 \pm 0 \cdot 12) \times 10^{-4}$	$(16 \cdot 41 \pm 0 \cdot 87) \times 10^{-4}$

At high magnetic fields the relaxation rates τ_n^* and τ_e may be derived. Results obtained are given in Table 18 A.15. The ratios τ_n^*/τ_e are closely in accord with the theoretical values 18 and 8 for ^{85}Rb and ^{87}Rb respectively. Notes on further theoretical aspects of these results are given on p. 3606.

Ackermann, Laulainen, zu Putlitz, and Weber[†] have measured cross-sections for disorientation of Sr^+ ions in the $^2S_{\frac{1}{2}}$ state, by rare-gas atoms. The ions, in concentration varying from 10^7 to 10^{10} cm^{-3}, were generated by a d.c. discharge. Otherwise the experiment was of the standard type using transient probing pulses to monitor the recovery time of the polarization after switching off a depolarizing r.f. field. No effects such as departure of the inverse relaxation time τ^{-1} from a linear relationship with the gas pressure or dependence on the magnetic field were observed.

TABLE 18 A.16

Observed cross-sections Q, in 10^{-16} cm^2, for disorientation of $5\,^2S_{\frac{1}{2}}$ Sr^+ ions

Disorienting gas	He	Ne	Ar	Kr	Xe
Q	2.0×10^{-5}	3.9×10^{-5}	5.7×10^{-3}	1.8×10^{-2}	4.0×10^{-2}

As the nuclear spin is zero the relaxation time observed is unique. The observed disorientation cross-sections are as given in Table 18 A.16. They are much larger than for the disorientation of Rb atoms by the same gases.

8. Recent theoretical work

8.1. *Depolarization cross-sections for* Hg ($6\,^3P_1$) *in rare gas collisions*

Calculations of cross-sections for disorientation and disalignment of $^{202}Hg(6\,^3P_1)$ by impact with rare-gas atoms have been carried out by Faroux.[‡] The analysis of Chapter 18, §§ 8.3.2 and 8.3.3.2 assumes that the long range attractive force varying as R^{-6}, where R is the nuclear separation, is dominant. If this is so the depolarization cross-sections should vary with temperature T as $T^{-\frac{1}{5}}$. As described on p. 3565, Faroux found that this applies to the heavier rare gases but not to helium and neon for which little variation with T over the range 20–500 °K was observed. This suggests that for these gases the short range repulsion, which falls off much more rapidly with R, is important. Faroux showed that this was plausible on theoretical grounds by estimating the short range interactions, assumed to be of (12, 6) form (see p. 1335).

For the heavier rare gases the quantities β_1, β_2, and $\overline{\Delta E}$ of (424) p. 1901

[†] ACKERMANN, H., LAULAINEN, N. S., ZU PUTLITZ, G., and WEBER, E. W., *7th int. conf. on the physics of electronic and atomic collisions, Amsterdam. Abstracts*, North Holland, 1971, p. 669.

[‡] loc. cit., p. 3565.

were evaluated with improved wave functions and the calculated disorientation cross-sections $\bar{Q}_{or}$ agree remarkably well with those observed (see Table 18 A.17). In the table less accurate theoretical

TABLE 18 A.17

Calculated and observed disorientation cross-sections $\bar{Q}_{or}$ (in 10^{-16} cm^2) for Hg($6\,^3P_1$) in collision with rare-gas atoms, obtained by Faroux

Rare gas	He	Ne	Ar	Kr	Xe
$\bar{Q}_{or}$ obs.	44·0±1·0	56·2±2·5	100·2±5·0	134±7	179±8
$\bar{Q}_{or}$ calc.	(33)	(45)	104	137	178

TABLE 18 A.18

Observed and calculated depolarization cross-sections for ^{199}Hg($6\,^3P_1$) in collision with rare gases, expressed as a ratio to $\bar{Q}_{or}$ for ^{202}Hg($6\,^3P_1$)

		He	Ne	Ar	Kr	Xe
$\bar{Q}_{he}/\bar{Q}_{or}$	obs.	0·464±0·015	0·455±0·015	0·463±0·015	0·463±0·015	0·461±0·015
	calc.	0·47	0·47	0·47	0·47	0·47
$^{\frac{3}{2}}\bar{Q}_{al}/\bar{Q}_{or}$	obs.	0·95	0·94	0·92	0·96	0·95
	calc.	0·955	0·945	0·94	0·96	0·97
$^{\frac{3}{2}}\bar{Q}_{or}/\bar{Q}_{or}$	obs.	0·59	0·60	0·59	0·61	0·595
	calc.	0·62	0·62	0·615	0·62	0·62
$^{\frac{1}{2}}\bar{Q}_{or}/\bar{Q}_{or}$	obs.	0·89	0·87	0·82	0·87	0·90
	calc.	0·87	0·84	0·83	0·88	0·90
$^{\frac{1}{2},\frac{3}{2}}\bar{Q}_{or}/\bar{Q}_{or}$	obs.	0·125	0·135	0·14	0·12	0·095
	calc.	0·145	0·155	0·16	0·14	0·135

values are given for helium and neon based on the estimated (12, 6) interaction. For these cases the theoretical value based on the attractive interaction alone would be about 10 per cent smaller.

Little evidence is available about the interaction from comparison of observed and calculated values of $\bar{Q}_{or}/\bar{Q}_{al}$. For the attractive and repulsive interactions the calculated ratios are 1·11 and 1·05 respectively according to the 'simple' approximation (396) p. 1896. Neither these, nor the experimental ratios, are accurate enough to discriminate —the ratios observed by Faroux are 1·15, 1·20, 1·22, 1·12, and 1·09 for He, Ne, Ar, Kr, and Xe respectively (cf. Table 18.34, p. 1906).

Calculated ratios of the various cross-sections for the odd isotope ^{199}Hg($6\,^3P_1$) to $\bar{Q}_{or}$ for the even isotope are compared with the ratios observed by Faroux in Table 18 A.18. The agreement is very good throughout.

In Table 18 A.19 we give values for the corresponding ratios of cross-sections for ^{201}Hg($6\,^3P_1$) calculated by Faroux for helium. These also were found to agree well with observation.

TABLE 18 A.19

Calculated depolarization cross-sections for ^{201}Hg($6\,^3P_1$) in collision with He, expressed as a ratio to $\bar Q_{or}$ for ^{202}Hg($6\,^3P_1$)

$^{1}_{2}\bar Q_{or}/\bar Q_{or}$	$^{3}_{2}\bar Q_{or}/\bar Q_{or}$	$^{5}_{2}\bar Q_{or}/\bar Q_{or}$	$^{1\,3}_{2\,2}\bar Q_{or}/\bar Q_{or}$	$^{3\,5}_{2\,2}\bar Q_{or}/\bar Q_{or}$	$^{1\,5}_{2\,2}\bar Q_{or}/\bar Q_{or}$
0·83	0·75	0·57	0·08	0·25	0·21

$^{3}_{2}\bar Q_{al}/\bar Q_{or}$	$\bar Q_{al}/\bar Q_{or}$	$^{3\,5}_{2\,2}\bar Q_{al}/\bar Q_{or}$
0·92	0·70	0·029

8.2. Depolarization of 2p levels of Ne

Carrington and Corney[†] have calculated cross-sections for disalignment of the $2p$ levels of neon (in Paschen notation) by collisions with neon and with helium atoms, assuming that the long range attractive forces are dominant. While the absolute cross-sections for neon agree reasonably well with those observed for all but the $2p_2$ and $2p_6$ states, those for helium are considerably too small, being on the average about 0·5 of those observed. Apart from the anomalous behaviour of the $2p_2$ and $2p_6$ states in exhibiting a rapid increase of cross-section with temperature,[‡] the main source of discrepancy is probably neglect of the repulsive short range interaction which is of major importance for helium, though less so for neon. The anomalous behaviour of $2p_2$ and $2p_6$ is probably due to depolarization through inelastic collisions which deactivate the level concerned.

8.3. Theory of Penning ionization

Referring to (20) the cross-sections for Penning ionization can be expressed in terms of the probability $W(R)$ of an autoionizing transition taking place, at a nuclear separation R, between a state i arising from the interaction between a helium metastable atom and the target atom and one f arising from interaction between a normal helium atom and the ionized atom. $W(R)$ has been estimated approximately by Fujii, Nakamura, and Mori,[§] by Bell,[||] and by Miller, Slocomb, and Schaeffer[††]

[†] CARRINGTON, C. G. and CORNEY, A., *J. Phys. B* (GB) **4** (1971) 869.
[‡] CARRINGTON, C. G., CORNEY, A., and DURRANT, A. V., loc. cit., p. 3568.
[§] FUJII, H., NAKAMURA, H., and MORI, M., *J. phys. Soc. Japan* **29** (1970) 1030.
[||] BELL, K. L., *J. Phys. B* (GB) **53** (1970) 1308.
[††] MILLER, W. H., SLOCOMB, C. A., and SCHAEFFER, H. F., *J. chem. Phys.* **56** (1972) 1347.

for the ionization of H atoms by impact of He($2\,^3S$). The last authors carried out an elaborate calculation using wave-functions for the initial and final states which fully included configuration interaction.† They found an attractive interaction for the initial state ($^2\Sigma$) concerned, the well-depth being 1·77 eV and the classical turning point for 0·06 eV impact energy occurring at $R_c = 2\cdot3a_0$. This agrees much better with the values (2·44 eV and $2\cdot15a_0$) derived from measurements of the velocity distribution of the ejected electrons (see p. 3581 and Table 18 A.13) than do the results of the less elaborate calculations of Fujii et al. and of Bell which gave smaller depths of around 0·55 and 0·5 eV respectively, with $R_c = 1\cdot34$ and $1\cdot36a_0$.

The total ionization cross-section was found by Miller et al. to range from about $4\cdot0\times10^{-15}$ cm² at a relative impact energy of 0·01 eV to $3\cdot3\times10^{-15}$ cm² at 0·03 eV and $1\cdot2\times10^{-15}$ cm² at 0·8 eV. The value at 0·03 eV is not very different from that, $2\cdot2\pm0\cdot6\times10^{-15}$ cm², measured by Shaw et al. (see p. 3577). The calculated energy distribution of the ejected electrons at an impact energy of 0·06 eV is qualitatively in agreement with that observed by Hotop et al. (see p. 3581). Miller et al. also calculated the contribution to the total cross-section from associative ionization and found it to be about 22 per cent of the total in the limit of zero impact energy. At 0·03 eV it is about 18 per cent, which is not inconsistent with the rough experimental estimate made by Shaw et al. (p. 3577).

For ionization by He($2\,^1S$) less detailed theoretical data are available but Miller and Schaeffer found for the upper ($^2\Sigma$) state an attractive well-depth of 0·39 eV with $R_c = 3\cdot6a_0$ for 0·06 eV impact energy. This agrees well with values (0·46 eV and $3\cdot4a_0$) derived from analysis of observed energy distribution of ejected electrons (see p. 3582 and Table 18 A.13). In contrast Fujii et al. obtained a repulsive curve for the upper state.

8.4. *Impact excitation of fine structure transitions in alkali-metal atoms*

An alternative approach to the problem of calculating cross-sections for excitation of transitions between the doublet levels in alkali-metal atoms to that described in Chapter 18, p. 1926 has been developed by Nikitin and his collaborators.‡ This treats the problem as one of transitions between molecular potential energy curves of the molecule M*S, where M is the alkali-metal atom and S the foreign gas atom.

† MILLER, W. H. and SCHAEFFER, H. F., loc. cit., p. 3582.
‡ NIKITIN, E. E., *Optika Spektrosk.* **19** (1965) 91; **22** (1967) 379; DASHEVSKAYA, E. I. and NIKITIN, E. E., *Optics Spectrosc.* **22** (1967) 473.

The molecular states concerned are the $A\,^2\Pi_{\frac{1}{2}}$, $A\,^2\Pi_{\frac{3}{2}}$, and $B\,^2\Sigma$. In the limit $R \to \infty$, where R is the nuclear separation, the first of these tends to the $\mathrm{M}(^2P_{\frac{1}{2}})$ state of the alkali-metal atom, and the ground state of S, whereas the other two tend to the $\mathrm{M}(^2P_{\frac{3}{2}})$ state. Coupling which produces transitions significant for the problem concerned is limited to two regions. The first, outermost region, arises from partial breakdown of the atomic spin–orbit coupling by exchange forces between M* and S which will occur when the separation ΔV of the $A\,^2\Pi$ and $B\,^2\Sigma$ curves becomes comparable with the spin–orbital separation $\Delta\epsilon$. This region, for the cases considered, occurs at separations R_1 from 12 to $14 a_0$ and the order of magnitude of the cross-section is

$$Q_1 = 4\pi R_1^2 (\tfrac{1}{3}\pi)^{\frac{1}{2}} \gamma \exp(-3\gamma^{\frac{2}{3}}), \tag{50}$$

where $\gamma = (\tfrac{2}{3}\pi)(\Delta\epsilon/\alpha\hbar)(M/2\kappa T)^{\frac{1}{2}}$. α is such that ΔV falls off as $\mathrm{e}^{-\alpha R}$, M is the reduced mass of the colliding systems, and T the temperature.

Because of the exponential factor Q_1 varies very much with $\Delta\epsilon$, ranging from 3×10^{-15} cm² for K–He to 10^{-17} cm² for Rb–He and $0{\cdot}7 \times 10^{-21}$ cm² for Cs–He, with a reasonable choice of the parameters determining R_1 and γ, in general agreement with the observed results (see pp. 1734–9 and Table 18.9 of Chap. 18).

However, when Q_1 is very small account must be taken of the inner interaction region which arises from rotational coupling between the $^2\Pi_{\frac{1}{2}}$ and $^2\Pi_{\frac{3}{2}}$ molecular states. For this the separation R_2 is much smaller than R_1 (near 5–6 a.u.) but the transition probability within this region may be much higher because the potential energy curves are rising much more steeply.

Using the interactions calculated by Baylis,[†] Dashevskaya, Nikitin, and Reznikov[‡] evaluated cross-sections for Rb–He, Rb–Ar, Rb–Xe, and Cs–He collisions as functions of temperature for comparison with the observed results obtained by Gallagher.[§] Allowing for the rather drastic approximations which must be made in deriving numerical data from the theory, reasonably good agreement is obtained when account is taken of the contributions from both interaction regions.

A detailed numerical calculation has been carried out by Masnou-Seeuws[‖] for sodium using a similar formulation to that of Mandelberg (Chap. 18, p. 1926) but replacing the van der Waals interaction with the

[†] Baylis, W. E., *J. chem. Phys.* **51** (1969) 2665.
[‡] Dashevskaya, E. I., Nikitin, E. E., and Reznikov, A. I., ibid. **53** (1970) 1175.
[§] loc. cit., p. 3569.
[‖] Masnou-Seeuws, F., *J. Phys. B (GB)* **3** (1970) 1437.

exchange interaction used by Nikitin *et al.* In this case three coupled equations result. These were solved without further approximation. The calculated cross-sections for collisions at 400 °K of Na with He, Ne, Ar, Kr, and Xe are, in 10^{-16} cm^2, 78, 68, 62, 57, and 55 respectively. The agreement with experiment (see p. 3569) is good for He, Ne, and Ar but for the two heavier rare gases the observed values are considerably larger than the calculated. This may be because of the neglect of the van der Waals interaction which is likely to be relatively more important for heavier colliding systems and at low velocities.

The theoretical discussion in Chapter 18, pp. 1926–8 is based on effects due to van der Waals forces only. Part of the discrepancy between theory and experiment shown in Table 18.40 is undoubtedly due to the use by Mandelberg of the so-called unitary approximation (see p. 1927) to solve the coupled equations involved. Kumar and Callaway[†] have checked this by obtaining accurate numerical solutions of Mandelberg's equations for Na–He and Rb–He collisions. For the former the result was quite close to that calculated by Mandelberg but for Rb–He the accurate solution was six times smaller and so quite close to the experimental value.

Reid and Dalgarno[‡] have carried out a close-coupling calculation for Na–He collisions in which they assume the interaction given by Baylis which includes both exchange and van der Waals terms. Their results, which cover the temperature range from threshold to 600 °K, agree very well with those of Masnou-Seeuws, and hence with experiment, at 400°, and show clearly that exchange forces are important and indeed dominant in this case. For the heavier atoms van der Waals forces may be relatively more important but it is difficult to check this because of the small cross-sections in these cases.

8.5. *Theoretical analysis of observed cross-sections for spin exchange*

Pritchard and Chu[§] have analysed the results observed by Pritchard, Carter, Chu, and Kleppner on the differential cross-section for elastic collisions between alkali-metal atoms with and without spin exchange (see p. 3592 for notes on the experiments). Of particular interest is the light thrown on the magnitude of the long-range interactions (see Chap. 16, p. 1432 and also p. 3531).

Spin-exchange collisions may be analysed in a very similar way to charge-exchange collisions (see Chap. 23, p. 2532). In terms of the

[†] KUMAR, L. and CALLAWAY, J., *Phys. Lett.* **28A** (1968) 385.
[‡] REID, R. H. G. and DALGARNO, A., *Chem. Phys. Lett.* **6** (1970) 85.
[§] PRITCHARD, D. E. and CHU, F. J., *Phys. Rev.* A2 (1970) 1932.

amplitudes $f_s(\theta)$, $f_t(\theta)$ for singlet and triplet scattering the probability of spin exchange P_{ex} is given by

$$P_{ex} = \frac{1}{2}\frac{|f_s(\theta)-f_t(\theta)|^2}{|f_s(\theta)|^2+3|f_t(\theta)|^2} \quad (51\,a)$$

$$= (1+x^2-2x\cos\delta)/(6+2x^2) \quad (51\,b)$$

where
$$x(\theta) = |f_s(\theta)|/|f_t(\theta)|, \quad (51\,c)$$

$$\delta(\theta) = \arg f_s(\theta) - \arg f_t(\theta). \quad (51\,d)$$

Under the experimental conditions the scattering is semi-classical. The triplet scattering is due to an interaction which is largely repulsive, only becoming attractive at large distances. Rainbow scattering occurs but, at angles smaller than the rainbow angle, almost all the contributions to the amplitude arises from a single impact parameter $p_t(\theta)$. The singlet scattering arises from an interaction with a relatively deep, chemical attraction. Orbiting occurs (see Chap. 16, p. 1310) but the dominant contribution to the amplitude, especially at small angles, is again associated with a single impact parameter $p_s(\theta)$.

Following a similar analysis to that on p. 2534 we now have, approximately,

$$\delta(\theta) = (\hbar v)^{-1}\int_{-\infty}^{\infty}\Delta V\{(p^2+z^2)^{\frac{1}{2}}\}\,dz \quad (52)$$

where p is a mean impact parameter $\frac{1}{2}(p_s+p_t)$ and ΔV is the difference V_s-V_t between the singlet and triplet interactions. Again, it may be shown that, if v is the relative velocity of collision,

$$v\delta = b_0(\tau)+E^{-1}b_1(\tau)+\ldots \quad (53)$$

where E is the relative collision energy and τ is the reduced scattering angle $E\theta$. Under the experimental conditions we would expect the first term of the series to be dominant so $v\delta$ should be a function of $E\theta$.

To separate δ from x in (51 b) advantage was taken of the fact that, when P_{ex} has an extreme value, δ is an integral multiple of π while it is an odd integral multiple of $\frac{1}{2}\pi$ when P_{ex} passes through its average value. At small angles $x \to 1\cdot 0$ and this enables δ to be determined at these angles. Knowing δ at an extreme value, x may be obtained there and is found to be $\simeq 0\cdot 8$. In view of the slow variation of x with θ this enables further values of δ to be determined near the extrema.

It was found that $v\delta$ determined in this way is indeed a function of $E\theta$ only; at least within the experimental and analytical uncertainties. Comparison was then made with values calculated from (52) on the assumption of different interactions V_s and V_t at the relevant large atomic separations. The mean interaction $\frac{1}{2}(V_s+V_t)$ is determined by the long-range dispersion force while the difference ΔV falls off exponentially.

Calculations were carried out using four different estimates of the long-range interactions, combined with ΔV as calculated by Dalgarno and Rudge (see Chap. 18, p. 1889). The four assumed long-range interactions were as follows.

(a) An interaction C_6/r^6 with C_6 derived from the experiments of Buck and Pauly on alkali–alkali scattering (see Chap. 16, p. 1433).
(b) An interaction C_6/r^6 with C_6 calculated by Dalgarno and Kingston.
(c) Interaction (a) together with terms $C_8 r^{-8}+C_{10} r^{-10}$, C_8/C_6, C_{10}/C_6 being as calculated by Davison† and by Fontana‡ respectively.
(d) Interaction (b) together with terms $C_8 r^{-8}+C_{10} r^{-10}$ with the same ratios C_8/C_6 and C_{10}/C_6 as in (c).

It was found that a good fit with the experimental data for all the atom pairs investigated could be obtained by combining interaction (d) with the difference interaction of Dalgarno and Rudge. Other mean interactions, combined with the same ΔV, give too large values for $v\delta$, increasing in the order (b), (c), (a). Any increase in ΔV will require a larger mean interaction than (d) if agreement with experiment is preserved. Although Dalgarno and Rudge obtained reasonably good agreement with observed values of total cross-sections for spin exchange (see Chap. 18, p. 1890) the check is only semi-quantitative. In view of the reliance which must therefore be placed on the theoretical value of ΔV in order to determine the dispersion forces, the latter cannot yet be regarded as well known. They do, however, appear to be stronger than those derived from analysis of the experiments of Buck and Pauly.

8.6. *Theory of spin-reversal collisions*

Bouchiat (C. C.), Bouchiat (M. A.), and Pottier§ have carried out a detailed analysis of the experimental data on the disorientation of polarized Rb atoms in collision with Kr atoms discussed in § 7.

The central interaction between Rb and Kr atoms was taken to be of the $(12, 6)$ form with the parameters $\epsilon = 10^{-2}$ eV and $r_m = 4\cdot 53$ Å (see p. 3530). Spin relaxation can occur through spin–orbit interaction of the form
$$\gamma(r)\mathbf{S}.\mathbf{L}, \qquad (54)$$
where **S** is the spin angular momentum of the Rb atoms, **L** the angular momentum of relative motion of the Rb and Kr atoms and r their

---
† DAVISON, W. D., *J. Phys. B (GB)* **1** (1968) 139.
‡ FONTANA, P. R., *Phys. Rev.* **123** (1961) 1865.
§ BOUCHIAT, C. C., BOUCHIAT, M. A., and POTTIER, L. C. L., ibid. **181** (1969) 144.

separation. A second possible interaction which could be effective is the hyperfine interaction

$$\beta(r)[\{(\mathbf{S}.\mathbf{r})(\mathbf{I}.\mathbf{r})/r^2\}-\tfrac{1}{3}\mathbf{S}.\mathbf{I}\}, \tag{55}$$

$\mathbf{I}$ being the nuclear spin. The data of Table 18 A.15, which refer to relaxation arising in sudden (non-sticking) collisions, discriminate between these possibilities. The ratio of τ_n^* to τ_e for both Rb isotopes agrees well with that predicted on the assumption of the form (54) but is in strong disagreement with that predicted from (55).

To analyse the data further $\gamma(r)$ was taken to be of the form

$$\gamma(r) = \gamma_a \exp\{-\kappa^2(r^2-r_0^2)\}+\gamma_b(r_0/r)^{2s} \tag{56}$$

combining a short-range and long-range term. Using the interaction (56) and an impact parameter formulation in which the trajectory of relative motion of the colliding atoms was taken to be that for two hard spheres of radius r_0 equal to that at which the (12, 6) interaction vanishes ($2^{-\frac{1}{6}}r_m$), the relaxation time τ in (40) is given by

$$\tau^{-1} = nQ\bar{v}M^2 r_0^4 \hbar^{-4}\{\gamma_a^2 F_a(\kappa r_0)+\gamma_b^2 F_b(s)\}(r_0/r_m)^2, \tag{57}$$

where M is the reduced mass of the two atoms, v their mean relative velocity, n is the concentration of Kr atoms, and Q is the total collision cross-section which is $\simeq \pi r_m^2$. The most probable value of s is 4 and $F_b(4) = 0.105$. $F_a(x)$ decreases rapidly as x increases.

To give agreement with the observed value of τ^{-1} at 300 °K

$$\gamma_a^2 F_a(\kappa r_0)+\gamma_b^2 F_b(s) = 1.05\times 10^{13}\hbar^2 \text{ s}^{-2}. \tag{58}$$

A second relation can be obtained from the analysis of the relaxation arising from sticking collisions. Referring to p. 3597 we see that $\bar{\gamma}\bar{L}/\gamma_s = 9.59\pm0.28$ G where $\bar{\gamma}$ is a mean value of $\gamma(r)$ and $\bar{L}$ of L. The latter may be calculated statistically from the allowed vibrational and rotational energy levels in the (12, 6) interaction assumed between Rb and Kr and is found to be $38\hbar$. $\bar{\gamma}$ may be written $\xi\gamma(r_m)$ where ξ depends on the shape of $\gamma(r)$. We then have

$$\bar{\gamma}^2 = \xi^2\{\gamma_a e^{-\kappa^2(r_m^2-r_0^2)}+\gamma_b(r_0/r_m)^{2n}\}^2$$
$$= 0.403\times 10^{12}\hbar^2 \text{ s}^{-2}. \tag{59}$$

If the short-range interaction is omitted, s taken as 4, and ξ calculated classically by averaging $\gamma(r)$ over the classical trajectory, two values for γ_b may be obtained from (58) and (59). These disagree by a factor of 10. On the other hand if γ_b is taken as zero the equations (58) and (59) are satisfied with $\hbar^{-1}\gamma_a = 35$ MHz and $\kappa r_0 = 4$ if ξ is taken as unity which is not far from that, 0.92, estimated as before. The values of γ_a and κr_0

are reasonable and suggest that the short-range contribution to γ is the most effective in producing spin relaxation.

From the ratio τ/τ_f, as given on p. 3597, the equilibrium constant K for the reaction
$$\text{Rb} + \text{Kr} + \text{Kr} \rightleftharpoons \text{RbKr} + \text{Kr} \qquad (60)$$
may be derived. It is found to be 1.7×10^{-22} cm^3/molecule at 300 °K corresponding to a proportion of about one bound to 1.8×10^5 free Rb atoms when the Kr pressure is 1 torr. The value estimated for K from the statistical formula in terms of the vibrational and rotational energies of RbKr, calculated from the assumed (12, 6) interaction, is 1.46×10^{-22} cm^3/molecule, which is in remarkably good agreement.

19

COLLISIONS UNDER GAS-KINETIC CONDITIONS—IONIC MOBILITIES AND IONIC REACTIONS

SINCE Vol. 3 was written most of the new work within the ambit of this chapter has been concerned with ionic reactions. However, data have now become available on the transverse and longitudinal diffusion coefficients of ions drifting in a gas under the influence of a uniform electric field. The results show reasonably good agreement with Wannier's theory. As far as mobilities are concerned the most important new result is the pressure dependence of the mobility of K^+ ions in rare gases, observed by Elford (see p. 3611), a result which, at the time of writing, still defies explanation.

The flowing afterglow technique continues to be of major importance for the measurement of the rates of ionic reactions and has been adopted for use over a considerable temperature range and for a wide variety of bimolecular and termolecular reactions. Nevertheless, other techniques, some only introduced recently, have made valuable contributions. Thus Biondi and his collaborators (see p. 3613) have developed further the use of the drift tube with mass-analyser while Good, Durden, and Kebarle (see p. 3616) have developed a technique for measurement of rates of hydration of ions which has proved fruitful.

The use of ion cyclotron-resonance for measurement of ionic mobilities and reaction rates has not developed as fast as might have been expected but the recent introduction of transient rather than equilibrium techniques is showing promise.

Very many of the new results on ionic reactions which we summarize below are concerned with reactions of interest for the interpretation of the behaviour of the terrestrial and other planetary ionospheres. Considerable progress has been made in the selection of the reactions which lead to production of the dominant clustered positive and negative ions in the D region. Incidental to this much new information is available about the rates of termolecular associative reactions as well as of cluster transfer reactions.

1. Transport processes associated with positive ions—transverse and longitudinal diffusion (see Chap. 19, § 2.1)

For a swarm of ions drifting through a gas at pressure p, under the influence of a uniform electric field F, the mean energy $\bar{\epsilon}$ per ion is given, under certain assumptions, by Wannier's formula

$$\bar{\epsilon} = \tfrac{1}{2}M_1 u^2 + \tfrac{1}{2}M_2 u^2 + \tfrac{3}{2}\kappa T. \tag{1}$$

M_1, M_2 are the masses of the ions and gas molecules respectively, u the drift velocity, and T the absolute temperature. If $\bar{\epsilon} \simeq \tfrac{3}{2}\kappa T$, at small F/p, so that the mean energy acquired from the field is small compared with the thermal energy, the diffusion coefficient D of the ions in the gas is related to their (zero-field) mobility μ by

$$\mu = eD/\kappa T. \tag{2}$$

However, when F/p increases so that the energy acquired by the field becomes comparable with or greater than the thermal the directed component of motion of the ions becomes important. As a result the diffusion can no longer be specified by a single quantity D but rather by a tensor involving longitudinal and transverse diffusion coefficients D_L and D_T respectively. With the development of drift tubes and other devices with high space and time resolution it is possible to measure these separate coefficients. In fact, in experiments at high F/p, the fact that $D_L \neq D_T$ must be taken into account. Thus, referring to the description on p. 2013 of the combined drift tube and mass spectrometer developed by McDaniel and his collaborators (see Fig. 19.34) the formula (102) must be modified to read

$$\phi(z,t) = \{A\sigma_0/4(\pi D_L t)^{\frac{1}{2}}\}(u+z/t)\exp\{-(z-ut)^2/4D_L t\} \times$$
$$\times \{1-\exp(-r_0^2/4D_T t)\}. \tag{3}$$

Under the same assumptions as those made in deriving (1), namely that there is a constant mean free time between collisions and that the angular distribution of the scattering is isotropic, Wannier† obtained the following formulae for D_L and D_T:

$$D_T(F) = D(0) + \zeta_T\left(\frac{M_1 M_2}{M_1+M_2}\right)u^3/eF, \tag{4a}$$

$$D_L(F) = D(0) + \zeta_L\left(\frac{M_1 M_2}{M_1+M_2}\right)u^3/eF, \tag{4b}$$

with $\quad \zeta_T = \dfrac{1}{3}\dfrac{(M_1+M_2)^2}{M_1(M_2+2M_1)}, \quad \zeta_L = \dfrac{1}{3}\dfrac{(M_1+M_2)(M_2+4M_1)}{M_1(M_2+2M_1)}$

and $D(0)$ satisfies (2).

† WANNIER, G. H., *Bell System. Tech. J.* **32** (1953) 170.

Very nearly the same results were obtained on the assumption that the scattering was determined by polarization forces. Skullerrud[†] has considered two further cases, one in which the collision cross-section is independent of velocity and the other in which the collisions involve symmetrical charge exchange. In the former case the product DN of the diffusion coefficient with the neutral molecule concentration N is again found to be proportional to $u^3(F/N)^{-1}$ just as with (4a) and (4b). For the second case the semi-empirical formula of Firsov and Demkov ((387) Chap. 23) was used for the collision cross-section so that it could be written

$$Q(v) = Q_0(\ln V/v)^2.$$

D_L was then obtained in the form

$$ND_L = (F/N)^{\frac{1}{2}}(eQ_0^3 M_1)^{-\frac{1}{2}} f\left(\frac{F}{N} \frac{e}{M_1 Q_0 V^2}\right)$$

with the function $f(x)$ calculated numerically over a range of $\log x$ from -14 to -2. Again it seems that ND_L is approximately proportional to $u^3(F/N)^{-1}$.

Referring to (3) it may be seen how D_T and D_L may be measured in drift tube experiments. (3) gives the flux of ions from a source which creates at time $t = 0$ a thin uniform disc of ions of radius r_0 which drift in the z-direction, normal to the disc, under the action of a uniform electric field, with a drift velocity u. The ion flux is received on an area A normal to the electric field at a distance z from the source at time t. If the experimental arrangement permits of the variation of z this gives a direct method of determining D_L. D_T may be obtained by choosing it to give the best fit of observed data to (3). Alternatively, as in the experiments of Dutton and Howells,[‡] D_T may be measured in a similar way to the corresponding measurements for electrons. D_L may also be measured in any shutter device for mobility measurements. If the effective collecting aperture is large enough the 'line width' of the ion current at the aperture is determined by D_L as in (3).[§]

McDaniel and Moseley[∥] have carried out a number of measurements of D_L and D_T, using apparatus essentially the same as that described on p. 2013. Results were obtained for N^+ and K^+ ions in N_2, O^-, and K^+ ions in O_2, and K^+ and H_3^+ ions in H_2. All measurements were made at room temperature and for F/N ranging from a few times 10^{-17} V cm^{-1} to up to a few times 10^{-15} V cm^{-1}. Good agreement, on a log–log scale,

[†] SKULLERRUD, H. R., *J. Phys. B (GB)* **2** (1969) 86.
[‡] DUTTON, J. and HOWELLS, P., ibid. **1** (1968) 1160.
[§] MILLOY, H. B. *Phys. Rev.* A**7** (1973) 1182.
[∥] McDANIEL, E. W. and MOSELEY, J. T., ibid. A**3** (1971) 1040.

was obtained with (4a) and (4b) in all cases. This also applied to observations by Fleming, Tunnicliffe, and Rees† (D_T; K^+–H_2, K^+–N_2) and by Dutton, Llewellyn-Jones, Rees, and Williams‡ (D_T; H_3^+–H_2).

Milloy§ has used a drift tube, developed by Elford,‖ to measure precisely ionic drift velocities by the Bradbury–Nielsen method (see p. 58), to make accurate observations of D_L for K^+ in N_2 over the range of F/N from 2·83 to 28·3×10^{-17} V cm^{-2}. The results differ significantly from those of McDaniel and Moseley and show a maximum departure of 11 per cent from the predictions of Wannier's theory. They do tend to the predicted limit (2) as $F/N \to 0$, within 4 per cent. The discrepancy at higher F/N may be due to neglect of inelastic collisions in the theory, which might be serious with molecular targets.

2. Mobilities of alkali-metal ions in rare gases

2.1. *Pressure dependence of mobility*

The most important new results are those of Elford‖ who has carefully remeasured the mobilities of K^+ ions in He, Ne, Ar, H_2, and N_2 at 291 °K as functions of F/N at different gas pressures. The method used was that of Bradbury and Nielsen (Chap. 2, § 3.1).

Two pressure gauges were used, an MKS Baratron (type 90H–3E) manometer for pressures less than 3 torr and a quartz-spiral manometer at higher pressures. The absolute error in pressure measurements was estimated to be 0·1 per cent in the range $1·0 < p < 3·0$ torr rising to 0·4 per cent for the range $0·2 < p < 0·5$ torr.

The potential across the drift chamber was directly measured to less than 0·05 per cent and the electric field strength F was estimated to be correct to less than 0·1 per cent, apart from effects of contact potential differences. To ensure temperature stability the envelope of each drift tube was immersed in a water bath so the change in gas temperature was less than 0·1 °K per hour. Thermal gradients which render uncertain the determination of the gas density were reduced to negligible proportions by heat shielding.

A series of checks were carried out on the performance of the equipment. Three experimental systems with different geometry were used. This made it possible to change the drift chamber length, the diameter of the ion entry hole, the diameter and type of collector (grid wires or

---

† FLEMING, I. A., TUNNICLIFFE, R. J., and REES, J. A., *J. Phys. B (GB)* **2** (1969) 780.
‡ DUTTON, J., LLEWELLYN-JONES, F., REES, W. D., and WILLIAMS, E. M., *Phil. Trans.* A259 (1966) 339.
§ loc. cit., p. 3610.   ‖ ELFORD, M. T., *Aust. J. Phys.* **24** (1971) 705.

plate), the geometry of the guard ring system, the aperture of the shutters, and the distance above the upper shutter in which the ions moved in the same electric field as in the drift space.

No significant differences in the measured values resulted from these changes. In particular, the independence of drift length eliminates any effects of contact potential differences between the shutter as well as diffusive corrections (Chap. 2, § 3.4) and other end effects.

Equally so, changes in magnitude of the sinusoidal potential applied to the shutters when sine wave gating was used, or of the pulse height and duty cycle when pulse gating was used, produced no variations in measured values. Increasing the ion current from 10^{-11} to 10^{-9} A and change of potential between the filament and mount electrode also had no effect.

Moreover, the data were highly reproducible and, except in argon at pressures as high as 100 torr, there was no trace of a second ion present.

The remarkable result obtained was that the mobility, defined as on p. 1934, for gas at 1 atmosphere pressure and 0 °C, is not a function only of F/p at fixed gas temperature but also varies with the gas pressure (or density). For example, in argon with $F/p = 3.0$ V cm^{-1} torr^{-1} ($F/N = 8.48 \times 10^{-17}$ V cm^{-2}) the mobility in cm^2 s^{-1} V^{-1}, varies from 2·673 to 2·624 as the pressure increases from 0·937 to 18·8 torr. The proportional dependence on p is independent of the value of F/p.

In addition there was some evidence that the mobility does not become independent of F/p at small F/p.

No explanation of these results is forthcoming at the time of writing. It is noteworthy that an equally inexplicable pressure dependence has been observed for the mobilities of electrons in H_2 and D_2 but not in He (see p. 3441).

2.2. *Cluster formation* (for hydration of atmospheric ions see p. 3627)

The equilibrium constants of hydration reactions of the K$^+$ ion

$$K^+(H_2O)_{n-1} + H_2O \rightleftharpoons K^+(H_2O)_n \qquad (5)$$

have been measured by Searles and Kebarle.† K$^+$ ions from a hot coated platinum filament were caused to drift through water vapour in the filament housing, under the action of an electric field, to enter the field-free reaction chamber in which the ions came into thermal equilibrium with the water vapour. Ions which diffused out through the exit orifice were accelerated and mass-analysed. From the relative currents of different clustered ions the equilibrium constants were obtained. The

† SEARLES, S. K. and KEBARLE, P., *Can. J. Chem.* **47** (1969) 2619.

reaction chamber could be electrically heated or cooled by circulation of a refrigerant mixture through the walls so that measurements could be obtained over a range of temperatures.

For the reactions $(n-1, n)$ the enthalpy differences $H_n^\circ - H_{n-1}^\circ$ between the n and $n-1$ hydrated clusters given in Table 19 A.1 were obtained.

TABLE 19 A.1

Observed values of the enthalpy differences for hydration reactions $(n-1, n)$ of K^+

Reaction	0, 1	1, 2	2, 3	3, 4	4, 5	5, 6
$H_n^\circ - H_{n-1}^\circ$ (eV)	0·77	0·69$_5$	0·57	0·51	0·46	0·43
(kcal/mole)	17·9	16·1	13·2	11·8	10·7	10·0

Clementi and Popkie[†] have applied the Hartree–Fock–Roothaan method to calculate the energy surface for Li^+–H_2O keeping the geometry of H_2O undistorted. The most stable configuration is planar with the Li–O and O–H bond lengths 1·84 and 0·96 Å respectively and the Li–O–H and H–O–H angles are 127·0° and 106·1° respectively. The calculated binding energy of the molecule to the ion is 1·53 eV.

3. Notes on techniques for measuring ionic reaction rates

In Chapter 19, § 3.4 we described a number of methods for measuring the rates of ionic reactions under thermal and superthermal conditions. We mention here some new features which have been developed since that section was written.

An interesting and important development of the drift tube technique with mass analysis has been made by Biondi and his collaborators.[‡] Suppose that the rate of a reaction is required in which parent ions A^+ produce ions B^+. A pulse of ions, A^+, produced by a suitable source, is admitted to the reactant gas in the drift tube where it drifts under the action of an electric field F while undergoing reactions with the gas. To reduce loss by diffusion a buffer gas, usually helium, is added in the drift tube.

The time rate of decrease of concentration $[A^+]$ of A^+ ions through reaction is determined by one or other of two techniques.

(a) If the mobilities of the ions A^+ and B^+ in the buffer gas are sufficiently different, the rate of decrease of $[A^+]$ may be derived from the

[†] CLEMENTI, E. and POPKIE, H., *J. chem. Phys.* **57** (1972) 1077.
[‡] HEIMERL, J., JOHNSEN, B., and BIONDI, M. F., ibid. **51** (1969) 5041.

so-called characteristic arrival spectrum of ions B⁺. Suppose the pulse of A⁺ ions enters the drift tube at time $t = 0$. Let D_p, D_r be the diffusion coefficients of A⁺ and B⁺ ions respectively in the buffer gas. Since the diffusion equation may be written

$$\frac{\partial n}{\partial (Dt)} = \nabla^2 n,$$

the time appears in the solution only in the product Dt. The density distribution of a product ion group produced at time t' and arriving at the collector at time t_{obs} will therefore be determined by

$$D_p t' + D_r(t_{obs}-t') = D_p\{t' + (D_r/D_p)(t_{obs}-t')\}. \qquad (6)$$

Also if v_r, v_p are the drift velocities of A⁺ and B⁺, t_{obs} is given by

$$t' + (v_r/v_p)(t_{obs}-t') = L/v_p, \qquad (7)$$

where L is the length of the drift tube. If the relation $D = \kappa T\mu/e$ is valid, where μ is the mobility, $v_r/v_p = D_r/D_p$ so the right-hand side of (6) becomes simply $D_p L/v_p$ independent of t'. Thus the diffusion broadening of a product ion group is the same no matter when it is produced. Hence, except for t_{obs} close to L/v_r or L/v_p, the arrival spectrum of ions produced at time t' is given by

$$N(t') = C\exp\{-k[R]t'\}, \qquad (8)$$

where k is the rate constant for the reaction, [R] the concentration of reactant gas, and C is constant. Eliminating t' from (8) we find

$$N(t_{obs}) = C'\exp\left\{-[R]k\left(\frac{v_r}{v_r-v_p}\right)t_{obs}\right\}. \qquad (9)$$

The validity of this result, which would not be expected to hold for too high values of F/p, where p is the buffer gas pressure, because of the failure of the relationship between D and μ, can be tested experimentally.

(b) If the mobilities of A⁺ and B⁺ in the buffer gas are nearly equal the following method may be used. The residence time of the ions A⁺ in the drift tube may be increased by modifying the electric field. Thus if the field is made to vanish for a time Δt, the ions spend this additional time in the reacting gas, at thermal energy. If it is desired to increase the residence time by Δt while leaving the ion energy unchanged, the field is reversed for a time $\tfrac{1}{2}\Delta t$. By observing the total flux of A⁺ received at the collector in successive cycles with and without the field switching, the loss of A⁺ due to the reaction in time Δt at the chosen mean energy, may be obtained.

Problems arise through Penning ionization by metastable helium atoms not only when the parent ions are He⁺ but in most other cases in which helium is deliberately introduced into the ion source to facilitate the production of the wanted ions. In the former case, with N_2 as the reactant gas, the contribution from Penning ionization is obtained by reversing the drift field for a few μs immediately after the discharge pulse. This destroys all ions by wall recombination but the metastable atoms survive this period so that all ions observed subsequently when the accelerating field is restored must arise from Penning ionization. With O_2 as reactant gas, for which the contribution from Penning ionization is relatively greater, the drift field is made to vanish for a controlled interval when the parent He^+ ions reach the centre of the drift tube. This superposes peaks on the O_2^+ and O^+ arrival spectrum which are produced only by He^+ ions at some distance from the ion source. From these the ratio of the rate production of O^+ and O_2^+ from He^+ may be obtained. Since Penning ionization has little effect in producing O^+ this enables the total reaction cross-section to be obtained.

A further procedure used for studying reactions involving other parent ions is to reduce the electric field to zero for several hundred μs during which time the metastable atoms are destroyed and the ion pulse is still sufficiently sharp for the additional residence time technique to be applied.

Results obtained by application of these techniques are described on pp. 3621–5.

By heating or cooling the flow tube, the flowing afterglow technique (see Chap. 19, § 3.4.5, p. 2021) has been extended to measure reaction rates over a temperature range from 82 to 600 °K.† It is important to remember that, whereas with a drift tube it is possible to measure reaction rates at different ion energies, the gas temperature and hence the vibrational distribution in reactant molecules remains that at room temperature. With a heated afterglow tube not only is the mean impact energy increased but so also is the vibrational temperature.

Twiddy and his collaborators‡ have measured reaction rates for Ne^+ ions in a flowing afterglow system with helium as the streaming gas. Neon gas is first produced at a side-entry port so Ne^+ ions and Ne^* metastable atoms are produced from He_2^+ and He^*. He^+ ions are at first unaffected but would gradually be converted into He_2^+ along the flow tube giving rise to a distributed source of Ne^+ which would be unusable. To obviate this He^+, He_2^+, He^*, and Ne^* are removed by injection of N_2

† DUNKIN, D. B., FEHSENFELD, F. C., SCHMELTEKOPF, A. L., and FERGUSON, E. E., J. chem. Phys. **49** (1968) 1365.

‡ HEMSWORTH, R. S., BOLDEN, R. C., SHAW, M. J., and TWIDDY, N. D., Chem. Phys. Lett. **5** (1970) 237.

or CO through a second side port. Both these gases react very slowly with Ne^+. Results obtained are given on p. 3632.

A further application of the flowing afterglow technique has been to measure the rates of three-body reactions as functions of temperature and pressure. Measurements of the variation with pressure at a fixed ion temperature are difficult with drift tube or other methods in which an electric field is present. In typical experiments using the apparatus of Dunkin et al.[†] rate coefficients for ionic association reactions with helium as the third body were measured over the temperature range from 82 to 280 °K at helium pressures from 0·1 to 3·0 torr. Results obtained are described on p. 3630.

Because of the growth of observational study of the D-region of the ionosphere using rocket-borne instruments, particularly mass-spectrometers to determine the positive and negative ion composition, there has been a rapid increase of interest in cluster formation by ions. The flowing afterglow technique has been adapted[‡] to the measurement of hydration rates, either bimolecular or termolecular, with such third bodies as He, N_2, and Ar. In some experiments the H_2O inlet was movable so the reaction rates concerned could be determined either as a function of water vapour concentration with a fixed reaction length or as a function of reaction length with fixed water vapour concentration. The water vapour concentration was determined by measuring, with a capacitance manometer, the increase in pressure due to addition of the water. It was verified that this is valid under the dynamical conditions of the afterglow experiments.

The measurement of rates of hydration of ions of atmospheric interest have also been carried out by Good, Durden, and Kebarle[§] using a quite different technique similar to that of Tal'roze and Frankevich (see Chap. 19, p. 2006). An electron beam of 4 keV energy was fired through a narrow slit into a cell containing the gas mixture under investigation. Ions formed could escape by diffusion through an exit slit in the centre of the cell, about 1 mm below the electron beam, to enter a quadrupole mass analyser through a set of focusing electrodes. The electron beam was pulsed, the duration being 10 μs, and ions observed during selected 10 μs intervals at different times t after the cessation of the electron pulse. The residence time in the reacting gas mixture is equal to $t-t'$

[†] loc. cit., p. 3615.
[‡] BOHME, D. K., DUNKIN, D. B., FEHSENFELD, F. C., and FERGUSON, E. E., J. chem. Phys. 51 (1969) 863.
[§] DURDEN, D. A., KEBARLE, P., and GOOD, A., ibid. 50 (1969) 805; GOOD, A., DURDEN, D. A., and KEBARLE, P., ibid. 52 (1970) 212 and 222.

where t' is the time for the ions to pass from the exit slit to the mass analyser. In experimental conditions $t' \ll t$ and could be estimated with sufficient accuracy from the electrode geometry and the applied voltage gradients. The total pressure in the ion source was measured with a capacitance manometer. Usually the reacting mixture included a known small concentration of water (0.3–10×10^{-3} torr) in a concentration of N_2 or O_2 of 1–4 torr pressure. To avoid disappearance of water vapour on the container walls a flow system was used in which the water vapour was fed into the main gas stream through a capillary from a small bulb containing liquid water. By controlling the bulb temperature and hence the water vapour pressure the flow rate through the capillary could be controlled and measured by determining at suitable intervals the loss of weight of the bulb. With the same apparatus it was also possible to study other ion association reactions. Results obtained, which usually agree well with those obtained using the flowing afterglow technique, are discussed on pp. 3628–30.

In Chapter 19, § 2.2.6, p. 1954 the application of ion cyclotron resonance techniques to the determination of ionic mobilities of ions was discussed. Further progress has been made recently through the development of transient rather than steady-state techniques.

Consider the motion of an average ion in a cell containing gas at a pressure p under the action of combined electric and magnetic fields $\mathbf{E}$, $\mathbf{H}$ respectively, where

$$\mathbf{E} = \{E_\mathrm{d} + E_\omega \sin(\omega t + \phi)\}\mathbf{i} + E_\mathrm{t}(z)\mathbf{k},$$

$$\mathbf{H} = H\mathbf{k}.$$

E_d is the field producing drift motion, E_t that which traps the ions. We suppose that the ions do not take part in any ionic reactions (apart from symmetrical charge transfer) and that loss to the walls is negligible. Under these conditions, if $\bar{\mathbf{v}}$ is the ion velocity then

$$\frac{\mathrm{d}\bar{\mathbf{v}}}{\mathrm{d}t} = \{e\mathbf{E}(t)/M_1\} + (e\bar{\mathbf{v}} \times \mathbf{H}/M_1) - \nu\bar{\mathbf{v}},$$

where M_1 is the ion mass. ν is the rate of momentum relaxation of the ion which is given by

$$\nu = \{M_2/(M_1 + M_2)\}\nu_\mathrm{c},$$

where ν_c is the momentum transfer collision frequency. When ν_c is independent of $\bar{\mathbf{v}}$, ν is simply equal to $e/M_1\mu$ where μ is the ion mobility in the gas under the experimental conditions.

It may then be shown that the instantaneous power absorption by

the ions is given by
$$P(t) = \bar{P} + P_1 e^{-\nu t}\sin\{(\omega_H - \omega)t + \gamma\}, \tag{10}$$
where
$$\bar{P} = \tfrac{1}{4}N^+ e^2 E_\omega^2 (\nu/M_1)\{\nu^2 + (\omega_H - \omega)^2\}^{-1} \tag{11a}$$
as in (25) p. 1955 and
$$P_1 = \tfrac{1}{2}N^+ e^2 E_\omega^2 (\omega_H/M_1)\{(\nu^2 + \omega_H^2 - \omega^2)^2 + 4\nu^2\omega^2\}^{-\frac{1}{2}}, \tag{11b}$$
$$\tan(\gamma - \pi) = 2\nu\omega_H(\nu^2 - \omega_H^2 + \omega^2)^{-1}. \tag{11c}$$

ω_H is the cyclotron angular frequency eH/M_1 and N^+ the number of ions in the cell.

If the ions are lost by ionic reactions at a rate k_r and to the walls at a rate k_w then we need only replace $P(t)$ in (10) by
$$P = e^{-kt}P(t) \tag{12}$$
where $k = k_r + k_w$.

At resonance when $\omega = \omega_H$ we have, when $\omega_H \gg \nu$,
$$P_r(t) = (N^+ e^2 E_r^2/4M_1\nu)e^{-kt}(1 - e^{-\nu t}). \tag{13a}$$

For short times t, P_r increases linearly with t so
$$P_r(t) = N^+ e^2 E_r^2 t/4M_1 = P'_r(0)t; \tag{13b}$$
while for t large, when k is negligible,
$$P_r(t) \sim N^+ e^2 E_r^2/4M_1\nu = P_r(\infty). \tag{13c}$$

Three techniques using these results have been developed. The first, due to Dunbar,[†] takes advantage of the form of (10), which is directly observed. Although, in principle, it is possible then to determine $\nu + k$ and k separately from the rate of decay of the envelope of the decaying oscillations and of the base line respectively, in practice the accuracy is only high enough to determine $\nu + k$.

The second, due to Huntress,[‡] measures the power absorption as a function of time under resonance conditions following the injection of a pulse of ions into the cyclotron resonance cell. This makes it possible to determine ν from the ratio $P'_r(0)/P_r(\infty)$ where $P'_r(0)$, $P_r(\infty)$ are as given in (13b) and (13c) respectively.

In both experiments the ions are injected over a finite time Δt which must be taken into account. However, under these circumstances $P(t)$ retains the form (10) but with $\bar{P}$, P_1, and γ replaced by different quantities independent of the time. Similarly (13b) is unchanged for $t > \Delta t$ and so also is (13c) provided $\nu\Delta t \ll 1$.

[†] DUNBAR, R. C., *J. chem. Phys.* **54** (1971) 711.
[‡] HUNTRESS, W. T., ibid. **55** (1971) 2146.

The third method, due to Lieder, Wien, and McIver,† depends on the direct measurement of the relaxation time for momentum transfer of the ions. A few μs after a pulse of ions is trapped in the cell an r.f. pulse is applied to the upper plate of the cell establishing a linear r.f. electric field perpendicular to the magnetic field. At the angular cyclotron frequency ω_H ions are accelerated in phase with the r.f. field so that, if this field is maintained for a time τ the ion will acquire a kinetic energy $(eE_r\tau)^2/8M_1$. After the field is removed the ion velocity relaxes according to the equation

$$\frac{d\bar{v}}{dt} = \frac{e}{M_1}(\bar{v}\times H) - \nu\bar{v}.$$

If the plates of the analyser cell are incorporated as a capacitance in a parallel LC circuit, resonant at the cyclotron frequency, the forced response of the circuit is an exponentially damped sinusoidal voltage with angular frequency ω_H and decay constant ν.

In all three methods the ion kinetic energy resulting from heating by the r.f. field can be calculated so that the measured values of ν and hence of ν_c apply to particular ion energies. An important advantage of the cyclotron resonance method is that the measurements can be made at ion energies very close to thermal.

All three of the methods outlined above have been applied to N_2^+ ions in N_2. Dunbar,‡ with the first method, determined ν as a function of pressure and thence obtained the collision frequency at s.t.p. for an average ion energy of about 0·02 eV, equivalent roughly to a value of F/p of 25 V cm^{-1} torr^{-1}. The derived value of the mobility was $1\cdot33\pm0\cdot2$ cm^{-2} V^{-1} s^{-1}, somewhat smaller than that, $1\cdot70$ cm^{-2} V^{-1} s^{-1}, observed by Moseley et al. (see p. 2036) by drift tube methods. Huntress§ obtained a mobility at thermal energies of $1\cdot90\pm0\cdot02$ cm^{-2} V^{-1} s^{-1} which agrees quite well with the drift measurement $1\cdot87$ cm^{-2} V^{-1} s^{-1} at zero field. Lieder et al.† obtained results in reasonably good agreement with Huntress who also measured the zero field mobilities of CO_2^+ in CO_2 and H_3^+ in H_2. No results obtained by other methods are available for CO_2 but the result obtained for H_3^+, $10\cdot9\pm0\cdot2$ cm^{-2} V^{-1} s^{-1}, agrees very well with the drift tube measurements of Albritton et al. (see p. 2030) which give $11\cdot1$ cm^{-2} V^{-1} s^{-1}.

These new techniques show considerable promise but need further application before they can be established effectively.

† LIEDER, C. A., WIEN, R. W., and McIVER, R. T., J. chem. Phys. 56 (1972) 5184. ‡ loc. cit., p. 3618.
§ loc. cit., p. 3618.

Flowing afterglow techniques[†] have been used to measure rates of such reactions as

$$Mg^{++} + A \to Mg^+ + A^+.$$

The helium buffer gas flowed past a furnace containing a sample of Mg or Ca. Metal atoms were carried downstream past an electron gun producing 100 eV electrons which created singly and doubly charged metallic ions. The decay of the latter due to addition of various reactants could then be obtained as usual. Results obtained are given on p. 3633.

Merging beam techniques are being applied to the measurement of ionic reaction rates (see p. 3631).

4. New results on mobilities and reaction rates of positive ions

4.1. *Ions in oxygen and in helium–oxygen mixtures* (see Chap. 19, § 3.5.3)

Further experiments on the mobilities of positive oxygen ions in oxygen have been carried out by McDaniel and his collaborators[‡] using equipment similar to that discussed on p. 2012. They identified O_4^+ in addition to O_2^+ and obtained mobilities at zero field of $2 \cdot 16 \pm 0 \cdot 07$ and $2 \cdot 24 \pm 0 \cdot 07$ cm^2 V^{-1} s^{-1} for the respective ions. The latter value agrees well with earlier results. No mass-analysed data are available on the mobility of O_4^+ but evidence for the presence of a second species of positive ions in oxygen has been deduced by Fleming and Rees[§] and by Dutton and Howells[||] from observations of the pressure dependence of the drift mobilities they observe. However, in both cases the mobility deduced for the second ion was slightly larger rather than smaller than that for the main ion. This may not be inconsistent because the observations were made at higher values of F/p and do not refer to the zero field mobility.

The rate of the association reaction

$$O_2^+ + 2O_2 \to O_4^+ + O_2, \tag{14}$$

which plays a special role in the reaction sequence which leads to $H^+(H_2O)_n$ ions in the D region of the ionosphere (see p. 3628), has been measured by Good, Durden, and Kebarle[††] at 300 °K as $2 \cdot 4 \times 10^{-30}$ cm^6 s^{-1}. The existence of O_4^+ ions had been established some time before by Yang and Conway[‡‡] using a mass spectrometer with a high pressure source.

[†] SPEARS, K. G., FEHSENFELD, G. C., McFARLAND, M., and FERGUSON, C. E., *J. chem. Phys.* **56** (1972) 2562.
[‡] SNUGGS, R. M., VOLZ, O. J., SCHUMMERS, J. H., MARTIN, D. W., and McDANIEL, E. W., *Phys. Rev.* **A3** (1971) 477.
[§] FLEMING, I. A. and REES, J. A., *J. Phys. B (GB)* **2** (1969) 423.
[||] DUTTON, J. and HOWELLS, P., ibid. **1** (1968) 1160.
[††] loc. cit., p. 3616.
[‡‡] YANG, J. H. and CONWAY, D. C., *J. chem. Phys.* **40** (1964) 1729.

The equilibrium constant for the reaction which they determined† agrees very well with that obtained by Good et al. and yields a dissociation energy of O_4^+ into $O_2^+ + O_2$ close to 0·4 eV.

Bohme et al.,‡ using the flowing afterglow method, measured the rate coefficient at 82 °K of the reaction in which O_2 was replaced by helium as third body and obtained $3·1 \times 10^{-29}$ cm^6 s^{-1} at pressures below 1·6 torr. Saturation was observed at higher pressures at an apparent bimolecular rate near 5×10^{-12} cm^3 s^{-1}. With large inlet flow of O_2 into the reaction tube heavy complex ions O_6^+ and O_{10}^+ could be detected.

The reaction
$$O_4^+ + O \rightarrow O_2^+ + O_3 \tag{15}$$
has been studied by Fehsenfeld and Ferguson,§ using the afterglow technique and is found to have a rate constant $3 \pm 2 \times 10^{-10}$ cm^3 s^{-1}, which is significant in connection with the height distribution of $H_5O_2^+$ ions in the lower ionosphere (see p. 3628).

New data on the rates of the reactions
$$He^+ + O_2 \rightarrow O^+ + O + He, \tag{16a}$$
$$\rightarrow O_2^+ + He, \tag{16b}$$
have been obtained by Heimerl, Johnsen, and Biondi,‖ using the drift tube technique described on p. 3613. As the mobility of O^+ ions in He is considerably higher than that of He^+ (due to charge exchange in the latter case) the arrival spectrum technique can be applied. Production of O^+ ions was estimated to occur in 80 per cent of the reactions. The total rate coefficient was measured at 300 °K as $8·5^{+2·5}_{-2·0} \times 10^{-10}$ cm^3 s^{-1}, agreeing reasonably well with earlier data (see Chap. 19, pp. 2049–50) and with new flowing afterglow measurements by Bolden et al.†† (9×10^{-10} cm^3 s^{-1}) and by Dunkin et al.‡‡ ($10·5 \times 10^{-10}$ cm^3 s^{-1}). No variation with ion energy was detected over the range from 0·04 to 0·12 eV.

4.2. Ions in nitrogen and in nitrogen–helium mixtures (see Chap. 19, § 3.5.4)

New measurements of the rates of the association reactions (see Chap. 19, pp. 2041–2)
$$N_2^+ + N_2 + N_2 \rightarrow N_4^+ + N_2, \tag{17}$$
$$N^+ + N_2 + N_2 \rightarrow N_3^+ + N_2, \tag{18}$$

† Conway, D. C. and Yang, J. H., ibid. **43** (1965) 2900. ‡ loc. cit., p. 3616.
§ Fehsenfeld, F. C. and Ferguson, E. E., *Radio Science* **7** (1972) 113.
‖ loc. cit., p. 3613.
†† Bolden, R. C., Hemsworth, R. S., Shaw, M. J., and Twiddy, N. D., *J. Phys. B* (*GB*) **3** (1970) 45.
‡‡ Dunkin, D. B., Fehsenfeld, F. C., Schmeltekopf, A. L., and Ferguson, E. E., *J. chem. Phys.* **49** (1968) 1365.

have been made by Good et al.† They find, at 300 °K, 8×10^{-29} and 5×10^{-29} cm^{-6} s^{-1} for the respective rates. The first agrees well with the earlier observations of Warneck (p. 2042). The second is somewhat larger than observed by Moseley et al. (p. 2042) from drift tube measurements but is closer to the more recent measurements, with similar technique, of McKnight, McAffee, and Dipler‡ (3×10^{-29} cm^6 s^{-1}).

Bohme et al. measured the rates of the reactions (17) and (18) with N_2 replaced as third body by He. At 82 °K they found $1 \cdot 2 \times 10^{-28}$ and $7 \cdot 2 \times 10^{-29}$ cm^6 s^{-1} for formation of N_4^+ and N_3^+ respectively. The first saturated at a pressure of 0·6 torr to an apparent bimolecular rate of 7×10^{-12} cm^3 s^{-1} but the second remained termolecular up to 2 torr. At 280 °K the termolecular rates fell off to $1 \cdot 9 \times 10^{-29}$ and $8 \cdot 6 \times 10^{-30}$ cm^6 s^{-1}.

New measurements of rate coefficients for the reactions

$$He^+ + N_2 \rightarrow He + N + N^+, \tag{19a}$$

$$\rightarrow He + N_2^+, \tag{19b}$$

have been made by Dunkin et al.§ and Bolden et al.,∥ using the flowing afterglow technique, and by Heimerl et al.†† and by Ong and Hasted,‡‡ using drift tube techniques. At thermal energies the values found for the overall rate coefficient by these respective authors are, in 10^{-10} cm^3 s^{-1}, 12·0, 12·5, 10·0, and 15·0, which fall within the range of earlier measurements (see Table 19.16). The branching ratio, in favour of production of N^+, is 55:45 according to Heimerl et al., which is consistent with the earlier measurement of Warneck (see p. 2051) but disagrees with Ong and Hasted who find a ratio of 1:2.

The evidence concerning the variation of the rates with ion energy and with gas temperature remains obscure. Heimerl et al., using the techniques described on p. 3613, find little dependence of the reaction rate on ion energy whereas Ong and Hasted find that the rate of (19) increases by a factor of about 3 as the ion energy increases from 0·04 to 0·1 eV. Again, Dunkin et al., who obtained data for gas temperatures up to 600 °K using a heated flow tube, observed no appreciable change of rate constant k with temperature, in conflict with the earlier measurements of Sayers and Smith with a static afterglow system (see p. 2052),

† loc. cit., p. 3616.
‡ McKnight, L. C., McAffee, K. B., and Dipler, D. P., *Phys. Rev.* **164** (1967) 62.
§ Dunkin, D. B., Fehsenfeld, F. C., Schmeltekopf, A. L., and Ferguson, E. E., *J. chem. Phys.* **49** (1968) 1365.
∥ Bolden, R. C., Hemsworth, R. S., Shaw, M. J., and Twiddy, N. D., *J. Phys. B (GB)* **3** (1970) 45. †† loc. cit., p. 3613.
‡‡ Ong, P. P. and Hasted, J. B., *J. Phys. B (GB)* **2** (1969) 91.

who observed k to decrease from 1·75 to $1·03 \times 10^{-9}$ cm^3 s^{-1} as the temperature rose from 200 to 500 °K.

4.3. Other reactions of He$^+$ ions (see Chap. 19, § 3.5.5)

Bolden et al.† have measured rate coefficients for reactions of He$^+$ with CO, NO, and CO$_2$, obtaining values, in 10^{-10} cm^3 s^{-1}, of 14·0, 12·5, and 8·5 as compared with earlier flowing afterglow results (see p. 2052) of 17, 15, and 11. The same authors have also measured rate coefficients for reactions with CH$_4$ and NH$_3$ (13·0 and $11·5 \times 10^{-10}$ cm^3 s^{-1} respectively).

4.4. Ions in nitrogen–oxygen mixtures (see Chap. 19, § 3.5.6)

New data on the reactions

$$N^+ + O_2 \to NO^+ + O, \qquad (20\,a)$$
$$\to N + O_2^+, \qquad (20\,b)$$
and
$$N_2^+ + O_2 \to N_2 + O_2^+, \qquad (21)$$

have been obtained by Johnsen, Brown, and Biondi‡ using the drift tube technique described on p. 3613. As the mobilities of the ions involved are not very different in the buffer gas the additional residence time procedure was adopted.

Comparable rates were found for the two branches of (20). The total rate coefficient obtained was 5×10^{-10} cm^3 s^{-1}, independent of ion energy over the observed range 0·03 to 0·8 eV. This agrees well with the flowing afterglow experiments of Dunkin et al. who obtained a mean rate constant of 6×10^{-10} cm^3 s^{-1}, nearly constant over a temperature range from 300 to 500 °K. These results are also consistent with earlier measurements (see Table 19.16) with the likely exception of the considerably smaller value measured by Warneck for an energy of 0·15 eV.

For (21) the measurements are difficult because of the slow rate. Johnsen et al.‡ observed a rate coefficient decreasing from 6×10^{-11} to 8×10^{-12} cm^3 s^{-1} as the ion energy increased from 0·039 to 1 eV. Dunkin et al.,§ using their temperature-controlled flowing afterglow tube, found a decrease with increasing temperature from 77 to 550 °K, and there seems to be a reasonable consistency between the results when account is taken of the different velocity distribution of the ions in the two experiments. The absolute value of the rate coefficient at 300 °K observed by Johnsen et al. (6×10^{-11} cm^3 s^{-1}) is somewhat larger than

† loc. cit., p. 3622.
‡ JOHNSEN, R., BROWN, H. L., and BIONDI, M. A., J. chem. Phys. **52** (1970) 5080.
§ loc. cit., p. 3615.

that measured by Dunkin et al. (4.7×10^{-11} cm³ s⁻¹). The consistency with earlier experiments, which agree well among themselves, is poor. Even as regards variation with ion energy, Warneck† observed no change in rate coefficient at temperatures between 400 and 800 °K.

The rate of the reaction

$$N_2^+ + O \to NO^+ + N \tag{22}$$

has been remeasured by Fehsenfeld, Dunkin, and Ferguson‡ who find a rate coefficient of 1.4×10^{-10} cm³ s⁻¹, somewhat smaller than in their earlier afterglow measurements (2.5×10^{-10} cm³ s⁻¹, see Table 19.16).

Dunkin et al.,§ using the apparatus referred to on p. 3615, have measured the rate constant for

$$O^+ + N_2 \to NO^+ + N, \tag{23}$$

as a function of temperature between 300 and 600 °K. At 300 °K they obtain a rate coefficient k of 1.2×10^{-12} cm³ s⁻¹ which is a little smaller than any previous value (see Table 19.16). k was found to fall gradually to about one-half at 600 °K.

4.5. *Reactions of oxygen and nitrogen ions with* NO (see Chap. 19, § 3.5.7)

$$O_2^+ + NO \to O_2 + NO^+. \tag{24}$$

Johnsen et al.∥ observed a rate coefficient of 3×10^{-10} cm³ s⁻¹ for (24) independent of ion energy over the range 0·039 to 1·5 eV. Fehsenfeld et al.,‡ in a remeasurement of the coefficient, find 6.3×10^{-10} cm³ s⁻¹. Earlier measurements (see Table 19.16) gave somewhat smaller values.

$$O^+ + NO \to NO^+ + O \tag{25}$$

is a very slow reaction. Dunkin, McFarland, Fehsenfeld, and Ferguson†† have placed a smaller limit, 1.3×10^{-12} cm³ s⁻¹, on its value than available from earlier experiments (see Table 19.16).

$$N_2^+ + NO \to NO^+ + N_2. \tag{26}$$

The rate of this reaction has been remeasured by Fehsenfeld, Dunkin, and Ferguson‡ as 3.3×10^{-10} cm³ s⁻¹, a little smaller than in earlier measurements (see Table 19.16).

† loc. cit., p. 2057.
‡ FEHSENFELD, F. C., DUNKIN, D. B., and FERGUSON, E. E., *Planet. Space Sci.* **18** (1970) 1267.
§ DUNKIN, D. B., FEHSENFELD, F. C., SCHMELTEKOPF, A. L., and FERGUSON, E. E., *J. chem. Phys.* **49** (1968) 1365.
∥ loc. cit., p. 3623.
†† DUNKIN, D. B., MCFARLAND, M., FEHSENFELD, F. C., and FERGUSON, E. E., *J. geophys. Res.* **76** (1971) 3820.

4.6. *Reactions of oxygen with* CO_2 (see Chap. 19, § 3.5.9), N_2O, *and* NO_2
Johnsen et al.† have measured the rate of
$$O^+ + CO_2 \rightarrow O_2^+ + CO, \tag{27}$$
using the drift tube technique described on p. 36, and obtained a value of 1×10^{-9} cm³ s⁻¹ at 300 °K. Assuming a very sharp velocity distribution of the ions the rate coefficient varied as $v^{-1 \cdot 6 \pm 0 \cdot 2}$, where v is the mean ion velocity.

With the temperature-controlled flowing afterglow equipment Dunkin et al.‡ found $1 \cdot 1 \times 10^{-9}$ cm³ s⁻¹ at 300 °K, decreasing to about $0 \cdot 8 \times 10^{-9}$ at 600 °K. An earlier flowing afterglow measurement (see Table 19.17) gave $1 \cdot 2 \times 10^{-9}$ at 300 °K.

Dunkin, McFarland, Fehsenfeld, and Ferguson§ have used the flowing afterglow technique to measure the rates of
$$O^+ + N_2O \rightarrow N_2O^+ + O, \tag{28}$$
$$O^+ + NO_2 \rightarrow NO_2^+ + O, \tag{29}$$
at 393 °K, obtaining rate coefficients of $6 \cdot 3 \times 10^{-10}$ and $1 \cdot 6 \times 10^{-9}$ cm³ s⁻¹ respectively.

4.7. *Other reactions of atmospheric interest*

4.7.1. *Reactions of* H^+ *with* O, NO, *and* CO_2. Fehsenfeld and Ferguson‖ have used the flowing afterglow technique to measure rate coefficients for the reactions
$$H^+ + O \rightarrow O^+ + H, \tag{30}$$
$$H^+ + NO \rightarrow NO^+ + H. \tag{31}$$

The H^+ ions were produced by adding a large flow of H_2 to the flow of the argon buffer gas past the electron gun. Many other ions are produced in addition to H^+ but none of these would be expected to yield H^+ in any reactions down the tube. The atomic oxygen was produced as described on p. 2024.

The rate coefficient was measured as $3 \cdot 75 \times 10^{-10}$ cm³ s⁻¹ for (30) and $1 \cdot 9 \times 10^{-9}$ for (31). For the inverse reaction
$$O^+ + H \rightarrow H^+ + O \tag{32}$$
the rate is deduced, through calculation of the thermodynamic equilibrium constant, to be $6 \cdot 8 \times 10^{-10}$ cm². The corresponding mean cross-section $2 \cdot 7 \times 10^{-15}$ cm² is consistent with extrapolation of the measurements of Stebbings and Rutherford using a beam technique (see Chap. 24, p. 2786).

† loc. cit., p. 3623.   ‡ loc. cit., p. 3622.   § loc. cit., p. 3624.
‖ FEHSENFELD, F. C. and FERGUSON, E. E., *J. chem. Phys.* **56** (1972) 3066.

The rate of the reaction

$$H^+ + CO_2 \rightarrow COH^+ + O \tag{33}$$

has been measured by Fehsenfeld and Ferguson[†] as 3×10^{-9} cm^3 s^{-1} at 300 °K.

4.7.2. *Associative reactions of* NO^+ *with* N_2, O_2, *and* CO_2. Dunkin, Fehsenfeld, Schmeltekopf, and Ferguson[‡] have applied the flowing afterglow technique to measure the rates of a number of associative reactions of NO^+ with different molecules. Results obtained are given in Table 19 A.2.

TABLE 19 A.2

Rates of NO^+ *associative reactions measured by Dunkin et al.*

	T (°K)	Rate constant (cm^6 s^{-1})
$NO^+ + N_2 + He \rightarrow NO^+ . N_2 + He$	200	$< 5 \times 10^{-33}$
$NO^+ + O_2 + He \rightarrow NO^+ . O_2 + He$	200	$< 6 \times 10^{-34}$
$NO^+ + CO_2 + He \rightarrow NO^+ . CO_2 + He$	197	$1 \cdot 0 \pm 0 \cdot 3 \times 10^{-29}$
$NO^+ + CO_2 + N_2 \rightarrow NO^+ . CO_2 + N_2$	200	$2 \cdot 5 \pm 1 \cdot 5 \times 10^{-29}$
$NO^+ + CO_2 + Ar \rightarrow NO^+ . CO_2 + Ar$	196	$3 \cdot 1 \pm 1 \cdot 0 \times 10^{-29}$

4.7.3. *Reaction of* CO^+ *with* O *and* NO. Fehsenfeld and Ferguson[§] have measured the rate coefficients of the reaction

$$CO^+ + O \rightarrow O^+ + CO + 0 \cdot 53 \text{ eV}, \tag{34}$$

$$CO^+ + NO \rightarrow NO^+ + CO - 4 \cdot 75 \text{ eV}, \tag{35}$$

obtaining the respective values $1 \cdot 4 \times 10^{-10}$ and $3 \cdot 3 \times 10^{-10}$ cm^3 s^{-1}.

4.7.4. *Reactions of* CO_2^+ *with* O, O_2, NO, *and* H. Fehsenfeld, Dunkin, and Ferguson,[||] using the flowing afterglow technique, have measured rate coefficients for reactions of CO_2^+ with O, O_2 and NO.

For the reactions
$$CO_2^+ + O \rightarrow O_2^+ + CO, \tag{36a}$$
$$\rightarrow O^+ + CO_2, \tag{36b}$$

a total rate coefficient of $2 \cdot 6 \times 10^{-10}$ cm^3 s^{-1} was measured. The branching ratio between the two possible reactions was difficult to determine but the ratio of O_2^+ to O^+ production was found to be around 5:3.

The rate coefficient for the charge-transfer reaction with O_2 was remeasured as $5 \cdot 0 \times 10^{-11}$ cm^3 s^{-1}, considerably smaller than in earlier measurements.

[†] FEHSENFELD, F. C. and FERGUSON, E. E., *J. geophys. Res.* **76** (1971) 8453.
[‡] DUNKIN, D. B., FEHSENFELD, F. C., SCHMELTEKOPF, A. L., and FERGUSON, E. E., *J. chem. Phys.* **54** (1971) 3817.
[§] loc. cit., p. 3625.   [||] loc. cit., p. 3624.

For the charge-transfer reaction with NO the measured rate coefficient was $1 \cdot 2 \times 10^{-10}$ cm^3 s^{-1}.

The reactions
$$CO_2^+ + H \to HCO^+ + O, \qquad (37\text{a})$$
$$\to H^+ + CO_2, \qquad (37\text{b})$$

have been studied by Fehsenfeld and Ferguson,[†] who obtained an overall rate coefficient of 6×10^{-10} cm^3 s^{-1} at 300 °K.

4.7.5. *Charge-transfer reaction between NO_2^+ and NO.* Fehsenfeld, Ferguson, and Mosesman,[‡] using the flowing afterglow technique, measured the rate coefficient of the charge-transfer reaction between NO_2^+ and NO at 300 °K as $2 \cdot 9 \times 10^{-10}$ cm^3 s^{-1}. This shows the reaction to be exothermic so the ionization potential of NO_2 must be greater than that of NO.

4.7.6. *Water cluster ions.* Rocket experiments[§] have shown that the main positive ion present in the D region of the terrestrial ionosphere at altitudes below 82 km is the doubly-hydrated proton, $H_3O^+H_2O$, while H_3O^+ and $H_3O^+(H_2O)_2$ have also been observed.

The reaction sequence leading to the production of these ions is not immediately obvious. In the laboratory the initial reaction which leads to formation of such clustered ions is

$$H_2O^+ + H_2O \to H_3O^+ + OH \qquad (38)$$

which has a rate constant $\simeq 10^{-9}$ cm^3 s^{-1}.[||] However, in the ionosphere there is insufficient H_2O^+ for this to be effective. This is because the reaction

$$H_2O^+ + O_2 \to O_2^+ + H_2O \qquad (39)$$

is quite fast, with a rate constant 2×10^{-10} cm^3 s^{-1}[††] so that, because of the great excess of O_2 as compared with H_2O, the reaction (38) would be very improbable.

Independently, Fehsenfeld and Ferguson[‡‡] and Good, Durden, and Kebarle[§§] suggested that the reaction chain which leads to the hydrated protons begins either with

---

[†] loc. cit., p. 3626.
[‡] FEHSENFELD, F. C., FERGUSON, E. E., and MOSESMAN, M., *Chem. Phys. Lett.* **4** (1969) 73.
[§] NARCISI, R. S. and BAILEY, A. D., *J. geophys. Res.* **70** (1965) 3687.
[||] TAL'ROZE, V. L. and FRANKEVICH, E. L., *Zh. Fiz. Khim.* **34** (1960) 2709; see also p. 2006.
[††] FEHSENFELD, F. C., SCHMELTEKOPF, A. L., and FERGUSON, E. E., *J. chem. Phys.* **46** (1967) 2802.
[‡‡] FEHSENFELD, F. C. and FERGUSON, E. E., *J. geophys. Res.* **74** (1969) 2217.
[§§] GOOD, A., DURDEN, D. A., and KEBARLE, P., *J. chem. Phys.* **52** (1970) 222.

	(a)	(b)	
$O_2^+ + O_2 + O_2 \to O_4^+ + O_2,$	—	$2\cdot4 \times 10^{-30}$	(40)
$O_4^+ + H_2O \to O_2^+ \cdot H_2O + O_2,$	$2\cdot2 \times 10^{-9}$	$1\cdot3 \times 10^{-9}$	(41)

or

	(a)	(b)	
$O_2^+ + H_2O + M \to O_2^+ \cdot H_2O + O_2,$	$0\cdot9, 2\cdot8, 2\cdot0 \times 10^{-28}$		(42)

followed by

	(a)	(b)	
$O_2^+ \cdot H_2O + H_2O \to H_3O^+ + OH + O_2$	$\leqslant 0\cdot3 \times 10^{-9}$	$0\cdot3 \times 10^{-9}$	(43)
$\to H_3O^+ \cdot OH + O_2$	$1\cdot9 \times 10^{-9}$	$0\cdot9 \times 10^{-9}$	(44)

and

	(a)	(b)	
$H_3O^+ \cdot OH + H_2O \to H_3O^+ \cdot H_2O + OH$	3×10^{-9}	$\geqslant 10^{-9}.$	(45)

A further three-body association

$$H_3O^+ \cdot H_2O + H_2O + M \to H_3O^+ \cdot (H_2O)_2 + M, \qquad (46)$$

produces triply-hydrated protons.

The rates of the reactions (40) to (45) have been investigated by Fehsenfeld, Mosesman, and Ferguson,† using the flowing afterglow technique, and by Good, Durden, and Kebarle‡ with the method described on p. 3616. Their results are summarized under (a) and (b) respectively above. The units are cm³ s⁻¹ for two-body and cm⁶ s⁻¹ for three-body reactions. For (42) the values given under (a) are for He, N_2, and Ar as third bodies respectively.

Provided there is a sufficiently strong source of O_2^+ ions, the measured rates of the different reactions, which agree remarkably well, are consistent with $H_3O^+ \cdot H_2O$ being the major component of the positive ionization in the D region. The rapid fall-off in the concentration of this ion above 80 km is probably due to the fast reaction of O_4^+ with atomic oxygen (see (15), p. 3621).

Fehsenfeld, Mosesman, and Ferguson† have also used the flowing afterglow technique to investigate the reaction sequence which occurs in ionized $NO-H_2O$ mixtures which is as follows:

$$NO^+ + H_2O + M \to NO^+ \cdot H_2O + M, \qquad k_1 \qquad (47)$$

$$NO^+ \cdot H_2O + H_2O + M \to NO^+ \cdot (H_2O)_2 + M, \qquad k_2 \qquad (48)$$

$$NO^+ \cdot (H_2O)_2 + H_2O + M \to NO^+ \cdot (H_2O)_3 + M, \qquad k_3 \qquad (49)$$

$$NO^+ \cdot (H_2O)_3 + H_2O \to H_3O^+ \cdot (H_2O)_2 + HNO_2, \qquad k_4 \qquad (50)$$

† Fehsenfeld, F. C., Mosesman, M., and Ferguson, E. E., *J. chem. Phys.* **55** (1971) 2115.
‡ Good, A., Durden, D. A., and Kebarle, P., loc. cit., p. 3627.

leading ultimately to the production of nitrous acid and triply-hydrated protons. These reactions may well be important in the D region in addition to the oxygen sequence (40–5) because NO$^+$ is the major unclustered ion produced. Reaction rates were measured with He, Ar, and N$_2$ as third bodies. Lineberger and Puckett† and later Puckett and Teague‡ have studied the same reaction sequence in a static afterglow (see Chap. 19, p. 2018) and have identified HNO$_2$ as a product from their mass analysis. In this work the third body M is NO which should not behave very differently in this respect from N$_2$.

Table 19 A.3 summarizes the measured values of the rate coefficients.

TABLE 19 A.3

Rates of reactions in NO hydration sequence

Third body	He	Ar	N$_2$	NO
k_1 (cm^6 s^{-1})	$3 \cdot 6 \times 10^{-29}$	$7 \cdot 8 \times 10^{-29}$	$1 \cdot 6 \times 10^{-28}$	$1 \cdot 6 \times 10^{-28}$
	$(3 \cdot 2 \times 10^{-29})$	$(6 \cdot 5 \times 10^{-29})$	$(1 \cdot 4 \times 10^{-28})$	
k_2 (cm^5 s^{-1})	$3 \cdot 0 \times 10^{-28}$	$8 \cdot 0 \times 10^{-28}$	$1 \cdot 0 \times 10^{-27}$	$1 \cdot 1 \times 10^{-27}$
k_3 (cm^6 s^{-1})	$4 \cdot 0 \times 10^{-28}$	$1 \cdot 5 \times 10^{-27}$	$2 \cdot 0 \times 10^{-27}$	$1 \cdot 9 \times 10^{-27}$
k_4 (cm^3 s^{-1})	8×10^{-11}	8×10^{-11}	8×10^{-11}	7×10^{-11}

Values for He, Ar, and N$_2$ as third bodies, from Fehsenfeld, Mosesman, and Ferguson (loc. cit) for NO from Puckett and Teague (loc. cit.). Bracketed values for k_1 are taken from HOWARD, C. J., RUNDLE, H. W., and KAUFMAN, F., *J. chem. Phys.* **55** (1971) 4772.

A further possibility of producing NO$^+$.H$_2$O in the lower ionosphere is through the pair of reactions

$$\text{NO}^+ + \text{CO}_2 + \text{M} \rightarrow \text{NO}^+.\text{CO}_2 + \text{M}, \tag{51}$$

$$\text{NO}^+.\text{CO}_2 + \text{H}_2\text{O} \rightarrow \text{NO}^+.\text{H}_2\text{O} + \text{CO}_2. \tag{52}$$

The first of these is quite fast (see Table 19 A.2), while Dunkin et al.§ find that the second is also fast with a rate constant of order 10^{-9} cm^3 s^{-1}. It is likely that NO$^+$.H$_2$O is produced at a faster rate from these two reactions than through the direct association (47).

Good, Durden, and Kebarle‖ have also studied hydration reactions associated with N$_2^+$. Beginning with the formation of N$_4^+$ through the reaction (17) (see p. 3621) they determined the rates of

$$\text{N}_4^+ + \text{H}_2\text{O} \rightarrow \text{H}_2\text{O}^+ + 2\text{N}_2, \tag{53}$$

$$\text{H}_2\text{O}^+ + \text{H}_2\text{O} \rightarrow \text{H}_3\text{O}^+ + \text{OH}, \tag{54}$$

$$\text{H}_3\text{O}^+ + \text{H}_2\text{O} + \text{N}_2 \rightarrow \text{H}^+(\text{H}_2\text{O})_2 + \text{N}_2, \tag{55}$$

† LINEBERGER, W. C. and PUCKETT, L. J., *Phys. Rev.* **187** (1969) 286.
‡ PUCKETT, L. J. and TEAGUE, M. W., *J. chem. Phys.* **54** (1971) 2564.
§ loc. cit., p. 3626.   ‖ loc. cit., p. 3616.

$$H^+.(H_2O)_2+H_2O+N_2 \to H^+.(H_2O)_3+N_2, \qquad (56)$$
$$H^+.(H_2O)_3+H_2O+N_2 \to H^+.(H_2O)_4+N_2. \qquad (57)$$

For (53) and (54) they found rate coefficients of $1 \cdot 9 \times 10^{-9}$ and $1 \cdot 8 \times 10^{-9}$ cm^3 s^{-1} respectively and for (55), (56), and (57): $3 \cdot 4 \times 10^{-27}$, $2 \cdot 3 \times 10^{-27}$, and $2 \cdot 4 \times 10^{-27}$ cm^6 s^{-1} respectively. Although N_2 is the most abundant neutral species in the lower atmosphere this sequence is of no importance there because N_2^+ initially produced is rapidly converted to O_2^+ by the charge-exchange reaction

$$N_2^+ + O_2 \to N_2 + O_2^+. \qquad (58)$$

4.7.7. *Other cluster reactions.* Adams, Bohme, Dunkin, Fehsenfeld, and Ferguson† have applied their temperature-controlled flowing afterglow equipment to study a number of termolecular cluster-forming reactions of O_2^+ and of O_4^+ ions with He as the third body. The rate coefficients observed for production of $O_2^+.H_2$, $O_2^+.N_2$, $O_2^+.CO_2$, $O_2^+.N_2O$, $O_2^+.SO_2$, and $O_2^+.H_2O$ clusters in collision with the appropriate neutral molecule were measured, in 10^{-29} cm^6 s^{-1}, as $0 \cdot 074$, $1 \cdot 9$, $2 \cdot 3$, $5 \cdot 2$, 60, and $8 \cdot 5$ respectively, the temperature being 80 °K for the first two reactions and 200 °K for the others.

For formation of $O_4^+.N_2$ and $O_4^+.O_2$ by collision of O_4^+ with N_2 and O_2 respectively, in the presence of He at 80 °K, coefficients of 10^{-29} and $5 \cdot 0 \times 10^{-30}$ cm^6 s^{-1} were measured while that for the reaction

$$O_2^+.N_2 + N_2 + He \to O_2^+.N_4 + He, \qquad (59)$$

again at 80 °K, was found to be $\simeq 10^{-29}$ cm^6 s^{-1}.

In addition to these cluster-formation reactions, rates were also measured for termolecular cluster-exchange reactions, the observed rate coefficient being as given in Table 19 A.4. These results show that the order of increasing strength of bonding of different clusters to O_2^+ is N_2, O_2, N_2O, H_2O. This is not the same order as for clusters to O_2^- (see p. 3639).

4.7.8. *Reactions of* Si$^+$ *and* SiO$^+$ *in atmospheric gases.* Fehsenfeld‡ has used the flowing afterglow technique to investigate the rates of the reactions

$$Si^+ + O_2 \to SiO^+ + O, \qquad k_1 \qquad (60)$$
$$SiO^+ + N \to Si^+ + NO, \qquad k_{2a} \qquad (61\,a)$$
$$\to NO^+ + Si, \qquad k_{2b} \qquad (61\,b)$$
$$SiO^+ + O \to Si^+ + O_2. \qquad k_3 \qquad (62)$$

---

† ADAMS, N. G., BOHME, D. K., DUNKIN, D. B., FEHSENFELD, F. C., and FERGUSON, E. E., *J. chem. Phys.* **52** (1970) 1951.

‡ FEHSENFELD, F. C., *Can. J. Chem.* **47** (1969) 1808.

SiO⁺ ions were produced by passing buffer argon gas over a few grammes of SiO placed in a furnace upstream of the electron gun. Downstream, argon ions and metastable atoms produced SiO⁺ from the SiO. If a helium buffer is used Si⁺ is produced from the dissociative ionization processes

$$SiO + He^+ \rightarrow Si^+ + O + He, \qquad (63)$$

$$SiO + He \rightarrow Si^+ + O + He + e. \qquad (64)$$

k_1 was found to be $< 10^{-11}$ cm³ s⁻¹, $k_{2a} + k_{2b} \simeq 3 \times 10^{-10}$ cm³ s⁻¹, $k_{2a} \simeq 2k_{2b}$, and $k_3 \simeq 2 \times 10^{-10}$ cm³ s⁻¹, all measured at room temperature.

TABLE 19 A.4
Observed rate constants for cluster exchange reactions involving O_2^+ ions

Cluster exchange	$N_2 \rightarrow O_2$	$O_2 \rightarrow N_2O$	$O_2 \rightarrow SO_2$	$N_2O \rightarrow SO_2$		$N_2O \rightarrow H_2O$	$SO_2 \rightarrow H_2O$
Rate coefficent (10^{-10} cm³ s⁻¹)	⩾ 0·5	2·5	8·2	4·2	5·6	> 1·0	> 1·0
Temperature (°K)	80	200	200	300	200	200	200

4.7.9. Reactions between metal atoms and atmospheric ions. Merging beam techniques are being used to investigate the rates of reaction of Ba atoms with O_2^+, N_2^+, and NO⁺ ions[†] and of Na with CO⁺ ions.[‡] In interpreting these experiments account has to be taken of the fact that the vibrational distribution in the reacting molecular ions depends on their mode of production and, in the case of O_2^+, metastable ions in the $^4\Pi_u$ state are also likely to be present.

For Ba the reaction cross-sections are of order 10^{-15} cm² at a C.M. energy E of 0·1 eV, and decrease at a rate between $E^{-\frac{1}{2}}$ and E^{-1} as the energy is increased. The measured energy distributions of the product ions suggest that most of the product ions are projected in the direction of the reactant Ba atoms. This shows that the reactions do not proceed primarily through formation of a collision complex.

A number of rate coefficients for reactions of metallic ions with atmospheric gases have been measured at thermal energies.

[†] NEYNABER, R. H., TRUJILLO, S. M., and MAGNUSON, G. D., *7th int. conf. on the physics of electronic and atomic collisions*, Amsterdam. Abstracts, North Holland, 1971, p. 987.
[‡] WENDELL, K. L. and ROL, P. K., ibid. p. 989.

Farragher, Peden, and Fite[†] have studied the charge-transfer reactions between Na atoms and N_2^+, NO^+, and O_2^+ ions, obtaining rate coefficients, in 10^{-10} cm³ s⁻¹, of 5·8, 0·7, and 6·7 respectively.

Ferguson and Fehsenfeld[‡] have measured the rates of reactions with ozone
$$M^+ + O_3 \to MO^+ + O_2, \qquad (65)$$
where M is Mg, Ca, Fe, Na, and K, the rate coefficients found being, in 10^{-10} cm³ s⁻¹, 2·3, 1·6, 1·5, < 0·1, and < 0·1 respectively. The same experimenters have measured rates for the three-body reaction
$$M^+ + O_2 + Ar \to MO_2^+ + Ar, \qquad (66)$$
where M is Mg, Ca, Fe, Na, and K. For these respective cases the rate coefficients, in 10^{-30} cm⁶ s⁻¹, were found to be 2·5, 6·6, 1·0, < 0·2, and < 0·2. For Ca⁺, with Ar replaced by He as third body, the rate coefficient fell to 2×10^{-30} cm⁶ s⁻¹. Keller and Beyer,[§] using a drift tube technique, find that, with O_2 as third body and M = Na, the rate coefficient is small, around 5×10^{-32} cm⁶ s⁻¹.

Keller and Beyer[§] have investigated the rates of associative reactions of Na⁺ and K⁺ with CO_2, with CO_2 as third body, and find coefficients 2×10^{-29} and 4×10^{-30} cm⁶ s⁻¹ respectively.

4.8. *Reactions of* Ar⁺ *and* Ne⁺ *ions* (see Chap. 19, § 3.5.11)

A number of new measurements of rate coefficients for reactions of Ar⁺ with CO, N_2, O_2, and NO have been made by flowing afterglow (Adams et al.[||]), static afterglow (for Ar⁺–O_2, Smith et al.[††]), and drift tube (Ong and Hasted,[‡‡] Kaneko et al.,[§§] Birkinshaw and Hasted[||||]) techniques. Considerable divergences still exist (see Birkinshaw and Hasted).[||||]

Hemsworth et al.[†††] have measured rate coefficients for reactions of Ne⁺ (see p. 3615) with Xe, N_2, O_2, H_2, CO, NO, CO_2, NH_3, CH_4, C_2H_6, C_3H_8, and C_4H_{10}, obtaining the respective values, in 10^{-10} cm³ s⁻¹, of $< 5 \times 10^{-3}$, $< 10^{-2}$, 0·55, $\leqslant 3 \times 10^{-3}$, $< 10^{-3}$, 1·2, 0·5, 2·0, 0·09, 6·0, 9·5, and 8·7.

[†] FARRAGHER, A. L., PEDEN, J. A., and FITE, W. L., *J. chem. Phys.* **50** (1969) 287.
[‡] FERGUSON, E. E. and FEHSENFELD, F. C., *J. geophys. Res.* **73** (1968) 6215.
[§] KELLER, G. E. and BEYER, R. A., ibid. **76** (1971) 289.
[||] ADAMS, N. G., BOHME, D. K., DUNKIN, D. B., and FEHSENFELD, F. C., *J. chem. Phys.* **52** (1970) 1951.
[††] SMITH, D., GOODALL, L. V., ADAMS, N. G., and DEAN, A. G., *J. Phys. B (GB)* **3** (1970) 34. [‡‡] ONG, P. P., and HASTED, J. B., ibid. **2** (1969) 91.
[§§] KANEKO, Y., KOBAYASHI, N., and ICHIRO, K., *J. Phys. Soc. Japan* **27** (1969) 992.
[||||] BIRKINSHAW, K. and HASTED, J. B., *J. Phys. B (GB)* **4** (1971) 1711.
[†††] HEMSWORTH, R. S., BOLDEN, R. C., SHAW, M. J., and TWIDDY, N. D., *Chem. Phys. Lett.* **5** (1970) 237.

4.9. Reactions of He_2^+, Ne_2^+, and Ar_2^+ with rare gas atoms and with molecules

Bohme, Adams, Mosesman, Dunkin, and Ferguson[†] have applied the flowing afterglow technique to measure the rates of reactions at 200 °K of the molecular ions of He, Ne, and Ar with Ne, Ar, and Kr and with NO, O_2, CO, N_2, and CO_2.

In most cases the main reaction is one of charge transfer. The rate coefficient for these reactions with rare-gas atoms increases with decrease in the internal energy change ΔE involved. Thus, measuring the efficiency in terms of the ratio of the rate coefficient to that calculated on the orbiting model of Gioumousis and Stevenson (see Chap. 19, p. 2005) we find that, for He_2^+, the ratio varies from 1·0 to 0·2 to 0·01 for reactions with Ne, Ar, and Kr for which the values of ΔE are, in eV, $3·02 - D$, $8·83 - D$, and $10·58 - D$ where D is the dissociation energy of He_2^+. For Ne_2^+ collisions with Ar and Kr the ratios are very small, $< 7 \times 10^{-5}$ and $< 7 \times 10^{-4}$ respectively, the values of ΔE being, in eV, $5·80 - D(Ne_2^+)$ and $7·56 - D(Ne_2^+)$. On the other hand for Ar_2^+ collisions with Kr, for which $\Delta E = 1·759 - D(Ar_2^+)$, the ratio is as high as 1·3.

For collisions with molecules, in most cases, the efficiency ratios are close to unity, the most marked exception being that of Ar_2^+ with NO for which it is only 4×10^{-3}.

Rates were also measured for reactions of HeNe+ with Ne and of ArKr+ with Kr, the efficiency ratios being found to be 0·3 and 0·6 respectively.

4.10. Partial charge transfer from doubly-charged ions

Spears et al.[‡] have applied the flowing afterglow technique to measure the rate constants at room temperature for the partial charge-transfer reactions

$$Ca^{++} + NO \rightarrow Ca^+ + NO^+, \quad (67)$$

and

$$Mg^{++} + X \rightarrow Mg^+ + X^+, \quad (68)$$

with X = Xe, NO, O_2, N_2O, CO_2, SO_2, NH_3, and NO_2. The method used to obtain the doubly-charged ions was described on p. 3620.

For the Ca^{++} reaction the rate constant was measured as $9·9 \times 10^{-10}$ cm^3 s^{-1} and for the Mg^{++} reactions with respective reaction partners X, 5×10^{-13}, $5·6 \times 10^{-10}$, $6·5 \times 10^{-10}$, $5·4 \times 10^{-10}$, $1·6 \times 10^{-11}$, $1·4 \times 10^{-9}$, $1·8 \times 10^{-9}$, and around $1·6 \times 10^{-9}$. The corresponding values of the

[†] BOHME, D. K., ADAMS, N. G., MOSESMAN, M., DUNKIN, D. B., and FERGUSON, E. E., J. chem. Phys. 52 (1970) 5094.
[‡] SPEARS, K. G., FEHSENFELD, G. C., McFARLAND, M., and FERGUSON, E. E., ibid. 56 (1972) 2562.

internal energy change ΔE involved are, in eV, 2·89, 5·77, 2·96, 2·13, 1·23, 2·68, 4·87, and 5·24. There is clearly no close correlation between measured rates and ΔE. The data were, however, analysed in terms of long range pseudo-crossing of potential energy curves (see Chap. 23, p. 2597) but considerable problems of interpretation remain. In particular, why is the cross-section for O_2 10^3 times larger than for Xe when the crossing-point is at nearly the same separation (4·7 Å for Xe 4·87 Å for O_2)? If these problems could be resolved, light might be thrown on other long range pseudo-crossing cases as in the reactions of A^+ with B^- ions.

5. Mobilities of negative ions in oxygen (see Chap. 19, § 4.1)

New measurements of the mobilities of mass-identified negative ions in oxygen have been made by Snuggs et al.,[†] using a drift tube essentially similar to that described on pp. 2012–14, and by McKnight.[‡] The former investigators obtained zero field mobilities, in $cm^2\ V^{-1}\ s^{-1}$ at 300 °K, of $3·20\pm0·09$, $2·16\pm0·07$, $2·55\pm0·08$, and $2·14\pm0·08$ for O^-, O_2^-, O_3^-, and O_4^-. McKnight's measurements covered the range of F/N between 2 and 400×10^{-17} V cm^{-1}. Some quantitative discrepancies still remain between the results obtained in these and earlier investigations. Thus, for O^- at $F/N = 5\times10^{-17}$ V cm^{-1}, McKnight finds a mobility a little below 3·4 $cm^2\ V^{-1}\ s^{-1}$ as compared with 3·2 and 3·0 according to Snuggs et al. and to Rees (see p. 2070) respectively. For O_2^- at the same F/N the two most recent measurements agree quite well but Rees obtains a value about 5 per cent higher. At high F/N the behaviour observed is qualitatively the same by most observers but, particularly for O^-, there are considerable differences in absolute values.

Especial interest attaches to the O_4^- ion. Because its mobility is so close to that of O_2^- it was not observed in most earlier work. Voshall, Pack, and Phelps (see p. 2082) did ascribe certain features of the temperature variation of the mobility of O_2^- ions in O_2 to the setting up of a statistical equilibrium

$$O_2^- + O_2 \rightleftharpoons O_4^-, \qquad (69)$$

and estimated the dissociation energy of O_4^- as 0·06 eV. Also Moruzzi and Phelps did not observe O_4^- ions in their mass analyses of different negative ions in O_2–CO_2 (see Fig. 19.88) and O_2–H_2O (see Fig. 19.90) mixtures. However, a little later Conway and Nesbitt§ carried out a mass analysis of negative ions produced when O_2 was irradiated with

[†] SNUGGS, R. M., VOLZ, D. J., SCHUMMERS, J. H., MARTIN, D. W., and MCDANIEL, E. W., Phys. Rev. A3 (1971) 477.   [‡] MCKNIGHT, L. G., ibid. A2 (1970) 762.
§ CONWAY, D. C. and NESBITT, L. E., J. chem. Phys. 48 (1968) 509.

a 2 Ci tritium source. They observed O_2^-, O_3^-, $O_2^-.H_2O$, O_4^-, and $^{16}O_3{}^{15}O^-$ ions, the mass scale being checked by addition of a trace of D_2O which introduced additional peaks due to $HOD.O_2^-$ and $DOD.O_2^-$. From the relative abundance of the ions over a temperature range from 273 to 353 °K and pressures of 6, 9, and 12 torr equilibrium constants for the reaction (69) were obtained. These led to a value of the enthalpy of -13.55 ± 0.16 heat/mole (0.59 eV), much higher than estimated by Voshall, Pack, and Phelps. Conway and Nesbitt give theoretical reasons why the dissociation energy of O_4^- should be comparable with that for O_4^+ (see p. 3620).

McKnight and Sawina[†] studied particularly the behaviour of O_4^- ions in their drift tube. They verified that the low field mobility of these ions is nearly equal to that of O_2^- and their measured value agrees well with that observed by Snuggs et al. It seems fairly certain that O_4^-, with a binding energy close to that obtained by Conway and Nesbitt, does exist and this must be taken into account in considering negative ion reactions which occur in the lower ionosphere. As will be seen below, measurements of the rates of reactions involving O_4^- have recently been made. The reason why Moruzzi and Phelps failed to observe O_4^- in their experiments remains obscure.

6. Negative ion reactions

6.1. *Reactions of atmospheric interest*

Much of the experimental work on negative ion reaction rates has been directed towards those reactions which are likely to be of significance in the negative ion chemistry of the lower ionosphere.

The expectation that NO_3^- ions either free or hydrated are likely to be the major negative constituents in the D region (see Chap. 20, p. 2303) has been confirmed by recent rocket flights.[‡] This has provided a further stimulus to experimental study of the rates of reactions which lead to production or loss of these ions.

The first proposed reaction sequence (see p. 2305) involved the reactions

$$e + O_2 + X \rightarrow O_2^- + X, \tag{70}$$

$$O_2^- + O_3 \rightarrow O_3^- + O_2, \tag{71}$$

$$O_3^- + CO_2 \rightarrow CO_3^- + O_2, \tag{72}$$

$$O_3^- + NO \rightarrow NO_3^- + O, \tag{73}$$

$$CO_3^- + NO \rightarrow NO_2^- + CO_2. \tag{74}$$

[†] McKnight, L. G. and Sawina, J. M., *Phys. Rev.* **A4** (1971) 1043.
[‡] Narcisi, R. S., Bailey, A. D., Della Lucca., L., Sherman, C., and Thomas, D. M., *J. atmos. terrest. Phys.* **33** (1971) 1147.

Fehsenfeld and Ferguson,† using the flowing afterglow technique, have since shown that NO_2^- may be converted to NO_3^- through the reaction

$$NO_2^- + O_3 \to NO_3^- + O_2, \qquad (75)$$

for which they measured a rate constant of $1\cdot 8 \times 10^{-11}$ cm³ s⁻¹ at 300 °K.

A little later Fehsenfeld, Ferguson, and Bohme‡ investigated the possibility that NO_3^- might be lost by reactions with O, N, and NO. For reactions with O and N the rate constant was found to be 10^{-11} cm³ s⁻¹ and with NO less than 10^{-12} cm³ s⁻¹. In fact the reaction

$$NO_2^- + NO_2 \to NO_3^- + NO \qquad (76)$$

was observed to occur with a rate constant near 4×10^{-12} cm³ s⁻¹. For (76) to be exothermic the electron affinity of NO_3 must exceed that of NO_2 by more than 0·9 eV.

Further more definite information about the electron affinities of NO_2 and NO_3 has been obtained by Ferguson, Dunkin, and Fehsenfeld.§ They studied the reactions of NO_2^- and NO_3^- ions with HCl and HBr. For NO_2^-, reactions with HCl and HBr, producing the halogen negative ions and HNO_2, were found to be clearly exothermic with rate constants of 1·4 and $1\cdot 9 \times 10^{-9}$ cm³ s⁻¹ respectively. Taking the heat of formation of HNO_2, which is not well known,∥ as $3\cdot 37 \pm 0\cdot 03$ eV the electron affinity $A(NO_2)$ of NO_2 must be $< 2\cdot 6$ eV.

The reaction $\qquad NO_3^- + HCl \to Cl^- + HNO_3 \qquad (77)$

was not observed, the rate constant being less than 10^{-12} cm³ s⁻¹. Assuming the reaction to be endothermic $A(NO_3) > 3\cdot 5 \pm 0\cdot 2$ eV, the uncertainty being that of the heat of formation of NO_3.∥ More evidence was obtained from the reaction with HBr. When HBr was added to the flow tube the concentration of NO_3^- decreased initially as expected for a reaction with a rate constant of $6\cdot 3 \times 10^{-10}$ cm³ s⁻¹ but, as the input flow of HBr increased the rate of loss of NO_3^- decreased rapidly. The most likely interpretation of this is that the equilibrium

$$NO_3^- + HBr \rightleftharpoons Br^- + HNO_3 \qquad (78)$$

is rapidly set up, implying that the reactions are nearly thermoneutral. In this case $A(NO_3)$ would be $3\cdot 9 \pm 0\cdot 2$ eV which would exceed $A(NO_2)$ by more than $1\cdot 3 \pm 0\cdot 2$ eV, consistent with the evidence from the reaction

---

† FEHSENFELD, F. C. and FERGUSON, E. E., *Planet. Space Sci.* **16** (1968) 701.
‡ FEHSENFELD, F. C., FERGUSON, E. E., and BOHME, D. K., ibid. **17** (1969) 1759.
§ FERGUSON, E. E., DUNKIN, D. B., and FEHSENFELD, F. C., *J. chem. Phys.* **57** (1972) 1459.
∥ Natl. Std. Ref. Data. Ser. Natnl. Bur. Std. (US) **37** (1971).

(77). It is also consistent with the limits derived by Berkowitz, Chupka, and Gutman[†] from their experiments on the threshold energy for the reaction
$$I^- + HNO_3 \rightarrow NO_3^- + HI. \qquad (79)$$
They find $2 \cdot 68$ eV $< A(NO_3) < 4 \cdot 34$ eV. If $A(NO_3)$ is as large as $3 \cdot 9$ eV all possible reactions of NO_3^- with O are endothermic.

In the course of their experiments Ferguson, Dunkin, and Fehsenfeld[‡] also measured the rate coefficients for the reactions of O_2^- and NO^- with HCl obtaining the same value, $1 \cdot 6 \times 10^{-9}$ cm^3 s^{-1}, for each.

Fehsenfeld and Ferguson have measured the total rate constant for
$$NO_2^- + H \rightarrow OH^- + NO, \qquad (80\,a)$$
$$\rightarrow HNO_2 + e, \qquad (80\,b)$$
as 3×10^{-10} cm^3 s^{-1} at 300 °K. This may be of some importance in reducing the NO_3^- concentration in the lower ionosphere.

A further new set of measurements relating to the negative ion reaction sequence in the D region concerns O_4^-. McKnight and Sawina,[§] by the drift tube technique, measured the rates of the forward and backward reactions
$$O_2^- + O_2 + O_2 \rightleftharpoons O_4^- + O_2 \qquad (81)$$
as 3×10^{-31} cm^6 s^{-1} and 2×10^{-14} cm^3 s^{-1} respectively at 310 °K. The forward reaction will produce O_4^- at a significant rate in the D region.

Adams et al.[||] have applied the flowing afterglow technique to measure the rates of a number of cluster-forming termolecular reactions at 200 °K involving O_2^- in which the third body is a He atom. For
$$O_2^- + O_2 + He \rightarrow O_4^- + He \qquad (82)$$
they find a rate coefficient of $3 \cdot 4 \times 10^{-31}$ cm^6 s^{-1}, close to the drift tube results with O_2 as third body. The corresponding coefficients for reactions in which the $O_2^- \cdot N_2$ and $O_2^- \cdot CO_2$ clusters are formed were found to be $4 \cdot 0 \times 10^{-32}$ and $4 \cdot 7 \times 10^{-29}$ cm^6 s^{-1}. Measurements were also made for reactions involving cluster formation by O^-,
$$O^- + N_2 + He \rightarrow O^- \cdot N_2 + He, \qquad (83)$$
$$O^- + CO_2 + He \rightarrow O^- \cdot CO_2 + He, \qquad (84)$$
the resulting coefficients being $4 \cdot 0 \times 10^{-32}$ and $2 \cdot 6 \times 10^{-28}$ cm^6 s^{-1}.

[†] BERKOWITZ, J., CHUPKA, W. A., and GUTMAN, D., J. chem. Phys. **55** (1971) 2733.
[‡] FERGUSON, E. E., DUNKIN, D. B., and FEHSENFELD, F. C., J. chem. Phys. **57** (1972) 1459.
[§] loc. cit., p. 3635.
[||] ADAMS, N. G., BOHME, D. K., DUNKIN, D. B., FEHSENFELD, F. C., and FERGUSON, E. E., loc. cit., p. 3630.

Fehsenfeld, Ferguson, and Bohme[†] have used the afterglow technique to measure the rates of the reactions

$$O_4^- + NO \rightarrow NO_3^- + O_2, \qquad (85)$$

$$O_4^- + CO_2 \rightarrow CO_4^- + O_2, \qquad (86)$$

$$CO_4^- + NO \rightarrow NO_3^- + CO_2. \qquad (87)$$

Values of $2 \cdot 5 \times 10^{-10}$, $4 \cdot 3 \times 10^{-10}$, and $4 \cdot 8 \times 10^{-11}$ cm^3 s^{-1} were found for the respective rate coefficients. These are high enough to make it likely that (81) followed by (85) or by (86) and (87) will be more effective than the sequence initially proposed.

O_4^- also reacts with atomic oxygen:

$$O_4^- + O \rightarrow O_3^- + O_2, \qquad (88\,a)$$

$$\rightarrow O^- + 2O_2, \qquad (88\,b)$$

with a rate coefficient of 4×10^{-10} cm^3 s^{-1}, (88 a) being the more probable process.

CO_4^- was also found to react with O with a coefficient of $1 \cdot 5 \times 10^{-10}$ cm^3 s^{-1}, CO_3^- being the most likely product.

There is little doubt that hydrated negative ions will be of importance in the D region of the ionosphere. Much remains to be done in the measurement of the relevant reaction rates. The rate coefficient for

$$O_4^- + H_2O \rightarrow O_2^- \cdot H_2O + O_2 \qquad (89)$$

has been estimated[‡] to be around 10^{-9} cm^3 s^{-1}. Puckett and Lineberger[§] have used their technique of observing the decay of the concentration of mass-analysed negative ions diffusing through an orifice from an afterglow at such a time that the plasma is controlled by ambipolar diffusion of positive and negative ions (see also p. 3452). Working with pure NO and with NO–H$_2$O mixtures they measured the rate constant for

$$NO_2^- + H_2O + NO \rightarrow NO_2^- \cdot H_2O + NO \qquad (90)$$

as $1 \cdot 3 \pm 0 \cdot 3 \times 10^{-28}$ cm^6 s^{-1}. Multiply-hydrated NO_2^- ions were observed as well as clusters with HNO$_2$.

The rate coefficients of the cluster-exchange reactions

$$O_2^- \cdot H_2O + CO_2 \rightarrow O_2^- \cdot CO_2 + H_2O, \qquad (91)$$

$$O_2^- \cdot H_2O + NO \rightarrow O_2^- \cdot NO + H_2O \qquad (92)$$

have been measured at 300 °K by the flowing afterglow technique by Adams et al.[∥] as 5·8 and $3 \cdot 1 \times 10^{-10}$ cm^3 s^{-1} respectively. On the basis of

[†] loc. cit., p. 3636.   [‡] FERGUSON, E. E., Annls Géophys. **26** (1970) 589.
[§] PUCKETT, L. J. and LINEBERGER, W. C., Phys. Rev. A**1** (1970) 1635.
[∥] ADAMS, N. G., BOHME, D. K., DUNKIN, D. B., FEHSENFELD, F. C., and FERGUSON, E. E., loc. cit., p. 3630.

their and other measurements referred to earlier, it appears that the order of increasing strength of binding of clusters to O_2^- is O_2, H_2O, CO_2, and NO, which is not the same as for clusters to O_2^+ (see p. 3630).

McFarland, Dunkin, Fehsenfeld, Schmeltekopf, and Ferguson† have made a detailed study of detachment reactions of NO^- using the flowing afterglow technique. The electron affinity of NO is very low (0·02 eV) as measured from photodetachment (see p. 3520) so that it is difficult to choose a mixture of buffer gas and a source of NO^- ions in which detachment is not serious. In these experiments the NO^- was produced from the reaction

$$N_2O + e \to O^- + N_2, \qquad (93)$$
$$O^- + N_2O \to NO^- + NO \qquad (94)$$

with argon as buffer gas. Otherwise the technique was as usual, the apparatus being that referred to on p. 3615 with which observations could be made over a considerable temperature range.

Detachment rate coefficients k_d were measured at temperatures of 193, 285, 382, and 506 °K for collisions with Ne, He, H_2, CO, NO, CO_2, N_2O, and NH_3. It was found that over this temperature range k_d could be represented by

$$k_d = A \exp(-E_0/\kappa T), \qquad (95)$$

where A and E_0 have the values given in Table 19 A.5. It will be seen

TABLE 19 A.5

Values of parameters A and E_0 fitting observed detachment rate coefficients for impact of NO^- with various atoms and molecules

Target	He	Ne	H_2	CO	NO	CO_2	N_2O	NH_3
E_0 (mV)	74·8	98·0	82·8	45·5	105	81·2	107·2	59·0
A (10^{-10} cm^3 s^{-1})	0·523	0·0147	0·0629	0·0346	3·28	2·2	4·26	2·32

from the wide spread in values of E_0 that the reaction is not simple. Furthermore, E_0 exceeds the detachment energy (20 mV) in every case.

By using the values of the electron affinity of NO and rotational constant B for NO^- measured from photodetachment (see p. 3520) the equilibrium constant for the reaction

$$NO^- + X \to NO + e + X \qquad (96)$$

could be calculated and hence from the measured detachment rate coefficients those for attachment could be derived. In units of 10^{-31} cm^6 s^{-1} the values obtained for He, Ne, H_2, CO, NO, CO_2, N_2O, and NH_3 were 2·97, 0·0342, 0·262, 0·606, 5·80, 9·76, 6·95, and 24·2 respectively.

† McFarland, M., Dunkin, D. B., Fehsenfeld, F. C., Schmeltekopf, A. L., and Ferguson, E. E., *J. chem. Phys.* **56** (1972) 2358.

These results do not seem to be consistent with an interpretation of the detachment process as arising from vibrational activation of NO⁻ to an autodetaching vibrational level (see p. 3445). This would require an activation energy of 0·160 eV which is considerably larger than any observed. If detachment does not proceed in this way then attachment cannot arise through a similar mechanism to that for O_2 (see pp. 3443–5). It is far from clear why this is not so.†

As seen from Table 19 A.5 the detachment rate coefficients for the polyatomic molecules are relatively high. This suggests that internal energy of these molecules may be effective in producing detachment.

It appears from day airglow measurements‡ that metastable $O_2(^1\Delta_g)$ molecules are present in appreciable concentration (mixing ratio $\simeq 10^{-5}$) in the D region of the ionosphere. They may therefore play an important part in causing detachment from O_2^- through the reaction

$$O_2(^1\Delta_g) + O_2^- \to 2O_2 + e. \qquad (97)$$

Fehsenfeld, Albritton, Burt, and Schiff§ have applied the flowing afterglow technique to measure the rate coefficient for (97) and also for the reaction

$$O_2(^1\Delta_g) + O^- \to O_3 + e. \qquad (98)$$

The O_2^- and O^- ions were produced by flowing helium buffer gas with an admixture of O_2 past an electron gun. $O_2(^1\Delta_g)$ molecules were produced by a microwave discharge in O_2, the discharge products being passed through a glass wool plug containing mercuric oxide to remove atomic oxygen. The concentration of $O_2(^1\Delta_g)$ in the molecular mixture was determined by passing the mixture through a cell of diameter 6 cm and length 20 cm and measuring the intensity of the 1·27 μ emission from $O_2(^1\Delta_g)$ with a PbS photometer. From the cell the molecules were admitted to the afterglow tube through a side arm.

Rate coefficients $\simeq 2 \times 10^{-10}$ and 3×10^{-10} cm³ s⁻¹ were measured for the respective reactions (97) and (98) for both of which the kinetic energy released is 0·6 eV.

6.2. *Reactions involving* C⁻

A further application of the flowing afterglow technique has been made by Fehsenfeld and Ferguson,‖ who studied thermal reactions of C⁻ ions

---

† See, however, PARKES, D. A., *J Chem. Soc., Far. Trans. I* **68** (1972) 2121.

‡ EVANS, W. F. J., HUNTEN, D. M., LLEWELLYN, E. J., and JONES, A. V., *J. geophys. Res.* **73** (1968) 2885.

§ FEHSENFELD, F. C., ALBRITTON, D. L., BURT, J. A., and SCHIFF, H. I., *Can. J. Chem.* **47** (1969) 1793.

‖ FEHSENFELD, F. C. and FERGUSON, E. E., *J. chem. Phys.* **53** (1970) 2614.

with O_2, H_2, CO, N_2O, and CO_2. The C^- ions were produced by impact of 100 eV electrons in CH_4 contained in a large excess of helium. Observed values for the respective rate coefficients, in 10^{-10} cm^3 s^{-1}, are 4·0, $< 10^{-3}$, 4·1, 9·0, and 0·47. For O_2, CO_2, and CO the reactions involved appear to be

$$C^- + O_2 \to O^- + CO, \tag{99}$$

$$C^- + CO_2 \to 2CO + e, \tag{100}$$

$$C^- + CO \to C_2O + e. \tag{101}$$

For N_2O there are a number of possible associative detachment processes which could contribute.

20

RECOMBINATION

Since Vol. 4 was written the most important further experimental work on the recombination of electrons has been concerned, on the one hand, with the further study of electron loss processes in a helium discharge at a few tens of torr pressure and, on the other hand, with the variation of the dissociative recombination coefficient to O_2^+ and N_2^+ ions with the temperature of the neutral gas and of the electrons, when these are equal.

1. Electron loss in helium afterglows at high pressures

In Chapter 20, p. 2220 the experiments of Berlande, Cheret, Deloche, Gonfalone, and Manus[†] on electron recombination in a helium afterglow at pressures p in the range 10 to 100 torr are described. The electron concentrations n_e concerned were between 10^9 and 5×10^{11} cm^{-3}. Evidence was obtained of a dependence of the recombination coefficient on both p and n_e.

More recent experiments, carried out under nearly the same conditions of pressure and electron concentration, disagree in some important respects with these results and with each other so the nature of the processes involved is somewhat obscure.

Collins, Hicks, and Wells[‡] used an optical technique to determine the effective loss rates through recombination to He_2^+. This consists in measuring the absolute intensity of emission of the He_2 bands in the visible, as well as the electron concentration, as a function of time in the afterglow. Then, if I_j is the total number of photons emitted per second in transitions from the jth molecular state the recombination coefficient α satisfies

$$\alpha \geqslant \Sigma_j I_j / K n_e^2, \qquad (1)$$

where K is the ratio of the concentration of He_2^+ ions to that of the electrons. If one and only one photon is detected for each ion recombined the inequality becomes an equality. To determine K, Collins et al. used the intensity I_q emitted in an observable transition from a Rydberg state of such high total quantum number q that the concentration N_q of atoms

[†] loc. cit., p. 2165.
[‡] Collins, C. B., Hicks, H. S., and Wells, W. E., Phys. Rev. A2 (1970) 797.

in this state was determined by the Saha equilibrium equation. We then have
$$I_q = A_q n_e n_+ c(q, T_e), \quad (2)$$
where $c(q, T_e)$ is calculable in terms of the ionization energy of the state q and the atomic constants and A_q is the probability for the radiative transition concerned. Substitution of (2) in (1) gives
$$\alpha \geqslant A_q c(q, T_e) \Sigma_j I_j / I_q. \quad (3)$$

Unfortunately A_q is not known and some means of calibration of the observed values of α must be introduced. Collins et al. assumed that the rate of change of concentration n_+^m of the He_2^+ ion satisfies
$$\frac{dn_+^m}{dt} = -\alpha n_+^m n_e, \quad (4)$$
in which case we find on integration and use of (2) that
$$A_q c(q, T_e) = I_q/n_e \int_t^\infty \alpha n_+^m n_e \, dt$$
$$= I_q/n_e \Sigma_j \int_t^\infty I_j \, dt. \quad (5)$$

Measurements were made at a pressure of 44·6 torr with a quartz afterglow cell excited by pulses of 5 kV with duration not greater than 10 μs. The intensities of bands originating from Rydberg states with $q \geqslant 16$ were found to decay at the same rate and this was taken to indicate that these states were in collisional equilibrium with the electrons. Assuming that, for the visible bands, one and only one photon is emitted for each recombination event, the data were analysed to give
$$\alpha \simeq 2{\cdot}8 \times 10^{-11} n_e^{0{\cdot}185} \text{ cm}^3 \text{ s}^{-1} \quad (6)$$
over a range of n_e from 10^{10} to 10^{12} cm³ s⁻¹. Over the range 10^{10}–10^{11} cm³ s⁻¹ common to both experiments there is good agreement with the earlier work of Berlande et al. at the same pressure p. However, the variation with n_e according to (6), over a ten times larger range of n_e, is in marked disagreement with expectation based on the assumption that collisional radiative recombination is dominant.

Johnson and Gerardo[†] have also studied the variation of electron concentration in helium afterglows over the range 15 to 56 torr of p and 3×10^{11} to 3×10^{10} cm⁻³ of n_e. They were particularly concerned with investigating the possibility that, under these conditions, the interpretation of the results is complicated by the existence of a source of free

[†] JOHNSON, A. W. and GERARDO, J. B., Phys. Rev. A5 (1972) 1410.

electrons presumably due to ionizing collisions between pairs of metastable atoms and molecules. If such a source exists producing S electrons cm^{-3} s^{-1} then, if only one kind of ion is present,

$$\frac{dn_e}{dt} = -\alpha n_e^2 + S. \tag{7}$$

Under a wide range of experimental conditions it was found that

$$n_e^{-1} = n_{eo}^{-1} + \alpha_{\text{eff}} t, \tag{8}$$

where n_{eo} and α_{eff} are independent of time t, α_{eff} being an apparent recombination coefficient. This being so we must have

$$S = (\alpha - \alpha_{\text{eff}})/(n_{eo}^{-1} + \alpha_{\text{eff}} t)^2. \tag{9}$$

To determine α the following procedure was used. At a suitable time in the afterglow the electrons were heated by a microwave pulse of about 200 μs duration. This reduced the rate of recombination so that the decay of n_e was not only checked but reversed during application of the pulse. After cessation of the pulse the electron concentration n_{eh} satisfies

$$\frac{dn_{eh}}{dt} = -\alpha n_{eh}^2 + (\alpha - \alpha_{\text{eff}})/(n_{eho}^{-1} + \alpha_{\text{eff}} t)^2, \tag{10}$$

where n_{eho} now refers to the electron concentration at a short time, about 10 μs, after pulse cessation and it is assumed that (9) still applies. Using previously measured values of α_{eff}, α was determined so as to give the best fit of (10) to the observed variation of n_{eh} with time. Quite good fits were obtained with α independent of n_e but varying linearly with pressure so that

$$\alpha = \alpha_0 + kp \tag{11}$$

with $\alpha = 1 \cdot 3 \times 10^{-8}$ cm^3 s^{-1}, $k = 4 \cdot 08 \times 10^{-10}$ cm^3 s^{-1} torr^{-1}. α_0 is considerably larger than in previous measurements, suggesting failure in those experiments to allow for a significant source of free electrons under the experimental conditions.

A number of checks were carried out to verify the assumptions made and to provide further evidence about the absence of any dependence of α on n_e. In particular measurements were made of the intensity of the 4650 Å band of He$_2$. It was found to vary very closely as n_e^2 which would be expected if it arose from a definite fraction of recombination events the rates of which were independent of n_e. Similar results were obtained from a study of the 3690 and 5730 Å bands.

The intensity of molecular radiation was also used to check that no serious error was introduced by non-uniformity of the electron distribution under the experimental conditions.

A more detailed analysis of the reactions occurring in the afterglow was made on the assumption that collisions involving metastable He atoms and He_2 molecules provided the source of the free electrons. If no distinction is made between the rates of reactions between each pair of these metastable species then it may be shown that, if $dn_e/dt = -\alpha_{eff} n_e^2$, then

$$(n_m/n_e)^2 = 2(\alpha - \alpha_{eff})^2/\beta \tag{12}$$

and

$$\alpha_m/\alpha = 2\left[1 - \frac{\alpha_{eff}}{\alpha}\left\{1 + \left(\frac{\alpha - \alpha_{eff}}{2\beta}\right)^{\frac{1}{2}}\right\}\right]. \tag{13}$$

Here n_m is the concentration of metastable atoms and molecules, α_m is the coefficient for recombination of electrons to He_2^+ which produces either metastable atoms or molecules and β is the rate coefficient for the metastable–metastable ionizing collisions. There is evidence from measurements at lower pressures, where metastable atoms rather than molecules are dominant, that a relation of the form (12) is indeed satisfied. Taking for β a value close to 2×10^{-9} cm^3 s^{-1} (corresponding to a reaction cross-section of 10^{-14} cm^2, see pp. 1803–6) $\alpha_m/\alpha \simeq 0.7$.

As far as the effect of the heating pulse on the free electron source is concerned it was noted that, after cessation of the pulse, both the electron concentration and the intensity of molecular bands returned asymptotically to the same value which each attained in the absence of the pulse. This showed that no permanent change in the strength of the source was produced.

The electron temperature in an unheated afterglow was estimated to be within 33° of room temperature at 15 torr pressure and closer at higher pressures. During the heating pulse the electron temperature reached about 1500 °K.

The recombination coefficients obtained in these experiments are difficult to interpret. The absence of any dependence on n_e and the quite strong dependence on pressure is inconsistent with the assumption that the process responsible is collisional radiative recombination (see pp. 2136–44).

Monchicourt† has made a detailed experimental study of the electron temperature in helium afterglows using a microwave radiometer. He finds that, working at a gas temperature of 300 °K the electron temperature is apparently above the gas density even at pressures up to 80 torr for electron concentrations above 10^{11} cm^{-3}. At the latter density observed electron temperatures of 380 and 330 °K were found at pressures of 10·15 and 20·25 torr respectively. At 10·15 torr good agreement is

† MONCHICOURT, P., Thesis, Paris 1972.

found with measurements made by Miller, Verdeyen, and Cherrington[†] at electron concentrations of 1, 2, 3, and 4×10^{11} cm^{-3}. These authors determined the temperature from measurement of the electron–helium collision frequency (see Chap. 2, p. 81). Theoretical estimates suggest that the production of superthermal electrons through collisions between metastable atoms is responsible for the elevated temperatures, again indicating the importance of these collisions (see also below for argon).

2. Further measurements of recombination in argon

In the course of an experimental study of electron and ion decay processes in afterglows in Ar–O$_2$ mixtures, Smith (D.), Goodall, Adams, and Dean[‡] have made further measurements of the recombination coefficient α of electrons to Ar$_2^+$. They used the Langmuir probe technique operated in the orbital-limited region (see Chap. 20, p. 2168) to measure the electron concentration. At pressures near the working pressure 0·45 torr, no dependence of α on pressure was observed but the values found decreased as the power associated with the ionizing pulse was increased, the range of variation being from 4·7 to $12·6 \times 10^{-7}$ cm^3 s^{-1} at 300 °K.

The measurements, discussed in Chapter 20, p. 2233, give values between 6·7 and $8·5 \times 10^{-7}$ cm^3 s^{-1}. It is not clear how the dependence on pulse power observed by Smith *et al.* arises but it may be associated with the production of metastable atoms which in turn modify the effective recombination coefficient by providing an electron source and/or modifying the electron temperature. There is need to clarify this matter further.

3. Further measurements of recombination to O$_2^+$ and O$_4^+$

The Langmuir probe technique, operated in the orbital limited region (see Chap. 20, p. 2168) has been further applied (see § 2 above) to study recombination in mixtures of O$_2$ dilute in Ar and Kr (see Chap. 20, p. 2248).

In Ar at 295 °K under conditions in which O$_2^+$ would be expected to be the dominant ion, Smith, D., *et al.*[‡] obtained a recombination coefficient $2·1 \pm 0·3 \times 10^{-7}$ cm^3 s^{-1} which agrees well with earlier measurements.

Plumb, Smith (D.), and Adams§ have carried out similar measurements in Kr. They worked particularly with the walls of the discharge

[†] MILLER, P. A., VERDEYEN, J. T., and CHERRINGTON, B. E., *Phys. Rev.* A4 (1971) 692.
[‡] SMITH, D., GOODALL, G. V., ADAMS, N. G., and DEAN, A. G., *J. Phys. B* (*GB*) 3 (1970) 34.
§ PLUMB, I. C., SMITH, D., and ADAMS, N. G., ibid. 5 (1972) 1762.

vessel cooled to 180 °K by surrounding it with solid CO_2. This made it possible to observe recombination to O_2^+ in the earlier part of the afterglow and to O_4^+ in the latter part. They obtained values of $3\cdot5\pm1\cdot0\times10^{-7}$ and $1\cdot8\pm0\cdot06\times10^{-6}$ cm^3 s^{-1} for O_2^+ and O_4^+ respectively agreeing well with values obtained by Kasner and Biondi at 205 °K (see Chap. 20, p. 2251). Measurements for O_2^+ were also made at 300 °K giving again very good agreement with earlier measurements.

4. Dissociative recombination to Kr_2^+, O_2^+, and N_2^+ at high gas and electron temperatures

Cunningham and Hobson[†] have applied the shock wave technique described on pp. 2174–8 to measure recombination of electrons in a krypton afterglow at gas and electron temperatures ranging from 800 to 2500 °K. They find that the recombination coefficient varies closely as $T^{-\frac{3}{2}}$ where T is the gas and electron temperature. Referring to the theory outlined on pp. 2124–6 this is the behaviour expected if the vibrational temperature of the recombining ions is the same as the translational and if vibrationally excited molecules are ineffective in recombination. For molecules as heavy as Kr_2^+ the vibrational quanta are sufficiently small for the vibrational relaxation time to be relatively short (see Chap. 17, § 3.1). Estimates indicate that in the experimental conditions sufficient time was available for the vibrational energy to come to full thermal equilibrium at the shock temperature.

Similar experiments were carried out to obtain corresponding results for recombination to O_2^+ [‡] and N_2^+.[§] In these cases care had to be taken to choose conditions so that it can be confidently assumed that the diatomic ion is indeed the dominant one. For O_2^+ Cunningham and Hobson used dilute mixtures of about 1 per cent oxygen in three different rare gases, He, Ne, and Ar—the total pressure being in the range 2 to 6 torr. Under these conditions (see Chap. 20, pp. 2248–51) O_2^+ is the dominant ion but there is some doubt about the degree of vibrational excitation and the possible presence of metastable excited O_2^+. However, Cunningham and Hobson obtained essentially the same results with all three gas mixtures. At 625 °K the value obtained for the recombination coefficient was $1\cdot14\times10^{-7}$ cm^3 s^{-1} and the variation with temperature T over the range 600–2500 °K followed a $T^{-0\cdot63}$ law. The extrapolated value at room temperature $1\cdot83\times10^{-7}$ cm^3 s^{-1} agrees well

[†] CUNNINGHAM, A. J. and HOBSON, R. M., ibid. 1773.
[‡] idem, 2320.
[§] idem, 2328.

with $1 \cdot 95 \pm 0 \cdot 2 \times 10^{-7}$ cm^3 s^{-1} measured directly by Mehr and Biondi (see Chap. 20, pp. 2248–51 and also p. 3646 of these notes).

Measurements for N_2^+ were carried out in a mixture of 1 per cent N_2 with 99 per cent Ne at pressures between 2 and 6 torr. Previous experiments (see Chap. 20, pp. 2241–8) indicate that under these conditions N_2^+ will be the dominant ion. At temperatures T between 700 and 2700 °K the observed recombination coefficient varied as $T^{-0 \cdot 37}$. The extrapolated value at 300 °K agrees well with the directly measured value of Mehr and Biondi.

The relatively slow decrease of recombination coefficient α with temperature for O_2^+ and N_2^+ suggests that, in these cases, the vibrational excitation has not reached full thermal equilibrium during the time scale of the experiment. This is not surprising because for these relatively light molecules vibrational relaxation is very slow (cf. Chap. 17, Fig. 17.31 for O_2 and N_2). Further evidence that the variation with T is mainly due to the dependence on T_e is provided from the fact that Mehr and Biondi found for $T_g = 300°$ that α varied as $T_e^{-0 \cdot 39}$.

Experiments at much higher temperatures have been carried out by Dunn and Lordi[†] who worked with gas at several atmospheres pressure heated in a reflected shock wave and expanded through a conical nozzle. With T between 2000 and 7000 °K for O_2 and between 3500 and 7200 °K for N_2 they found a recombination coefficient varying as $T^{-\frac{3}{2}}$ as would be expected for molecular ions with vibrational excitation in full thermal equilibrium. At these temperatures vibrational relaxation is much faster (see Chap. 17, Fig. 17.31) so these results are not inconsistent with the results at lower temperatures.

5. Dissociative recombination to H_2^+—theoretical calculation

On p. 3493 we referred to the calculations of Nielsen and Berry[‡] on the rates of autoionization of excited molecular states through coupling between vibrational and electronic motion. They have applied the same method, which is essentially a perturbed stationary state method (see Chap. 23, § 7.1), to calculate the rates of associative ionization on impact between excited and normal H atoms and also for the inverse process of dissociative recombination, in which curve crossing does not occur. They find for the rate coefficients of the latter process associated with H_2^+ ions in the vibrational states with vibrational quantum numbers v from 4 to 8, and arising from interaction with the $4s\sigma$ repulsive potential

[†] DUNN, M. G. and LORDI, J. A., *AIAA J.* **8** (1970) 614.
[‡] NIELSEN, S. E. and BERRY, R. S., *Phys. Rev.* **A4** (1971) 865.

energy curve of H_2, values no greater than 10^{-10} cm^3 s^{-1}. For H_2^+ ions in lower vibrational states considerably smaller values were found but significant contributions could also be made from interactions with other repulsive states of H_2. An upper bound of 5×10^{-11} cm^3 s^{-1} was estimated for H_2^+ ions in the ground vibrational state. This low value is not inconsistent with experiment (see Chap. 20, p. 2254).

6. Reactions in the earth's ionosphere

Timothy (A. F.), Timothy (J. G.), Willmore, and Wager† have published a very comprehensive study of the ion chemistry of the E and lower F regions of the daytime ionosphere based on observations made in a single rocket flight of the densities of the major neutral atmospheric constituents, the electron concentration, the intensity and spectrum of the ultraviolet and X-radiation from the sun, and the temperature of the neutral atmosphere and of the electrons, as functions of height over the range 120–270 km. The rocket used was a sun-stabilized Skylark launched at Woomera, Australia, on 3 April 1969.

Because the various quantities concerned were measured in the same flight, complications of interpretation due to time variation were eliminated. Unfortunately it was not possible to measure in the same flight the composition of the positive ions. However, with full knowledge of all the reaction rates involved as functions of gas and electron temperature complete analysis is possible without this information. In practice full information about the relevant reaction rates is not yet available. Using the best data available Timothy et al. determined the effective recombination coefficient of electrons as 1×10^{-7} cm^3 s^{-1} at 120 km. At 130 km the ion concentration calculated from the measured intensity of the solar radiation, the measured concentration of the different atmospheric constituents and their photoionization cross-sections, agrees with the measured electron concentration to within 30 per cent but becomes too low at lower altitudes.

The most important progress in the understanding of reaction rates in the lower ionosphere has already been discussed on pp. 3627 and 3635.

† TIMOTHY, A. F., TIMOTHY, J. G., WILLMORE, A. P., and WAGER, J. H., *J. atmos. terrest. Phys.* **34** (1972) 969.

22

ELASTIC COLLISIONS BETWEEN PARTICLE BEAMS AND NEUTRAL ATOMS AND MOLECULES

1. Collisions between neutral systems

CONSIDERABLE progress has been made, since this chapter was written, in the development and application of sputter sources of neutral atoms to obtain measurements of 'total' cross-sections for elastic scattering of atoms by various atomic and molecular targets. Experimental studies of glory oscillations in collisions between alkali-metal and mercury atoms have been extended to energies of several hundred eV and the results used to obtain information about the interactions concerned. On the theoretical side an analysis of the combining rule for repulsive potentials has been carried out yielding an improved formula for the case in which the intersecting systems are unlike.

1.1. *The glory effect in collisions at relatively high energies*

Neumann and Pauly† have measured the velocity dependence of total cross-sections for collisions of the alkali-metal atoms with Hg and for Cs–Na collisions over an energy range extending from a few eV to several hundred eV. For these systems the product ϵr_m, of the depth ϵ of the potential well in the atomic interaction with the separation r_m at the potential minimum, is large so that the energy at which the first glory maximum occurs (see Chap. 16, p. 1337 (146)) is quite high. As all of the collisions with mercury atoms had previously been studied at thermal energies (see pp. 3528–9) the analysis of the glory structure provides a valuable source of additional information about the interactions concerned.

The measurements were carried out by a crossed beam method—the alkali-metal atomic beams being obtained by the usual charge exchange technique. Glory oscillations were observed in all cases. Thus with the best experimental angular resolution three successive maxima and two minima were found for Cs–Hg over the energy range 8–500 eV. Even

† NEUMANN, W. and PAULY, H., *J. chem. Phys.* **52** (1970) 2548.

for the lightest collision partner, Li, one minimum and one maximum were clearly present in the interval 6–1000 eV.

In analysing the results account was taken of the finite angular resolution and of the presence of rainbow structure in the differential cross-section.

As in the case of infinitely fine angular resolution (see Chap. 16, p. 1327) the observed total cross-section Q may be written in the form

$$\tilde{Q} = \tilde{Q}_0 + \Delta\tilde{Q}, \qquad (1)$$

where $\tilde{Q}_0$ varies monotonically with relative impact velocity v and $\Delta\tilde{Q}$ is an oscillatory function of v. $\tilde{Q}_0$ and $\Delta\tilde{Q}$ may be expressed in terms of an effective minimum scattering angle θ_0 given by

$$\theta_0 = 0{\cdot}8(dF/l), \qquad (2)$$

where d is the detector width, l the distance from the detector to the scattering centre, and F the ratio of the beam profile cross-section to the surface area of the detector. We then have

$$\tilde{Q}_0 = 2\pi \int_{\theta_0}^{\pi} \langle I(\theta)\rangle \sin\theta\, d\theta, \qquad (3)$$

$\langle I(\theta)\rangle\, d\omega$ being the differential cross-section averaged over the interference effects. The velocity variation of $\tilde{Q}_0$, while still approximately of the form $v^{-2/(s-1)}$, is best fitted by a value of s less than 6 which is valid when $\theta_0 = 0$.

$\Delta\tilde{Q}$ is given to a close approximation by

$$\Delta\tilde{Q} = -2\pi U \int_{\theta_0}^{\theta_r} \cos\{2\eta_0(1-\theta/\theta_r)^2 + \tfrac{1}{4}\pi\}\, d\theta, \qquad (4)$$

where U is a slowly varying quantity, η_0 is the zero-order phase shift, and θ_r the rainbow angle. Provided $2\eta_0(1-\theta/\theta_r)^2$ is not too small

$$\Delta\tilde{Q} \simeq W\cos\{2\eta_0(1-\theta_0/\theta_r)^2 - \tfrac{1}{4}\pi\}, \qquad (5)$$

where W is slowly varying. The effect of the rainbow scattering is to change the values of the relative velocity v at which glory extrema occur so that, if v'_m is an observed value, the limiting value for $\theta_0 = 0$ is given by

$$v_m = v'_m(1-\theta_0/\theta_r)^{-2}. \qquad (6)$$

From the observed locations of the glory extrema v_m was derived and hence $a_1\epsilon r_m$ where a_1 depends on the shape of the interaction.

An alternative procedure used by Neumann and Pauly was to calculate $\tilde{Q}$ with different assumed interactions and determine $a_1\epsilon r_m$ from that which gave the best fit. Values obtained were, in 10^{-22} erg cm, 26·5,

20·8, 20·6, 20·5, and 20·3 for Li, Na, K, Rb, and Cs respectively. Values obtained from thermal measurements (see Chap. 16, pp. 1377–81) are available for Li and K and agree well, being 27·7 and 19·2 respectively in the same units.

For Na–Hg accurate information is available on the shape as well as the magnitude of the interaction (see p. 3528). In this case $a = 0.51$. Assuming the same shape of interaction for the other cases and using calculated values of the van der Waals constant ϵ and r_m may be determined separately for these cases also. Results so obtained agree well with those observed from thermal measurements.

The analysis of the Na–Cs case is complicated by the fact that the observed cross-section is a weighted mean of cross-sections for scattering by two separate interactions, the predominantly repulsive triplet and the singlet which includes a chemical attraction. However, the glory structure arising from these interactions can be distinguished quite readily because the values of ϵr_m are very different. From the observations $a_1 \epsilon r_m$ was found to be, in 10^{-22} erg cm, 136 and 15·5 for the singlet and triplet interactions respectively.

1.2. *The use of sputter sources of neutral atom beams*

The general design of sputter sources of neutral atoms with energies ranging from a few eV to a few tens of eV has been described in Chapter 21, p. 2323. An early application to the measurement of 'total' cross-sections (see p. 1317 for definition of 'total') was made by Politiek, Los, and Schiffer[†] for collisions of K atoms with energy between 0·35 and 16·5 eV with He atoms. We draw attention here to an ingenious technique using artificial radioactive isotopes which was introduced by Thompson and Farmery[‡] and applied to the elastic scattering of gold atoms with energy between 5 and 90 eV by argon atoms.[§]

The technique is essentially a time-of-flight method in which the velocity analysis is carried out mechanically. The gold atoms are produced as a pulsed beam by bombarding a radioactive gold surface with a pulsed beam of inert gas ions of energy 25 keV. The sputtered atoms, after a flight path of 1·14 m, are collected on a stainless steel rotor which is electromagnetically supported *in vacuo* and spins at 157 Hz. The deposit which they produce will have a density distribution determined by the time-of-flight spectrum of the gold atoms. As the atoms are

---

[†] POLITIEK, J., LOS, J., and SCHIFFER, J. J. M., *Physica*, **49** (1970) 165.
[‡] THOMPSON, M. W. and FARMERY, B. W., *Phil. Mag.* **18** (1968) 361.
[§] FOSTER, C., WILSON, I. H., and THOMPSON, M. W., *J. Phys. B (GB)* **5** (1972) L140.

radioactive the density distribution in the deposit may be determined by standard techniques of autoradiography.

1.3. *Combining rule for repulsive interactions*

In various parts of Chapter 22 reference has been made to the so-called geometrical combining rule for repulsive atomic potentials (see, for example, pp. 1416, 2359, 2364–6). This is an empirical rule with little but experience in application to justify it. However, Smith (F. T.)† has analysed the matter in physical terms and shown that, while the rule is reasonably justified if the intersecting systems are similar, it needs some modification if it is to be effective when they are very different.

The starting point of Smith's analysis is the fact that, because of the exclusion principle, the electron clouds in each of a pair of colliding atoms A and B are largely prevented from penetration. This may be represented approximately by supposing that the respective wave functions vanish at a plane normal to the nuclear axis which it cuts at some point P to be determined.‡ Let r_a be the distance of the point P from the nucleus A. The repulsive energy between A and B will then be the sum of the energies due to the distortion of the atomic charge clouds. We may therefore write

$$V_{AB}(r, r_a) = \tfrac{1}{2}V_{AA}(2r_a) + \tfrac{1}{2}V_{BB}\{2(r-r_a)\}, \tag{7}$$

for the interaction between A and B at a separation r, V_{AA} and V_{BB} being the interactions between the nuclei AA and BB respectively. It remains to determine r_a. This will be such as to minimize the total energy, i.e. so that the restoring forces in the two atoms are equal and opposite. For this

$$\{dV_{AA}(r)/dr\}_{r=2r_a} = \{dV_{BB}(r)/dr\}_{r=2r_b}. \tag{8}$$

If the interactions between like atoms have the simple exponential form

$$V_{ii} = W_{ii}\exp(-r/\rho_{ii}), \tag{9}$$

then according to (7) and (8) the interaction V_{ij} between unlike systems is given by

$$V_{ij} = W_{ij}\exp(-r/\rho_{ij}) \tag{10}$$

where

$$\rho_{ij} = \tfrac{1}{2}(\rho_{ii}+\rho_{jj}), \tag{11a}$$

$$(W_{ij}/\rho_{ij})^{2\rho_{ij}} = (W_{ii}/\rho_{ii})^{\rho_{ii}}(W_{jj}/\rho_{jj})^{\rho_{jj}}. \tag{11b}$$

If $\rho_{ii} \simeq \rho_{jj}$ this agrees with the geometric rule

$$W_{ij} = (W_{ii}W_{jj})^{\tfrac{1}{2}}$$

but otherwise it differs appreciably.

† SMITH, F. T., *Phys. Rev.* **A5** (1972) 1708.
‡ MATCHA, R. L. and NESBET, R. K., ibid. **160** (1967) 72.

2. Collisions between ionized and neutral systems

A number of measurements of 'total' and differential cross-sections for scattering of ions by neutral atoms and molecules have been made since Chapter 22 was written. Of particular interest are a number of observations of differential cross-sections for scattering of protons by rare-gas atoms. Comparative observations, at energies of a few tens of eV, of scattering by N_2 have been made to study the damping of interference effects by the anisotropy of the basic interaction.

Some outstanding problems in the interpretation of Li^+–He scattering data have been cleared up by observations at higher energies.

Most of the interpretation of the data on elastic scattering is linked with that of the observed inelastic scattering which is discussed in notes on new material within the purview of Chapters 23, 24, and 25.

New measurements have been made for collisions of K^+ ions with various gases while 'total' cross-sections have been determined for collision of ions of noble elements and of period II elements, with Ne and Ar atoms.

2.1. Elastic scattering of H^+ ions

2.1.1. *By rare gas atoms.* The elastic scattering of H^+ ions by rare gas atoms has been investigated recently by different experimenters over a wide energy range.

Between 4 and 10 eV differential cross-sections for collisions with He, Ne, and Ar have been made by Champion, Doverspike, Rich, and Bobbio[†] and analysed to determine interatomic potentials. Mittmann, Weise, Ding, and Henglien[‡] have used a crossed-beam method to measure differential cross-sections for collisions with all the rare-gas atoms, the energy range being between 3 and 80 eV. Measurements at much higher energies (0·5 to 3 keV) have been carried out by Abignoli, Barat, Baudon, Fayeton, and Houver[§] using the experimental arrangement described in Chapter 22, p. 2378 so that elastic and inelastic processes could be distinguished. Crandall, McKnight, and Jaecks[||] have carried out measurements over a similar energy range without distinguishing elastic

---

[†] CHAMPION, R. L., DOVERSPIKE, L. D., RICH, W. G., and BOBBIO, S. M., *Phys. Rev.* A2 (1970) 2327.

[‡] MITTMANN, H. U., WEISE, H. P., DING, A., and HENGLIEN, A., *Z. Naturf.* **26A** (1971) 1112; WEISE, H. P., MITTMANN, H. U., DING, A., and HENGLIEN, A., ibid. 1122.

[§] ABIGNOLI, M., BARAT, M., BAUDON, J., FAYETON, J., and HOUVER, J. C., *J. Phys. B (GB)* **5** (1972) 1533.

[||] CRANDALL, D. H., McKNIGHT, R. H., and JAECKS, D. H., *Phys. Rev.* A7 (1973) 1261.

from total scattering. Finally, Loster† has carried out similar measurements up to 30 keV.

Analysis of the data at the low energy end to derive the scattering potential has been discussed by Bobbio et al.‡ In experiments at low resolution the rainbow angle is not well determined and this may lead to appreciable errors in analysis. However, they found that for H⁺–Ne collisions, the interaction calculated by Peyerimhoff§ gave good agreement.

Mittmann et al. found that their results could not be fitted either with a Lennard-Jones (n, m) potential even with additional terms of the form introduced by Düren, Ribe, and Schlier (see p. 3529). They did, however, find empirical forms which gave a good fit. The values of the depth (ϵ)

TABLE 22 A.1

Parameters ϵ and r_m for interactions of H^+ with rare-gas atoms derived from the experiments of Mittmann et al.

Rare gas	He	Ne	Ar	Kr	Xe
ϵ (eV)	2·0	2·28	4·04	4·45	6·75
	(1·94)	(2·21)			
r_m (Å)	0·77	0·99	1·31	1·47	1·74
	(0·77)	(0·97)			

Bracketed values are calculated from the theoretical interactions given by Peyerimhoff.

and separation (r_m) at the potential minimum which they obtained with these forms are given in Table 22 A.1. Theoretical interactions obtained by Peyerimhoff§ for He and Ne give values for ϵ and r_m which agree quite well.

At the much higher energies used in the experiments of Abignoli et al. and of Crandall, McKnight, and Jaecks the effective interaction is predominantly repulsive though at small values of τ, the reduced scattering angle, rainbow scattering will occur.

For collisions with He there is good agreement between the elastic differential cross-section as observed by Abignoli et al. and that for total scattering observed by Crandall, McKnight, and Jaecks for 6 keV deg $\leqslant \tau \leqslant$ 30 keV deg. At higher values of τ the total scattering exceeds

---

† LOSTER, W., *7th int. conf. on the physics of electronic and atomic collisions, Amsterdam.* Abstracts, North Holland, 1971, p. 274.
‡ BOBBIO, S. M., RICH, W. G., DOVERSPIKE, L. D., and CHAMPION, R, L., *Phys. Rev.* A4 (1971) 957.
§ PEYERIMHOFF, S., *J. chem. Phys.* 43 (1965) 998.

the elastic to an increasing extent. The results of Abignoli *et al.* agree well with those calculated for an interaction of the form

$$Ar^{-1}\exp(-r/a) \qquad (12)$$

with $a = 0.91a_0$ and $A = 2e^2$. Only at separation $r > 2a_0$ does this compare well with that calculated for the ground $X\,^1\Sigma$ state of HeH$^+$ by Michels.† For smaller r it leads to quite close agreement with that for the first excited $^1\Sigma$ state.

For Ne collisions good agreement is obtained if the elastic scattering is calculated from an interaction derived by Sidis‡ which at small distances is of the form (12) with $A = 13.5e^2$, $a = 0.425a_0$. Evidence for the presence of a rainbow effect for $\tau < 800$ eV deg is obtained.

The observed elastic and total scattering agree for argon up to values of τ of only 5 keV deg. Values calculated classically for the elastic differential§ cross-section from the interaction obtained theoretically§ for the ground state of ArH$^+$ agree quite well for $1\text{ keV deg} \leqslant \tau < 2\text{ keV deg}$. For smaller τ rainbow effects are important and, for larger, inelastic processes must be allowed for.

The inelastic processes become significant at considerably lower values of τ (800 eV deg) for collisions with Kr and Xe. There is evidence from the observed elastic scattering of oscillations arising from diffraction by the absorbing core as in He$^+$–Kr and He$^+$–Xe collisions (see Chap. 22, p. 2410).

A considerable amount of information about inelastic collisions including charge transfer was also obtained by Abignoli *et al.* This will be discussed on p. 3671.

2.1.2. *By molecular nitrogen.* Interesting measurements of differential cross-sections for scattering of protons of a few eV energy by N_2 molecules have been made which show clear evidence of damping of interference effects by the anisotropic character of the interaction. Thus because of the anisotropy the location of the extrema will depend on the molecular orientation and so will tend to be smoothed out in the observed average over all orientations.

Udseth, Giese, and Gentry‖ have compared differential cross-sections for collisions of 10 eV protons with N_2 and with Ar, the measurements being made from a static gas target at 78 °K. The energy and angular

† MICHELS, H. H., *J. chem. Phys.* **44** (1966) 3834.
‡ SIDIS, V., private communication to Abignoli *et al.*
§ SIDIS, V., *J. Phys. B (GB)* **5** (1972) 1517.
‖ UDSETH, H., GIESE, C. F., and GENTRY, W. R., *7th int. conf. on the physics of electronic and atomic collisions, Amsterdam.* Abstracts, North Holland, 1971, p. 264.

resolutions were 0·2 eV and 1·5° respectively in terms of the full width at half maximum. Both distributions show rainbow undulations at nearly the same angles (five minima are clearly visible in the range from 4° to 40°) but those for N_2 are of much smaller amplitude. Similar measurements made by Doverspike and Champion[†] for 6 eV protons show the same effect. The latter authors carried out an energy analysis of the scattered protons which showed that vibrational excitation is negligible (see p. 3682), thus ruling out inelastic collisions of this kind as a significant source of damping. It is not yet possible to rule out the possibility that rotational inelastic collisions may be significant.

2.2. Elastic collisions of Li+ with He

The measurements made by Zehr and Berry of the 'total' cross-section for scattering of Li+ ions by He are described in Chapter 22, § 4.3.6 together with the analysis in terms of scattering by a central potential. This potential was found to disagree markedly with the theoretical potential due to Fischer at nuclear separations below $0·5a_0$. No fresh light was thrown on the reason for this discrepancy by the differential cross-section measurements of Lorents and Aberth described in Chapter 22, p. 2413. However, these results were of interest in that they showed no interference effects and that no inelastic collisions were observed. A more refined calculation of the Li+–He interaction made by Junker and Browne[‡] agreed well with the earlier work of Fischer and it was suggested that the discrepancy with experiment arose because of the onset of inelastic collisions at impact parameters round $0·5a_0$ or less. A further refined calculation by Catlow et al.[§] again agreed with Fischer's earlier work.

To investigate the possible importance of inelastic collisions as suggested, experiments to measure differential cross-sections at energies higher than those covered by Aberth and Lorents have recently been carried out.

The investigation not only of the elastic but also of the inelastic scattering is of interest not only in clearing up the discrepancy with theory but also because the Li+–He system is the simplest one in which the colliding systems comprise two filled K shells and no other electrons are involved.

[†] DOVERSPIKE, L. D. and CHAMPION, B. L., ibid., p. 267.
[‡] JUNKER, B. R. and BROWNE, J. C., Proc. 6th int. conf. on the physics of electronic and atomic collisions, Cambridge, Mass., 1969, p. 220.
[§] CATLOW, G. W., McDOWELL, M. R. C., KAUFMAN, J. J., SACHS, L. M., and CHANG, E. S., J. Phys. B (GB) **3** (1970) 833.

Lorents and Conklin† used essentially the same apparatus as that of Aberth and Lorents (see Chap. 22, p. 2373) and extended the energy range to 2 keV. In the experiment of François, Dhuicq, and Barat‡ the ions were extracted from a pure Li^{7+} ion emitter, focused by an einzel lens and collimated by defining slits before entering the collision chamber. Ions and neutral atoms scattered at a chosen angle could be separated and velocity-analysed by an electrostatic analyser. This could be switched on and off so that the full scattered flux and the neutral component could be separately measured by electron multiplier detectors.

The results of both experiments, when comparable, agree quite well and provide confirmation that the discrepancy between theory and experiment is really due to the onset of inelastic processes for collisions with small impact parameter. Thus François et al. found that, over the full range of the reduced scattering angle τ ($= E\theta$, see p. 2386) from 1 to 16 keV deg the *total* scattered flux agreed quite well with that calculated, due to loss to the inelastic channels. Similar results were obtained by Lorents and Conklin, after correcting for the effect of their slit geometry on the behaviour at small τ.

The absence of interference effects in the differential cross-sections seems to be due to the mechanism of the inelastic processes. These do not arise from interactions over a limited region as with curve crossing but mainly from rotational (angular momentum) coupling as will be discussed in more detail on pp. 3662 and 3681. Other aspects of the inelastic scattering including charge exchange which was measured by François et al. and confirmed by Lorents and Conklin are discussed on p. 3662.

2.3. *Collisions of ions of noble metals of period II elements with* Ne *and* Ar

Foster, Wilson, and Thompson§ have measured 'total' cross-sections for collisions of Na^+, Mg^+, Al^+, P^+, C^+, Cu^+, Ag^+, Au^+, Ni^+, and Zn^+ with energies in the range 8–25 keV, with Ne and Ar. The procedure was essentially the same as in the experiments described in Chapter 22, § 2. Subsidiary experiments were carried out to establish that under the experimental conditions inelastic collisions were unimportant. Results were analysed in terms of an inverse power interaction (see (3) of Chap. 22) rather as described in Chapter 22, § 2.

† LORENTS, D. C. and CONKLIN, G. M., *J. Phys. B* (*GB*) **5** (1972) 950.
‡ FRANÇOIS, R., DHUICQ, D., and BARAT, M., ibid. 963.
§ FOSTER, C., WILSON, I. H., and THOMPSON, M. W., *J. Phys. B* (*GB*) **5** (1972) 1332.

2.4. Collisions of K⁺ ions with various gases

Amdur, Jordan, Fung, Hance, Hulpke, and Johnson[†] have measured 'total' cross-sections for scattering of K⁺ ions of energies between 150 and 2350 eV by Ar, CO, N_2, and O_2.

[†] AMDUR, I., JORDAN, J. E., FUNG, L. W.-M., HANCE, R. L., HULPKE, E., and JOHNSON, S. E., *7th int. conf. on the physics of electronic and atomic collisions, Amsterdam.* Abstracts, North Holland, 1971, p. 955.

23

INELASTIC COLLISIONS BETWEEN ATOMIC SYSTEMS—THEORETICAL CONSIDERATIONS

There have been comparatively few developments since this chapter was written. Notes are given below on some recent calculations for slow collisions of He$^+$ and of Li$^+$ with He and for some new results on H$^+$–H collisions at intermediate energies which show some disagreement with earlier calculations carried out with the same binuclear close-coupling approximation.

1. Slow collisions of He$^+$ with He

In § 6.2.2 of Chapter 23 we discussed the experimental and theoretical evidence about the scattering of He$^+$ ions by He atoms. Although this referred particularly to elastic collisions, inelastic collisions were also concerned through the perturbations they produced in the elastic differential cross-section. The most explicit effects are those which arise from excitation of He($2\,^3S$) through the crossing of the diabatic potential energy curves for the $(1\sigma_g)(1\sigma_u)^2$ ground state with that for the $(1\sigma_g)^2(2\sigma_g)$ state. More detailed calculations have since been carried out by Olson† for excitation of He($2\,^3S$) by He$^+$ ions with energies 27·5, 30, and 50 eV close to the threshold.

Account was taken only of the interaction between the ground and excited states and the coupled equations solved electronically. The diabatic interactions V_{11}, V_{22} in the initial and final states respectively were obtained by fitting to the theoretical adiabatic potentials, using the calculations of Bardsley to estimate the coupling potential V_{12}. Some modifications were made near the crossing point to obtain better agreement with the observed scattering at small angles.

An interesting feature of this case is that V_{22}, representing the even interactions between He$^+$ and He($2\,^3S$), has a maximum at a separation larger than that for crossing. This is similar to the maximum which occurs in the interaction between He and He($2\,^3S$) (see Chap. 18, p. 1876). The existence of this maximum introduces additional interference effects at small angles and energies below 32 eV. This is because the occurrence

† Olson, R. E., *Phys. Rev.* A5 (1972) 2094.

of regions of both attractive and repulsive potential leads to both positive and negative angles of deflexion under these conditions. Hence for certain angles of scattering there are two impact parameters and so two amplitudes with different phase which have to be combined to give the total scattered amplitude (cf. Chap. 16, p. 1319).

Interference effects of this kind appear in the calculated differential cross-section as well as (a) the much broader interference peaks due to interference between the two possible paths in and out through the crossing point, and (b) the oscillations of smaller wavelength due to the nuclear symmetry.

Qualitative agreement is found with the observed results which are hardly yet of sufficient resolution to show the small angle interference effects due to the maximum in V_{22}.

McCarroll and Piacentini† have used close-coupling expansions to investigate the importance of rotational coupling in producing transitions between molecular states of He_2^+ and hence in leading to different excitation processes in He⁺–He collisions. This follows on similar lines to their calculations for H⁺–H collisions which involve excitation of the $2p$ state of H_2^+ (see Chap. 23, p. 2537).

They considered collisions in which the ion kinetic energy, in the centre-of-mass system, ranged from 0·6 to 3 keV. Account was taken of the coupling between the ground $^2\Sigma_u$ and the lowest $^2\Pi_u$ state of He_2^+ and between the lowest $^2\Sigma_g$, $^2\Pi_g$, and $^2\Delta_g$ states. By considering the correlation diagram of molecular states in detail they showed that rotational coupling can lead to the following transitions:

$$He^+ + He \rightarrow (^2\Pi_u, {}^2\Pi_g) \rightarrow He^+(1s) + He(1s, 2p)^{1,3}P \quad (1\,a)$$

$$\rightarrow {}^2\Pi_g' \quad \rightarrow He^+(2p) + He(1s^2)^1S \quad (1\,b)$$

$$\rightarrow {}^2\Delta_g \quad \rightarrow He^+(1s) + He(2p^2)^1S, {}^1D \quad (1\,c)$$

$$\rightarrow He^+(2p) + He(1s, 2p)^{1,3}P. \quad (1\,d)$$

The processes (1 a) may occur either through the $^2\Pi_u$ or the $^2\Pi_g$ state so we would expect interference effects to arise between the corresponding scattered amplitudes. Excitation of the $^{1,3}P$ states can also arise from radial coupling due to crossing between the $^2\Sigma_g$ potential energy curves referred to earlier. However, this is of importance only at relatively small values of reduced scattering angle τ ($\leqslant 0\cdot 6$ keV deg) whereas rotational coupling is most effective at considerably larger values of τ ($\geqslant 4$ keV deg).

† McCarroll, R. and Piacentini, R. D., *J. Phys. B* (*GB*) **4** (1971) 1026.

In carrying out detailed calculations the potential energy curves for $^2\Sigma_u$ and $^2\Sigma_g$ were taken to be those given by Marchi and Smith and by Lichten (see Chap. 23, p. 2347). For $^2\Pi_u$ the calculated curve due to Michels† was used for separations $r \geqslant 0{\cdot}75a_0$. In the absence of calculations for smaller r the curve was extrapolated smoothly to the united atom limit. For $^2\Pi_g$ and $^2\Delta_u$ extrapolation between united and separated atom limits was all that could be done.

Comparison of calculated results with data on charge exchange probability and on differential cross-sections for inelastic scattering is, on the whole, quite satisfactory.

2. Slow collisions of Li⁺ with He

McCarroll and Piacentini‡ have investigated rotational coupling effects in Li⁺–He collisions by methods similar to those which they used to investigate similar effects in He⁺–He impacts (see § 1). In the Li⁺–He case they considered specifically rotational coupling between $^1\Sigma$, $^1\Pi$, and $^1\Delta$ states of Li–He⁺. Through $^1\Sigma$–$^1\Pi$ coupling the following reactions are possible

$$\text{Li}^+ + \text{He} \to \text{Li}^+(1s^2)^1S + \text{He}(1s, 2p)^1P \tag{2a}$$
$$\to \text{Li}(1s^2\, 2p)^2P + \text{He}^+(1s)^2S, \tag{2b}$$

and through $^1\Sigma$–$^1\Delta$

$$\text{Li}^+ + \text{He} \to \text{Li}^+(1s^2)^1S + \text{He}(2p^2)^1D, {}^1S$$
$$\to \text{Li}(1s^2\, 2p)^2P + \text{He}^+(2p)^2S.$$

The charge transfer process

$$\text{Li}^+ + \text{He} \to \text{Li}(1s^2\, 2s) + \text{He}^+(1s)$$

can occur through crossing of Σ–Σ potential curves but as this crossing occurs close to the united atom limit the near symmetry of the Li–He⁺ system could well make the interaction weak.

In detailed calculations the ground state $X\,{}^1\Sigma$ potential energy curve was taken from the calculations of Catlow *et al.*§ and that for the $^1\Pi$ state from those of Junker and Browne‖ extrapolated smoothly to the united atom limit. For $^1\Delta$ an estimated curve was drawn between the united and separated atom limits.

Discussion of the results obtained is deferred to p. 3681 where they are considered in relation to recent experimental data.

† MICHELS, H. H., National Bureau of Standards (USA) Technical Note 438.
‡ MCCARROLL, R. and PIACENTINI, R. D., *J. Phys. B* (*GB*) **5** (1972) 973.
§ CATLOW, G. W., MCDOWELL, M. R. C., KAUFMAN, J. J., SACKS, L. M., and CHANG, E. S., ibid. **3** (1970) 833.
‖ JUNKER, B. F. and BROWNE, J. C., *7th int. conf. on the physics of electronic and atomic collisions, Amsterdam.* Abstracts, North Holland, 1971, p. 130.

3. Unsymmetrical charge transfer

Bottcher and Oppenheimer[†] have applied the perturbed stationary state method (see Chap. 23, § 7.1.1) to calculate cross-sections for charge transfer between Na^+ and Li for energies in the C.M. system in the range 100 eV–12 keV. A model potential method, in the distorted wave approximation (Chap. 23 (431)), was used to determine the required potential energy curves and wave-functions for Na–Li^+.

The calculated cross-sections show oscillatory structure which can be correlated quite well with similar structure observed by Perel and Daly (see Chap. 24, p. 2796) although the absolute magnitude of the cross-section is about 30 per cent higher than the observed. The relation of the oscillations to the existence of a crossing point between the diabatic potential energy curves is discussed.

4. Excitation of vibration in slow collisions of ions with molecules

Fremerey and Toennies[‡] have calculated cross-sections for excitation of vibration in H_2 molecules by impact of He atoms, the kinetic energy in the C.M. system being 1·09 eV.

For the He–H_2 interaction they used the form given by Krauss and Mies (see p. 2362) with and without an added van der Waals potential obtained from analysis of total cross-section measurements (see p. 3538). In the initial calculations the H_2 molecule was taken to be non-rotating but coupling between channels with vibrational quantum numbers 0, 1, and 2 was included.

5. Collisions at intermediate energies. H^+–H collisions

Rapp, Dinwiddie, Storm, and Sharp[§] have recomputed cross-sections for symmetrical charge transfer and for excitation of the $2s$ and $2p$ states, with or without charge transfer, using a binuclear expansion including the 1 and 2 quantum states and allowing for electron momentum transfer. The new results disagree quite appreciably in some cases with earlier calculations using the same approximation. Of these the most accurate are probably those of Cheshire, Gallaher, and Taylor who in the course of calculations using pseudo-state expansions (see Chap. 23, p. 2613) recalculated cross-sections obtained earlier by Wilets and Gallaher. The results of Cheshire et al. are included in Figs. 23.84–8.

[†] BOTTCHER, C. and OPPENHEIMER, M., J. Phys. B (GB) 5 (1972) 492.
[‡] FREMEREY H. and TOENNIES, J. P., 7th int. conf. on the physics of electronic and atomic collisions, Amsterdam. Abstracts North Holland, 1971, p. 249.
[§] RAPP, D., DINWIDDIE, D., STORM, D., and SHARP, T. E., Phys. Rev. A5 (1972) 1290.

Rapp *et al.* find reasonably good agreement with the results of Cheshire *et al.* for symmetrical charge transfer (Fig. 23.84) but for excitation of $2s$, either directly or with charge transfer, there are quite serious discrepancies. Less serious but quite considerable differences occur also for excitation of $2p$. Tables 23 A.1 and 2 give some comparative values.

TABLE 23 A.1

Cross-sections for excitation of $H(2s)$ *by proton impact, calculated* (a) *by Rapp et al. and* (b) *by Cheshire, Gallaher, and Taylor using a four-state binuclear expansion*

Proton energy (keV)	1·0	2·0	4·0	6·0	7·0	9·0	10·0	15·0	16·0	25·0	40·0	100·0
Cross-sections (10^{-17} cm^2)												
Direct excitation												
(a)	0·125	0·800	0·402	0·265	—	0·584	—	—	1·142	0·967	1·334	1·165
(b)	0·222	0·554	0·246₅	—	0·747₅	—	1·025	1·343	—	1·130	1·591	1·371
Excitation with transfer												
(a)	0·136	0·789	0·230	0·637	—	1·93	—	—	2·90	3·46	2·30	0·240
(b)	0·226	0·528	0·202₅	—	1·840	—	2·985	3·123	—	3·655	2·48	0·263

TABLE 23 A.2

Cross-sections for excitation of $H(2p)$ *by proton impact, calculated* (a) *by Rapp et al. and* (b) *by Cheshire, Gallaher, and Taylor using a four-state binuclear expansion*

Proton energy (keV)	1·0	2·0	4·0	6·0	7·0	9·0	10·0	15·0	16·0	25·0	40·0	100·0
Cross-sections (10^{-17} cm^2)												
Direct excitation												
(a)	2·547	2·841	4·458	4·299	—	2·341	—	—	1·490	4·051	7·034	9·441
(b)	2·439	3·105	4·543	—	3·370	—	1·584	1·141	—	3·440	6·905	8·618
Excitation with transfer												
(a)	2·533	2·850	4·330	4·464	—	3·469	—	—	1·777	1·136	0·696	0·062
(b)	2·436	3·128	4·228	—	3·320	—	2·610	1·522	—	0·924	0·572	0·053

The reasons for the discrepancies are not clear. Rapp *et al.* also raise questions about the importance of including pseudo-states (see Chap. 23, p. 2613) as they consider that at not too high energies the convergence of the 4-state ($1s$, $2s$, $2p_0$, $2p_{\pm 1}$) binuclear expansion should be adequate for calculating cross-sections for excitation of $2s$ and $2p$ states.

24

CHARGE CHANGING AND IONIZING COLLISIONS—EXPERIMENTAL METHODS AND RESULTS INCLUDING THEORETICAL ANALYSIS

WE include here notes of some recent new results obtained using techniques already described in Chapter 24, § 1. These are mostly concerned with the study of the charge-changing collisions of metastable species, inner shell ionization and dissociation of molecules by ion impact, all subjects which are actively developing.

1. Energy and angular distributions of electrons emitted in ionizing collisions (see Chap. 24, § 2.1.5)

Further measurements have been carried out on the energy and angular distributions of electrons ejected in ionizing collisions of protons with neutral systems. Stoltenfoht† has used a crossed-beam technique with 200–500 keV protons colliding with He, N_2, and CH_4 while Toburen‡ has extended the proton energy range to 2 MeV for He and 1·5 MeV for H_2 targets. In the latter experiments, which used a static gas target, the electron energy spectra were obtained at angles from 20 to 130° with respect to the incident direction of the protons. One of the aims of the work was to investigate further the enhanced intensity of electrons emitted near the forward direction attributed by Macek (see pp. 2485 and 2766–7) to interaction with the incident proton. Some of the discrepancies between this theory and observation are still present at the higher incident energies.

Barat, Dhuicq, François, McCarroll, Piacentini, and Salin,§ in the course of measurements of differential cross-sections for scattering of He^+ ions with definite chosen energy loss by He (see also p. 3660), were able to measure energy distributions of electrons ejected at a fixed angle with respect to the neutral ion. Considering these distributions as a

---

† STOLTENFOHT, N., 7th int. conf. on the physics of electronic and atomic collisions, Amsterdam. Abstracts, North Holland, 1971, p. 1123.
‡ TOBUREN, L. H., ibid., p. 1121; Phys. Rev. A3 (1971) 216.
§ BARAT, M., DHUICQ, D., FRANÇOIS, R., MCCARROLL, K., PIACENTINI, R. D., and SALIN, A., J. Phys. B (GB) 5 (1972) 1343.

function of reduced scattering angle τ ($= E\theta$, see p. 2386) for incident ion energies between 1 and 3 keV it was found that the maximum of the distribution shifts to larger electron energies as τ increases. Eventually with increasing τ a second maximum appears which also shifts to larger energies with τ. At the largest observed value of τ three such maxima were observed. An explanation is suggested in terms of the crossing of the diabatic curve for the $\{(1s\sigma_g)(2p\sigma_u)^2\}\,^2\Sigma_g$ state of He_2^+ with the continuum $^2\Sigma_g$ states of the system He_2^{++} +free electron.

Gerber et al.† have studied the energy and angular distributions of electrons emitted in 100–5000 eV He^+ and He collisions with Ne, Ar, Kr, and Xe. As in collisions at higher impact energies‡ the energy spectra exhibit discrete peaks arising from autoionization superimposed on a continuous background. The latter is more pronounced for collisions involving He^+, a feature which, it is suggested, is more consistent with autoionization of the quasi-molecule formed during the collisions rather than the Russek§ statistical model which is valid at higher impact energies.

2. Total cross-sections for charge transfer

2.1. *Charge transfer involving triply-charged positive ions* (see Chap. 24, § 2.2.4.7)

Zwally and Cable‖ have studied one-electron capture in B^{3+}–He collisions, a reaction which enables the Landau–Zener theory of pseudo-crossing to be compared with experiment. The potential energy curves for the two final states corresponding to the formation of B^{2+} in the $2s$ and $2p$ states cross the initial state at separations of $4\cdot 18 a_0$ and $4\cdot 37 a_0$ respectively. The large separation of these crossings allows them to be treated individually. The sum of the cross-sections for capture into the $2s$ and $2p$ states, calculated using the Landau–Zener theory, although somewhat smaller, is shown to be in good general accord with the total capture cross-section measured over the range 400 eV–30 keV.

2.2. *Charge transfer involving halogen negative ions*

Dimov and Rosljakov†† have measured charge transfer cross-sections Q_{tr} for Cl^-, Br^-, and I^- ions with energies between 200 and 2500 eV in collision with O_2, Cl_2, I_2, CO_2, NO_2, SF_6, C_6F_6, and CCl_4. The ions were

---

† GERBER, G., MORGENSTERN, R., and NIEHAUS, A., *J. Phys. B* (*GB*) **5** (1972) 1396.
‡ RUDD, M. E., JORGENSEN, T., and VOLZ, D. J., *Phys. Rev.* **151** (1966) 28.
§ RUSSEK, A. and MELI, J., *Physica 's Grav.* **46** (1970) 222.
‖ ZWALLY, H. J. and CABLE, P. G., *Phys. Rev.* A**4** (1971) 2301.
†† DIMOV, G. I. and ROSLJAKOV, G. V., *7th int. conf. on the physics of electronic and atomic collisions, Amsterdam.* Abstracts, North Holland, 1971, p. 800.

produced in a surface ionization source and the slow negative ions produced were mass-analysed. Charge transfer in O_2 and NO_2 leads primarily to O_2^- and NO_2^- ions respectively but in CO_2, SF_6, C_6F_6, and CCl_4 the main products are atomic negative ions O^-, F^-, Cl^-. In I_2 and Cl_2 the yields of atomic and diatomic ions are comparable.

2.3. *Formation of excited negative ions by charge transfer*

Oparin, Il'in, Serenkov, Solov'ev, and Fedorenko[†] have used electric field ionization to observe the presence of excited negative ions resulting from charge transfer. The ions were passed through an electric field $\leqslant 4 \times 10^5$ V cm^{-1}. If W_n is the probability of field ionization of ions in the nth excited state and f_n the fraction of ions in that state, the ion current I after passing through the field will be given in terms of the initial current I_0 by

$$I/I_0 = \Sigma_n f_n \exp(-W_n t), \qquad (1)$$

where t is the flight time of ions in the field (typically 3×10^{-10} s). W_n is given by

$$W_n = \alpha_n E^{k_n} \exp(-6 \cdot 83 \times 10^7 E^{\frac{3}{2}}/F), \qquad (2)$$

where E is the binding energy of the electron in eV, F the electric field strength in V cm^{-1}, and α_n and k_n are constants depending on the ion state. In the experiments I/I_0 and also dI/dF were measured as functions of F.

The first application of this technique was to He^- in which the binding energy of the attached electron relative to the $He(2\,^3S)$ state was found to be 0·076 eV. Later measurements[‡] have been made for C^- and Si^-. In the former case evidence was obtained of an excited state with detachment energy 0·035 eV and relative weight consistent with a $2p^3\,^2D$ state. In Si^- preliminary evidence suggested the existence of an excited state and binding energy 0·03 eV.

3. Charge-changing collisions involving fast excited species

3.1. *Metastable hydrogen atoms* (see Chap. 24, § 2.2.8.4)

Electron loss cross-sections $Q_{2s,1}$ for the process

$$H(2s)+X \rightarrow H^+ + X(\Sigma n, l) + e$$

have been measured for various target gases by Dose and Gunz[§] (Ar,

[†] OPARIN, V. A., IL'IN, R. N., SERENKOV, I. T., SOLOV'EV, E. S., and FEDORENKO, N. V., *Soviet Phys. JETP Lett.* **12** (1970) 162.

[‡] OPARIN, V. A., IL'IN, R. N., SERENKOV, I. T., SOLOV'EV, E. S., and FEDORENKO, N. V., *7th int. conf. on the physics of electronic and atomic collisions, Amsterdam.* Abstracts, North Holland, 1971, p. 796.

[§] DOSE, V. and GUNZ, R., *J. Phys. B (GB)* **5** (1972) 636.

H_2, N_2, O_2), Hughes and Choe† (He, Ar, N_2), and Spiess et al.‡ (He, Ne, Ar, Kr, Xe, H_2, N_2) in the respective energy ranges 2–60 keV, 20–120 keV, and 2·5 keV. These results are in general accord with the earlier data of Gilbody et al.§ although the cross-sections measured by Dose and Gunz are significantly larger. Calculated cross-sections based on the classical impulse approximation of Bates and Walker‖ tend to overestimate the loss cross-sections for the heavier targets, particularly at the lower impact energies, although agreement is generally well within a factor of two.

Bruckmann et al.†† have studied the formation of deuterons by electron loss in the passage of 1 keV $D(2s)$ atoms through iodine vapour. A large deuteron yield of up to about five times that observed for $D(1s)$ atoms was explained in terms of the near resonant process

$$D(2s)+I_2 \to D^++I_2^- -0\cdot3 \text{ eV}.$$

Electron capture cross-sections $Q_{2s,\bar{1}}$ for the process

$$H(2s)+X \to H^-+X^+$$

measured by Dose and Gunz‡‡ for targets of argon, nitrogen, and oxygen in the energy range 1·4–8 keV indicate that this particular collisional destruction mode for $H(2s)$ atoms becomes relatively unimportant at impact energies above a few keV. The results of these workers show that the ratio $Q_{2s,\bar{1}}/Q_{2s,1}$ decreases from about 25, 8, and 33 per cent at 2 keV to 6, 6, and 6 per cent at 6 keV for argon, nitrogen, and oxygen respectively.

Total collisional destruction cross-sections

$$Q_D = (Q_{2s,1}+Q_{2s,1s}+Q_{2s,\bar{1}})$$

measured by Krotkov et al.§§ for 0·25–30 keV $H(2s)$ atoms in helium and argon are uncorrected for elastic scattering, a fact which may explain why these values are significantly larger than those obtained by Gilbody et al.‖‖ Values of Q_D for helium calculated by Krotkov et al. using a pseudopotential and eikonal approximation are a factor of four lower than their experimental values.

† HUGHES, R. H. and CHOE, S. S., Phys. Rev. **A6** (1972) 1413.
‡ SPIESS, G., VALANCE, A., and PRADEL, P., ibid. **A6** (1972) 746.
§ GILBODY, H. B., BROWNING, R., REYNOLDS, R. M., and RIDDELL, G. I., J. Phys. B (GB) **4** (1971) 94.
‖ BATES, D. R. and WALKER, J. C. G., Planet. Space Sci. **14** (1966) 1367.
†† BRUCKMANN, H., FINKEN, D., and FRIEDRICH, L., Nucl. Inst. Meth. **87** (1970) 155.
‡‡ DOSE, V. and GUNZ, R., J. Phys. B (GB) **5** (1972) 1412.
§§ KROTKOV, R. V., BYRON, F. W., MEDEIROS, J. A., and YANG, K. H., Phys. Rev. **A5** (1972) 2078.
‖‖ GILBODY, H. B., BROWNING, R., REYNOLDS, R. M., and RIDDELL, G. I., J. Phys. B (GB) **4** (1971) 94.

3.2. Metastable helium atoms (see Chap. 24, § 2.2.9)

Peterson and his collaborators† have studied the excitation and de-excitation of $2\,^1S$ and $2\,^3S$ metastable He atoms in collisions with helium and nitrogen in the range 150 eV–2200 eV.

Cross-sections Q_D for the de-excitation of $2\,^1S$ atoms in helium are from 20 to 75 per cent larger than the corresponding triplet cross-sections; both cross-sections have a velocity dependence in accord with the relation $Q_D^{\frac{1}{2}} = a - b \ln v$ characteristic of a symmetric excitation transfer process. There is also evidence of an oscillatory structure superimposed on the monotonic velocity dependence.

In the case of nitrogen targets, values of Q_D decrease with increasing energy above 300 eV, and $Q_D(2\,^1S)$ is between 20 and 50 per cent greater than $Q_D(2\,^3S)$. In addition $Q_D(2\,^1S) \simeq 2 \times 10^{-15}$ cm² at 300 eV, becoming smaller at lower energies. At these energies metastable He atoms are believed to be de-excited predominantly by the Penning ionization process,
$$\mathrm{He}(2\,^1S \text{ or } 2\,^3S) + \mathrm{N_2} \to \mathrm{He}(1\,^1S) + \mathrm{N_2^+} + e,$$
although at energies above 10 keV the results of Browning et al.‡ and Gilbody et al.§ show that helium metastable atoms are destroyed predominantly by collisions involving electron loss.

The results of Peterson and his collaborators also show that metastable helium atoms are collisionally excited to produce HeI emissions much more readily than ground-state atoms. In addition Gilbody et al.∥ have observed a strong enhancement of both projectile and target $4\,^3S$ and $4\,^3D$ emissions in 70 keV He–He collisions when the metastable population of the helium atom beam was increased. For the projectile emissions the following processes were considered
$$\mathrm{He}(2\,^3S) + \mathrm{He}(1\,^1S) \to \mathrm{He}(4\,^3S \text{ or } 4\,^3D) + \mathrm{He}$$
$$\mathrm{He}(1\,^1S) + \mathrm{He}(1\,^1S) \to \mathrm{He}(4\,^3S \text{ or } 4\,^3D) + \mathrm{He}$$
with respective cross-sections $Q_m(p)$ and $Q_g(p)$. Target emissions were

---

† HOLLSTEIN, M., SHERIDAN, J. R., PETERSON, J. R., and LORENTS, D C., Phys. Rev. **187** (1969) 118; Can. J. Chem. **47** (1969) 1858. HOLLSTEIN, M., LORENTS, D. C., and PETERSON, J. R., Bull. Am. phys. Soc. **13** (1968) 197. PETERSON, J. R., LORENTS, D. C., and MOSELY, J. T., 7th int. conf. on the physics of electronic and atomic collisions, Amsterdam. Abstracts, North Holland, 1971, p. 1089.

‡ BROWNING, R., LATIMER, C. J., and GILBODY, H. B., J. Phys. B (GB) **3** (1970) 667.

§ GILBODY, H. B., DUNN, K. F., BROWNING, R., and LATIMER, C. J., ibid. **3** (1970) 1105.

∥ GILBODY, H. B., BLAIR, W. G. F., SIMPSON, F. R., and McCULLOUGH, R. W., J. Phys. B (GB) **5** (1972) L101; ibid. **6** (1973) 1265.

considered to arise through electron exchange in the reactions

$$He(2\,^3S)+He(1\,^1S) \to He+He(4\,^3S \text{ or } 4\,^3D)$$

and

$$He(1\,^1S)+He(1\,^1S) \to He+He(4\,^3S \text{ or } 4\,^3D)$$

with respective cross-sections $Q_m(t)$ and $Q_g(t) = Q_g(p)$. For the $4\,^3S$ emissions, it was shown that $Q_m(p)/Q_g(p) \simeq 2 \cdot 9$ and $Q_m(t)/Q_g(t) \simeq 5 \cdot 6$ while for the $4\,^3D$ emissions $Q_m(p)/Q_g(p) \simeq 7 \cdot 5$ and $Q_m(t)/Q_g(t) \simeq 12$.

Dunn and Gilbody,[†] by using fast He atom beams in which the metastable population could be controlled, have obtained estimates of the cross-sections for the He⁻ formation process

$$He(2\,^3S)+X \to He^-(^4P)+X^+$$

for targets of He, Kr, and N_2 in the energy range 60–200 keV. He⁻ formation in collisions involving ground-state He atoms was found to be negligibly small in each case. The formation of He⁻ in the passage of He⁺ ions through vapours of Na and Mg has been studied by Il'in et al.[‡] and Baragiola and Salvatelli.[§] The He⁻ yield exhibits a quadratic pressure dependence consistent with the two-step formation process observed previously by Schlachter et al.[||] for Cs and Gilbody et al.[††] for H_2. However, Baragiola and Salvatelli have observed a linear pressure dependence in Pb vapour which they have noted, unlike the other targets, does not violate the Wigner total electron spin conservation rule for the two-electron capture process

$$He^+(^2S)+Pb(^3P) \to He^-(^4P)+Pb^{2+}(^1S).$$

3.3. Excited hydrogen molecules

The formation and collisional destruction of H_2 molecules in the $c\,^3\Pi_u$ long-lived electronically excited state (p. 880) has been studied by Morgan et al.[‡‡] The molecules were formed in electron capture by 2–84 keV H_2^+ or D_2^+ in Mg vapour. In this energy range, total electron loss cross-sections for $c\,^3\Pi_u$ hydrogen molecules in hydrogen are shown

---

[†] DUNN, K. F. and GILBODY, H. B., 7th int. conf. on the physics of electronic and atomic collisions, Amsterdam. Abstracts, North Holland, 1971, p. 1085.

[‡] IL'IN, R. N., OPARIN, V. A., SERENKOV, I. T., SOLOV'EV, E. S., and FEDORENKO, N. V., ibid., p. 793.

[§] BARAGIOLA, R. and SALVATELLI, E., ibid., p. 791.

[||] SCHLACHTER, A., LLOYD, D., BJORKHOLM, P., ANDERSON, L., and HAEBERLI, W., Phys. Rev. 174 (1968) 201.

[††] GILBODY, H. B., BROWNING, R., DUNN, K. F., and McINTOSH, A. I., J. Phys. B (GB) 2 (1969) 465.

[‡‡] MORGAN, T. J., BERKNER, K. H., and PYLE, R. V., Phys. Rev. A5 (1972) 1591.

to be up to a factor of five larger than the corresponding cross-sections for $X\,^1\Sigma_g^+$ ground-state molecules.

The formation of highly excited states ($n \geqslant 10$) of hydrogen molecules formed in electron capture by 50–450 keV H_2^+, HD^+, and D_2^+ ions in hydrogen has been studied by Barnett et al.† using an electric field ionization technique. The fraction of molecules in states $n \geqslant 10$ was about 10^{-3} with D_2^* yield approximately twice that for H_2^*. Barnett and Ray‡ have also observed highly excited H_3 molecules formed during the passage of fast H_3^+ ions through hydrogen.

4. Differential scattering in charge transfer collisions involving H^+ (see Chap. 24, § 2.2.12.3)

Measurements of differential cross-sections for charge transfer collisions of protons with rare-gas atoms, the impact energy being in the range 0·5 to 3 keV, have been made by Abignoli, Barat, Baudon, Fayeton, and Houver§ in the course of a detailed study of energy loss spectra in H^+–rare-gas-atom collisions as a function of scattering angle and impact energy. The elastic scattering data have been discussed already on p. 3654 and that which refers to excitation on p. 3680.

The only case in which the diabatic potential energy curve for the initial state crosses that for the charge-exchanged state is that for an Xe target. In that case the charge transfer probability, as a function of reduced scattering angle τ ($= E\theta$, see p. 2386), exhibits the typical oscillations expected. For the other cases less pronounced oscillatory behaviour is observed, showing that the range of nuclear separation over which the probability of transfer is appreciable is quite localized.

5. Dissociation of fast molecular ions on impact with neutral targets

Stearns, Berkner, Pyle, Briegleb, and Warren‖ have measured cross-sections for the separate dissociation processes which can occur on impact of HeH^+ ions, with energies of 520, 765, and 1025 keV, with H_2 and N_2. These processes result in the respective pairs of atomic ions and neutral atoms as follows: (He, H) (He, H^+), (He^+, H) (He^+, H^+), (He^{++}, H) (He^{++}, H^+).

† BARNETT, C. F., RAY, J. A., and RUSSEK, A., Phys. Rev. A5 (1972) 2110.
‡ BARNETT, C. F. and RAY, J. A., ibid. A5 (1972) 2120.
§ loc. cit., p. 3654.
‖ STEARNS, J. W., BERKNER, K. H., PYLE, R. V., BRIEGLEB, B., and WARREN, L., 7th int. conf. on the physics of electronic and atomic collisions, Amsterdam. Abstracts, North Holland, 1971, p. 419.

The beam and its collision products emerging from the gas-filled collision chamber were magnetically analysed and the separate products detected by surface barrier detectors. Cross-sections for production of different pairs were determined by coincidence-counting methods. The total break-up cross-section in H_2 varied from about $1·5 \times 10^{-16}$ cm² at the lowest impact energy to 10^{-16} cm² at the highest. Of the partial cross-sections that for (He^+, H^+) production was the largest at all energies and that for (He^+, H) next. At the lowest energy that for (He, H) slightly exceeded that for (He, H^+) as the third strongest but at the highest energy (He, H) was nearly an order of magnitude smaller, it being the only process for which the cross-section fell rapidly with increasing impact energy.

Similar measurements† have been carried out for H_3^+ with energies of 410, 900 and 1800 keV in H_2 and N_2. In this case the order of importance of the partial cross-sections was found to be (H, $2H^+$), (H^+, 2H), (H^+, H_2), and (H, H_2^+) and at the lowest energy (H^+, H_2) and (3H). However, the cross-section for (3H) fell rapidly with energy by a factor of nearly 400 in contrast with the slow change of the other cross-sections.

Dong, Bizot, Durup, Fourmann, and Ozenne,‡ using the method described in Chapter 24, p. 2918 have measured the velocity distributions of N^+ and O^+ ions resulting from dissociation of NO^+ ions, with energy between 2 and 5 keV, on impact with He. From analysis of these data it was deduced that N^+ is produced in the ^3P state in conjunction with $O(^3P)$ while O^+ is produced in the ^4S state in conjunction with $N(^4S^0)$.

6. Inner shell ionization by heavy particles

It has been pointed out in the text (§ 2.3.3.7) that plane-wave Born approximation is inadequate to account for the observed inner shell ionization cross-sections by H, He, Li ions on heavier target atoms at low energies but that good agreement can be obtained when allowance is made for the Coulomb distortion of the incident ions and for their polarization effect on the target atoms, leading to an effective increase in the inner shell ionization energy. By introducing appropriate universal variables, Basbas et al.§ showed that measurements for many different

---

† BERKNER, K. H., MORGAN, T. J., PYLE, R. V., and STEARNS, J. W., *7th int. conf. on the physics of electronic and atomic collisions, Amsterdam.* Abstracts, North Holland, 1971, p. 422.

‡ PHAM DONG, BIZOT, M., DURUP, J., FOURMANN, B., and OZENNE, J.-B., ibid., p. 427.

§ BASBAS, G., BRANDT, W., and LAUBERT, R., *Bull Am. phys. Soc.* **15** (1970) 555; see also BRANDT, W., *Proc. int. conf. on inner shell ionization phenomena and future applications, Atlanta, Georgia, April 1972,* published by USAEC, Technical Information Center, Oak Ridge, Tenn., 1973, p. 948.

projectiles and targets in the low velocity region were in good agreement with the absolute K ionization cross-sections, Q_K^I, calculated on the basis of plane-wave Born approximation corrected for Coulomb distortion of the incident projectile according to the method of Bang and Hansteen† and for the increased ionization energy arising from the presence of the positively charged projectile according to the method of Brandt et al.‡

In this universal plot the quantity

$$\frac{\epsilon \theta_K}{9 E_{10}(\pi d q_0 \epsilon)} (Q_K^I / Q_0)$$

is plotted against $\eta / \epsilon^2 \theta_K^2$ where

$$\theta_K = E_K / Z_K^2 \, \mathrm{Ry},$$
$$Q_0 = 8\pi (a_0 z / Z_K^2)^2,$$
$$\eta = (v/v_K)^2 = 40 E \text{ (in MeV)} / M Z_K^2,$$

ze, M, v, E are respectively the charge, mass number, velocity, and kinetic energy of the projectile; $Z_K e$, E_K, v_K respectively the effective nuclear charge, ionization energy, and K electron velocity of the target atom; a_0, Ry the Bohr radius and the Rydberg constant ($= 13 \cdot 6$ eV); $2d$ ($= z Z_K e^2 / E$) the minimum distance of approach in the collision, $\hbar q_0$ ($= E_K / v$) the minimum momentum transfer for K shell ionization; $E_{10}(x) = \int_1^\infty e^{-xt} t^{-10} \, dt$; and ϵ is a calculable parameter which measures the effective ratio by which the K ionization energy is increased by the presence of the projectile. Provided $Z \gg z$, the universal curve does not depend on the degree of stripping of the incident ion because the K shell radius of the projectile is so much larger than that of the target atom that effectively the target electrons see the projectile nucleus unscreened by electrons. For example, the measured variation with energy of the K shell ionization cross-sections of Al by α-particles of energy 1–5 MeV obtained by Sellers et al.§ fits smoothly to the variation of this cross-section for ionization by $^4\mathrm{He}^+$ ions of energy up to 200 keV obtained by Brandt et al.‡

At higher energies, but still corresponding to ion velocities below the orbital velocities of the inner electrons of the target, the observed inner shell ionization cross-sections are larger than those predicted by Born approximation corrected as described above, the effect increasing with

† BANG, J. and HANSTEEN, J. M., Kgl. Danske Videnskab. Selskab. Mat. Fys. Medd. **31** (1959), No. 13.
‡ BRANDT, W., LAUBERT, R., and SELLIN, I., Phys. Rev. **151** (1966) 56.
§ SELLERS, B., HANSER, F. A., and WILSON, H. H., ibid. **182** (1969) 90.

the atomic number of the incident ion. This effect has been interpreted by Ashley et al.† in terms of the presence of a z^3 term in the dependence of the ionization cross-section on the atomic number of the incident ion. Classically they show such a term would arise from distant collisions of the incident ion with an atom containing bound electrons that can be set in oscillation by the passing ion. The quantal counterpart of this model would be given by Born's second approximation but quantal calculations of inner shell ionization of this kind have not been made.

The need for a term such as this had been suspected already as a result of atomic stopping power measurements of Andersen et al.‡ They found that for H$^+$ and He^{++} ions passing through tantalum, the stopping powers, S_{He}, S_H, were related by

$(S_{He}-4S_H)/S_{He} = 0 \cdot 01 \pm 0 \cdot 003$ at the velocity of 7 MeV protons

and $= 0 \cdot 033 \pm 0 \cdot 003$ at the velocity of 2 MeV protons.

Similarly Barkas§ had speculated on a difference in the range–energy relation for positive and negative particles. For π^+ and π^- of velocity corresponding to an energy of $1 \cdot 2$ MeV/a.m.u., the ranges $R_{\pi^\pm}$ determined by Heckman‖ are related by $(R_{\pi^-}-R_{\pi^+})/R_{\pi^+} = 1 \cdot 14$ compared with a value of $1 \cdot 09$ estimated by Ashley et al.

A similar difference in the range in nuclear emulsion of Σ^- and Σ^+ hyperons of energy $12 \cdot 44$ MeV has been observed by Tovee et al.†† They infer that for Σ's of this energy $R_{\Sigma^-}-R_{\Sigma^+} = 17 \cdot 5 \pm 2 \cdot 5$ μm in emulsion compared with the estimate of 17 ± 5 μm from the theory of Ashley et al.

Considerable attention has been paid to the characteristic X-radiation emitted in collisions of very heavy ions with various targets where the condition $z \ll Z$ is not satisfied. Under low resolution it was observed that the K_α and K_β X-radiation emitted in such collisions was shifted toward higher energy compared with the energies of these radiations excited by electron impact or by fluorescence. The magnitude of this shift was found to depend on the energy of the incident ions. Thus Betz et al.‡‡ observed such shifts for Cl, Br, and I ions incident on targets

† ASHLEY, J. C., RITCHIE, R. H., and BRANDT, W., Phys. Rev. B5 (1972) 2393.
‡ ANDERSEN, H. H., SIMONSEN, H., and SØRENSEN, H., Nucl. Phys. A125 (1969) 171.
§ BARKAS, W. H., Nuovo Cim. 8 (1958) 201; BARKAS, W. H., DYER, J. N., and HECKMAN, H. H., Phys. Rev. Lett. 11 (1963) 26.
‖ HECKMAN, H. H., Proc. int. conf. on hypernuclear physics, Argonne, 1969, Vol. 1, p. 199.
†† TOVEE, D. N., DAVIS, D. H., SIMONOVIC, J., BOHM, G., KLABUHN, J., WYSOTZKI, F., CSEJTHEY-BARTH, M., WICKENS, J. H., CANTWELL, T., NI GHOGAIN, C., MONTWILL, A., GARBOWSKA-PNIEWSKA, K., PNIEWSKI, T., and ZAKRZEWSKI, J., Nuc. Phys. B33 (1971) 493.
‡‡ BETZ, H. D., DELVAILLE, J. P., KALATA, K., SCHNOPPER, H. W., SOHVAL, A. R., JONES, K. W., and WEGNER, H. E., Proc. int. conf. on inner shell ionization phenomena

of Cu, Ni, Ti, Zr, Ag. Experiments carried out using higher resolution detection of the X-rays have established that the observed shift is due to strong excitation of satellite lines,† the relative intensities of satellites arising from higher states of multiple inner shell ionization being dependent on the incident ion energy. The intensities of satellite lines produced in such collisions are, under some conditions, much greater than that of the parent line. This is a situation which is not met with in inner shell ionization produced by electron or photon impact.

Strong emission of Auger satellite lines had already been observed in heavy ion collisions by Rudd and his colleagues.‡

It has been pointed out (p. 2982) that some of the multiple inner shell ionization in heavy ion collisions giving rise to strong satellite Auger electron or characteristic X-ray emission can be accounted for by the Fano–Lichten theory of pseudo-molecule formation, with electron promotion occurring between near-crossing electron orbitals. Larkins§ has applied the approach of Lichten and Fano to a number of systems, estimating the energies of single molecular orbitals for different projectile ion–target atom separations using a Hartree–Fock field. He found reasonable agreement with the observed phenomena in the case of collisions between the symmetrical systems, (Ar^+–Ar), (Ne^+–Ne). The model is inadequate to account for the observed phenomena for collisions between asymmetrical systems however. For example in Ne^+–Ar collisions Ne K emission is observed at energies less than 5 keV/a.m.u., but electron promotion cannot take place from the Ne K shell to the Ar L shell since the latter is completely filled. Similarly Al K emission is observed for N^+, O^+, Ne^+ ions of energy less than 15 keV/a.m.u. but electron promotion from the Al K shell cannot be accounted for on such a model.

That the Fano–Lichten model cannot provide the complete picture is evidenced by the high probability of double inner shell ionization produced in Al due to He^+ bombardment.‖ The ratio of KL (double) to K (single) ionization of Al is found to be ~ 0.1 for electrons or photons,

and future applications, Atlanta, Georgia, April 1972, published by USAEC, Technical Information Center, Oak Ridge, Tenn., 1973, p. 1374.

† RICHARD, P., ibid., p. 1641.

‡ RUDD, M. E., JORGENSON, T., and VOLZ, D. J., Phys. Rev. **151** (1966) 28. EDWARDS, A. K. and RUDD, M. E., ibid. **170** (1968) 140. VOLZ, D. J. and RUDD, M. E., ibid. A**2** (1970) 1395. CACAK, R. K. and JORGENSON, T., ibid. A**2** (1970) 1322. CACAK, R. K., KESSEL, Q. C., and RUDD, M. E., ibid. A**2** (1970) 1327.

§ LARKINS, F. P., J. Phys. B (GB) **5** (1972) 571; Proc. int. conf. on inner shell ionization phenomena and future applications, Atlanta, Georgia, April 1972, published by USAEC, Technical Information Center, Oak Ridge, Tenn., 1973, p. 1543.

‖ KNUDSON, A. R., BURKHALTER, P. G., and NAGLE, D. J., preprint, 1973.

and is also small for He⁺ ions of energy 1 MeV/a.m.u. For He⁺ ions of energy 0·4 MeV/a.m.u., however, the ratio is $\sim$ 2·2. A He⁺ ion of energy 1 MeV/a.m.u. has a velocity equal to that of an electron of energy 60 eV, i.e. insufficient to ionize the K shell of Al. It is known that double inner shell ionization produced by electron impact can be understood in terms of 'shake-up' or 'shake-off' processes accompanying the production of an initial K shell vacancy. Some other process must be responsible for the much greater proportion of double ionization observed in ion collisions. Knudson et al. suggest that, at the energies at which it is large, ion orbitals which are most likely to produce K ionization are also particularly likely to produce L ionization so that the double ionization takes place as a single event.

K Auger electron and X-ray yields of Ne have been measured simultaneously in 30 MeV O^{7+} collisions with Ne.† In both cases the spectra were found to be dominated by satellite transitions. The Ne K-shell fluorescence yield was found to be 0·043±0·009, 2·2 times the value (0·018) found for single ionization by electron or photon excitation.

Bhalla and Hein‡ have shown that multiple ionization of the $n = 2$ shell of Ne accompanying single ionization of the $n = 1$ shell could increase the fluorescence yield by the amount observed.

7. Ion pair formation in neutral–neutral collisions

Cross-sections for reactions of the type

$$A+B \to A^+ + B^-$$

have been measured at impact energies not far above the threshold which, in the C.M. system, is given by $E_i(A) - E_a(B)$, where E_i is the ionization energy of A and $E_a(B)$ the electron affinity of B. Much of the work has been devoted to determining $E_a(B)$ from observations of the threshold energy, with A chosen as an alkali metal atom. Energetic beams of neutral atoms are obtained either by low-energy charge transfer or from sputter sources.

Measurements of electron affinities of halogen molecules§ and of a number of other molecules∥,†† including NO_2 and SF_6 have been carried out as well as measurements of cross-sections as functions of collision energy.

† Burch, D., Ingalls, W. B., Risley, J. S., and Heffner, R., Phys. Rev. Lett. **29** (1972) 1719.
‡ Bhalla, C. P. and Hein, M., ibid. **30** (1973) 39.
§ Baede, A. P. M., Auerbach, D. J., and Los, J., Physica 's Grav. **64** (1973) 134; Helbing, R. K. B. and Rothe, E. W., J. chem. Phys. **51** (1969) 1607.
∥ Leffert, C. B., Jackson, W. M., and Rothe, E. W., ibid. **58** (1973) 5801.
†† Lacmann, K. and Herschbach, D. R., Chem. Phys. Lett. **6** (1970) 106.

25

COLLISIONS INVOLVING EXCITATION— EXPERIMENTAL METHODS AND DISCUSSION OF RESULTS OBTAINED

WE include here notes of some recent results obtained using techniques already described in Chapter 25, § 1. In the last section we refer to the interesting new results obtained on vibrational excitation of H_2 by impact of H^+ and of Li^+ ions, which mark the beginning of an extensive new subject for study.

1. Excitation in collisions involving hydrogen and helium beams
(see Chap. 25, § 2.2)

Morgan et al.† have used a modulated crossed-beam technique to determine cross-sections for the four reactions

$$H^+ + H(1s) \rightarrow H(2s \text{ or } 2p) + H^+ \quad \text{(electron capture)}$$
$$H^+ + H(1s) \rightarrow H^+ + H(2s \text{ or } 2p) \quad \text{(excitation)}$$

within the energy range 2·25–26 keV. The $H(2p)$ formation cross-sections are consistent with those of Stebbings et al.‡ while the $H(2s)$ capture cross-sections are in excellent agreement with Bayfield.§ The previously unmeasured $H(2s)$ excitation cross-section is in good general accord with the close-coupling calculations of Cheshire et al. ∥

McNeal and Bireley†† have measured cross-sections for emission of Ly α radiation in impact of H^+ and H beams with energy between 1 and 30 keV on H_2. In these collisions fast excited atoms are produced by charge transfer and slow by molecular dissociation. These were distinguished from the Doppler shift of the radiation emitted by the fast atoms (see Chap. 25, p. 3041). Results obtained for H^+ impact agree well with previous results (see Chap. 25, § 2.2.1.3).

† MORGAN, T. J., GEDDES, J., and GILBODY, H. B., J. Phys. B (GB), 6 (1973) 2118.
‡ STEBBINGS, R. F., YOUNG, R. A., OXLEY, C. L., and EHRHARDT, H., Phys. Rev. 138 (1965) A1312.
§ BAYFIELD, J. E., Phys. Rev. 185 (1969) 105.
∥ CHESHIRE, I. M., GALLAHER, D. F., and TAYLOR, A. J., J. Phys. B (GB) 3 (1970) 813.
†† MCNEAL, R. J. and BIRELEY, J. H., 7th int. conf. on the physics of electronic and atomic collisions, Amsterdam. Abstracts, North Holland, 1971, p. 815.

1.1. Excitation of helium

Gray, Haselton, Krause, and Soltysik[†] have carried out experiments to detect and measure the very weak excitation of triplet states of helium by proton impact—excitation which can only arise through spin–orbit coupling. For this purpose they observed the light emitted as a function of distance z from the point at which the beam is composed entirely of protons. The apparent cross-section for excitation can be written in the form

$$Q_{\text{app}} = Q_{\text{p}} + B(z+z_1)N,$$

where Q_{p} denotes the true proton cross-section, BzN the contribution due to direct excitation by H atoms produced from the beam by charge transfer, and $Bz_1 N$ that due to excitation transfer, N being the concentration of neutral atoms. Plots of Q_{app} against z for different N should all converge to Q_{p} as $z \to 0$. Observed results conformed to these requirements and it was found that

$$Q_{\text{p}}(3\,^3S)/Q_{\text{p}}(3\,^1S) = 0 \pm 0.0005,$$
$$Q_{\text{p}}(3\,^3P)/Q_{\text{p}}(3\,^1P) = 0.0028 \pm 0.0005,$$
$$Q_{\text{p}}(3\,^3D)/Q_{\text{p}}(3\,^1D) = 0.06 \pm 0.005.$$

Park and his collaborators[‡] have used an energy analysis of fast forward scattered ionic collision products to determine cross-sections for excitation of various atomic and molecular species (see also p. 3681).

In this way absolute cross-sections have been measured[§] for excitation of the $2\,^1S$, $2\,^3S$, $2\,^1P$, and $2\,^3P$ states of He by impact of He⁺ ions of energy 30–60 keV on He. The cross-sections for excitation of the corresponding singlet and triplet states are of comparable order of magnitude although the former, in contrast with the latter, can be excited directly as well as by electron exchange.

A photon-coincidence technique has been used by Rahmat et al.[‖] to study the process

$$\text{He}^+(1s) + \text{He}(1s^2) \to \text{He}^+(1s) + \text{He}(1s, 3p, \,^3P)$$

in the energy range 200–350 eV for He⁺ scattering angles between 5 deg

---

[†] GRAY, R. L., HASELTON, N. H., KRAUSE, D., and SOLTYSIK, E. A., ibid., p. 833.

[‡] PARK, J. T. and SCHOWENGERDT, F. D., *Rev. Sci. Instrum.* **40** (1969) 753; PARK, J. T. and SCHOWENGERDT, F. D., *Phys. Rev.* **185** (1969) 152; PARK, J. T., SCHOONOVER, D. R., and YORK, G. W., ibid. A2 (1970) 2304; SCHOONOVER, D. R. and PARK, J. T., ibid. A3 (1971) 228; CRANDALL, D. H., YORK, G., POL, V., and PARK, J. T., *Phys. Rev. Lett.* **28** (1972) 397; YORK, G. W., PARK, J. T., POL, V., and CRANDALL, D. H., *Phys. Rev.* A6 (1972) 1497.

[§] SCHOONOVER, D. R. and PARK, J. T., *7th int. conf. on the physics of electronic and atomic collisions, Amsterdam.* Abstracts, North Holland, 1971, p. 839.

[‖] RAHMAT, G., VASSILEV, G., BAUDON, J., and BARAT, M., ibid., p. 113.

and 18 deg. Differential cross-sections exhibit an oscillatory structure consistent with a curve-crossing excitation process.

1.2. *Excitation of* Ne

Jaecks et al.† have determined cross-sections for the processes

$$H^+ + Ne \to H(2p, 2s) + Ne^+$$

by recording the Lyman α photons emitted either spontaneously or in the presence of an electric quenching field in delayed coincidence with the scattered H atom. Observations were made at a scattering angle of 3 deg over the energy range 2–20 keV. The electron capture probabilities $P(2p)$ and $P(2s)$ exhibit an oscillatory dependence on impact energy with the two curves out of phase. The separations of the maxima are shown to be consistent with a simple two-state description of electron capture.

1.3. *Excitation of* N_2 (see Chap. 24, § 2.3.2)

Wehrenberg and Clark‡ have carried out coincidence experiments to determine the relative importance of charge transfer and direct ionization in exciting the first negative 3914 Å band of N_2^+ by impact of protons in N_2. This was done by coincident detection of the photon and the fast neutralized scattered atom which distinguishes the charge transfer process. For scattering through angles up to 1·5° about $\frac{2}{3}$ of the excitation events due to 30 keV protons are estimated to arise through charge transfer.

2. Excitation of alkali-metal atoms in charge transfer collisions with alkali-metal ions

A crossed-beam method has been used by Aquilanti, Liuti, Vecchio-Cattivi, and Volpi§ to measure cross-sections for excitation of alkali-metal atoms by impact of alkali-metal ions with kinetic energies ranging from a few eV to 1·5 keV. The intensity of the neutral beam was determined by simultaneous measurement of resonance fluorescence stimulated by a photon beam of known intensity.

Quite large cross-sections were observed for production of the lowest ^2P state of the atom, either directly or by charge transfer. These cross-sections varied quite slowly with ion energy down to energies as low as 200 eV and even lower in some cases such as

$$Na^+ + Na(3\,^2S) \to Na^+ + Na(3\,^2P)$$

† Jaecks, D. H., de Rijk, W., and Martin, P. J., ibid., p. 605.
‡ Wehrenberg, P. J. and Clark, K. C., ibid., p. 387.
§ Aquilanti, V., Liuti, G., Vecchio-Cattivi, F., and Volpi, G. G., ibid., p. 600.

for which the cross-section at 100 eV was as high as 3×10^{-16} cm^2, rising to nearly 10^{-15} cm^2 at 1500 eV. The other excitation processes had smaller cross-sections, the smallest being for the charge transfer excitation

$$\text{Li}^+ + \text{Na}(3\,^2S) \rightarrow \text{Na}^+ + \text{Li}(2\,^2P),$$

the maximum observed value for which was 8×10^{-17} cm^2 at 1·5 keV. The corresponding process

$$\text{K}^+ + \text{Na}(3\,^2S) \rightarrow \text{Na}^+ + \text{K}(4\,^2P)$$

has a much larger cross-section, the observed value at 1·5 keV being about 2×10^{-16} cm^2. This is to be compared with 4×10^{-17} cm^2 for

$$\text{K}^+ + \text{Na}(3\,^2S) \rightarrow \text{K}^+ + \text{Na}(3\,^2P).$$

3. Differential cross-sections for collisions involving excitation

We have already referred to the experiments of Abignoli, Barat, Baudon, Fayeton, and Houver† in relation to their measurements of differential cross-sections for elastic scattering and for charge-exchange collisions of protons with rare-gas atoms. In addition to these measurements they also observed energy spectra of the scattered protons as a function of scattering angle for proton energies ranging from 0·5 to 3 keV.

Two main excitation processes are observed, corresponding to excitation of bound states or of autoionizing levels of the target atom. For He targets no excitation of any kind was observed within the experimental range of energies and scattering angles. Weak excitation is observed for Ne and is progressively more important as the mass of the target increases. When the energy resolution is sufficient to distinguish the excitation of individual states (incident energies below 500 eV) it is found that the strongest single excitation is not to the first excited state but to states in which one np orbital is excited to a higher s, p, or d orbital. The autoionizing levels are only observed at relatively high energies, 3, 2, and 1 keV for Ar, Kr, and Xe respectively.

The united–separated atom correlation of diabatic potentials shows that the lowest $X\,^1\Sigma^+$ curve for the H$^+$–rare-gas interaction does not cross any other curve except for Xe. In this case it crosses a curve which in the separated atom limit corresponds to charge transfer. For the other atoms the nearest curve, at large separations, to the $X\,^1\Sigma^+$ curve is the one corresponding to charge transfer. It is likely that excitation proceeds in two steps, the first involving a charge-exchange transition to this curve followed by a second transition to the final excited state. The latter transition may arise through rotational coupling.

† loc. cit., p. 3654.

Differential cross-sections for the elastic scattering of Li⁺ ions by He atoms have already been discussed on p. 3657. The same experiments, by Lorents and Conklin and by François et al. were used to obtain differential cross-sections for inelastic scattering via measurements of energy loss spectra. Two principal energy processes are observed. One, process B, corresponds to one-electron excitation of the He. Evidence was obtained which indicates that the most probable level excited is $(1s2p)^1P$. The other, process C, corresponds to double excitation of He to $(2s^2)^1S$, $(2s2p)^1P$, and $(2p^2)^1D$ (see Chap. 9, p. 619).

As both K shells are full, the orbitals of the two Li⁺ electrons correlate with the ground $1s$ orbitals of the united system B⁺ and so play no significant part in transitions to upper states of the LiHe⁺ system. In particular the probability that they will exchange with the He electrons will be very small so that only excitation of singlet states of He can be expected. The orbitals of the two He electrons correlate with the $2p$ excited orbitals of B⁺ and it is the interaction of these orbitals with higher excited orbitals that must be responsible for the inelastic processes observed. A brief account of the theory of the elastic and single excitation processes due to McCarroll and Piacentini, which places special emphasis on rotational coupling, has been given on p. 3662. It gives results in quite good agreement with the observations and suggest that the assumptions made in the theoretical treatment are valid.

Park and his collaborators† have measured the energy loss spectrum of 50 keV N⁺ ions in He. The results provide evidence for the presence of metastable $N^+(2p^2\,^1D)$ metastable ions in the beam. The cross-section for excitation of these ions to the $3s\,^1P$ and $2p^3\,^1D$ levels was estimated to be about $1·6\times10^{-17}$ cm². Cross-sections for excitation of the $2p^2\,^3P$ ground state ions to the $2p^3\,^3P$ and $2p^3\,^3D$ levels were $1·02\times10^{-17}$ cm² and $2·41\times10^{-17}$ cm² respectively.

Fournier et al.‡ have recently studied the energy spectra of H⁻ ions arising at small scattering angles from 4 keV proton collisions with molecules. Both the two-electron capture process

$$H^+ + M \to H^- + M^{2+}$$

(where the product M²⁺ may undergo subsequent dissociation) and the two-step H⁻ formation process

$$H^+ + M \to M_a^+ + H_a,$$
$$H_b + M \to M_b^+ + H^-,$$

† loc. cit., p. 3678.
‡ FOURNIER, P. G., APPELL, J., FEHSENFELD, F. C., and DURUP, J., *J. Phys. B (GB)* **5** (1972) L58.

where M_a^+, H_a, H_b, and M_b^+ allow for possible different excited states, were investigated. For a target of H_2, the H^- energy spectra were shown to correspond closely to the results expected on the basis of Franck–Condon transitions.

4. Excitation of vibration in slow collisions of ions with molecules

The energy resolution attainable in ion scattering experiments is now high enough to make it possible to observe energy losses in collisions with molecules which are due to vibrational excitation.

Udseth, Giese, and Gentry[†] have measured differential cross-sections for such excitation in collisions of protons, with energies between 4 and 16 eV, with H_2, HD, and D_2.

At a scattering angle of 22° and energy of 10 eV, in the C.M. system in H_2 the observed ratio of elastic scattering to inelastic, involving excitation of the $v = 1$ and 2 states respectively, is roughly as $10:7\cdot5:2\cdot7$ whereas for D_2 the corresponding numbers are more nearly $5:10:2$. Inelastic processes tend to be relatively more important as the scattering angle and as the impact energy increases.

Vibrational excitation associated with backward scattering (in the C.M. system) of Li^+ ions of 3 eV C.M. energy has been observed by David, Faubel, Marchand, and Toennies[‡] by a time-of-flight technique. This takes advantage of the fact that in the laboratory system the backward scattered ions in the C.M. system are moving slowly forward. The time-of-flight spectrum showed peaks corresponding to excitation of the $v = 1$, 2, and 3 states.

[†] UDSETH, H., GIESE, C. F., and GENTRY, W. R., *Phys. Rev.* A8 (1973) 2483.
[‡] DAVID, R., FAUBEL, M., MARCHAND, P., TOENNIES, P., *7th int. conf. on the physics of electronic and atomic collisions, Amsterdam.* Abstracts, North Holland, 1971, p. 252.

26

COLLISIONS OF SLOW POSITRONS AND OF SLOW POSITIVE MUONS IN GASES

The most important development in the experimental study of collision processes involving slow positrons has been the realization of the possibility of measuring total cross-sections for scattering of slow positrons, with quite closely defined energy, by various gases.

This work depended on the availability of a source of positrons of reasonably homogeneous energy and yet of sufficient strength to make transmission experiments practicable. Such a source was developed by Coleman, Griffith, and Heyland† who at the same time improved the efficiency of the positron detection system and were then able to carry out time-of-flight measurements.

The origin of the source dates back to experiments by Costello, Groce, Herring, and McGowan‡ which were directed towards the energy moderation of positrons produced through pair formation by γ-rays generated by fast electrons from a linear accelerator. Evidence was obtained which suggested that, if the positrons were slowed down by passage through a gold-coated mica sheet there emerged, on the far side, slow positrons with a narrow band of energies about a median value of a few eV. It was suggested by Tong§ that these arose through expulsion of thermalized positrons through the surface of the gold by the work function. Some preliminary measurements‖ of cross-sections for collisions with helium were made.

Although efforts to reproduce this effect using a radioactive source were initially unsuccessful Coleman et al. found that if positrons emerging from a moderator such as a thin scintillator covered by a suitable foil were caused to suffer back scattering from a gold-coated surface then slow positrons emerged in useful numbers with a narrow range of energies about a median of a few eV. Using a source of this kind with an

† Coleman, P. G., Griffith, T. C., and Heyland, G. R., Proc. R. Soc. A**331** (1973) 561.
‡ Costello, D. G., Groce, D. E., Herring, D. F., and McGowan, J. W., Phys. Rev. B**5** (1972) 1433.
§ Tong, B. Y., ibid. B**5** (1972) 1436.
‖ Costello, D. G., Groce, D. E., Herring, D. F., and McGowan, J. W., Can. J. Phys. **50** (1972) 23.

improved method of detecting positrons, time-of-flight transmission experiments could then be carried out.

The improvement in detection was based on the following considerations. In earlier work the passage of a positron through the system was indicated by a start signal arising from emission of the prompt γ-ray from the artificially radioactive source, and a stop signal from an annihilation γ-ray. The efficiency of detection of γ-rays with crystal scintillators is less than 10 per cent so Coleman et al. replaced the start signal by one caused by the direct passage of the positron through a thin scintillator on its way into the experimental system. This increases the efficiency by at least an order of magnitude.

The arrangement used in the first transmission experiments[†] was as follows. Positrons from the ^{22}Na source after passage through a plastic scintillator 0·17 mm thick, a Melinex window of 0·075 mm thickness aluminized on the surface in contact with the scintillator, and an aluminium foil about 0·001 mm thick, entered a copper cylinder 20 mm long coated internally with gold to a thickness of about 0·025 mm. At the far end of the cylinder was a fine grid followed by an insulating ring and a second grid at earth potential. Positrons were accelerated to the desired energy by applying a positive potential to the moderator.

After acceleration the positrons entered the time-of-flight tube, consisting of a straight section 70 cm long followed by a sector 15 cm long curved in an arc of 25 cm radius. By application of a suitable magnetic field the positrons were confined to helical paths close to the axis throughout the full flight path, to be detected by the annihilation radiation produced in an Al foil by an NaI counter. The final curved section of the flight path ensures that the detector is not exposed directly to the source, so reducing the number of background events.

Improved performance of the source was obtained by evaporating MgO on to the gold surface. In later experiments[‡] the gold-coated copper cylinder was replaced by a series of vanes as in a photomultiplier. About 1 in 10^5 of the positrons emitted from the source entered the flight tube as slow positrons.

By delayed coincidence observations the energy distribution of the positrons emerging from the source could be measured and was found to have a full width at half height of about 1 eV down to the lowest usable

[†] COLEMAN, P. G., GRIFFITH, T. C., and HEYLAND, G. R., *Proc. R. Soc.* A**331** (1973) 561.

[‡] CANTER, K. F., COLEMAN, P. G., GRIFFITH, T. C., and HEYLAND, G. R., *J. Phys. B (GB)* **5** (1972) L167 and **6** (1973) L201; COLEMAN, P. G., GRIFFITH, T. C., and HEYLAND, G. R., *Applied Physics*, vol. 3 (1974), in course of publication.

median energy (2 eV). Transmission cross-sections for different gases were then measured by observing the variation with gas pressure of the positron counting rate.

Results have been obtained in this way for positrons with energy extending from 2 to 20 eV for He, Ne, Ar, Kr, H_2, N_2, and CO. For He the cross-section falls from $0.28\pi a_0^2$ at 20 eV to $0.17\pi a_0^2$ at 9 eV, then varies only slowly to 6 eV after which it falls quite rapidly to about $0.08\pi a_0^2$ at 2 eV where there is evidence that a Ramsauer minimum may exist. The shape of the variation with energy is consistent with Drachman's theory (see Chap. 26, p. 3173) but allowance must be made for the rather poor angular resolution in the experiments which will lead to some underestimation of the total cross-section. Further experimental and theoretical work is in progress.†

† See, for example, JADUSZLIWER, B., KEEVER, W. C., and PAUL, D. A. L., *Can. J. Phys.* **50** (1972) 1414; HUMBERSTON, J. W., *J. Phys.* B6 (1973) L305.

ERRATA†

Volume 1

p. 43, 2nd last line. Replace 'measuring' by 'mercury'.

p. 111, 3rd last line. Replace 'MeV' by 'keV'.

p. 147, line 8. Replace $\frac{1}{4}$ by $\simeq 0.45$.

p. 149, line 8. Omit 'and He^+'.

p. 237,‡ lines 8 and 9 should read 'Measurements of excitation functions of highly excited long-lived Rydberg states of inert gas atoms have been made by Kuprianov'.

p. 253, Fig. 4.56, underline. After 'Milatz' (2^3S); change the legend to read 'and of highly excited long-lived Rydberg states, ... (Kuprianov)'.

p. 253, last four lines. Omit 'and of Kuprianov ... metastable atoms'.

p. 255, 3rd last line. Replace '§ 3.2' by '§ 2.2'.

p. 257, Fig. 4.60, underline. After 'Hadeishi', change underline to read 'and of highly excited long-lived Rydberg states, ... (Kuprianov)'.

p. 259, Fig. 4.62, underline. Replace present underline by 'Measurements of Kuprianov of excitation functions of highly excited long-lived Rydberg states of (1) Kr, (2) Xe, (3) Ar'.

p. 259, line 5. Replace last sentence on this page by 'Fig. 4.62 gives the results obtained by Kuprianov for excitation of highly excited long-lived Rydberg states of Ar, Kr, Xe. Fig. 4.63 shows the structure near the threshold in excitation curves for metastable states of Ar, Kr, and Xe'.

p. 270, Fig. 4.71, underline, last line. Change 'hydrogen' to 'helium' and Fig. 1.11 to Fig. 1.10.

p. 271, 6th last line. Change 'Dolder et al.' to 'Dance et al. (Chapter 3, § 4.1, p. 157)'.

† We wish to express our thanks to Dr. A. C. H. Smith for pointing out some of the corrections given above for Vols. 1 and 2.

‡ In Chap. 4 several references have been made to a paper by S. E. Kuprianov, Optika Spektrosk. 20 (1966), 163 (English translation, Optics Spectrosc., N.Y. 20 (1966), 85). We were able to consult only the English translation of this paper and it appears that we have misinterpreted what was actually measured. It appears that his results refer to the excitation of highly excited long-lived Rydberg states of inert gas atoms and not the low-lying metastable states of these atoms as stated. We are indebted to Dr. W. Shearer for pointing out to us this misrepresentation of Kuprianov's work, necessitating the corrections indicated on pp. 237–59.

ERRATA

p. 347, eleven lines from bottom. Replace 'Ehrhardt and Willmann' by 'Ehrhardt and Meister'.

p. 348, underline to Fig. 5.54. The reference to Ehrhardt and Meister should be given, viz. 'Ehrhardt, H., and Meister, G., *Phys. Lett.* **14** (1965), 200'.

p. 525, line 12. Replace, 'diagonal elements U_{nm}' by 'diagonal elements U_{nn}'.

p. 525, line 22. Replace 'electron scattering' by 'elastic scattering'.

p. 536, 5th last line. Replace '(40·5 eV)' by '(40·8 eV)'.

p. 536, 4th last line. Replace 'Harrison, Dance, and Smith' by 'Dance, Harrison, and Smith'.

p. 601, line 12. Replace 'of half-thickness' by 'with a full width at half maximum of'.

p. 643, line 17. Replace '$\left(\frac{\partial M}{\partial E}\right)_0$' by '$\left(\frac{\partial M}{\partial E}\right)_m$'.

p. 664. Last paragraph should read 'For n-fold ionization of an atom in a single collision the threshold behaviour might be expected to be given by

$$Q_i = C(E-E_i)^n \qquad (250)$$

but, as explained in Chapter 3 § 2.4.4, the complex structure near the threshold for these collisions makes it very difficult to prove'.

Volume 2

p. 739, Fig. 11.5, bottom left-hand scale. Replace '10, 20, 30, 40' by '1·0, 2·0, 3·0, 4·0'.

p. 791, Fig. 11.44, left-hand scale. Replace '10^{-14}' by '10^{-15}'.

p. 832, line 12. Replace 'maximum' by 'minimum'.

p. 879, line 4. Replace '$C^1\Sigma_u$' by '$C^1\Pi_u$'.

p. 919, Fig. 13.23, left-hand scale. Replace '10^{-16} cm^2' by '10^{-18} cm^2'.

p. 934, Fig. 13.35 (c), left-hand scale. Replace '0·005, 0·01' by '0·05, 0·1'.

p. 954, Fig. 13.45. The data plotted as observed by Dance, Harrison, Rundel, and Smith for electron energies greater than 200 eV refer to results which were later remeasured before publication. The published results agree well with those of Dunn and van Zyl.

p. 987, Fig. 13.69, left-hand scale. Replace '5, 4, 2' by '4, 2, 0'. Delete '0'.

p. 1127, Fig. 14.31, left-hand scale. Replace '10^{-8} cm^2' by '10^{-18} cm^2'.

ERRATA

Volume 3

p. 1535, Table 17.5. Replace 'Doubling dispersing' by 'Doubly dispersing'.

p. 1546, 2nd line below (148). Replace $m = 0$ by $m \neq 0$.

p. 1626, eqn (326). Delete 'i^l'.

p. 1693, Table 18.4, 9th column. Replace '10·6' by '50'.

p. 1733, Table 18·8 (b), 1st row. Replace 'Ne($^2P_4$)' by 'Ne($^2p_4$)'.

p. 1838, 6th line from bottom. Replace '$1·5 \times 10^{-36}$ cm^6 s^{-1}' by '$1·5 \times 10^{-34}$ cm^6 s^{-1}'.

p. 1984, last line. Replace '($0·52_5 \times 10^{-36}$ cm^6 s^{-1})' by '($0·52_5 \times 10^{-31}$ cm^6 s^{-1})'.

p. 1986, 3rd-last line. Replace '$0·84_5 \times 10^{-36}$ cm^6 s^{-1}' by '$0·84_5 \times 10^{-31}$ cm^6 s^{-1}'.

p. 1987, 1st line of second-last paragraph. Replace 'Mulcahy' by 'Hackam'.

p. 1987, 7th line from bottom. Replace '0·93 cm^6 s^{-1}' by '$0·93 \times 10^{-31}$ cm^6 s^{-1}'.

p. 1987, footnote reference. Replace 'Mulcahy, M. J.' by 'Hackam, R.' and '**80** (1962) 626' by '**84** (1964) 133'.

Volume 4

p. 2578. Replace '$Q_{tr} - Q_{tr}^0$' in (383) by '$Q_{tr}^0 - Q_{tr}$' and also in the equation three lines above.

AUTHOR INDEX

for Chapters 26 and 27

Aamodt, R. L., 3288
Adrianov, D. G., 3248
Alberigi-Quaranta, A., 3305, 3324, 3325, 3326, 3327, 3331, 3333, 3335, 3336, 3344, 3347
Ali-Zade, S. A., 3257
Alvarez, L. W., 3306, 3308, 3309, 3310
Amato, J. J., 3242, 3266
Anderson, E. W., 3266, 3312, 3319, 3321
Anderson, L. W., 3247
Armstead, R. L., 3178
Ashmore, A., 3308

Babayev, A. I., 3248, 3249, 3269, 3273, 3303
Bailey, J. M., 3238, 3239, 3264, 3266, 3274
Baird, J. C., 3247
Baker, G. A., 3279, 3280, 3287
Balats, M. Y., 3248, 3269
Barjal, J. S., 3287
Barker, M. I., 3216
Basiladze, S. G., 3327
Bates, D. R., 3281, 3300
Baumann, K., 3293
Beers, R. H., 3219
Belkic, D. S., 3261
Bell, R. E., 3143, 3203, 3204, 3209, 3219
Belyaev, V. B., 3321, 3337, 3341, 3342
Bennett, W., 3156, 3161
Berestetzki, V. B., 3148, 3168
Berg, H. G., 3269
Berko, S., 3218
Bertin, A., 3305, 3321, 3324, 3326, 3331, 3335
Bethe, H. A., 3278, 3286, 3290, 3291, 3292, 3293, 3295, 3297.
Bierman, E., 3296
Bleser, E., 3266, 3312, 3315, 3316, 3317, 3319, 3320 3321, 3324
Bloch, F., 3262
Block, M. M., 3298
Bradner, H., 3306
Brandt, W., 3158, 3159, 3161, 3164
Bransden, B. H., 3190, 3216
Breit, G., 3232, 3235
Breskman, D., 3261
Briscoe, C. V., 3148, 3200, 3201
Brouillard, F., 3304
Brown, R. A., 3247
Brown, S. C., 3149, 3156
Bruno, M., 3321
Budick, B., 3297

Buhler, A., 3257
Bukhvostov, A. P., 3301
Bunnell, K., 3298
Burhop, E. H. S., 3300
Burke, P. G., 3177, 3242
Burnstein, R. A., 3296
Byakov, V. M., 3269

Carboni, C., 3297, 3299, 3329
Carter, K., 3197, 3198, 3207, 3208, 3209, 3210, 3218, 3219, 3220, 3221, 3223
Casperson, D., 3242
Celitans, G. J., 3142, 3144, 3145, 3219
Chernagorova, V. A., 3303
Choi, S. I., 3200, 3201
Chuang, S. Y., 3228
Chultem, D., 3303, 3304
Clarke, G. A., 3247
Cleland, W. E., 3238, 3239, 3245, 3246, 3264, 3266, 3274
Cody, W. J., 3177
Cohen, S., 3321, 3334, 3335, 3339, 3341, 3342, 3343, 3344, 3345, 3347, 3350, 3351, 3352
Coleman, P. G., 3135
Condo, G. T., 3299
Conforto, G., 3312, 3320, 3321, 3323, 3325
Corben, H. C., 3125
Crane, P., 3242, 3243, 3244
Crane, T., 3242
Crawford, F. S., 3306
Crawford, J. A., 3306
Cresti, M., 3296
Crowe, K. M., 3253
Cuthbertson, C., 3187, 3189
Cuthbertson, M., 3187, 3189

Dalgarno, A., 3173
Dalpiaz, P., 3305, 3324, 3331, 3335
Davis, D. H., 3288
Day, T. B., 3286, 3292, 3295, 3297, 3298
De Benedetti, S., 3125, 3126, 3168
De Borde, A. H., 3286, 3300
Derrick, M., 3288, 3294, 3298
Deser, S., 3293
Deutsch, M., 3219, 3149, 3154, 3156, 3168, 3218
DeVoe, R., 3242, 3243
Diaz, J. A., 3287
Dicke, R. H., 3149
Dirac, P. A. M., 3124

Doede, J. H., 3296, 3306, 3307, 3308, 3310, 3311, 3319, 3321, 3322, 3323
Dombeck, T., 3288
Drachman, F. J., 3173, 3174, 3175, 3177, 3178, 3179, 3180, 3181, 3182, 3183, 3184, 3185, 3186, 3193, 3195, 3200, 3201, 3216, 3220, 3227
Drisko, R. M., 3150
Duclos, J., 3330
Duff, B. G., 3133, 3218, 3219
Dulit, E. P., 3149, 3154, 3218
Dzhelepov, V. P., 3310, 3312, 3320, 3321, 3322, 3330, 3331, 3333, 3335, 3336, 3344, 3347

Eckhause, M., 3239
Egan, P., 3242
Egorov, L. B., 3303, 3304
Ehrlich, R. D., 3242, 3244, 3246
Ensberg, E. S., 3247
Ermolov, P. F., 3310, 3312, 3321, 3322, 3327, 3330, 3335
Evseev, V. S., 3303

Falk, W. R., 3202, 3203, 3204, 3205, 3206, 3209
Falk-Vairant, P., 3306
Falomkin, I. V., 3298
Favart, D., 3243, 3244, 3245, 3246, 3304
Feibus, H., 3158, 3159, 3161, 3164
Fermi, E., 3287
Ferrell, R. A., 3168, 3259
Fetkovich, J. G., 3288, 3294, 3298, 3309
Fields, T. H., 3294, 3298, 3309
Fil'chenkov, V. V., 3310, 3312, 3322, 3330, 3335
Fillipov, A. J., 3298
Firsov, V. G., 3248, 3257, 3269
Fistul, V. I., 3248
Focardi, S., 3312, 3321, 3323
Frank, F. C., 3308
Fraser, P. A., 3182, 3214, 3215, 3216
Friml, M., 3310. 3312, 3321, 3335

Garwin, R. L., 3301
Gastaldi, U., 3297, 3299, 3329
Gershtein, S. S., 3282, 3287, 3305, 3318, 3319, 3321, 3326, 3327, 3328, 3334, 3335, 3336, 3339, 3341, 3342, 3346, 3347, 3348, 3349, 3350, 3352
Gittelman, B., 3154, 3218
Goldanskii, V., 3168, 3219, 3223, 3227
Goldberger, M. L., 3293
Good, N. L., 3306
Gorini, G., 3297, 3299, 3329
Gorodetzky, S., 3253, 3254, 3257
Gow, J. D., 3306
Graham, R. L., 31 43

Green, J. H., 3142, 3144, 3145, 3203, 3204, 3209, 3219, 3227
Grenacs, L., 3304
Griffith, T. C., 3135
Gurevich, I. I., 3250, 3252, 3257, 3258, 3262

Haas, F., 3242
Hadley, I., 3288
Hahn, Y., 3178
Halliday, D., 3149
Hatcher, C. R., 3228
Heinberg, M., 3152, 3153
Herman, R. M., 3247
Heyland, G. R., 3135
Heymann, F. F., 3133, 3137, 3138, 3139, 3140, 3141, 3168, 3218, 3219
Hildebrand, R. H., 3296, 3307
Hofer, H., 3242, 3244
Hogg, B. G., 3228
Houston, S. K., 3216, 3220, 3227
Huetter, T., 3296
Hughes, V. W., 3141, 3142, 3149, 3150, 3151, 3153, 3156, 3158, 3159, 3161, 3163, 3219, 3231, 3235, 3236, 3237, 3238, 3239, 3242, 3261, 3264, 3266, 3274
Humberston, J. W., 3180
Hutchinson, D. P., 3301
Hyman, L. G., 3298

Ignatenko, A. E., 3303, 3304
Igo-Kemenes, P., 3304
Israel, M. H., 3296
Ivanter, I. G., 3252, 3258, 3261, 3263, 3628
Izuyama, T., 3321, 3342

Jackson, J. D., 3322
Jakovleva, I. V., 3252
Janev, R. K., 3261
Jarecki, J., 3247
Jensen, S. H. P., 3289
Judd, D. L., 3321, 3322, 3334, 3335, 3339, 3342, 3344, 3350

Kanofsky, A., 3261
Kaplan, S. N., 3287
Katyshev, Y. V., 3310, 3321, 3322
Kaufman, S. L., 3247
Kelly, T. M., 3196, 3199, 3200
Keyes, G. J., 3288, 3298
Khalupa, B., 3303, 3304
Kleinman, C. J., 3178, 3179
Klem, R. D., 3304, 3306
Knop, R., 3296
Koller, E. L., 3296
Kolos, W., 3341
Kopelman, J. B., 3298
Koppe, H., 3259, 3289

Kraidy, M., 3182, 3215, 3216
Krumshteĭn, Z. V., 3288
Kulyukín, J. B., 3298

La Bahn, R. W., 3187, 3188, 3189, 3206
Lamkin, J. C., 3171, 3172, 3173, 3174, 3187
Landau, L. D., 3148, 3168
Langenberg, D. M., 3247
Lawson, J., 3177, 3187, 3206
Lederman, L. M., 3266, 3301, 3312, 3319, 3321
Ledsham, K., 3281
Lee, G. F., 3163, 3191, 3192, 3193, 3194, 3199, 3227
Leon, M., 3286, 3291, 3292, 3293, 3295, 3297
Leung, C. Y., 3163, 3185, 3191, 3192, 3193, 3194, 3195, 3211, 3212
Limentani, S., 3296
Lipnik, P., 3304
Lomnev, S. P., 3321, 3339
Loria, A., 3296
Lynn, N., 3173

McColm, D. W., 3236, 3261
McGervey, J. D., 3227
McIlwain, R. L., 3309
McIntyre, P. M., 3242, 3243
McKenzie, J., 3298
Macq, P. C., 3304
Magnon, A., 3242, 3244, 3330
Makar'ina, L. A., 3250, 3252, 3257, 3258
Mann, R. A., 3280, 3289, 3301, 3303
Marder, S., 3141, 3149, 3150, 3151, 3153, 3156, 3161, 3165, 3193
Martin, A. D., 3286
Massam, T., 3257
Massey, H. S. W., 3177, 3187, 3188, 3189, 3206, 3277, 3342
Matone, G., 3305, 3324, 3331, 3334, 3335
Matveenko, A. V., 3321, 3335, 3347
Mel'eshko, E. A., 3250, 3252, 3257, 3258
Melville, H. W., 3274
Meyer, J. L., 3266, 3312, 3319, 3321
Miller, D. B., 3190, 3202, 3204, 3206, 3208, 3209
Millett, W. E., 3228
Minaĭchev, E. V., 3248
Mizuno, Y., 3321, 3342
Mobley, R. M., 3239, 3242, 3264, 3266, 3274
Mohorovičić, S., 3168
Montgomery, R. E., 3187, 3188, 3189, 3206
Morgan, C. L., 3247
Moskalev, V. I., 3310, 3312, 3321, 3322, 3330, 3335
Mott, N. F., 3277, 3242
Mühlsehlegel, B., 3259, 3289

Muller, Th., 3253, 3257
Murphy, C. T., 3288
Myasishcheva, G. G., 3248, 3249, 3250, 3251, 3257, 3269, 3303

Neri, G., 3299
Nikol'skii, B. A., 3250, 3252, 3257, 3258
Nordhagen, B., 3308
Nosov, V. G., 3261, 3262, 3264

Obenshain, F. E., 3147, 3164, 3165
Obukhov, Yu. V., 3248, 3257, 3269, 3303
Oganesyan, K. O., 3327
Öre, A., 3125, 3128
Ormrod, J. H., 3227
Orth, P. H. R., 3133, 3156, 3163, 3190, 3191, 3202, 3203, 3204, 3205, 3206, 3207, 3208, 3209, 3219, 3223
Osmon, P. E., 3137, 3138, 3168, 3219
Øveras, H., 3289

Page, L. A., 3147, 3152, 3153, 3164, 3165
Palmonari, F., 3305, 3324, 3331, 3335
Panofsky, W. K. H., 3288
Parker, W. H., 3247
Paul, D. A. R., 3163, 3185, 3191, 3192, 3193, 3194, 3195, 3203, 3204, 3211, 3212
Pawlewicz, W. T., 3288, 3289
Penman, S., 3301
Percival, I. C., 3242
Peruzzo, L., 3296
Petch, H. E., 3143
Petruhnin, V. I., 3287, 3288, 3329
Pewitt, F. G., 3298
Picard, J., 3329
Pipkin, F. M., 3247
Pirenne, J., 3168
Placci, A., 3297, 3299, 3305, 3321, 3324, 3326, 3327, 3328, 3329, 3331, 3335
Polacco, E., 3297, 3299, 3329
Pond, T. A., 3140, 3149, 3153
Ponomarev, L. I., 3287, 3288, 3321, 3335, 3347
Port, M., 3253, 3257
Powell, J. L., 3125
Prentki, J., 3321
Prepost, R., 3236, 3238, 3239
Prokoshkin, Y. D., 3287, 3288
Pyaka, M. R., 3296
Pyle, R. V., 3287

Rabi, I. I., 3232
Ramsey, N. F., 3243, 3246
Ray, S., 3247
Riddell, R. J., 3321, 3334, 3335, 3339, 3341, 3342, 3344, 3350

Riley, B. R., 3298
Robb, J. C., 3274
Roellig, L. O., 3148, 3196, 3197, 3199, 3200, 3210, 3227
Roganov, V. S., 3248, 3252, 3257 3258, 3269, 3303
Roothaan, C. C. J., 3341
Rose, M. E., 3286, 3289, 3300, 3303
Rosen, J. L., 3266, 3312, 3319, 3321
Rosenfeld, A. H., 3306
Rothberg, J., 3266, 3312, 3319, 3321
Rothberg, J. E., 3239, 3242, 3264, 3266, 3274
Ruark, A. E., 3168
Rubbia, C., 3312, 3321, 3323
Russell, J. E., 3299

Sack, R. A., 3341
Sacton, J., 3288
St. Pierre, L., 3203, 3024
Salpeter, E. E., 3290
Samoilov, V. M., 3261
Santangelo, R., 3296
Savil'ev, G. I., 3248
Schenck, A., 3253
Schiff, M., 3310, 3312, 3315, 3321, 3331
Schneegans, H., 3257
Schwartz, C., 3171, 3172, 3174, 3175, 3177, 3178, 3180, 3181
Selivanov, V. I., 3250, 3252, 3257, 3258
Shantarovich, V. P., 3219
Shapiro, G., 3301
Shearer, J. W., 3168
Shestakov, V. D., 3252, 3258
Shimizu, M., 3321, 3342
Shmushkevich, I. M., 3301
Siegel, R. T., 3126, 3168
Smilga, V. P., 3252, 3258, 3261, 3262, 3263, 3268
Smith, K., 3177
Snow, C. A., 3292, 3296, 3297
Sokolov, B. V., 3250, 3252, 3257, 3258
Solmitz, F. T., 3306
Spruch, L., 3178
Stambaugh, R., 3242
Stamer, P., 3296
Stefanini, G., 3330
Stevenson, M. L., 3306
Stewart, A. L., 3281
Stewart, A. T., 3148, 3199, 3200, 3201, 3227
Stowell, D. Y., 3242, 3243, 3244
Strauch, K., 3308
Stueckelberg, E. C. G., 3348
Sucher, J., 3292, 3297
Sulyaev, R. M., 3298
Sun, C. R., 3298

Suvarov, V. M., 3329
Swanson, R. A., 3242, 3244, 3257, 3312

Tao, S. J., 3145, 3199, 3203, 3204, 3209, 3219, 3223, 3228
Taylor, B. N., 3247
Taylor, S., 3296
Telegdi, V. L., 3242, 3243
Teller, E., 3287
Temkin, A., 3171, 3172, 3173, 3174, 3187
Teutsch, W. B., 3156, 3158, 3159, 3163
Thirring, W., 3293
Thompson, D. G., 3187, 3188, 3206
Thompson, P. A., 3242, 3266
Ticho, H. K., 3306
Toraskar, J. R., 3297
Torelli, G., 3297, 3299, 3305, 3324, 3330, 3331, 3335
Tovee, D. M., 3288
Townes, B. M., 3308
Tripp, R. D., 3306
Tsupko-Sitnikov, V. M., 3298

Van der Velde-Wilquet, C., 3288
Veit, J. J., 3137, 3138, 3168, 3219
Vitale, A., 3321, 3326

Wallace, J. R. G., 3180
Wang, I. T., 3266, 3298, 3312, 3319, 3321
Wangler, T., 3288
Wangsness, R. K., 3262
Weinrich, M., 3301
Weinstein, R., 3149
Wheatley, J., 3149
Wheeler, J. A., 3168
Whiteside, H., 3296
Wickens, J. H., 3288
Wightman, A. S., 3278, 3280, 3283, 3284, 3285, 3286, 3287, 3297
Williams, W. F., 3137, 3138, 3168, 3219
Wolfenstein, L., 3318
Wu, C. S., 3141, 3149, 3150, 3151, 3156, 3161

Yaghoobia, I., 3297
Yakovleva, I. V., 3261, 3262, 3264
Yodh, G. B., 3294

Zaïmidoroga, O. A., 3298
Zakhar'ev, B. N., 3321, 3339, 3341, 3342
Zavattini, E., 3297, 3299, 3305, 3312, 3321, 3323, 3324, 3326, 3329, 3331, 3335
Zel'dovich, Y. B., 3308, 3335, 3336, 3339, 3342, 3350, 3352
Zichichi, A., 3253, 3257
Ziock, K., 3236, 3238, 3261, 3297, 3299
Zu Putlitz, G., 3242

SUBJECT INDEX

for Chapters 26 and 27

Al:
polarization of μ^+ stopping in, 3256
spin relaxation time of μ^+ stopping in, 3258

Al_2O_3:
muonium formation in, 3252
polarization of μ^+ stopping in, 3252

angular distribution of photons in 3-quantum annihilation of positrons, 3126
formula for, 3174
in He, theory of, 3186–7
in magnetic fields, 3149
observation of, 3150–2
observed results and comparison with theory, 3200–1
use of, to study effect of electric fields on Ps formation, 3164

angular distribution of photons in 2-quantum annihilation of positrons, 3144
effect of magnetic field on, 3152
in Ar, 3152
in O_2, 3152
measurement of, 3147–8

annihilation, of positrons in collisions with free-electrons, 3125
cross section for, 3125
effect of electric fields on, 3155
effect of impurities on, 3136, 3211
measurement of, 3147–8
pick-off, 3129
spectra:
analysis of, 3134
in Ar, 3202–7
at low temperatures, 3208–9
effect of N_2 on, 3211–13
in He, 3185–6, 3191–6
at low temperatures, 3197–3201
measurement of, 3133
three quantum, 3126
angular distribution of emitted photons in, 3126
two quantum:
analysis of observations of, 3146
angular correlation in, 3145
measurement of, 3147–8

antiprotons, properties of, 3277

Ar:
annihilation time spectra of positrons in:
at ordinary temperatures, 3202–7
at low temperatures, 3228–9

collisions of positrons in:
momentum transfer cross-sections in, 3189
observed results and comparison with theory, 3205
number of annihilation electrons, 3190
observed results and comparison with theory, 3202–9
theory of, 3188–9
cross-section for μ^- transfer from (μ^-p), (μ^-d) to, 3327, 3329
effect of magnetic field on angular correlation of annihilation quanta in, 3152
fraction of positrons forming Ps in, 3153
effect of electric field on, 3163–4
Mu microwave spectrum in, 3238–44
pressure shift of, 3247
observation of Mu precession frequency in, 3236–7
π^- transfer to, from (π^-p), 3329–30
pressure shift of microwave spectrum in:
for H, 3246–7
for Mu 3247
quenching of ortho Ps by, at ordinary temperature, 3219
at low temperatures, 3221, 3223
in presence of impurity of O_2, N_2, 3144

Auger effect, external, and de-excitation of (μ^-p), 3291–2

Auger process and mesic atom formation, 3278

Be, polarization of μ^+ stopping in, 3256
beats between hyperfine precession frequencies of Mu, 3252–3
benzene, formation of Mu, in, 3252
Bethe–Bloch energy loss formula, 3277
B_2O_3:
formation of Mu in, 3252
polarization of μ^+ stopping in, 3252
Boltzmann equation, for positrons in gases, 3136
in presence of electric field, 3155–6
bromoform ($CHBr_3$), polarization of μ^+ stopping in, 3255
precession of Mu in, 3250
bubble chamber method:
for measuring de-excitation time of mesic atoms, 3294–9
for studying atomic capture of negative particles, 3288

bubble chamber method (*cont.*):
for studying muon-catalysed nuclear fission, 3308–10
bubble formation, by ortho Ps in He and H_2, theory of, 3224–7

C, cross-section for μ^- transfer from (μ^-p), (μ^-d) to, 3327
polarization of μ^+ stopping in, 3256
catalysis of nuclear fusion by muons, 3306–23
basic processes in 3312–14
observation of, 3306–8
CCl_4, effect of, in inhibiting Ps formation, 3228–30
C_2H_4, electron spin exchange cross-section for Mu in, 3265, 3268
C_6H_6, C_6H_{10}, C_6H_{12}, C_8H_{18}, C_6H_5Br, $CHBr_3$, C_6H_5C, $C_2H_4Cl_2$, $CHCl_3$, CH_2I_2, CH_3OH, cross-sections for chemical reactions of Mu in, 3274–5
CCl_4, effect of, in inhibiting Ps formation, 3328–30
CO_2, solid, polarization of μ^+ stopping in, 3252
precession of Mu in, 3248, 3252
cluster formation, by positrons in He, 3199
coincidence counting techniques, for measurement of delayed coincidence annihilation spectra, 3133
for studying muon-catalysed nuclear fusion, 3308–9, 3312
for studying three photon annihilation, 3142
counters, scintillation, for measurement of delayed coincidence annihilation spectra, 3133
for studying positron annihilation, 3131–2

de-excitation times of mesic atoms, H, 3294–7
He, 3298–9
heavier atoms, 3299–3300
depolarization, of μ^- in atomic capture, 3289
of μ^+ in various materials, 3231–3, 3255–64
deuterium, cross-section and scattering lengths for (μ^-p), (μ^-d) collisions in, 3335
muonic (μ^-d), exchange collisions of in D, 3305–6
formation of mu-molecular ion $(d\mu^-d)^+$ by collisions in D, 3321–4

formation of mu-molecular ion $(p\mu^-d)^+$ by collisions in H, 3312, 3321
diffusion chamber, for studying muon-catalysed nuclear fusion, 3310–12
diffusion (momentum transfer) cross section, of positrons
calculated, in Ar, 3189
in He, 3183–5
in Ne, 3187
estimated from annihilation spectra in electric field, 3156
relation of, to effect of electric field on Ps formation, 3157–61
theoretical formula for, 3170
diffusion of (μ^-p) in H_2, 3305, 3330–3
diffusion time distribution of (μ^-p), (μ^-d) in H_2, D_2, 3331–4

energy loss of negatively charged particles in matter, 3277–8

Fermi–Teller Z law for negative particle capture, 3287
freon 12, use of, in analysing positron annihilation time spectra, 3135
fusion, nuclear, muon-catalysed, 3306–23
basic processes in, 3312–14
calculation of fusion rate, 3351–2
influence of hyperfine structure effects on fusion rate, 3318–21
influence of muon recycling and trapping on fusion rate, 3321–3
observation of, 3306–8
reaction rates and yields of, in liquid H_2 and D_2, 3320
time distribution of γ-ray emission in, 3315–18

gamma ray emission following nuclear fusion in $(p\mu^-d)$, 3313–14
yield of, 3314–15
time distribution of, 3315–18
gap size distribution for re-emitted muons in muon-catalysed nuclear fusion, 3330–2

H, capture of negative mesons by, 3279
collisions of ortho Ps with, theory of, 3213–15
momentum transfer and exchange quenching cross-sections, 3214–15
collisions of positrons with, theory of, 3171
number of annihilation electrons 3180–1

phase shift calculations for, 3171–9
Drachman's method for including atomic interaction, 3173–6, 3178
inclusion of virtual Ps formation, 3171–2
Kohn's variational method, 3171–2, 3178, 3180–1
polarized orbital method, 3171–2
formation of Mu in, by charge exchange, 3261
mesic, de-excitation of, 3290–8
Lyman α radiation from, 3297–8
muonic (μ^-p) diffusion of, in H, 3305
exchange collisions of, in D, 3313–14 in H, 3304–6
formation of mu-molecular ion $(p\mu^-p)^+$ in collisions of with H, 3312–14
hyperfine structure of, 3304
pionic (μ^-p), transfer of π^- to Ar from, 3328–9
pressure shift in microwave spectrum of, in Ar, He, Kr, 3246–7
H_2, bubble formation by ortho Ps in, theory of, 3224–5
comparison with experiment, 3225–7
capture of negatively charged particles by, 3286
cross-section and scattering length of (μ^-p), (μ^-d) collisions in, 3335
diffusion of (μ^-p) in, 3305
energy loss of negatively charged particles in, 3278
electron spin exchange cross-sections for Mu in, 3265
moderation time of negatively charged particles in, 3278
quenching of ortho Ps by, observed results, at low temperatures, 3220–1
at ordinary temperatures, 3219
He angular distribution in 2-quantum annihilation in, 3186–7, 3200–1
annihilation time spectrum observed in, at low temperatures, 3197–3201
at ordinary temperatures, 3191–6
collisions of ortho-Ps with:
bubble formation:
experimental results, 3225–7
theory of, 3224–5
pick-off quenching in:
observed results, at low temperatures, 3221
at ordinary temperatures, 3219
theory of, 3215–17
scattering length, 3227

collisions of positrons with, theory of, 3181
momentum transfer cross-sections, 3183, 3195–6
number of annihilation electrons, 3186, 3193–6
phase shift calculation for, 3182–3
Drachman's method for including atomic interaction, 3182
inclusion of virtual Ps formation, 3182
polarized orbital method, 3182
cross-section for μ^- transfer from (μ^-p), (μ^-d) to, 3325, 3327
formation of Mu in, by charge exchange, 3261
mesic, de-excitation time of, 3298–9
pressure shift of H microwave spectrum in, 3247

hydrocarbons, cross-sections for chemical reactions of Mu in, 3269–75
hyperfine, effects in (μ^-p) (μ^-d) scattering in H, D, 3345–7
splitting of Mu, 3231–3
structure of (μ^-p), (μ^-d), 3304
hyperons, Σ^-, properties of, 3277

I_2, effect of, in quenching Ps formation, 3228–30
ice, muonium formation in, 3245, 3251–2
polarization of μ^+ stopping in, 3252, 3256

kaons, properties of, 3277
Kr, cross-sections for μ^- transfer from (μ^-p), (μ^-d) to, 3327, 3329
mean number of annihilation electrons, observed, 3209
pressure shift of microwave spectrum in, of H, 3246–7
of Mu, 3247

mesic atoms, de-excitation processes of, 3290–3300
energy levels and orbital radii of, 3276
formation processes of, 3278
quantum number of capture orbit in, 3281–3
radius of capture orbit in, 3280
mesic H, cascade time of negative mesons in, 3294–8
de-excitation processes of, 3290–4
mesic He, cascade time of negative mesons in, 3298–9
mesons, negative, capture of, by atoms other than H, 3287
Fermi–Teller Z law, 3287
depolarization in the capture process, 3289

mesons, negative (*cont.*):
capture of by H, theory of, 3279–84
capture of by H_2, theory of, 3284–6
properties of, 3277
methods, experimental for investigating:
angular correlations in 2-quantum annihilation, 3147–8
angular distribution in 3-quantum annihilation, effect of magnetic field on, 3150
atomic capture of negatively charged particles, 3288
de-excitation time of mesic atoms, 3294–9
delayed coincidence annihilation spectra of positrons, 3133
elastic scattering cross-sections of $(\mu^- p)$, $(\mu^- d)$ in H_2, D_2, from gap size distribution, 3320–2
from diffusion time distribution, 3331–4
electron spin exchange cross-section for Mu, 3246–9
using microwave resonance spectra, 3264–6
using spin relaxation time, 3266–9
energy spectrum of single annihilation quanta, 3137–40
application to determination of Ps:
concentration, 3137
quenching ratio, 3138–40
muon-catalysed nuclear fusion, 3306–12
bubble chamber method, 3306–10
counter method, 3308–9, 3312
diffusion chamber method, 3310–12
muon transfer cross-sections from $(\mu^- p)$, $(\mu^- d)$ to heavier atoms, 3323–30
muonic hydrogen $(\mu^- p)$ and deuterium $(\mu^- d)$ collision cross-sections in H_2, D_2, 3315–18, 3330–6
muonium (Mu) formation, in condensed materials, 3248–64
in longitudinal magnetic field, 3253–5
in transverse magnetic field, 3248–53
muonium (Mu) formation in gases, 3236–48
by observing Mu precession frequency, 3236–8
by observing Mu microwave resonance spectrum, 3238–44
in strong magnetic fields, 3240–2
using 'magic' field, 3242
in weak magnetic fields, 3242–4
using Ramsey resonance method, 3243–4

polarization of μ^- meson decay at rest, 3300–3
positron and positronium (Ps) slow collisions, in gases, summary of, 3131
positronium (Ps) production
effect of electric field on, 3161, 3164
fraction of positrons forming Ps, 3140
reactions of positrons and Ps in gases, summary of, 3166
three-quantum positron annihilation
application to measurement of ortho-Ps concentration, 3142
microwave resonance spectrum of Mu, 3238–44
energy levels of, 3399–41
moderation time, of negatively charged particles in H_2, 3286
Mu-molecular ion $(d\mu^- d)^+$, formation of in $(\mu^- d)$ collision in D_2, 3321–3
theory of, 3349–51
Mu-molecular ion $(p\mu^- d)^+$:
energy levels of, 3339–41
formation of, in $(\mu^- d)$ collisions in H_2, 3313, 3321–4
theory of, 3349–51
nuclear fusion process in, 3313–14
calculation of the fusion rate, 3351–2
influence of hyperfine structure effects on, 3318–21
μ^- re-emission following fusion in, 3313–14
yield of re-emitted μ^-, 3314–15
γ-rays following nuclear fusion in, 3314–18
time distribution of γ-rays, 3315–18
yield of γ-rays, 3313–14
Mu-molecular ion $(p\mu^- p)^+$:
energy levels of, 3339–41
formation of in $(\mu^- p)$ collisions in H_2, 3312–14, 3321
theory of formation processes, 3349–51
muon catalysed nuclear fusion, 3306–23, 3336–45
basic processes in, 3312–14
cross-sections important in interpretation of, 3321
gap size distribution in, 3330–2
influence of hyperfine structure effects, 3318–21
influence of muon recycling and trapping, 3321–3
observation of, 3306–8
reaction rates and yields in liquid H_2, D_2, 3320
theory of, **3336–45**

time distribution of γ-ray emission in, 3315–18

muonic D (μ^-d):
  collisions in H_2, D_2:
    cross-sections and scattering length for, 3335
    theory of, 3336–43
      hyperfine effects in, 3345–7
      scattering phases for, 3343–5
  diffusion time distribution in H_2, D_2, 3331–4
  exchange collisions of, in H, 3305–6
    theory of, 3339–47
  formation of mu-molecular ions
    $(p\mu^-d)^+$, by collision of, in H_2, 3313–14, 3321
    $(d\mu^-d)^+$, by collision of, in D_2, 3321–3
    theory of, 3349–51
  Ramsauer effect in collisions in H, 3344
  transfer cross-section of μ^- to heavier atoms from, 3323–8
    theory of, 3347–9

muonic H (μ^-p), collisions in H_2, D_2, cross sections and scattering lengths for, 3335
  theory of, 3336–43
    scattering phases for, 3343–5
  diffusion time distribution in H_2, D_2, 3331–4
  exchange collisions of, in H, 3304–5, 3321
  exchange collisions of, in D, 3313–14, 3321
  formation of mu-molecular ion:
    $(p\mu^-p)^+$, by collisions of, in H, 3312–13, 3321
    $(p\mu^-d)^+$, by collisions of, in D, 3313–14, 3321
    theory of, 3349–51
  hyperfine structure of, 3304
  transfer cross-section of μ^- to heavier atoms from, 3323–8
    theory of, 3347–9

muonium (Mu), chemistry of, 3261, 3269–75
  cross-sections for substitution and capture reactions of, in hydrocarbons, 3274
  depolarization of muons in, in condensed materials, 3255–64
    interpretation of, 3259–64
  depolarized (Mu^D), 3234, 3270–1
  electron exchange collisions of, 3262–4
    measurement of cross-sections for, 3264–9, 3275
  energy levels and fine structure splitting of, 3231
  formation of, by charge exchange, 3261
  in condensed materials, 3248–53
    in bromoform, 3250
    in CO_2 (solid), 3248
    in ice, 3248–51
    in quartz, 3248, 3250–1, 3255–6
  in gases, 3236–48
  in longitudinal magnetic field, 3235
  hyperfine splitting of, 3231–3
  magnetic moment of, 3233
  microwave resonance spectrum of, in gases, 3238–44
    in strong magnetic fields, 3240–2
    in weak magnetic fields, 3242–4
  precession frequency of, in gases, 3236–8
  pressure shift in microwave resonance spectrum of, 3244–8
  singlet (Mu^s), 3235, 3260, 3263, 3270–1
  spin relaxation time of muons in, 3234, 3260
  triplet, 3234, 3248, 3263, 3270–1

muons, negative (μ^-), depolarization of in atomic capture, 3289
  polarization of, decaying from rest, 3300–4
  properties of, 3277

muons, positive (μ^+), asymmetric decay of, 3230
  changes in spin polarization of 3231
  interpretation of, 3259–64
  depolarization of, in condensed materials, 3255–64
  in Mu formation, 3233
  passage through matter of, 3230–1
  polarization of, at production, 3230
  stopping in various materials, 3256–7
  precession frequency of, in various magnetic fields, 3272
  in solid materials, 3250
  spin relaxation time of, stopping, 3257–9
  in Mu, 3234

N_2, effect of, on shoulder width in annihilation spectra, 3136, 3211–13
  formation of Mu in liquid, 3252
  mean number of annihilation electrons in, 3209
  polarization of μ^+ stopping in, 3252
  Ps formation in, effect of electric field on 3163–4

Ne, collisions of positrons with:
  mean number of annihilation electrons, 3187, 3209–10
  momentum transfer cross-section, 3189
  theory of, 3187–8

SUBJECT INDEX

Ne (*cont.*):
μ^- transfer to from $(\mu^- p)$, $(\mu^- d)$, cross-section for, 3323–5, 3327–9
Ps formation in, effect of electric field on, 3163
quenching of ortho-Ps by, at low temperatures, 3221, 3223
at ordinary temperatures, 3219
NO, conversion of ortho-Ps to para-Ps in, 3137
electron spin exchange cross-section in, for H, 3269
for Mu, 3265, 3267–9
fraction of positrons forming Ps in, 3140
quenching of ortho-Ps in, 3129
measurement of rate of, 3140
time spectra of annihilation of positrons, in, 3135
NO$_2$, electron spin exchange cross-section for Mu in, 3268
number, effective, of annihilation electrons, 3125, 3135
calculated values of, for Ar, 3189
H, 3180–1
He, 3186
Ne, 3188
relation to effect of electric field on Ps formation, 3158–61
theoretical formula for, 3170

O$_2$, angular distribution of annihilation quanta in, effect of magnetic field on, 3153
effect of, on shoulder width in annihilation spectra, 3136
electron spin exchange cross-sections in, of H 3269, of Mu, 3265, 3268–9
formation of Ps, effect of electric field on, 3165
μ^- transfer from $(\mu^- p)$, $(\mu^- d)$ to, cross section for, 3327
quenching of ortho-Ps in, 3129
measurement of rate of, 3140
Öre gap, 3128, 3140
ortho-positronium (o-Ps), 3125
bubble formation by, in He and H$_2$, 3222–7
concentration, measurement of:
from analysis of annihilation time spectra, 3135
from energy spectrum of single annihilation quanta, 3137
formation in radiative or three body capture, 3130
inhibition of formation of, 3130
lifetime ratio to that for para-Ps, 3149
quenching of, 3129
by chemical reactions, 3129, 3227–30
by electron exchange, 3129, 3213–15
by pick-off annihilation, 3129, 3215–16
measurement of rate of, 3138–40, 3142
observed results, 3218–20
at low temperatures, 3220–7

para-positronium (Ps), 3125
measurement of ratio of lifetime to that of ortho-Ps, 3149
paraffin, Mu formation in, 3252
photographic emulsion, polarization of μ^+ stopping in, 3257
pions, negative, (π^-), properties of, 3277
pionic hydrogen $(\pi^- p)$, transfer of π^- from, to Ar, 5328–9
plastic scintillator, polarization of μ^+ stopping in, 3257
polarization, of μ^- decaying from rest, 3300–4
of μ^+ stopping in condensed materials, 3252, 3256–7
polarized orbital method, application to positron collisions with Ar, 3188–9; H, 3171–2; He, 3182; Ne, 3187–8
polyethylene, Mu formation in, 3252
positronium (Ps), 3125
annihilation of:
effect of magnetic field on, 3149
three-quantum, of ortho-Ps, 3126
chemistry of, 3127
collisions of, with H, theory of, 3213–15
formation of:
cross-section for, 3158–61
in SF$_6$, 3161, 3164
effect of electric field on:
experimental observation of, 3161
in rare gases, H$_2$, D$_2$, N$_2$, 3163
theory of, 3151–6
effect of magnetic field on, 3148
in Öre gap, 3127
life-times of states of, 3125
ratio of, for para- and ortho-Ps, 3149
positrons (e$^+$), cluster formation by, in He, 3199
collisions of, with Ar, theory of, 3188–9
with H, theory of, 3171
with He, theory of, 3181
with Ne, theory of, 3187–8
fraction of forming Ps, estimated, 3128
method of measurement of, 3140, 3153, 3162
effect of electric fields on:
experimental observation of, 3161
in rare gases, H$_2$, D$_2$, N$_2$, 3163
in Ar, N$_2$, using angular correlations, 3164
theory of, 3157–61

SUBJECT INDEX

Positrons (e$^+$), free, annihilation of, with free electrons, 3124
  life history of, in a gas, 3129
  in Ps, 3125
  3-quantum annihilation of, 3126
  time distribution of, from μ^+ decay in Mu, 3237
  velocity distribution of, 3136
  in electric field, 3155

quartz (SiO$_2$), Mu formation in, 3248, 3250, 3252 3255–6
  polarization of μ^+ stopping in 3257
quenching, of ortho-Ps, 3129
  by Ar, at low temperatures, 3221, 3223
  at ordinary temperatures, 3219
  by Ar with small admixtures of O$_2$, N$_2$, 3144
  by H, theory of, 3213–15
  by He, at low temperatures, 3221–2
  at ordinary temperatures, 3219
  theory of, 3222–7
  by I$_2$, observed results, 3227–30
  by Ne, at low temperatures, 3221, 3223
  at ordinary temperatures, 3219
  measurement of rate of,
    from energy spectra of annihilation quanta, 3138–40
    from 3-photon coincidences, 3142

Ramsauer effect in collisions of (μ^-d) with H, 3344
Ramsey resonance method applied to Mu, 3243–4

retardation time of negatively charged particles in matter, 3278

S, polarization of μ^+ stopping in, 3256
  spin relaxation time of μ^+ stopping in, 3258
SF$_6$, formation of Ps in, effect of electric field on, 3161, 3164–5
shoulder, in positron annihilation time spectra, 3134
  analysis of, in Ar, 3204
  effect of CH$_4$ on, 3213
  effect of N$_2$ on, 3211–13
  analysis of, in He, 3193
  effect of impurities on, 3136
Si, polarization of μ^+ stopping in, 3256
Σ^- hyperons, properties of, 3277
SiO$_2$, Mu formation in, 3252
  polarization of μ^+ stopping in, 3252, 3257
spin relaxation time of μ^+, 3234, 3252, 3257–9
Stark effect, role of, in de-excitation of mesic atoms, 3292–4

variational method of Kohn, application to positron collisions in H, 3171
velocity distribution, of positrons in an electric field, 3155

water, polarization of μ^+ stopping in, 3256

Xe, cross-section for μ^- transfer from (μ^-p), (μ^-d) to, 3325–7, 3329